Molecular Electromagnetism

Molecular Electromagnetism

A Computational Chemistry Approach

Stephan P. A. Sauer
Department of Chemistry,
University of Copenhagen, Denmark

OXFORD
UNIVERSITY PRESS

Great Clarendon Street, Oxford OX2 6DP

Oxford University Press is a department of the University of Oxford.
It furthers the University's objective of excellence in research, scholarship,
and education by publishing worldwide in

Oxford New York

Auckland Cape Town Dar es Salaam Hong Kong Karachi
Kuala Lumpur Madrid Melbourne Mexico City Nairobi
New Delhi Shanghai Taipei Toronto

With offices in

Argentina Austria Brazil Chile Czech Republic France Greece
Guatemala Hungary Italy Japan Poland Portugal Singapore
South Korea Switzerland Thailand Turkey Ukraine Vietnam

Oxford is a registered trade mark of Oxford University Press
in the UK and in certain other countries

Published in the United States
by Oxford University Press Inc., New York

© Stephan P. A. Sauer 2011

The moral rights of the author have been asserted
Database right Oxford University Press (maker)

First published 2011

All rights reserved. No part of this publication may be reproduced,
stored in a retrieval system, or transmitted, in any form or by any means,
without the prior permission in writing of Oxford University Press,
or as expressly permitted by law, or under terms agreed with the appropriate
reprographics rights organization. Enquiries concerning reproduction
outside the scope of the above should be sent to the Rights Department,
Oxford University Press, at the address above

You must not circulate this book in any other binding or cover
and you must impose the same condition on any acquirer

British Library Cataloguing in Publication Data

Data available

Library of Congress Cataloging in Publication Data

Data available

Typeset by SPI Publisher Services, Pondicherry, India
Printed in Great Britain
on acid-free paper by
CPI Antony Rowe, Chippenham, Wiltshire

ISBN 978-0-19-957539-8

1 3 5 7 9 10 8 6 4 2

*To Sarah Rashmi,
David Ajit
and Hannah Jyotsna,
who grew up while this book was written.*

Preface

The interaction of molecules with static electric and magnetic fields or the time-dependent fields of electromagnetic radiation is often described in terms of so-called molecular electromagnetic properties. The outcome of many experiments involving electromagnetic fields are therefore interpreted with such properties. With modern quantum chemical computer software such as Dalton, CFOUR, Orca, Turbomole, Gaussian or Gamess, to name just a few, it is nowadays possible to calculate values of many of these electromagnetic properties for individual molecules or clusters of molecules with an accuracy comparable to experiment. Contrary to the experimental determination calculations allow to identify and analyse individual contributions to the properties and offer therefore an understanding of molecular properties, which cannot be obtained by their measurement alone. This creates a fruitful interplay between theory and experiment, prediction and experimental verification, measurement and theoretical rationale with applications ranging from the design of materials to the understanding of natural phenomena.

This book focuses on the definitions and quantum theory of molecular electromagnetic properties as well as on the theory of the computational methods for calculating them. It tries to treat both aspects equally thoroughly and differs therefore from typical textbooks on physical chemistry as well on computational chemistry. While electromagnetic properties like the polarizability or the nuclear magnetic shielding tensor are typically defined in physical chemistry textbooks, the quantum chemical methods for calculating them are at most superficially mentioned. In typical computational chemistry textbooks the situation is reversed, because computational methods and their application are discussed but the quantum theory underlying the definitions of molecular properties is typically not mentioned.

The list of molecular properties, which are discussed in this book, is not and hardly can be complete, because new experimental setups lead sometimes also to the definition of new electromagnetic properties. Instead, the most important properties of each type are described, i.e. one or two prototypical properties that are used in the description of the interaction with static or oscillating electric fields, with magnetic fields and with magnetic moments of nuclei. They are also the properties that most chemists will have heard of, like, e.g., the chemical shift of NMR spectroscopy or the electronic excitation energies that give rise to the absorption in UV/Vis spectra. Furthermore, the emphasis is more on the general concepts in the definition and derivation of molecular properties, on the interrelation between different types of properties and on detailed derivation of the selected molecular electromagnetic properties than on completeness.

This applies even more to the quantum chemical methods for calculating them. The list of methods is ever-growing and any attempt to give a comprehensive overview is bound to fail. The emphasis is therefore again on the general concepts, on the differences and similarities between methods and on detailed derivation of selected prototypical quantum chemical methods. Consequently, only so-called *ab initio* quantum chemical wavefunction methods based on the non-relativistic Schrödinger equation are discussed in this book.

Reading this book you will not only learn how the most important molecular properties are defined but will also learn how to derive molecular properties for new experimental setups. Furthermore, you can understand the relations between various molecular properties and how this can be used to predict the outcome of one experiment based on other measurements. In the third part of the book you acquire a thorough understanding of quantum chemical methods for the calculation of molecular properties. In particular, you find out how the various quantum mechanical methods are related to each other. At the same time you will become acquainted with different techniques for deriving computational methods and will learn how to apply these techniques to different types of wavefunctions. This will allow you to derive new methods on your own.

The book is aimed at graduate or senior undergraduate students and at PhD students or post-docs who want to embark on the calculation of molecular electromagnetic properties. It is expected that you have some basic knowledge of quantum mechanics corresponding to a second-year quantum mechanics or quantum chemistry course and of classical electromagnetism corresponding to a typical undergraduate physics course for non-physics students. Furthermore, you should be somewhat familiar with quantum chemical methods for the calculation of electronic energies such as the Hartree–Fock (HF) or self-consistent-field (SCF) method, Møller–Plesset (MP) perturbation and coupled cluster (CC) theory. However, the basic equations and notation of these methods are also discussed in the book.

The book is divided into three parts and preceded by an introduction. In *Part I, Quantum Mechanical Fundamentals*, the foundations are laid by deriving the Schrödinger equation for a molecule in the presence of electromagnetic fields and by presenting the perturbation theory tools for solving this Schrödinger equation approximately. Time-independent as well as time-dependent perturbation theory in the form of response theory are discussed. In *Part II, Definition of Properties*, many different molecular properties are defined and quantum chemical expressions for them are derived using perturbation theory. Finally, in *Part III, Computational Methods for the Calculation of Molecular Properties*, a selection of modern quantum chemical methods for the calculation of molecular properties is derived and discussed. Furthermore, some illustrative examples of calculated properties are presented. Exercises are included in all chapters that will allow you to test your understanding on some of the intermediate derivations. At the end of each chapter a *Further Reading* section is added with a list of books or review articles that I found useful while writing the corresponding chapter. The list is in no way complete but reflects much more my personal taste and preferences.

Acknowledgements

This book is the result of teaching a course on *Quantum Chemical Calculations of Molecular Electromagnetic Properties* for master and PhD students at the University of Copenhagen for more than 10 years. However, the first set of lecture notes was actually prepared for Professor Gustavo A. Aucar's *First and Second Mercosur Summerschool on Molecular Physics* that took place at Northeastern University in Corrientes, Argentina, February 1999 and 2000. During a sabbatical 2005/2006 in the group of Professor Dr. Walter Thiel at the Max-Planck-Institute für Kohlenforschung in Mülheim an der Ruhr, Germany, the notes have been completely restructured and largely rewritten. In the following years they were then continuously modified. The present version will hopefully not be the final version, but I feel that it is time to make something available to the wider public.

I ought to thank many people who have in some way or another contributed to this book, but here I want to thank in particular my two PhD thesis supervisors Professor Geerd H. F. Diercksen, Max-Planck-Institute for Astrophysics, Garching bei München, and Professor Jens Oddershede, University of Southern Denmark, Odense, who have taught me many things and not just about Quantum Chemistry.

Many colleagues and students have commented on previous versions. Among those I want to thank in particular Professor Michał Jaszuński, Dr. Martin J. Packer, Dr. Patricio F. Provasi, Dr. Andrea Ligabue and Dr. Kestutis Aidas.

Stephan P. A. Sauer
Copenhagen, Fall 2010

Contents

1	**Introduction**	**1**
Part I	Quantum Mechanical Fundamentals	
2	**The Schrödinger Equation in the Presence of Fields**	**5**
2.1	The Time-Dependent Schrödinger Equation	5
2.2	The Born–Oppenheimer Approximation	7
2.3	Electron Charge and Current Density	9
2.4	The Force due to Electromagnetic Fields	12
2.5	Minimal Coupling—Non-Relativistically	13
2.6	Minimal Coupling—Relativistically	17
2.7	Elimination of the Small Component	20
2.8	The Molecular Electronic Hamiltonian	23
2.9	Gauge Transformations	25
2.10	Further Reading	28
3	**Perturbation Theory**	**30**
3.1	The Hellmann–Feynman Theorem	31
3.2	Time-Independent Perturbation Theory	33
3.3	Time-Independent Response Theory	37
3.4	Second Derivatives of the Energy	38
3.5	Density Matrices	39
3.6	The Ehrenfest Theorem	41
3.7	The Off-Diagonal Hypervirial Theorem	42
3.8	The Interaction Picture	43
3.9	Time-Dependent Perturbation Theory	44
3.10	Transition Probabilities and Rates	47
3.11	Time-Dependent Response Theory	49
3.12	Matrix Representation of the Propagator	57
3.13	Pseudo-Perturbation Theory	64
3.14	Further Reading	66
Part II	Definition of Properties	
4	**Electric Properties**	**71**
4.1	Electric Multipole Expansion	71
4.2	Potential Energy in an Electric Field	75
4.3	Quantum Mechanical Expressions for Electric Moments	77

4.4	Induced Electric Moments and Polarizabilities	80
4.5	Quantum Mechanical Expressions for Polarizabilities	85
4.6	Molecular Electric Fields and Field Gradients	89
4.7	Further Reading	91

5 Magnetic Properties — 93

5.1	Magnetic Multipole Expansion	93
5.2	Potential Energy in a Magnetic Induction	96
5.3	Quantum Mechanical Expression for the Magnetic Moment	97
5.4	Induced Magnetic Moment, Magnetizability, and Nuclear Magnetic Shielding	100
5.5	Quantum Mechanical Expression for the Magnetizability	102
5.6	Molecular Magnetic Fields and ESR Parameters	105
5.7	Induced Magnetic Fields and NMR Parameters	109
5.8	Quantum Mechanical Expression for the NMR Parameters	112
5.9	Sum-over-States Expression for Diamagnetic Terms	118
5.10	The Gauge-Origin Problem	121
5.11	Further Reading	124

6 Properties Related to Nuclear Motion — 126

6.1	Molecular Rotation as Source for Magnetic Moments	126
6.2	Quantum Mechanical Expression for the Rotational g Tensor	128
6.3	Rotational g Tensor and Electric Dipole Moment	133
6.4	Rotational g Tensor and Electric Quadrupole Moment	135
6.5	Molecular Rotation as Source for Magnetic Fields	136
6.6	Quantum Mechanical Expression for the Spin Rotation Tensor	138
6.7	Non-Adiabatic Rotational and Vibrational Reduced Masses	141
6.8	Partitioning of the g Factors	148
6.9	Further Reading	152

7 Frequency-Dependent and Spectral Properties — 153

7.1	Time-Dependent Fields	153
7.2	Frequency-Dependent Polarizability	156
7.3	Optical Rotation	157
7.4	Electronic Excitation Energies and Transition Moments	161
7.5	Dipole Oscillator Strength Sums	166
7.6	van der Waals Coefficients	169
7.7	Further Reading	172

8 Vibrational Contributions to Molecular Properties — 174

8.1	Sum-over-States Treatment	175
8.2	Clamped-Nucleus Treatment	177
8.3	Vibrational and Thermal Averaging	179
8.4	Further Reading	184

Part III Computational Methods for the Calculation of Molecular Properties

9 Short Review of Electronic Structure Methods — 189
- 9.1 Hartree–Fock Theory — 191
- 9.2 Excited Determinants and Excitation Operators — 193
- 9.3 Multiconfigurational Self-Consistent Field Method — 196
- 9.4 Configuration Interaction — 197
- 9.5 Møller–Plesset Perturbation Theory — 198
- 9.6 Coupled Cluster Theory — 201
- 9.7 The Hellmann–Feynman Theorem for Approximate Wavefunctions — 203
- 9.8 Approximate Density Matrices — 207
- 9.9 Further Reading — 209

10 Approximations to Exact Perturbation and Response Theory Expressions — 210
- 10.1 Ground-State Expectation Values — 210
- 10.2 Sum-over-States Methods — 211
- 10.3 Møller–Plesset Perturbation Theory Polarization Propagator — 212
- 10.4 Multiconfigurational Polarization Propagator — 225
- 10.5 Further Reading — 226

11 Perturbation and Response Theory with Approximate Wavefunctions — 227
- 11.1 Coupled and Time-Dependent Hartree–Fock — 227
- 11.2 Multiconfigurational Linear Response Functions — 233
- 11.3 Second-Order Polarization Propagator Approximation — 235
- 11.4 Coupled Cluster Linear Response Functions — 236
- 11.5 Further Reading — 242

12 Derivative Methods — 243
- 12.1 The Finite-Field Method — 243
- 12.2 The Analytic Derivative Method — 245
- 12.3 Time-Dependent Analytical Derivatives — 248
- 12.4 Further Reading — 252

13 Examples of Calculations and Practical Issues — 253
- 13.1 Basis Sets for the Calculation of Molecular Properties — 253
- 13.2 Reduced Linear Equations — 259
- 13.3 Examples of Electron Correlation Effects — 260
- 13.4 Examples of Vibrational Averaging Effects — 266
- 13.5 Further Reading — 267

Part IV Appendices

Appendix A Operators — 271
- A.1 Perturbation Operators — 271
- A.2 Other Electronic Operators — 277

Appendix B	Definitions of Properties	278
Appendix C	Perturbation Theory Expressions for Properties	280
References		282
Index		297

1
Introduction

All experiments carried out on molecules can be understood as the interaction of a molecule with an electromagnetic field. If we exclude situations or fields strengths where the molecules change their identity, i.e. undergo chemical reactions, or where it becomes difficult to distinguish between the molecule and e.g. an intense laser field, we can treat the fields as perturbations that slightly modify the nuclear and electronic structure of molecules. These interactions with weak electromagnetic fields and the corresponding changes in the molecular structure are often described in terms of so-called **molecular properties**. Some of the best-known molecular properties are the electric dipole moment, the frequency-dependent polarizability, the chemical shift and indirect nuclear spin-spin coupling constant of nuclear magnetic resonance (NMR) spectroscopy or the hyperfine coupling constant of electron spin resonance (ESR) spectroscopy. Molecular properties are intrinsic properties of a particular state of a molecule, which means that they are independent of the strength of the fields. Therefore, they can be used to describe the **response** of a molecule to an arbitrary field within the above-described limits.

The electromagnetic fields discussed here can be the static external electric field in a capacitor or the electric field due to another molecule nearby, the static external magnetic field in an NMR or ESR spectrometer but also the internal magnetic fields due to the magnetic moments of nuclei with spin or the internal electric field due to the electric quadrupole moment of a nucleus. Furthermore it can be the oscillating electric and magnetic fields of electromagnetic radiation. Molecular properties thus play an important role in the interpretation of numerous experimental phenomena such as the refractive index, the Stark and Zeeman effects, the Kerr effect, nuclear and electric magnetic resonance spectra and many more. Even long-range interactions between molecules can be understood in terms of molecular electric moments.

Although many of these properties can be determined to a high accuracy in experiments it is also important to be able to calculate them. Unknown compounds or molecular configurations can be identified by their calculated properties. Candidates for new materials with desired properties can be screened fast and inexpensively by calculating the respective properties instead of synthesizing them first. The value of an electromagnetic property of a particular molecule or the changes of a property in a series of molecules can often be explained by analysing all the terms that contribute to it. The information about the structure and nature of a molecule that is contained in the measured properties can only in this way be fully unfolded. Theoretical calculations of molecular electromagnetic properties can therefore supplement experiments in many ways in addition to calculation of the energetics and structure of molecules.

2 Introduction

This book is concerned with the quantum chemical methods for the calculations of electromagnetic properties of molecules. However, in detail only so-called *ab initio* quantum chemical methods will be discussed in Part III. As *ab initio* methods one normally describes those quantum chemical methods that start *from the beginning*, i.e. methods that require the evaluation of all the terms in the Schrödinger or Dirac equation and do not include other experimentally determined quantities than the nuclear charges, nuclear masses, nuclear dipole and quadrupole moments and maybe positions of the nuclei. This is in contrast to the so-called *semi-empirical* methods where many of the integrals over the operators in the Hamiltonian are replaced by experimentally or otherwise determined constants. However, in the case of density functional theory (DFT) methods the classification is somewhat debatable.

In this book the interaction between fields and molecules is treated in a semi-classical fashion. Quantum mechanics is used for the description of the molecule, whereas the treatment of the electromagnetic fields is based on classical electromagnetism. A complete quantum mechanical description using quantum electrodynamics is beyond the scope of this presentation, although we will make use of the correct value of the electronic g-factor as given by quantum electrodynamics. Furthermore, only *ab initio* methods derived from the non-relativistic Schrödinger equation are discussed. Nevertheless, the Dirac equation is briefly discussed in order to introduce the electronic spin via the Pauli Hamiltonian.

Part I

Quantum Mechanical Fundamentals

2
The Schrödinger Equation in the Presence of Fields

A complete quantum mechanical treatment of the interaction of molecules and fields would require quantum electrodynamics, which is probably the most successful theory, that was ever derived. Using it, one can reproduce very accurately the Lamb shift in the hydrogen atom spectrum and the g-factor of the free electron $g_e \approx 2.002$. Nevertheless it is not yet regularly[1] employed in the calculation of electromagnetic properties of molecules and the effects are expected to be very small.

Therefore, we will make a series of approximations to this approach. First, we will only use quantum mechanics for the description of the molecule and use classical electrodynamics for the electromagnetic fields. In this **semi-classical approach** the perturbing fields and nuclear moments are considered to be unaffected by the molecular environment, the **so-called minimal coupling** approximation.

Secondly, the exposition will be restricted to non-relativistic quantum mechanics, i.e. to the Schrödinger equation.[2] This approach is justified, if we restrict ourselves to atoms of the first three rows of the periodic table, for which relativistic effects are generally unimportant. However, if we are interested in discussing properties, which include interactions with the spin of the electrons such as NMR and ESR coupling constants, the Schrödinger equation is not sufficient alone, because it is in principle a spin-free theory contrary to the Dirac equation. The necessary operators for the interaction with the electron spin are therefore derived from the Dirac equation and then added to the Schrödinger Hamiltonian in an *ad hoc* fashion.

Finally, the Born-Oppenheimer approximation is applied in order to separate the nuclear and electronic wavefunctions.

2.1 The Time-Dependent Schrödinger Equation

The total Hamiltonian for a molecule with N electrons and M nuclei in the absence of fields[3] is, in non-relativistic quantum mechanics, given as

---

[1] See Romero and Aucar (2002) for an example of the calculation of quantum electrodynamics effects on molecular properties.
[2] See Saue (2001) for a discussion of *ab initio* methods for the calculation of electromagnetic properties based on the Dirac equation.
[3] When talking about fields we do not include the electric field arising from the charges of the electrons and nuclei in the molecule, unless stated otherwise.

6 The Schrödinger Equation

$$\hat{H}^{(0)}_{nuc,e} = \frac{1}{2}\sum_K^M \frac{\hat{\vec{p}}_K^2}{m_K} + \frac{e^2}{4\pi\epsilon_0}\sum_{K<L}\frac{Z_K Z_L}{|\vec{R}_K - \vec{R}_L|}$$

$$+ \frac{1}{2m_e}\sum_i^N \hat{\vec{p}}_i^2 - \frac{e^2}{4\pi\epsilon_0}\sum_{iK}^{NM}\frac{Z_K}{|\vec{r}_i - \vec{R}_K|} + \frac{e^2}{4\pi\epsilon_0}\sum_{i<j}\frac{1}{|\vec{r}_i - \vec{r}_j|} \quad (2.1)$$

where $\vec{R}_K$ is the position vector of nucleus K with mass m_K and atomic number Z_K and $\vec{r}_i$ is the position vector of electron i. The operators for the canonical momentum of the nuclei, $-i\hbar\hat{\vec{\nabla}}_K$, and electrons, $-i\hbar\hat{\vec{\nabla}}_i$, are denoted $\hat{\vec{p}}_K$ and $\hat{\vec{p}}_i$, respectively. All the other constants have their usual meaning (see e.g. Mills et al., 1993).

The first axiom of quantum mechanics states that the state of a molecule is completely described by the time-dependent wavefunction $\Phi^{(0)}(\{\vec{R}_K\},\{\vec{r}_i\},t)$, where $\{\vec{R}_K\}$ and $\{\vec{r}_i\}$ stand collectively for the position vectors of all nuclei and electrons and t is the time. Another axiom states that the average measured value $<P>_t$ of a physical observable P obtained in a series of measurements on a large ensemble of molecules, which are all in the same state $|\Phi^{(0)}(\{\vec{R}_K\},\{\vec{r}_i\},t)\rangle$, can be calculated as the **expectation value** of the corresponding quantum mechanical operator $\hat{P}$

$$<P>_t \equiv \frac{\langle\Phi^{(0)}(\{\vec{R}_K\},\{\vec{r}_i\},t)|\hat{P}|\Phi^{(0)}(\{\vec{R}_K\},\{\vec{r}_i\},t)\rangle}{\langle\Phi^{(0)}(\{\vec{R}_K\},\{\vec{r}_i\},t)|\Phi^{(0)}(\{\vec{R}_K\},\{\vec{r}_i\},t)\rangle} \quad (2.2)$$

Finally, the last postulate states that the time dependence of the wavefunction is governed by the **time-dependent Schrödinger equation**

$$\hat{H}^{(0)}_{nuc,e}|\Phi^{(0)}(\{\vec{R}_K\},\{\vec{r}_i\},t)\rangle = i\hbar\frac{\partial}{\partial t}|\Phi^{(0)}(\{\vec{R}_K\},\{\vec{r}_i\},t)\rangle \quad (2.3)$$

When the Hamiltonian does not depend explicitly on time like the one given in Eq. (2.1), we can apply the separation of variables technique and separate the time variable t from the spatial coordinates $\{\vec{R}_K\}$ and $\{\vec{r}_i\}$. We write therefore the time-dependent wavefunction $\Phi^{(0)}(\{\vec{R}_K\},\{\vec{r}_i\},t)$ as the product of a time-independent wavefunction $\Phi^{(0)}(\{\vec{R}_K\},\{\vec{r}_i\})$ and a time-dependent phase factor $\vartheta(t)$.

$$\Phi^{(0)}(\{\vec{R}_K\},\{\vec{r}_i\},t) = \Phi^{(0)}(\{\vec{R}_K\},\{\vec{r}_i\})\,\vartheta(t) \quad (2.4)$$

Inserting this trial solution in Eq. (2.3), the time-dependent Schrödinger equation separates in two equations: the **time-independent Schrödinger equation**

$$\hat{H}^{(0)}_{nuc,e}|\Phi^{(0)}(\{\vec{R}_K\},\{\vec{r}_i\})\rangle = E^{(0)}|\Phi^{(0)}(\{\vec{R}_K\},\{\vec{r}_i\})\rangle \quad (2.5)$$

and an equation for the time-dependent phase factor $\vartheta(t)$.

The time-independent Schrödinger equation in Eq. (2.5) is a second-order partial differential equation. However, it can also be interpreted as an eigenvalue equation. The time-independent wavefunctions $\Phi^{(0)}(\{\vec{R}_K\},\{\vec{r}_i\})$ are then the eigenfunctions of the Hamiltonian with the energy $E^{(0)}$ as eigenvalue.

For the **time-dependent phase factor** one obtains [see Exercise 2.1]

$$\vartheta(t) = e^{-\frac{i}{\hbar}E^{(0)}t} \quad (2.6)$$

Exercise 2.1 Derive Eq. (2.6) for the time-dependent phase factor.

The time dependence of the time-dependent wavefunction in Eq. (2.4) corresponds therefore simply to a rotation in the complex plane and the probability density of the time-dependent wavefunction, $|\Phi(\{\vec{R}_K\},\{\vec{r}_i\},t)|^2$, given as

$$|\Phi^{(0)}(\{\vec{R}_K\},\{\vec{r}_i\},t)|^2 = \left(\Phi^{(0)}(\{\vec{R}_K\},\{\vec{r}_i\})e^{-\frac{i}{\hbar}E^{(0)}t}\right)^* \left(\Phi^{(0)}(\{\vec{R}_K\},\{\vec{r}_i\})e^{-\frac{i}{\hbar}E^{(0)}t}\right)$$

$$= |\Phi^{(0)}(\{\vec{R}_K\},\{\vec{r}_i\})|^2 \qquad (2.7)$$

is consequently constant in time. The solutions of the time-independent Schrödinger equation are therefore called **stationary states**. For these states the **expectation value** of the corresponding quantum mechanical operator[4] $\hat{P}$, and thus the average measured value $<P>$ of the physical observable P becomes also time-independent

$$<P> \equiv \frac{\langle\Phi^{(0)}(\{\vec{R}_K\},\{\vec{r}_i\})|\hat{P}|\Phi^{(0)}(\{\vec{R}_K\},\{\vec{r}_i\})\rangle}{\langle\Phi^{(0)}(\{\vec{R}_K\},\{\vec{r}_i\})|\Phi^{(0)}(\{\vec{R}_K\},\{\vec{r}_i\})\rangle} \qquad (2.8)$$

2.2 The Born–Oppenheimer Approximation

The masses of the nuclei, m_K, are at least three orders of magnitude larger than the mass of an electron. We can therefore assume that the electrons will instantaneously adjust to a change in the positions of the nuclei and that we can find a wavefunction for the electrons for each arrangement of nuclei. In the **Born–Oppenheimer approximation** the total molecular Hamilton operator $\hat{H}^{(0)}_{nuc,e}$ from Eq. (2.1) is thus partitioned in the kinetic energy operator of the nuclei, $\frac{1}{2}\sum_K^M \frac{\hat{p}_K^2}{m_K}$, and a **molecular field free electronic Hamiltonian** $\hat{H}^{(0)}$, defined as

$$\hat{H}^{(0)} = \frac{1}{2m_e}\sum_i^N \hat{p}_i^2 - \frac{e^2}{4\pi\epsilon_0}\sum_{iK}^{NM}\frac{Z_K}{|\vec{r}_i-\vec{R}_K|} + \frac{e^2}{4\pi\epsilon_0}\sum_{i<j}\frac{1}{|\vec{r}_i-\vec{r}_j|} \qquad (2.9)$$

$$+ \frac{e^2}{4\pi\epsilon_0}\sum_{K<L}\frac{Z_K Z_L}{|\vec{R}_K-\vec{R}_L|}$$

where the set of nuclear coordinates $\{\vec{R}_K\}$ is held fixed.

Furthermore, setting up a time-independent Schrödinger equation with this operator we obtain the **time-independent field-free electronic Schrödinger equation** for a given set of nuclear coordinates $\{\vec{R}_K\}$

$$\hat{H}^{(0)}|\Psi_k^{(0)}(\{\vec{r}_i\};\{\vec{R}_K\})\rangle = E_k^{(0)}(\{\vec{R}_K\})|\Psi_k^{(0)}(\{\vec{r}_i\};\{\vec{R}_K\})\rangle \qquad (2.10)$$

The solution $\Psi_k^{(0)}(\{\vec{r}_i\};\{\vec{R}_K\})$ of this equation is called the **electronic wavefunction** of the kth electronic state and is a function of the electronic coordinates $\{\vec{r}_i\}$ but

---

[4] It is assumed that the operator $\hat{P}$ is also independent of time.

depends only parametrically on the nuclear coordinates $\{\vec{R}_K\}$. This means that for each set of nuclear coordinates $\{\vec{R}_K\}$ the electronic wavefunction shows a different functional dependence on the electronic coordinates $\{\vec{r}_i\}$. Strictly speaking, the nuclear repulsion energy, $\frac{e^2}{4\pi\epsilon_0}\sum_{K<L}\frac{Z_K Z_L}{|\vec{R}_K - \vec{R}_L|}$, which is a constant for space-fixed nuclei, is not part of the "pure" electronic Hamiltonian, as it does not include any electronic variables. However, it is often added to it such that the eigenvalues of this operator, the electronic energies $E_k^{(0)}(\{\vec{R}_K\})$, become the total energy of a molecule for fixed nuclei.

In the Born–Oppenheimer approximation the total time-independent wavefunction is then approximated by the product of the electronic wavefunction $\Psi_k^{(0)}(\{\vec{r}_i\};\{\vec{R}_K\})$ for the given electronic state k and a **nuclear wavefunction** $\Theta_{v,J}^{(0)}(\{\vec{R}_K\})$, which depends only on the nuclear coordinates $\{\vec{R}_K\}$, i.e.

$$\Phi_{k,v,J}^{(0)}(\{\vec{R}_K\},\{\vec{r}_i\}) = \Psi_k^{(0)}(\{\vec{r}_i\};\{\vec{R}_K\})\,\Theta_{v,J}^{(0)}(\{\vec{R}_K\}) \tag{2.11}$$

The quantum numbers v and J stand collectively for the quantum numbers of all the vibrational modes and for the quantum numbers of the rotational motion of the whole molecule. When we insert this approximate trial wavefunction in the time-independent Schrödinger equation, Eq. (2.5), neglect two small terms [see Exercise 2.2] and make use of the time-independent field free electronic Schrödinger equation, Eq. (2.10), we obtain the **nuclear Schrödinger equation**

$$\left[\frac{1}{2}\sum_K^M \frac{\hat{\vec{p}}_K^2}{m_K} + E_k^{(0)}(\{\vec{R}_K\})\right]|\Theta_{v,J}^{(0)}(\{\vec{R}_K\})\rangle = E_{k,v,J}^{(0)}|\Theta_{v,J}^{(0)}(\{\vec{R}_K\})\rangle \tag{2.12}$$

We can see that the total electronic energy, $E_k^{(0)}(\{\vec{R}_K\})$ as a function of the nuclear coordinates $\{\vec{R}_K\}$ fulfills the role of the potential energy for the motion of the nuclei and is therefore often called a potential-energy surface.

Exercise 2.2 Derive the nuclear Schrödinger equation, Eq. (2.12), and the two terms that are neglected in the Born–Oppenheimer approximation.

In the following, we will consider neither the nuclear Schrödinger equation nor the parametrical dependence of the electronic wavefunction on the nuclear coordinates until Chapter 8. We will therefore also omit the dependence on the nuclear coordinates, $(\{\vec{R}_K\})$, in the notation for the electronic energies and wavefunctions until that chapter.

On the other hand, we will later on use the time-dependent version of Eq. (2.10), i.e. the **time-dependent electronic Schrödinger equation**,

$$\hat{H}^{(0)}\,|\Psi_k^{(0)}(\{\vec{r}_i\},t)\rangle = \imath\hbar\frac{\partial}{\partial t}\,|\Psi_k^{(0)}(\{\vec{r}_i\},t)\rangle \tag{2.13}$$

whose solutions are the electronic stationary states

$$|\Psi_k^{(0)}(\{\vec{r}_i\},t)\rangle = e^{-\frac{\imath}{\hbar}E^{(0)}t}|\Psi_k^{(0)}(\{\vec{r}_i\})\rangle = e^{-\frac{\imath}{\hbar}\hat{H}^{(0)}t}|\Psi_k^{(0)}(\{\vec{r}_i\})\rangle \tag{2.14}$$

2.3 Electron Charge and Current Density

According to Max Born's interpretation of the wavefunction the square of the absolute value of the wavefunction of an electron, $|\psi(\vec{r})|^2 = \psi^*(\vec{r})\,\psi(\vec{r})$, gives the probability density of finding the electron at the point $\vec{r}$. Generalizing to the N-electron case we can say that

$$\left|\Psi_k^{(0)}(\vec{r}_1, \vec{r}_2, \cdots, \vec{r}_N)\right|^2 d\vec{r}_1 d\vec{r}_2 \cdots d\vec{r}_N$$
$$= \Psi_k^{(0)*}(\vec{r}_1, \vec{r}_2, \cdots, \vec{r}_N)\Psi_k^{(0)}(\vec{r}_1, \vec{r}_2, \cdots, \vec{r}_N)\, d\vec{r}_1 d\vec{r}_2 \cdots d\vec{r}_N \quad (2.15)$$

is the probability of finding electron 1 in the volume element $d\vec{r}_1$ at the point $\vec{r}_1$ and at the same time electron 2 in the volume element $d\vec{r}_2$ at point $\vec{r}_2$ and so forth for a system in a state $|\Psi_k^{(0)}(\vec{r}_1, \vec{r}_2, \cdots, \vec{r}_N)\rangle$. If we are only interested in knowing the probability for finding electron 1 in the volume element $d\vec{r}_1$ at point $\vec{r}_1$, i.e. $P_1(\vec{r}_1)\,d\vec{r}_1$, we have to integrate over the coordinates of the other electrons

$$P_1(\vec{r}_1)\,d\vec{r}_1 = \left(\int_{\vec{r}_2}\cdots\int_{\vec{r}_N}\left|\Psi_k^{(0)}(\vec{r}_1, \vec{r}_2, \cdots, \vec{r}_N)\right|^2 d\vec{r}_2\cdots d\vec{r}_N\right) d\vec{r}_1 \quad (2.16)$$

where e.g. $d\vec{r}_2$ stands for $dx_2\,dy_2\,dz_2$ and $\int_{\vec{r}_2}$ denotes a triple integral $\int\int\int$ over the volume element $d\vec{r}_2$. Of course electrons are indistinguishable and the probability of finding electron 2 in a volume element at the same point is identical. The **probability of finding an electron** in the volume element $d\vec{r}$ at a point $\vec{r}$ is therefore N times the probability of finding electron 1 at this point[5] and is thus given as

$$P(\vec{r})\,d\vec{r} = N\left(\int_{\vec{r}_2}\cdots\int_{\vec{r}_N}\left|\Psi_k^{(0)}(\vec{r}, \vec{r}_2, \cdots, \vec{r}_N)\right|^2 d\vec{r}_2\cdots d\vec{r}_N\right) d\vec{r} \quad (2.17)$$

where $P(\vec{r})$ is the **electron density**. The integral over $N-1$ electrons in Eq. (2.17) is inconvenient for actual calculations. Using the properties of the **Dirac δ function**, $\delta(\vec{r}_0)$, in three dimensions

$$\int_{\vec{r}} \delta(\vec{r} - \vec{r}_0) f(\vec{r})\,d\vec{r} = f(\vec{r}_0) \quad (2.18)$$

we can extend the integration to include all electrons and obtain for the electron density

$$P(\vec{r}) = N\int_{\vec{r}_1}\cdots\int_{\vec{r}_N} \delta(\vec{r}_1 - \vec{r})\left|\Psi_k^{(0)}(\vec{r}_1, \cdots, \vec{r}_N)\right|^2 d\vec{r}_1\cdots d\vec{r}_N \quad (2.19)$$

Making use of the indistinguishability of the electrons again we can alternatively write the electron density as an expectation value

$$P(\vec{r}) = \langle \Psi_k^{(0)}(\{\vec{r}_i\})|\hat{D}(\vec{r})|\Psi_k^{(0)}(\{\vec{r}_i\})\rangle \quad (2.20)$$

---

[5] The subscript 1 that indicates coordinates of electron 1 is therefore dropped in the following.

10 The Schrödinger Equation

of the **density operator** $\hat{D}(\vec{r})$, which is defined as

$$\hat{D}(\vec{r}) = \sum_{i}^{N} \delta(\vec{r}_i - \vec{r}) \tag{2.21}$$

This expression will be used later in Part III, when we want to calculate the electron density for various approximate wavefunctions.

Sometimes, it is also necessary to generalize the definition of the electron density $P(\vec{r})$ by defining a **reduced one-electron density matrix**

$$P(\vec{r}, \vec{r}\,') = N \int \cdots \int \Psi_k^{(0)}(\vec{r}, \vec{r}_2, \cdots, \vec{r}_N) \, \Psi_k^{(0)*}(\vec{r}\,', \vec{r}_2, \cdots, \vec{r}_N) \, d\vec{r}_2 \cdots d\vec{r}_N \tag{2.22}$$

Finally, we can define now **the charge density** $\rho^{el}(\vec{r})$ of an N-electron system in a state described by the wavefunction $\Psi_k^{(0)}(\vec{r}_1, \cdots, \vec{r}_N)$ as

$$\rho^{el}(\vec{r}) = -e \, \langle \Psi_k^{(0)}(\{\vec{r}_i\}) | \hat{D}(\vec{r}) | \Psi_k^{(0)}(\{\vec{r}_i\}) \rangle \tag{2.23}$$

In order to derive an expression for the current density of the electrons, $\vec{j}^{\,el}(\vec{r})$, i.e. the flux of the electronic charges, we have to start from the continuity equation of classical electromagnetism

$$\frac{\partial \rho(\vec{r}, t)}{\partial t} = -\vec{\nabla} \cdot \vec{j}(\vec{r}) \tag{2.24}$$

which relates the rate of change of a conserved charge density to the divergence of the current density. This implies taking the time derivative of the time-dependent charge density $\rho^{el}(\vec{r}, t)$ of the electrons, i.e. the generalization of Eq. (2.23) for time-dependent electronic wavefunctions. However, here we will call the variable $\vec{r}_1$ and we will start the derivation from the probability density as given implicitly in Eq. (2.17) but generalized for a time-dependent wavefunction $\Psi_k^{(0)}(\{\vec{r}_i\}, t) = \Psi_k^{(0)}(\vec{r}_1, \vec{r}_2, \cdots, \vec{r}_N, t)$

$$\frac{\partial \rho^{el}(\vec{r}_1, t)}{\partial t} = -eN \frac{\partial}{\partial t} \int_{\vec{r}_2} \cdots \int_{\vec{r}_N} \left| \Psi_k^{(0)}(\vec{r}_1, \vec{r}_2, \cdots, \vec{r}_N, t) \right|^2 d\vec{r}_2 \cdots d\vec{r}_N \tag{2.25}$$

The time derivative of the wavefunction and its complex conjugate is given by the time-dependent Schrödinger equation Eq. (2.13) leading to [see Exercise 2.3]

$$\frac{\partial \rho^{el}(\vec{r}_1, t)}{\partial t} = \frac{-\imath e N}{2 m_e \hbar} \int_{\vec{r}_2} \cdots \int_{\vec{r}_N} \left\{ \Psi_k^{(0)}(\{\vec{r}_i\}, t) \sum_{i=1}^{N} \hat{p}_i^2 \Psi_k^{(0)*}(\{\vec{r}_i\}, t) \right. \tag{2.26}$$

$$\left. - \Psi_k^{(0)*}(\{\vec{r}_i\}, t) \sum_{i=1}^{N} \hat{p}_i^2 \Psi_k^{(0)}(\{\vec{r}_i\}, t) \right\} d\vec{r}_2 \cdots d\vec{r}_N$$

Exercise 2.3 Derive Eq. (2.26) for the time derivative of the electronic charge density using the time-dependent electronic Schrödinger equation, Eq. (2.13).

We split the summation over i now up in the term $i=1$ and the remaining terms $i=2\cdots N$, because the latter terms can be shown to vanish [see Exercise 2.4]. Using, furthermore, that

$$\hat{\vec{\nabla}}\cdot\left[\Psi_k^{(0)}(\{\vec{r}_i\},t)\hat{\vec{\nabla}}\Psi_k^{(0)*}(\{\vec{r}_i\},t)\right]$$
$$=\left[\hat{\vec{\nabla}}\cdot\Psi_k^{(0)}(\{\vec{r}_i\},t)\right]\left[\hat{\vec{\nabla}}\cdot\Psi_k^{(0)*}(\{\vec{r}_i\},t)\right]+\Psi_k^{(0)}(\{\vec{r}_i\},t)\hat{\nabla}^2\Psi_k^{(0)*}(\{\vec{r}_i\},t) \quad (2.27)$$

and a corresponding equation for the expression where $\Psi_k^{(0)}(\{\vec{r}_i\},t)$ and $\Psi_k^{(0)*}(\{\vec{r}_i\},t)$ are interchanged we arrive at

$$\frac{\partial \rho^{el}(\vec{r}_1,t)}{\partial t} = \frac{i\hbar e N}{2m_e}\hat{\vec{\nabla}}_1\cdot\int_{\vec{r}_2}\cdots\int_{\vec{r}_N}\left[\Psi_k^{(0)}(\{\vec{r}_i\},t)\hat{\vec{\nabla}}_1\Psi_k^{(0)*}(\{\vec{r}_i\},t)\right. \quad (2.28)$$
$$\left.-\Psi_k^{(0)*}(\{\vec{r}_i\},t)\hat{\vec{\nabla}}_1\Psi_k^{(0)}(\{\vec{r}_i\},t)\right]\,d\vec{r}_2\cdots d\vec{r}_N$$

Exercise 2.4 Show that the terms for $i=2\cdots N$ in Eq. (2.26) are all zero.

Changing the variable $\vec{r}_1$ to $\vec{r}$ and introducing the Dirac δ function again we can continue in analogy to the derivation of Eq. (2.20) and obtain

$$\frac{\partial \rho^{el}(\vec{r},t)}{\partial t} = \hat{\vec{\nabla}}\cdot\left(\frac{e}{2m_e}\langle\Psi_k^{(0)}(\{\vec{r}_i\},t)|\sum_i^N \delta(\vec{r}_i-\vec{r})\hat{\vec{p}}_i|\Psi_k^{(0)}(\{\vec{r}_i\},t)\rangle\right.$$
$$\left.+\frac{e}{2m_e}\langle\Psi_k^{(0)}(\{\vec{r}_i\},t)|\sum_i^N \delta(\vec{r}_i-\vec{r})\hat{\vec{p}}_i|\Psi_k^{(0)}(\{\vec{r}_i\},t)\rangle^*\right) \quad (2.29)$$

Comparison of this expression with the continuity equation in Eq. (2.24) leads us finally to the desired expression for the **current density**

$$\vec{j}^{el}(\vec{r}) = \frac{e}{2m_e}\langle\Psi_k^{(0)}(\{\vec{r}_i\},t)|\sum_i^N \delta(\vec{r}_i-\vec{r})\hat{\vec{p}}_i|\Psi_k^{(0)}(\{\vec{r}_i\},t)\rangle$$
$$+\frac{e}{2m_e}\langle\Psi_k^{(0)}(\{\vec{r}_i\},t)|\sum_i^N \delta(\vec{r}_i-\vec{r})\hat{\vec{p}}_i|\Psi_k^{(0)}(\{\vec{r}_i\},t)\rangle^* \quad (2.30)$$

of an N-electron system in a state described by the time-dependent wavefunction $\Psi_k^{(0)}(\vec{r}_1,\cdots,\vec{r}_N,t)$. For a stationary state, i.e. a wavefunction of the form of Eq. (2.4), this reduces to

$$\vec{j}^{el}(\vec{r}) = \frac{e}{2m_e}\left(\langle\Psi_k^{(0)}(\{\vec{r}_i\})|\sum_i^N \delta(\vec{r}_i-\vec{r})\hat{\vec{p}}_i|\Psi_k^{(0)}(\{\vec{r}_i\})\rangle\right.$$
$$\left.+\langle\Psi_k^{(0)}(\{\vec{r}_i\})|\sum_i^N \delta(\vec{r}_i-\vec{r})\hat{\vec{p}}_i|\Psi_k^{(0)}(\{\vec{r}_i\})\rangle^*\right) \quad (2.31)$$

2.4 The Force due to Electromagnetic Fields

Before we study the interaction between the electrons in a molecule and classical electromagnetic fields, we should review briefly the relevant equations from classical electromagnetism.

The force $\vec{F}$ of an electromagnetic field on a particle with charge q and mass m is called **the Lorentz force** and is given as

$$\vec{F}(\vec{r},t) = q\left[\vec{\mathcal{E}}(\vec{r},t) + \vec{v}\times\vec{\mathcal{B}}(\vec{r},t)\right] \qquad (2.32)$$

where $\vec{v} = \frac{d\vec{r}}{dt}$ is the velocity of the particle and $\vec{\mathcal{E}}(\vec{r},t)$ and $\vec{\mathcal{B}}(\vec{r},t)$ are the vectors of the electric field and the magnetic induction or flux density, respectively.

In electromagnetism it is often more convenient to work with two potentials instead of the electric and magnetic fields directly. These so-called **scalar potential**, $\phi(\vec{r},t)$, and **vector potential**, $\vec{A}(\vec{r},t)$, are indirectly defined by their relations to the electric and magnetic fields

$$\vec{\mathcal{E}}(\vec{r},t) = -\vec{\nabla}\phi(\vec{r},t) - \frac{\partial \vec{A}(\vec{r},t)}{\partial t} \qquad (2.33)$$

$$\vec{\mathcal{B}}(\vec{r},t) = \vec{\nabla}\times\vec{A}(\vec{r},t) \qquad (2.34)$$

These relations can be derived from Maxwell's equations in vacuum

$$\vec{\nabla}\cdot\vec{\mathcal{E}}(\vec{r},t) = 0 \qquad (2.35)$$

$$\vec{\nabla}\cdot\vec{\mathcal{B}}(\vec{r},t) = 0 \qquad (2.36)$$

$$\vec{\nabla}\times\vec{\mathcal{E}}(\vec{r},t) = -\frac{\partial \vec{\mathcal{B}}(\vec{r},t)}{\partial t} \qquad (2.37)$$

$$\vec{\nabla}\times\vec{\mathcal{B}}(\vec{r},t) = \frac{1}{c^2}\frac{\partial \vec{\mathcal{E}}(\vec{r},t)}{\partial t} \qquad (2.38)$$

where c is the speed of light in vacuum. One should note that electric and magnetic phenomena are coupled in the case of a time-dependent vector potential, due to the time derivative in Eqs. (2.33), (2.37) and (2.38).

One possible solution to Eqs. (2.33) and (2.34) for the case of a static and homogeneous electric field $\vec{\mathcal{E}}$ and magnetic induction $\vec{\mathcal{B}}$ are the potentials

$$\phi^{\mathcal{E}}(\vec{r}) = -\vec{r}\cdot\vec{\mathcal{E}} \qquad (2.39)$$

$$\vec{A}^{\mathcal{B}}(\vec{r}) = \frac{1}{2}\vec{\mathcal{B}}\times\vec{r} \qquad (2.40)$$

which we will meet many times in the following.

However, the scalar and vector potentials are not uniquely defined by Eqs. (2.33) and (2.34). Given an arbitrary scalar function $\chi(\vec{r},t)$, the following transformations of the time-dependent vector potential

$$\vec{A}(\vec{r},t) \to \vec{A}'(\vec{r},t) = \vec{A}(\vec{r},t) + \vec{\nabla}\chi(\vec{r},t) \qquad (2.41)$$

and simultaneously of the time-dependent scalar potential

$$\phi(\vec{r},t) \to \phi'(\vec{r},t) = \phi(\vec{r},t) - \frac{\partial \chi(\vec{r},t)}{\partial t} \tag{2.42}$$

leaves the fields, $\vec{\mathcal{E}}(\vec{r},t)$ and $\vec{\mathcal{B}}(\vec{r},t)$, unchanged [see Exercise 2.5]. The transformations in Eqs. (2.41) and (2.42) are so-called **gauge transformations** of the second kind and $\chi(\vec{r},t)$ is called the gauge function.[6] The consequences of this arbitrariness of the gauge function and thus the gauge transformations for the Schrödinger equation and molecular properties are discussed in more details in sections 2.9 and 5.10.

Exercise 2.5 Show that the electric field $\vec{\mathcal{E}}(\vec{r},t)$ and the magnetic induction $\vec{\mathcal{B}}(\vec{r},t)$ in Eqs. (2.33) and (2.34) are invariant to the gauge transformations in Eqs. (2.41) and (2.42).

Hint: Recall that the curl of a gradient vanishes, i.e. $\vec{\nabla} \times \vec{\nabla} \chi(\vec{r},t) = 0$.

Using the scalar and vector potential we can write the expression for the Lorentz force alternatively as

$$\vec{F} = q \left\{ -\vec{\nabla} \phi(\vec{r},t) - \frac{\partial \vec{A}(\vec{r},t)}{\partial t} + \vec{v} \times \left[\vec{\nabla} \times \vec{A}(\vec{r},t) \right] \right\} \tag{2.43}$$

2.5 Minimal Coupling—Non-Relativistically

The usual way to treat the interaction between electromagnetic fields or nuclear electromagnetic moments and molecules is a semi-classical way, where the fields or nuclear moments are treated classically and the electrons are treated by quantum mechanics. The fields or nuclear moments are thus not part of the system, which is treated quantum mechanically, but they are merely considered to be perturbations that do not respond to the presence of the molecule. They therefore enter the molecular Hamiltonian in terms of external potentials similar to the Coulomb potential due to the charges of the nuclei. This is therefore called the **minimal coupling** approach.

In order to reduce the number of indices and summation sign we derive here the Hamiltonian operator for the motion of a single electron in the presence of external fields. The final equations can then easily be generalized to the many-electron case in Section 2.8.

Before we can derive the additional terms in the Hamiltonian operator due to the interaction with external fields we should recall how in general one constructs the Schrödinger equation for a given system. The standard approach starts from the **classical Hamiltonian** $\mathcal{H}$ for the system that is a function of the position coordinates

---

[6] By choosing a particular form of the gauge function $\chi(\vec{r},t)$ we choose the gauge of the potentials, i.e. we calibrate the potentials. This is somewhat analogous to setting the zero point of a potential energy.

$\vec{r}$ and the canonical momenta $\vec{p}$ of the particles.[7] In the case of a field-free particle the classical Hamilton function is given as

$$\mathcal{H} = \frac{\vec{p}^{\,2}}{2m} \tag{2.44}$$

The time-dependent Schrödinger equation is then obtained if one replaces the functions by operators using the substitutions rules

$$\vec{p} \to \hat{\vec{p}} = -i\hbar \hat{\vec{\nabla}} \tag{2.45}$$

$$\mathcal{H} \to i\hbar \frac{\partial}{\partial t} \tag{2.46}$$

and lets both sides of Eq. (2.44) act on the wavefunction of the given system. In the example of the field-free particle this leads to the following time-dependent Schrödinger equation

$$i\hbar \frac{\partial}{\partial t} |\psi(\vec{r},t)\rangle = \frac{\hat{\vec{p}}^{\,2}}{2m} |\psi(\vec{r},t)\rangle \tag{2.47}$$

The problem for us is therefore to derive the classical Hamiltonian function for an electron in the presence of electromagnetic fields, which is normally done from the **classical Lagrangian**. Hamilton's and Lagrange's generalizations of classical mechanics are essentially the same theory as Newton's formulation but are more elegant and often computationally easier to use.[8] In our context, their importance lies in the fact that they serve as a springboard to quantum mechanics.

The classical Lagrangian $\mathcal{L}$

$$\mathcal{L}(\vec{r},\vec{v},t) = T(\vec{r},\vec{v}) - U(\vec{r},\vec{v},t) \tag{2.48}$$

is a function of generalized position coordinates $\vec{r}$ and their time derivatives, i.e. the generalized velocities $\vec{v}$. The generalized coordinates can, but need not, be cartesian coordinates. Alternatively, they could be any kind of polar coordinates or the set of independent coordinates in a system with one or more constraints. $T(\vec{r},\vec{v})$ is the kinetic energy and $U(\vec{r},\vec{v},t)$ is a **generalized potential**. The latter has to be chosen in such a way that on application of **Lagrange's equations of motion**, also called the **Euler–Lagrange equations**

$$\frac{d}{dt}\left[\frac{\partial \mathcal{L}}{\partial v_\alpha}\right] - \frac{\partial \mathcal{L}}{\partial r_\alpha} = 0 \tag{2.49}$$

Newton's second law

$$\vec{F} = m\frac{d\vec{v}}{dt} \tag{2.50}$$

is recovered.[9]

---

[7] The canonical momentum is often also called generalized or conjugate momentum. Its precise definition is given later in this section.

[8] References to more detailed discussions and to derivations of the Lagrangian and Hamiltonian formulations of classical mechanics can be found in the *Further Reading* section.

[9] Contrary to Newton's second law the Lagrangian and Hamiltonian formulations of classical mechanics are form invariant under a change of coordinates, i.e. the Euler–Lagrange equations have

In our case of the motion of an electron in external fields the respective force is the Lorentz force given in Eq. (2.43) and we therefore have to find a generalized potential U that will reproduce the following relation

$$-e\left\{-\vec{\nabla}\phi(\vec{r},t)-\frac{\partial \vec{A}(\vec{r},t)}{\partial t}+\vec{v}\times\left[\vec{\nabla}\times\vec{A}(\vec{r},t)\right]\right\}=m_e\frac{d\vec{v}}{dt} \qquad (2.51)$$

Let us try to use
$$U(\vec{r},\vec{v},t)=-e\,\phi(\vec{r},t)+e\,\vec{v}\cdot\vec{A}(\vec{r},t) \qquad (2.52)$$
as generalized potential U, which gives the following expression for the Lagrangian

$$\mathcal{L}(\vec{r},\vec{v},t)=\frac{m_e\vec{v}^{\,2}}{2}+e\,\phi(\vec{r},t)-e\,\vec{v}\cdot\vec{A}(\vec{r},t) \qquad (2.53)$$

The potential U, in Eq. (2.52), is called the generalized potential because it depends not only on the position of the electron and on time but also on the velocity of the electron. When we insert this Lagrangian in Lagrange's equations, Eq. (2.49), we obtain Newton's second law, Eq. (2.51) [see Exercise 2.6], which proves that our choice of the generalized potential in Eq. (2.52) was indeed correct.

Exercise 2.6 Show that the Lagrangian in Eq. (2.53) fulfills the Lagrange equations (2.49), which means that on inserting the Lagrangian into Lagrange equations one obtains Eq. (2.51).

From this Lagrangian one can then define a **classical Hamiltonian function**, which translated to operator form yields the desired Hamiltonian operator. The classical Hamiltonian $\mathcal{H}$ is a function of time, of the generalized position coordinates $\vec{r}$ and of their conjugated generalized momenta $\vec{p}$. It is defined as[10]

$$\mathcal{H}(\vec{r},\vec{p},t)=\vec{p}\cdot\vec{v}-\mathcal{L}(\vec{r},\vec{v},t) \qquad (2.54)$$

The components[11] of the generalized momentum vector $\vec{p}$, the canonical conjugate to $\vec{r}$, are given as

$$p_\alpha=\frac{\partial \mathcal{L}(\vec{r},\vec{v},t)}{\partial v_\alpha} \qquad (2.55)$$

For the Lagrangian in Eq. (2.53) this definition yields

$$\vec{p}=m_e\,\vec{v}-e\vec{A}(\vec{r},t) \qquad (2.56)$$

the same form for all types of coordinates. This is one of several advantages of the Lagrangian formulation.

---

[10] The relation between the classical Lagrangian and Hamiltonian is called a Legendre transformation and the Lagrangian and Hamiltonian are called Legendre transforms of each other. The purpose of this transformation is to exchange the role of the velocities and conjugated momenta as the independent variables. Legendre transformations are well known in thermodynamics where e.g. the enthalpy $H(S,P)$ and Gibbs free energy $G(T,P) = H(S,P) - TS$ are Legendre transforms of each other.

[11] Components of a vector or tensor are denoted by small greek subscripts $\alpha, \beta, \gamma, \ldots$. They will typically represent one of the cartesian components x, y, z.

16 The Schrödinger Equation

We should note that in the presence of electromagnetic fields the **canonical momentum** $\vec{p}$ is no longer equal to the product of mass and velocity. The latter is therefore often also called the **kinematical or mechanical momentum** $\vec{\pi}$

$$\vec{\pi} \equiv m_e \vec{v} = \vec{p} + e\vec{A}(\vec{r},t) \tag{2.57}$$

because the kinetic energy is defined as

$$T = \frac{1}{2} m_e \vec{v}^{\,2} = \frac{1}{2m_e} \vec{\pi}^{\,2} \tag{2.58}$$

Inserting the expression for the canonical momentum, Eq. (2.56), in the definition of the classical Hamiltonian, Eq. (2.54), we then obtain

$$\mathcal{H} = \frac{m_e \vec{v} \cdot \vec{v}}{2} - e\,\phi(\vec{r},t) \tag{2.59}$$

However, in order to use the usual substitution rule, Eq. (2.45), for the transition to quantum mechanics, the classical Hamiltonian has to be written in terms of the canonical momentum $\vec{p}$, and not the velocity $\vec{v}$. But with the help of Eq. (2.57) we can replace the mechanical momentum by the canonical momentum and get

$$\mathcal{H} = \frac{1}{2m_e} \left[\vec{p} + e\vec{A}(\vec{r},t) \right]^2 - e\,\phi(\vec{r},t) \tag{2.60}$$

When we apply now the substitution rules, Eqs. (2.45) and (2.46), and let both sides act on the time-dependent wavefunction of the electron, $|\psi(\vec{r},t)\rangle$, we obtain the time-dependent Schrödinger equation

$$i\hbar \frac{\partial}{\partial t} |\psi(\vec{r},t)\rangle = \hat{H}\, |\psi(\vec{r},t)\rangle \tag{2.61}$$

where the quantum mechanical **Hamiltonian operator** $\hat{H}$ for a single particle is given as

$$\hat{H} = \frac{1}{2m_e} \left[\hat{\vec{p}} + e\hat{\vec{A}}(\vec{r},t) \right]^2 - e\,\hat{\phi}(\vec{r},t) \tag{2.62}$$

In the **Coulomb gauge**, where one chooses $\vec{\nabla} \cdot \vec{A} = 0$ [see Exercise 2.7], this can then be written as

$$\hat{H} = \frac{1}{2m_e} \hat{\vec{p}}^{\,2} + \frac{1}{m_e} e\, \hat{\vec{A}}(\vec{r},t) \cdot \hat{\vec{p}} + \frac{1}{2m_e} e^2 \left[\hat{\vec{A}}(\vec{r},t)\right]^2 - e\,\hat{\phi}(\vec{r},t) \tag{2.63}$$

This is a non-relativistic, Schrödinger, Hamiltonian for a single, spin-less particle. In Section 2.8 it will be generalized to the case of many particles, electrons and nuclei.

Exercise 2.7 Show that in the static case the vector potential $\vec{A}$ can be chosen to be divergence free, i.e. $\vec{\nabla} \cdot \vec{A} = 0$, without effect on the magnetic induction. Secondly, investigate what consequences this choice has for time-dependent potentials and fields.

Hint: Remember that any vector field $\vec{F}(\vec{r},t)$ can be separated in two components

$$\vec{F}(\vec{r},t) = \vec{F}_T(\vec{r},t) + \vec{F}_L(\vec{r},t)$$

where $\vec{F}_T(\vec{r},t)$ and $\vec{F}_L(\vec{r},t)$ are the transverse or solenoidal and longitudinal or irrotational components that are defined by the following relations

$$\nabla \cdot \vec{F}_T(\vec{r},t) = 0$$

$$\nabla \times \vec{F}_L(\vec{r},t) = 0$$

2.6 Minimal Coupling—Relativistically

In the last section, a non-relativistic Hamiltonian for a spin-less particle was derived. However, electrons have spin and in general it would be desirable to use a Hamiltonian operator that fulfills the requirements of special relativity. The so-called **Dirac Hamiltonian operator** is such a relativistic operator for a single particle in the presence of an electromagnetic field. It can be derived in the same ways as the non-relativistic analogue was obtained in the previous section.

The Lorentz force in Eq. (2.43) is unchanged in special relativity, because electromagnetism in Maxwell's formulation fulfills the requirements of special relativity.[12] Newton's second law, on the other hand,

$$\vec{F} = \frac{d}{dt}(m^r \vec{v}) \tag{2.64}$$

is changed due to the velocity dependence of the relativistic mass m^r

$$m^r = \frac{m}{\sqrt{1 - \frac{\vec{v}^2}{c^2}}} \tag{2.65}$$

where m is the rest mass.

In complete analogy to the non-relativistic case we have to set up a Lagrangian again that, inserted in Lagrange's equations, Eq. (2.49), should yield Newton's second law, Eq. (2.64). The following Lagrangian

$$\mathcal{L}(\vec{r},\vec{v},t) = -m_e\, c^2 \sqrt{1 - \frac{\vec{v}^2}{c^2}} + e\, \phi(\vec{r},t) - e\, \vec{v} \cdot \vec{A}(\vec{r},t) \tag{2.66}$$

can be shown to have the correct form [see Exercise 2.8].

Exercise 2.8 Show that the relativistic Lagrangian in Eq. (2.66) also fulfills the Lagrange equations (2.49).

The components of the canonical momentum vector are again obtained as partial derivatives of the Lagrangian

---

[12] Maxwell's theory of electromagnetism was actually the first theory that fulfilled the requirements of special relativity (i.e. the equations are invariant under a Lorentz transformation), even before special relativity was formulated by Einstein.

18 The Schrödinger Equation

$$p_\alpha = \frac{\partial \mathcal{L}(\vec{r},\vec{v},t)}{\partial v_\alpha} = \frac{m_e\, v_\alpha}{\sqrt{1-\frac{\vec{v}^{\,2}}{c^2}}} - e\, A_\alpha(\vec{r},t) \tag{2.67}$$

With that, we have now all the necessary ingredients for the classical Hamiltonian according to Eq. (2.54) [see Exercise 2.9]

$$\mathcal{H} = \frac{m_e\, c^2}{\sqrt{1-\frac{\vec{v}^{\,2}}{c^2}}} - e\, \phi(\vec{r},t) \tag{2.68}$$

In terms of the canonical momentum the classical Hamiltonian can be written as

$$\mathcal{H} = \sqrt{m_e^2 c^4 + c^2\left[\vec{p} + e\,\vec{A}(\vec{r},t)\right]^2} - e\, \phi(\vec{r},t) \tag{2.69}$$

Exercise 2.9 Derive the expressions for the classical relativistic Hamiltonian in Eqs. (2.68) and (2.69).

Hint: In the second step you might want to use the following relation

$$\frac{m_e^2\, c^4}{1-\frac{\vec{v}^{\,2}}{c^2}} = m_e^2\, c^4 + c^2 \frac{m_e^2\, \vec{v}^{\,2}}{1-\frac{\vec{v}^{\,2}}{c^2}}$$

in order to replace $\vec{v}$ by $\vec{p}$.

Because of the square root, it is not possible to make the transition to quantum mechanics. However, if we write the term underneath the square root as a perfect square of something, we can continue. Therefore, Dirac proposed the following relation

$$m_e^2 c^4 + c^2\left[\vec{p}+e\,\vec{A}(\vec{r},t)\right]^2 = \left\{\beta\, m_e\, c^2 + c \sum_{\mu=x,y,z}\alpha_\mu\,[p_\mu + e\, A_\mu(\vec{r},t)]\right\}^2 \tag{2.70}$$

where the *a priori* unknown α's and β have to fulfill the conditions

$$\alpha_\mu^2 = \beta^2 = 1 \quad \text{for } \mu = x, y, z \tag{2.71}$$

$$\alpha_\mu \alpha_\nu + \alpha_\nu \alpha_\mu = 0 \quad \text{for } \mu \neq \nu \tag{2.72}$$

$$\alpha_\mu \beta + \beta \alpha_\mu = 0 \quad \text{for } \mu = x, y, z \tag{2.73}$$

in order for Eq. (2.70) to be fulfilled.

It turns out [see Exercise 2.10] that the simplest solution to these equations are a set of 4×4 matrices defined as

$$\beta = \begin{pmatrix} I_2 & 0_2 \\ 0_2 & -I_2 \end{pmatrix}, \quad \alpha_\mu = \begin{pmatrix} 0_2 & \sigma_\mu \\ \sigma_\mu & 0_2 \end{pmatrix} \tag{2.74}$$

where I_2 and 0_2 are the 2×2 unit and zero matrices

$$I_2 = \begin{pmatrix} 1 & 0 \\ 0 & 1 \end{pmatrix}, \quad 0_2 = \begin{pmatrix} 0 & 0 \\ 0 & 0 \end{pmatrix} \tag{2.75}$$

and the σ_μ are the **Pauli spin matrices**

$$\sigma_x = \begin{pmatrix} 0 & 1 \\ 1 & 0 \end{pmatrix}, \quad \sigma_y = \begin{pmatrix} 0 & -i \\ i & 0 \end{pmatrix}, \quad \sigma_z = \begin{pmatrix} 1 & 0 \\ 0 & -1 \end{pmatrix} \tag{2.76}$$

Exercise 2.10 Show that the α matrices in Eq. (2.74) fulfill the conditions (2.71), (2.72) and (2.73).

Hint: You may want to make use of the commutator and anti-commutator relations of the Pauli spin matrices

$$[\sigma_i, \sigma_j] = 2i\,\epsilon_{ijk}\,\sigma_k \quad \text{or} \quad \sigma_i\sigma_j = i\,\epsilon_{ijk}\,\sigma_k$$

and

$$[\sigma_i, \sigma_j]_+ = 2\,\delta_{ij}\,\boldsymbol{I}_2 \quad \text{or} \quad \sigma_i\sigma_i = \boldsymbol{I}_2$$

where ϵ_{ijk} is the Levi-Civita symbol (see e.g. Mills et al., 1993), defined as

$$\epsilon_{ijk} = \begin{cases} 1 & \text{if } i,j,k \text{ is an even permutation of } x,y,z \\ -1 & \text{if } i,j,k \text{ is an odd permutation of } x,y,z \\ 0 & \text{if any index is repeated} \end{cases}$$

The classical Hamiltonian can therefore be rewritten as

$$\mathcal{H} = \boldsymbol{\beta}\,m_e\,c^2 + c\sum_{\mu=x,y,z}\boldsymbol{\alpha}_\mu\left[p_\mu + e\,A_\mu(\vec{r},t)\right] - e\,\phi(\vec{r},t)\boldsymbol{I}_4 \tag{2.77}$$

where $\boldsymbol{I}_4$ is a 4×4 unit matrix.

Finally, we can now apply the substitution rules, Eqs. (2.45) and (2.46), let both sides act on the time-dependent wavefunction of the electron, $|\psi(\vec{r},t)\rangle$, and obtain in this way the time-dependent **Dirac equation**

$$i\hbar\frac{\partial}{\partial t}|\psi(\vec{r},t)\rangle = \left\{c\sum_{\mu=x,y,z}\boldsymbol{\alpha}_\mu\left[\hat{p}_\mu + e\,\hat{A}_\mu(\vec{r},t)\right] - e\,\hat{\phi}(\vec{r},t)\boldsymbol{I}_4 + \boldsymbol{\beta}\,m_e\,c^2\right\}|\psi(\vec{r},t)\rangle \tag{2.78}$$

Because $\boldsymbol{\beta}$ and the $\boldsymbol{\alpha}$'s are 4×4 matrices, the wavefunction ψ will consist of four components

$$\psi = \begin{pmatrix} \psi_L \\ \psi_S \end{pmatrix} = \begin{pmatrix} \psi_1 \\ \psi_2 \\ \psi_3 \\ \psi_4 \end{pmatrix} \tag{2.79}$$

which one calls a four-component spinor. The Dirac equation is therefore a set of four coupled differential equations that couple the four components of the wavefunction. The two-component spinors ψ_L and ψ_S are called the large and small component of the wavefunction, respectively, because ψ_L is the main component of the wavefunction for electrons. For a positron, on the other hand, the small component would be the main component of the wavefunction.

Substituting Eq. (2.74) for the $\boldsymbol{\alpha}$ matrices and Eq. (2.79) for the four-component wavefunction, the Dirac equation can alternatively be written as two coupled two-component equations

$$\left[c \sum_{\alpha=x,y,z} \sigma_\alpha \left(\hat{p}_\alpha + e\, \hat{A}_\alpha\right)\right] |\psi_S\rangle + \left(-e\, \hat{\phi} + m_e\, c^2\right) |\psi_L\rangle = i\hbar \frac{\partial}{\partial t} |\psi_L\rangle \quad (2.80)$$

$$\left[c \sum_{\alpha=x,y,z} \sigma_\alpha \left(\hat{p}_\alpha + e\, \hat{A}_\alpha\right)\right] |\psi_L\rangle + \left(-e\, \hat{\phi} - m_e\, c^2\right) |\psi_S\rangle = i\hbar \frac{\partial}{\partial t} |\psi_S\rangle \quad (2.81)$$

2.7 Elimination of the Small Component

We could continue now with the Dirac equation and derive expressions for the molecular properties using standard perturbation theory. However, as stated earlier, the exposition in these notes is restricted basically to a non-relativistic treatment with the exception that we want to include also interactions with the spin of the electrons. The appropriate operator can be found by reduction of the Dirac equation to a non-relativistic two-component form, which can be achieved by several approaches.[13] Here, we want to discuss only the simplest approach, the so-called elimination of the small component.

We assume that the potentials $\hat{\phi}(\vec{r})$ and $\hat{\vec{A}}(\vec{r})$ are time independent and collect the time dependence of the wavefunction in a phase factor

$$|\psi(\vec{r},t)\rangle = |\bar{\psi}(\vec{r})\rangle\, e^{-iEt/\hbar} \quad (2.82)$$

which implies that $\psi(\vec{r},t)$ is an eigenfunction of $i\hbar \frac{\partial}{\partial t}$ with eigenvalue E. Inserting this wavefunction in Eqs. (2.80) and (2.81) and rearranging we obtain

$$c \left\{\sum_{\alpha=x,y,z} \sigma_\alpha \left[\hat{p}_\alpha + e\, \hat{A}_\alpha(\vec{r})\right]\right\} |\bar{\psi}_S\rangle = \left(E + e\, \hat{\phi}(\vec{r}) - m_e\, c^2\right) |\bar{\psi}_L\rangle \quad (2.83)$$

$$c \left\{\sum_{\alpha=x,y,z} \sigma_\alpha \left[\hat{p}_\alpha + e\, \hat{A}_\alpha(\vec{r})\right]\right\} |\bar{\psi}_L\rangle = \left(E + e\, \hat{\phi}(\vec{r}) + m_e\, c^2\right) |\bar{\psi}_S\rangle \quad (2.84)$$

From Eq. (2.84) we can see that the small component of the wavefunction $|\bar{\psi}_S\rangle$ can be expressed in terms of the large component as

$$|\bar{\psi}_S\rangle = \frac{c}{E + e\, \hat{\phi}(\vec{r}) + m_e\, c^2} \left\{\sum_{\alpha=x,y,z} \sigma_\alpha \left[\hat{p}_\alpha + e\, \hat{A}_\alpha(\vec{r})\right]\right\} |\bar{\psi}_L\rangle \quad (2.85)$$

---

[13] See the references mentioned in the *Further Reading* section.

Inserting this expression in Eq. (2.83) we obtain a single two-component equation for the large component

$$\left\{\sum_{\alpha=x,y,z}\sigma_\alpha\left[\hat{p}_\alpha+e\,\hat{A}_\alpha(\vec{r})\right]\right\}\frac{c^2}{E+e\,\hat{\phi}(\vec{r})+m_e\,c^2}\left\{\sum_{\alpha=x,y,z}\sigma_\alpha\left[\hat{p}_\alpha+e\,\hat{A}_\alpha(\vec{r})\right]\right\}|\bar{\psi}_L\rangle$$
$$=\left(E+e\,\hat{\phi}(\vec{r})-m_e\,c^2\right)|\bar{\psi}_L\rangle \tag{2.86}$$

This equation together with the expression for the small component in Eq. (2.85) is still the Dirac equation. In order to reduce it to a non-relativistic expression we have to expand $\frac{c^2}{E+e\,\hat{\phi}(\vec{r})+m_e\,c^2}$. If we introduce the non-relativistic energy $E^{NR}=E-m_e\,c^2$, we can rewrite the denominator as

$$E+e\,\hat{\phi}(\vec{r})+m_e\,c^2=2\,m_e\,c^2+E^{NR}+e\,\hat{\phi}(\vec{r}) \tag{2.87}$$

Since $2\,m_e\,c^2$ is of the order of MeV we can assume that $E^{NR}+e\,\hat{\phi}(\vec{r})\ll 2\,m_e\,c^2$ and thus expand $\frac{c^2}{E+e\,\hat{\phi}(\vec{r})+m_e\,c^2}$ as

$$\frac{c^2}{E+e\,\hat{\phi}(\vec{r})+m_e\,c^2}=\frac{1}{2\,m_e}\left(\frac{1}{1+\frac{E^{NR}+e\,\hat{\phi}(\vec{r})}{2\,m_e\,c^2}}\right)$$
$$=\frac{1}{2\,m_e}\left(1-\frac{E^{NR}+e\,\hat{\phi}(\vec{r})}{2\,m_e\,c^2}+\cdots\right) \tag{2.88}$$

When we use only the first term, the equation for the large component reads

$$\frac{1}{2\,m_e}\left\{\sum_{\alpha=x,y,z}\sigma_\alpha\left[\hat{p}_\alpha+e\,\hat{A}_\alpha(\vec{r})\right]\right\}\left\{\sum_{\alpha=x,y,z}\sigma_\alpha\left[\hat{p}_\alpha+e\,\hat{A}_\alpha(\vec{r})\right]\right\}|\bar{\psi}_L\rangle$$
$$=\left[E^{NR}+e\,\hat{\phi}(\vec{r})\right]|\bar{\psi}_L\rangle \tag{2.89}$$

The left-hand side of this equation can be simplified, if we make use of a relation that holds for the Pauli spin matrices and two general, spin-free vector operators $\hat{\vec{C}}$ and $\hat{\vec{D}}$ with components $\hat{C}_\alpha$ and $\hat{D}_\alpha$ [see Exercise 2.11]

$$\left(\sum_{\alpha=x,y,z}\sigma_\alpha\hat{C}_\alpha\right)\left(\sum_{\beta=x,y,z}\sigma_\beta\hat{D}_\beta\right)=\left(\hat{\vec{C}}\cdot\hat{\vec{D}}\right)I_2+\imath\sum_{\alpha=x,y,z}\sigma_\alpha\left(\hat{\vec{C}}\times\hat{\vec{D}}\right)_\alpha \tag{2.90}$$

and a relation that holds for the gradient operator $\hat{\vec{\nabla}}$, a general vector operator $\hat{\vec{C}}$ and a scalar function ψ [see Exercise 2.12]

$$\hat{\vec{\nabla}}\times\left(\vec{C}\psi\right)=-\vec{C}\times\hat{\vec{\nabla}}\psi+\left(\hat{\vec{\nabla}}\times\vec{C}\right)\psi \tag{2.91}$$

Exercise 2.11 Prove relation (2.90).

Hint: You may want to make use again of the commutator and anti-commutator relations of the Pauli spin matrices given in Exercise 2.10.

Exercise 2.12 Prove relation (2.91).

The **non-relativistic Schrödinger-Pauli equation** can then finally be written as [see Exercise 2.13]

$$\left\{ \frac{1}{2m_e} \left[\hat{\vec{p}} + e\hat{\vec{A}}(\vec{r}) \right]^2 \mathbf{I}_2 + \frac{e\hbar}{2m_e} \sum_{\alpha=x,y,z} \boldsymbol{\sigma}_\alpha \left[\hat{\vec{\nabla}} \times \hat{\vec{A}}(\vec{r}) \right]_\alpha - e\hat{\phi}(\vec{r}) \mathbf{I}_2 \right\} |\bar{\psi}_L\rangle$$
$$= E^{NR} |\bar{\psi}_L\rangle \tag{2.92}$$

Exercise 2.13 Derive Eq. (2.92) from Eq. (2.89) using Eqs. (2.90) and (2.91).

On comparison with the Schrödinger Hamiltonian in Eq. (2.62) we can identify the additional term due to the interaction of the electron spin with a magnetic field, a so-called **Zeeman term**

$$\hat{H}^{Zeeman} = \frac{e\hbar}{2m_e} \sum_{\alpha=x,y,z} \boldsymbol{\sigma}_\alpha \left[\hat{\vec{\nabla}} \times \hat{\vec{A}}(\vec{r}) \right]_\alpha \tag{2.93}$$

The electron spin operator $\hat{\vec{s}}$ in units of Js is related to the Pauli spin matrices $\vec{\sigma}$ by

$$\hat{\vec{s}} = \frac{\hbar}{2} \vec{\sigma} \tag{2.94}$$

and the electron spin Zeeman operator becomes therefore

$$\hat{H}^{Zeeman} = \frac{2e}{2m_e} \hat{\vec{s}} \cdot \left[\hat{\vec{\nabla}} \times \hat{\vec{A}}(\vec{r}) \right] \tag{2.95}$$

However, from quantum electrodynamics we know that this should be written as

$$\hat{H}^{Zeeman} = \frac{g_e e}{2m_e} \hat{\vec{s}} \cdot \left[\hat{\vec{\nabla}} \times \hat{\vec{A}}(\vec{r}) \right] \tag{2.96}$$

where $g_e \approx 2.0023$ is the electron g-factor.

Including also the next term of the expansion, Eq. (2.88), gives rise to additional operators including the mass-velocity, Darwin and one-electron spin-orbit operators, which can be used in perturbation theory calculations of relativistic corrections to the non-relativistic results of the Schrödinger equation and molecular properties. However, the expansion is based on the assumption that the scalar potential $\hat{\phi}(\vec{r})$ is small, which is not fulfilled for the inner electrons of heavy atoms, because close to the nucleus they are exposed to the strong Coulomb potential of the nucleus. For this situation the expansion is then no longer valid. Alternative expansions exist, which circumvent this problem, like, e.g. the zeroth-order regular approximation (ZORA) by Chang et al. (1986) and van Lenthe et al. (1993).

2.8 The Molecular Electronic Hamiltonian

In the last three sections we have considered the effect of a time-dependent external electric field $\vec{\mathcal{E}}(\vec{r}, t)$ and a magnetic induction $\vec{\mathcal{B}}(\vec{r}, t)$ on the motion of an electron and denoted the corresponding potentials with $\phi(\vec{r}, t)$ and $\vec{A}(\vec{r}, t)$. In the present section we want to collect all the terms and derive our final expression for the molecular electronic Hamiltonian. However, we will not restrict ourselves to the case of external fields because in the following chapters we want to study also interactions with other sources of electromagnetic fields such as magnetic dipole moments and electric quadrupole moments of the nuclei, the rotation of the molecule as well as interactions with field gradients. Therefore, we do not include the superscripts $\mathcal{B}$ and $\mathcal{E}$ on the vector and scalar potential in this section. On the other hand, we will assume that the perturbations are time independent. The time-dependent case is considered in Section 3.9.

In the previous sections it was shown that in the minimal coupling approximation the vector potential enters the mechanical momentum of electron i

$$\hat{\vec{\pi}}_i = m_e \hat{\vec{v}}_i = \hat{\vec{p}}_i + e\hat{\vec{A}}(\vec{r}_i) \tag{2.97}$$

As we are working within the Born–Oppenheimer approximation the nuclei are fixed in space and there is thus no coupling between the momenta of the nuclei and the vector potential.

Secondly, terms consisting of the scalar potential times the charges of the particles have to be added to the Hamiltonian. Although we are only interested in the electronic Hamiltonian, one should also add the constant contribution from the interaction of the scalar potential with the nuclear charges. In total, the following terms due to the scalar potential have to be added

$$-e \sum_i^N \hat{\phi}(\vec{r}_i) + e \sum_K^M Z_K \hat{\phi}(\vec{R}_K) \tag{2.98}$$

The electronic Hamiltonian becomes then

$$\hat{H} = \frac{1}{2m_e} \sum_i^N \left[\hat{\vec{p}}_i + e\hat{\vec{A}}(\vec{r}_i) \right]^2 - \frac{e^2}{4\pi\epsilon_0} \sum_{iK}^{NM} \frac{Z_K}{|\vec{r}_i - \vec{R}_K|}$$
$$+ \frac{e^2}{4\pi\epsilon_0} \sum_{K<L} \frac{Z_K Z_L}{|\vec{R}_K - \vec{R}_L|} + \frac{e^2}{4\pi\epsilon_0} \sum_{i<j} \frac{1}{|\vec{r}_i - \vec{r}_j|} \tag{2.99}$$
$$- \sum_i^N e\hat{\phi}(\vec{r}_i) + e \sum_K^M Z_K \hat{\phi}(\vec{R}_K)$$

Thirdly, the interaction of the spin of the electrons with magnetic fields is introduced via the Zeeman term of the Pauli Hamiltonian, Eq. (2.96),

$$\hat{H}^{Zeeman} = \sum_i^N \frac{g_e e}{2m_e} \hat{\vec{s}}_i \cdot \left[\hat{\vec{\nabla}} \times \hat{\vec{A}}(\vec{r}_i)\right] \quad (2.100)$$

Collecting all terms we can finally write the molecular electronic Hamiltonian operator $\hat{H}$ in the presence of an electromagnetic field as

$$\hat{H} = \hat{H}^{(0)} + \hat{H}^{(1)} + \hat{H}^{(2)}$$
$$= \sum_i^N \hat{h}^{(0)}(i) + \sum_{i<j} \hat{g}(i,j) + \hat{H}^{(0)}_{nuc} + \sum_i^N \hat{h}^{(1)}(i) + \hat{H}^{(1)}_{nuc} + \sum_i^N \hat{h}^{(2)}(i) \quad (2.101)$$

$\hat{H}^{(0)}$ is the unperturbed Hamiltonian from Eq. (2.9) and contains one-electron $\hat{h}^{(0)}(i)$, two-electron $\hat{g}(i,j)$ and nuclear $\hat{H}^{(0)}_{nuc}$ contributions

$$\hat{h}^{(0)}(i) = \frac{1}{2m_e}\hat{\vec{p}}_i^2 - \frac{e^2}{4\pi\epsilon_0}\sum_K^M \frac{Z_K}{|\vec{r}_i - \vec{R}_K|} \quad (2.102)$$

$$\hat{g}(i,j) = \frac{e^2}{4\pi\epsilon_0}\frac{1}{|\vec{r}_i - \vec{r}_j|} \quad (2.103)$$

$$\hat{H}^{(0)}_{nuc} = \frac{e^2}{4\pi\epsilon_0}\sum_{K<L} \frac{Z_K Z_L}{|\vec{R}_K - \vec{R}_L|} \quad (2.104)$$

$\hat{H}^{(1)}$ includes all one-electron, $\hat{h}^{(1)}(i)$, and nuclear, $\hat{H}^{(1)}_{nuc}$, operators, which are linear in the perturbing field and thus first order

$$\hat{h}^{(1)}(i) = \frac{e}{m_e}\hat{\vec{A}}(\vec{r}_i) \cdot \hat{\vec{p}}_i + \frac{g_e e}{2m_e}\hat{\vec{s}}_i \cdot \left[\hat{\vec{\nabla}} \times \hat{\vec{A}}(\vec{r}_i)\right] - e\,\hat{\phi}(\vec{r}_i) \quad (2.105)$$

$$\hat{H}^{(1)}_{nuc} = e\sum_K^M Z_K \hat{\phi}(\vec{R}_K) \quad (2.106)$$

where we have assumed the Coulomb gauge, i.e. $\vec{\nabla} \cdot \vec{A} = 0$, again. Finally, $\hat{H}^{(2)}$ contains the one-electron operators quadratic in the perturbations and is thus second order

$$\hat{h}^{(2)}(i) = \frac{e^2}{2m_e}\hat{\vec{A}}^2(\vec{r}_i) \quad (2.107)$$

In Chapters 4, 5 and 6 explicit forms for these **perturbation Hamiltonian operators** will be derived by expressing the scalar and vector potentials in terms of components of the electric field $\mathcal{E}_\alpha$, the electric field gradient $\mathcal{E}_{\alpha\beta}$, the magnetic induction $\mathcal{B}_\alpha$, the nuclear moment m_α^K and the rotation of the molecule. The resulting operators are also collected in Appendix A.

In the meantime we will in Chapter 3 discuss perturbations by a general field with tensor components $\mathcal{F}_{\alpha\ldots}$. With the notation $\mathcal{F}_{\alpha\ldots}$ we will cover both vector fields,

$\vec{\mathcal{F}}$, with three components $\mathcal{F}_\alpha$ as well as second-rank tensor fields, $\mathcal{F}$, with nine components $\mathcal{F}_{\alpha\beta}$ like the electric-field gradient tensor $\mathcal{E}$. The first and second-order perturbation Hamiltonians $\hat{H}^{(1)} + \hat{H}^{(2)}$, in Eq. (2.101), will then be expressed as the scalar or tensor product of the perturbations, i.e. fields or magnetic moments, and two **interaction** or **perturbation operators** $\hat{O}^{\mathcal{F}}_{\alpha\ldots}$ and $\hat{O}^{\mathcal{FF}}_{\alpha\beta\ldots}$.

$$\hat{H}^{(1)} + \hat{H}^{(2)} = \sum_{\alpha\ldots} \hat{O}^{\mathcal{F}}_{\alpha\ldots}\, \mathcal{F}_{\alpha\ldots} + \sum_{\alpha,\beta,\ldots} \mathcal{F}_{\alpha\ldots}\, \hat{O}^{\mathcal{FF}}_{\alpha\beta\ldots}\, \mathcal{F}_{\beta\ldots} \qquad (2.108)$$

With $\hat{O}^{\mathcal{F}}_{\alpha\ldots}$ we denote cartesian components of a vector operator $\hat{O}_\alpha$ as well as components of a second-rank tensor operator $\hat{O}_{\alpha\beta}$, depending on the situation. Similarly, $\hat{O}_{\alpha\beta\ldots}$ stands for second- $\hat{O}_{\alpha\beta}$, third- $\hat{O}_{\alpha\beta\gamma}$ and fourth-rank tensor operators $\hat{O}_{\alpha\beta\gamma\delta}$. The superscripts $\mathcal{F}$ and $\mathcal{FF}$ are labels attached to the operators in order to associate them with their corresponding fields. In later chapters it will be convenient to express the perturbation operators as the sum over all electrons

$$\hat{O}^{\mathcal{F}}_{\alpha\ldots} = \sum_i^N \hat{o}^{\mathcal{F}}_{i,\alpha\ldots} \qquad (2.109)$$

$$\hat{O}^{\mathcal{FF}}_{\alpha\beta\ldots} = \sum_i^N \hat{o}^{\mathcal{FF}}_{i,\alpha\beta\ldots} \qquad (2.110)$$

where $\hat{o}^{\mathcal{F}}_{i,\alpha\ldots}$ and $\hat{o}^{\mathcal{FF}}_{i,\alpha\beta\ldots}$ are then the perturbation operators acting on electron i alone.

2.9 Gauge Transformations

In Section 2.4 it was mentioned that the vector and scalar potentials, $\vec{A}(\vec{r}_i, t)$ and $\phi(\vec{r}_i, t)$, are not uniquely determined by their relations to the fields, $\vec{\mathcal{E}}(\vec{r}_i, t)$ and $\vec{\mathcal{B}}(\vec{r}_i, t)$, in Eqs. (2.33) and (2.34). A simultaneous gauge transformation of the two potentials with a gauge function $\chi(\vec{r}, t)$, Eqs. (2.41) and (2.42), changes the potentials but leaves the fields $\vec{\mathcal{E}}(\vec{r}_i, t)$ and $\vec{\mathcal{B}}(\vec{r}_i, t)$ unchanged.

The fact that the observable fields, $\vec{\mathcal{E}}(\vec{r}_i, t)$ and $\vec{\mathcal{B}}(\vec{r}_i, t)$ do not change under such a gauge transformation, implies that all equations describing the physics of a system must be **form invariant** under this **gauge transformation**. This applies in particular to the time-dependent Schrödinger equation

$$i\hbar \frac{\partial}{\partial t} |\Psi(t)\rangle = \hat{H} |\Psi(t)\rangle \qquad (2.111)$$

with the Hamiltonian $\hat{H}$ given in Eq. (2.99).

Replacing the potentials $\phi(\vec{r}_i, t)$ and $\vec{A}(\vec{r}_i, t)$ in the Hamiltonian $\hat{H}$ by $\phi'(\vec{r}_i, t)$ and $\vec{A}'(\vec{r}_i, t)$, according to the gauge transformation of second kind in Eqs. (2.41) and (2.42), yields a new Hamiltonian $\hat{H}'$

$$\hat{H}' = \frac{1}{2m_e} \sum_i^N \left[\hat{\vec{p}}_i + e\hat{\vec{A}}(\vec{r}_i,t) + e\vec{\nabla}\chi(\vec{r}_i,t)\right]^2 - \sum_i^N \left[e\hat{\phi}(\vec{r}_i,t) - e\frac{\partial\chi(\vec{r}_i,t)}{\partial t}\right]$$

$$- \frac{e^2}{4\pi\epsilon_0} \sum_{iK}^{NM} \frac{Z_K}{|\vec{r}_i - \vec{R}_K|} + \frac{e^2}{4\pi\epsilon_0} \sum_{i<j} \frac{1}{|\vec{r}_i - \vec{r}_j|} \qquad (2.112)$$

where the terms in Eq. (2.99), which depend only on nuclear coordinates have been excluded. Their inclusion would require that we go beyond the Born–Oppenheimer approximation and also include the nuclear kinetic energy terms $\frac{1}{2}\sum_K \hat{\vec{\pi}}_K^2/m_K$ with kinematical momentum operators $\hat{\vec{\pi}}_K$ depending on the vector potential $\hat{\vec{A}}(\vec{R}_K,t)$ as given in Eq. (2.57). This **gauge-transformed Hamiltonian** $\hat{H}'$ can also be obtained directly by the following transformation [see Exercise 2.14]

$$\hat{H}' - i\hbar\frac{\partial}{\partial t} = e^{-i\sum_i \frac{e}{\hbar}\chi(\vec{r}_i,t)}\left(\hat{H} - i\hbar\frac{\partial}{\partial t}\right) e^{i\sum_i \frac{e}{\hbar}\chi(\vec{r}_i,t)} \qquad (2.113)$$

where the summation is over all electrons i.

Exercise 2.14 Prove Eq. (2.113) for a one-electron system, i.e. with the Hamiltonian in Eq. (2.62) and with a one-electron transformation operator $e^{i\frac{e}{\hbar}\chi(\vec{r},t)}$.

Form invariance of the time-dependent Schrödinger equation under the gauge transformations in Eqs. (2.41) and (2.42) or in Eq. (2.113) is therefore obtained if also the wavefunction $\Psi(t)$ is simultaneously transformed according to

$$|\Psi(t)\rangle \to |\Psi'(t)\rangle = e^{-i\sum_i \frac{e}{\hbar}\chi(\vec{r}_i,t)}|\Psi(t)\rangle \qquad (2.114)$$

which is called a gauge transformation of the first kind.

Similarly, form invariance of the time-independent Schrödinger equation under a gauge transformation is guaranteed by the simultaneous gauge transformation of the total Hamiltonian

$$\hat{H}' = e^{-i\sum_i \frac{e}{\hbar}\chi(\vec{r}_i)}\,\hat{H}\,e^{i\sum_i \frac{e}{\hbar}\chi(\vec{r}_i)} \qquad (2.115)$$

or the vector potential in the Hamiltonian

$$\vec{A}(\vec{r}_i) \to \vec{A}'(\vec{r}_i) = \vec{A}(\vec{r}_i) + \vec{\nabla}\chi(\vec{r}_i) \qquad (2.116)$$

and the gauge transformation of the time-independent wavefunction

$$|\Psi\rangle \to |\Psi'\rangle = e^{-i\sum_i \frac{e}{\hbar}\chi(\vec{r}_i)}|\Psi\rangle \qquad (2.117)$$

The form invariance of the Schrödinger equation will then lead to gauge invariant expectation values of the Hamiltonian. However, this will not be the case for an arbitrary operator. In particular, it turns out that expectation values of the canonical momentum operator, given in Eq. (2.45), are not gauge invariant, whereas expectation values of the mechanical or kinematical momentum operator, given in Eq. (2.97), are gauge invariant [see Exercise 2.15]

Gauge Transformations

$$\langle \Psi' | \sum_i \hat{\vec{\pi}}'_i | \Psi' \rangle = \langle \Psi | \sum_i \hat{\vec{\pi}}_i | \Psi \rangle \tag{2.118}$$

The mechanical or kinematical momentum operator is therefore sometimes also called the **gauge invariant momentum operator**.

Exercise 2.15 Prove equation (2.118) for a one-electron system, i.e. with the kinematical momentum operator in Eq. (2.57) and with a one-electron transformation operator $e^{i\frac{e}{\hbar}\chi(\vec{r},t)}$.

An important gauge transformation in the context of the calculation of static molecular properties is given by the following gauge function

$$\chi(\vec{r}_i) = -\frac{1}{2} \vec{B} \times \vec{R}_{GO} \cdot \vec{r}_i \tag{2.119}$$

where $\vec{R}_{GO}$ is the arbitrary **gauge origin**. This gauge function implies that

$$\vec{\nabla}\chi(\vec{r}_i) = -\frac{1}{2} \vec{B} \times \vec{R}_{GO} \tag{2.120}$$

and that the vector potential for a uniform magnetic induction

$$\vec{A}^{\vec{B}'}(\vec{r}_i) = \frac{1}{2} \vec{B} \times (\vec{r}_i - \vec{R}_{GO}) \tag{2.121}$$

previously given in Eq. (2.40), becomes a linear function of the arbitrary gauge origin $\vec{R}_{GO}$ under this gauge transformation. This has important consequences for all magnetic properties that will be discussed in Section 5.10.

For time-dependent properties three other gauge transformations will play an important role. Let us consider the case that the scalar potential $\phi(\vec{r},t)$ is zero and that the vector potential $\vec{A}(t)$ depends only on time. The latter assumption implies, according to Eq. (2.34), that the magnetic field vanishes. In the first transformation we choose now the gauge function to be

$$\chi(\vec{r}_i, t) = -\vec{A}(t) \cdot \vec{r}_i \tag{2.122}$$

According to Eqs. (2.41) and (2.42) the transformed potentials then become

$$\vec{A}'(\vec{r}_i, t) = \vec{A}(t) + \vec{\nabla}\chi(\vec{r}_i, t) = 0 \tag{2.123}$$

$$\phi'(\vec{r}_i, t) = \phi(\vec{r}_i, t) - \frac{\partial \chi(\vec{r}_i, t)}{\partial t} = \frac{\partial \vec{A}(t) \cdot \vec{r}_i}{\partial t} = -\vec{\mathcal{E}}(t) \cdot \vec{r}_i \tag{2.124}$$

where the last equality is due to the definition of the vector potential in Eq. (2.33). The effect of this gauge transformation is that the time-dependent electric field will enter the Hamiltonian via the scalar potential instead of via the vector potential and that it couples the time-dependent electric field with the position vectors of the electrons. This gauge is therefore called the **length gauge**.

28 The Schrödinger Equation

In the second transformation we choose the gauge function to be

$$\chi(t) = -\frac{e}{2m_e} \int_0^t \vec{A}^2(t')dt' \tag{2.125}$$

giving rise to a scalar potential

$$\phi'(t) = -\frac{\partial \chi(t)}{\partial t} = \frac{e}{2m_e}\vec{A}^2(t) \tag{2.126}$$

which cancels the second-order contribution, Eq. (2.107), to the molecular Hamiltonian. The time-dependent electric field enters in this gauge the molecular Hamiltonian only via the linear $\hat{\vec{A}}(t) \cdot \hat{\vec{p}}_i$ term in Eq. (2.105) and is thus coupled through its vector potential to the canonical momentum or velocity of the electrons. This gauge is called the **velocity gauge**.

Finally, in the **Lorenz gauge** the gauge function $\chi(\vec{r}_i, t)$ is chosen in such a way that

$$\vec{\nabla} \cdot \vec{A}'(\vec{r}_i, t) + \frac{1}{c^2}\frac{\partial \phi'(\vec{r}_i, t)}{\partial t} = 0 \tag{2.127}$$

With this gauge, the third and fourth Maxwell equations (2.37) and (2.38) take, in vacuum, a simple form in terms of the potentials

$$\nabla^2 \phi(\vec{r}, t) - \frac{1}{c}\frac{\partial^2 \phi(\vec{r}, t)}{\partial t^2} = 0 \tag{2.128}$$

$$\nabla^2 \vec{A}(\vec{r}, t) - \frac{1}{c}\frac{\partial^2 \vec{A}(\vec{r}, t)}{\partial t^2} = 0 \tag{2.129}$$

2.10 Further Reading

Time-Dependent Schrödinger Equation

P. W. Atkins and R. S. Friedman, *Molecular Quantum Mechanics*, 3rd edn, Oxford University Press, Oxford (1997): Chapter 1.12.

Electron Density

R. McWeeny, *Methods of Molecular Quantum Mechanics*, Academic Press, London, 2nd edn, (1992): Chapter 5.

I. N. Levine, *Quantum Chemistry*, Prentice Hall, Upper Saddle River, 5th edn, (2000): Chapter 13.14.

Classical Electromagnetism

V. D. Barger and M. G. Olsson, *Classical Electricity and Magnetism*, Allyn and Bacon, Boston (1987): Chapters 1 and 10.

Lagrangian and Hamiltonian Formulation of Classical Mechanics

R. Shankar, *Principles of Quantum Mechanics*, Plenum, New York (1980): Chapter 2.

R. E. Moss, *Advanced Molecular Quantum Mechanics*, Chapman and Hall, London (1973): Chapter 3.

V. D. Barger and M. G. Olsson, *Classical Mechanics*, McGraw Hill, New York (1995): Chapter 3.

H. Goldstein, C. Poole and J. Safko, *Classical Mechanics*, 3rd edn, Addison Wesley, San Francisco (2002): Chapters 1, 2 and 8.

Hamiltonian in Electromagnetic Fields

D. W. Davies, *The Theory of the Electric and Magnetic Properties of Molecules*, John Wiley and Sons, London (1967): Chapter 2.

A. Hinchliffe and R. W. Munn, *Molecular Electromagnetism*, John Wiley and Sons Ltd, Chichester (1985): Chapter 13.

R. McWeeny, *Methods of Molecular Quantum Mechanics*, 2nd edn, Academic Press, London (1992): Chapter 11.

R. Shankar, *Principles of Quantum Mechanics*, Plenum, New York (1980): Chapters 2.2 and 20.2.

H. Goldstein, C. Poole and J. Safko, *Classical Mechanics*, 3rd edn, Addison Wesley, San Francisco (2002): Chapter 1.5.

Dirac Equation and Elimination of the Small Component

R. Shankar, *Principles of Quantum Mechanics*, Plenum, New York (1980): Chapter 20.

R. E. Moss, *Advanced Molecular Quantum Mechanics*, Chapman and Hall, London (1973): Chapter 8.

J. J. Sakurai, *Advanced Quantum Mechanics*, Addison-Wesley, Reading (1967): Chapter 3–3.

K. G. Dyall and K. Fægri, *Introduction to Relativistic Quantum Chemistry*, Oxford University Press, Oxford (2007): Parts II and IV.

Gauge Transformations and Gauge Origin Problem

H. F. Hameka, *Advanced Quantum Chemistry: Theory of Interactions between Molecules and Electromagnetic Fields*, Addison-Wesley, Reading (1965): Chapters 1–4, 3–2 and 9–2.

D. W. Davies, *The Theory of the Electric and Magnetic Properties of Molecules*, John Wiley and Sons, London (1967): Chapter 2–3 and Appendix I–6.

3
Perturbation Theory

In the previous chapter we have derived the molecular electronic Hamiltonian $\hat{H}$ in the presence of static electromagnetic fields or fields due to nuclear moments. Throughout this chapter we will consider a general field and denote it as $\vec{\mathcal{F}}$ with components $\mathcal{F}_{\alpha\cdots}$. Examples for $\mathcal{F}_{\alpha\cdots}$ are one of the three components of the electric field $\mathcal{E}_\alpha$, of the magnetic induction $\mathcal{B}_\alpha$, of the nuclear moment m_α^K of a magnetic nucleus K or one of the nine components of the field gradient $\mathcal{E}_{\alpha\beta}$.[1]

Our first task in this chapter is to obtain expressions for the wavefunction $|\Psi_0(\vec{\mathcal{F}})\rangle$ of the ground state of our system in the presence of the components of such a static field and afterwards expressions for the energy $E_0(\vec{\mathcal{F}})$ and for the expectation value $\langle\Psi_0(\vec{\mathcal{F}})|\hat{P}|\Psi_0(\vec{\mathcal{F}})\rangle$ of an arbitrary operator $\hat{P}$ in the presence of the field. This means that we have to solve the time-independent Schrödinger equation for the system

$$\hat{H}(\vec{\mathcal{F}})\,|\Psi_0(\vec{\mathcal{F}})\rangle = E_0(\vec{\mathcal{F}})\,|\Psi_0(\vec{\mathcal{F}})\rangle \tag{3.1}$$

However, we consider only electromagnetic fields that cause a slight change in the nuclear and electronic structure of the molecules and we treat them as mere perturbations of the wavefunction and energy of our system. Furthermore, we are often more interested in these small changes in the wavefunction and energy of the system than in the final state of the system itself. Therefore, we will not attempt to solve Eq. (3.1) directly but rather apply **time-independent perturbation theory**[2] and **time-independent response theory**. We consider throughout the book only properties of molecules in their electronic ground state, because most of the computational methods discussed in Part III are restricted to the electronic ground state. Perturbation theory as discussed in the present chapter could, however, also be applied to non-degenerate excited states.

Our second task is to generalize this approach to the case of time-dependent fields and the solution of the time-dependent Schrödinger equation

$$i\hbar\frac{\partial}{\partial t}|\Psi_0(t,\vec{\mathcal{F}})\rangle = \hat{H}(t)\,|\Psi_0(t,\vec{\mathcal{F}})\rangle \tag{3.2}$$

---

[1] The subscript $\alpha\cdots$ indicates thus that we are dealing with tensor fields of different rank as discussed in Section 2.8.

[2] Sometimes, it is also called Schrödinger or Rayleigh–Schrödinger perturbation theory, because Erwin Schrödinger developed it for the calculation of the Stark effect on the hydrogen atom (Schrödinger, 1926).

There, we will employ **time-dependent perturbation theory** in the form of **time-dependent response theory** in order to derive terms in the expansion of the time-dependent wavefunction $|\Psi_0(t,\vec{\mathcal{F}})\rangle$ and of a time-dependent expectation value of a given operator $\hat{P}$, $\langle\Psi_0(t,\vec{\mathcal{F}})|\hat{P}|\Psi_0(t,\vec{\mathcal{F}})\rangle$. The time-dependent Schrödinger equation, i.e. Eq. (3.2), contrary to the time-independent version is not an eigenvalue equation with the energy as eigenvalue. Therefore, we will not consider the energy in Section 3.11 but only expectation values of the operators corresponding to physical observables. In Section 12.3, however, we will briefly discuss an alternative approach that defines a quasi- or pseudo-energy also for the time-dependent case.

Before starting properly with perturbation theory we will first introduce in the next section the Hellmann–Feynman theorem, which establishes a deep connection between the energy and molecular properties calculated as expectation values and that does not rely on perturbation theory.

3.1 The Hellmann–Feynman Theorem

In Part II we will see that all molecular properties can be defined as derivatives of the energy with respect to the strength of external or internal perturbations. A theorem, which will become very useful in this context, is the **Hellmann–Feynman theorem**.

Let us, for its derivation, consider a Hamiltonian $\hat{H}(\vec{\mathcal{F}})$ with eigenfunctions $|\Psi_0(\vec{\mathcal{F}})\rangle$ and eigenvalues $E_0(\vec{\mathcal{F}})$ that all depend on the general electromagnetic field $\vec{\mathcal{F}}$ with components $\mathcal{F}_{\alpha\dots}$

$$\hat{H}(\vec{\mathcal{F}})\,|\Psi_0(\vec{\mathcal{F}})\rangle = E_0(\vec{\mathcal{F}})\,|\Psi_0(\vec{\mathcal{F}})\rangle \tag{3.3}$$

We assume further that the eigenfunctions are normalized for all values of $\mathcal{F}_{\alpha\dots}$

$$\langle\Psi_0(\vec{\mathcal{F}})\,|\,\Psi_0(\vec{\mathcal{F}})\rangle = 1 \tag{3.4}$$

We are now interested in how the energy changes as a function of the strength of the component $\mathcal{F}_{\alpha\dots}$ of the field. Instead of the dependence on a field $\vec{\mathcal{F}}$ we could also consider any other parameter on which the Hamiltonian and therefore the energy and wavefunction depend. A typical example of such a parameter is the change in the position vector $\vec{R}_K$ of a nucleus K. However, here we want to know the first derivative of the energy with respect to a component $\mathcal{F}_{\alpha\dots}$ of a general electromagnetic field

$$\frac{dE_0(\vec{\mathcal{F}})}{d\mathcal{F}_{\alpha\dots}} = \frac{d}{d\mathcal{F}_{\alpha\dots}}\langle\Psi_0(\vec{\mathcal{F}})|\hat{H}(\vec{\mathcal{F}})|\Psi_0(\vec{\mathcal{F}})\rangle \tag{3.5}$$

$$= \langle\Psi_0(\vec{\mathcal{F}})|\frac{\partial\hat{H}(\vec{\mathcal{F}})}{\partial\mathcal{F}_{\alpha\dots}}|\Psi_0(\vec{\mathcal{F}})\rangle$$

$$+ \langle\frac{d\Psi_0(\vec{\mathcal{F}})}{d\mathcal{F}_{\alpha\dots}}|\hat{H}(\vec{\mathcal{F}})|\Psi_0(\vec{\mathcal{F}})\rangle + \langle\Psi_0(\vec{\mathcal{F}})|\hat{H}(\vec{\mathcal{F}})|\frac{d\Psi_0(\vec{\mathcal{F}})}{d\mathcal{F}_{\alpha\dots}}\rangle$$

The Hamiltonian depends only explicitly on the field, which allows us to replace the total by the partial derivative of the Hamiltonian. In the second and third terms we can make use of the fact that $\Psi_0(\vec{\mathcal{F}})$ is an eigenfunction of $\hat{H}(\vec{\mathcal{F}})$, Eq. (3.3), and that the eigenvalues are real, which allows us to write

$$\frac{dE_0(\vec{\mathcal{F}})}{d\mathcal{F}_{\alpha\cdots}} = \langle\Psi_0(\vec{\mathcal{F}})|\frac{\partial\hat{H}(\vec{\mathcal{F}})}{\partial\mathcal{F}_{\alpha\cdots}}|\Psi_0(\vec{\mathcal{F}})\rangle \quad (3.6)$$

$$+ E_0(\vec{\mathcal{F}})\left(\langle\frac{d\Psi_0(\vec{\mathcal{F}})}{d\mathcal{F}_{\alpha\cdots}}|\Psi_0(\vec{\mathcal{F}})\rangle + \langle\Psi_0(\vec{\mathcal{F}})|\frac{d\Psi_0(\vec{\mathcal{F}})}{d\mathcal{F}_{\alpha\cdots}}\rangle\right)$$

$$= \langle\Psi_0(\vec{\mathcal{F}})|\frac{\partial\hat{H}(\vec{\mathcal{F}})}{\partial\mathcal{F}_{\alpha\cdots}}|\Psi_0(\vec{\mathcal{F}})\rangle + E_0(\vec{\mathcal{F}})\frac{d}{d\mathcal{F}_{\alpha\cdots}}\langle\Psi_0(\vec{\mathcal{F}})|\Psi_0(\vec{\mathcal{F}})\rangle$$

The last term vanishes, because we have assumed that the eigenfunctions $\Psi_0(\vec{\mathcal{F}})$ are normalized for any value of $\mathcal{F}_{\alpha\cdots}$, Eq. (3.4). We thus arrive at the conclusion that the first derivative of the energy with respect to the component $\mathcal{F}_{\alpha\cdots}$ of the field can be obtained as an expectation value of the derivative of the Hamiltonian

$$\frac{dE_0(\vec{\mathcal{F}})}{d\mathcal{F}_{\alpha\cdots}} = \langle\Psi_0(\vec{\mathcal{F}})|\frac{\partial\hat{H}(\vec{\mathcal{F}})}{\partial\mathcal{F}_{\alpha\cdots}}|\Psi_0(\vec{\mathcal{F}})\rangle \quad (3.7)$$

Evaluating the derivative of the Hamiltonian with respect to the component $\mathcal{F}_{\alpha\cdots}$ of a field $\vec{\mathcal{F}}$ and therefore of the two perturbation Hamiltonians $\hat{H}^{(1)} + \hat{H}^{(2)}$ in Eq. (2.108) leads us to a possibly field-dependent operator, which we denote[3] with $\hat{P}(\vec{\mathcal{F}})$ and that is then given as

$$\hat{P}(\vec{\mathcal{F}}) = \frac{\partial\hat{H}(\vec{\mathcal{F}})}{\partial\mathcal{F}_{\alpha\cdots}} = \hat{O}^{\mathcal{F}}_{\alpha\cdots} + \sum_{\beta\cdots}\left(\hat{O}^{\mathcal{F}\mathcal{F}}_{\alpha\beta\cdots} + \hat{O}^{\mathcal{F}\mathcal{F}}_{\beta\alpha\cdots}\right)\mathcal{F}_{\beta\cdots} \quad (3.8)$$

In the presence of the field the operator $\hat{P}(\vec{\mathcal{F}})$ consists of a field-independent or zeroth-order term

$$\hat{P}^{(0)} = \hat{O}^{\mathcal{F}}_{\alpha\cdots} \equiv \hat{P} \quad (3.9)$$

for which we will mostly use the symbol $\hat{P}$ in the following, and a first-order, field-dependent contribution

$$\hat{P}^{(1)}(\vec{\mathcal{F}}) = \sum_{\beta\cdots}\left(\hat{O}^{\mathcal{F}\mathcal{F}}_{\alpha\beta\cdots} + \hat{O}^{\mathcal{F}\mathcal{F}}_{\beta\alpha\cdots}\right)\mathcal{F}_{\beta\cdots} \quad (3.10)$$

Often, one is interested in the derivative of the energy evaluated at zero field strength, i.e. for $|\vec{\mathcal{F}}| = 0$. This then gives

$$\frac{\partial E_0(\vec{\mathcal{F}})}{\partial\mathcal{F}_{\alpha\cdots}}\bigg|_{|\vec{\mathcal{F}}|=0} = \langle\Psi_0^{(0)}|\left(\frac{\partial\hat{H}(\vec{\mathcal{F}})}{\partial\mathcal{F}_{\alpha\cdots}}\right)_{|\vec{\mathcal{F}}|=0}|\Psi_0^{(0)}\rangle \quad (3.11)$$

$$= \langle\Psi_0^{(0)}|\hat{O}^{\mathcal{F}}_{\alpha\cdots}|\Psi_0^{(0)}\rangle = \langle\Psi_0^{(0)}|\hat{P}|\Psi_0^{(0)}\rangle \quad (3.12)$$

---

[3] In the derivations in this chapter it is convenient to distinguish between the operators of the perturbing fields, $\hat{O}^{\mathcal{F}}_{\alpha\cdots}$ and $\hat{O}^{\mathcal{F}\mathcal{F}}_{\alpha\beta\cdots}$, and the operator whose expectation value we are evaluating, although they are often the same operator. Therefore, we will call the latter $\hat{P}(\vec{\mathcal{F}})$ or $\hat{P}$.

which is called the the **generalized Hellmann–Feynman theorem**[4] and establishes
the equivalence between the first derivative of the energy and the expectation value
of an operator, given as the corresponding derivative of the Hamiltonian. One should
note that in the derivation of the Hellmann–Feynman theorem, we do not assume
that we have solved the Schrödinger equation for the system without the field $\vec{\mathcal{F}}$,
Eq. (2.10), as we are going to do in the following sections on perturbation theory.
This detail will become important later when we discuss approximate methods for
calculating molecular properties in Part III.

3.2 Time-Independent Perturbation Theory

Perturbation theory builds on the partitioning of the Hamiltonian from Eq. (2.101) in
an unperturbed Hamiltonian $\hat{H}^{(0)}$ and perturbation Hamiltonians $\lambda \hat{H}^{(1)} + \lambda^2 \hat{H}^{(2)}$,

$$\hat{H} = \hat{H}^{(0)} + \lambda \hat{H}^{(1)} + \lambda^2 \hat{H}^{(2)} \tag{3.13}$$

where λ is a dimensionless ordering parameter, which is introduced during the derivation of the perturbation theory expressions and is discarded afterwards. For $\lambda = 1$ the
perturbation is turned on, while $\lambda = 0$ corresponds to the unperturbed system. The
linear, quadratic or general mth-order dependence of the operators, energies and wavefunctions on the perturbing field $\vec{\mathcal{F}}$ is therefore shown twice: (a) by the superscripts
(m) and (b) explicitly by the powers of the ordering parameter λ.

We suppose now that the energies $E_n^{(0)}$ and wavefunctions $|\Psi_n^{(0)}\rangle$ ($n = 0, 1, 2, \cdots$) of
the unperturbed Hamiltonian $\hat{H}^{(0)}$ are known, i.e. that the unperturbed Schrödinger
equation

$$\hat{H}^{(0)} |\Psi_n^{(0)}\rangle = E_n^{(0)} |\Psi_n^{(0)}\rangle \tag{3.14}$$

has been solved exactly. This is Eq. (2.10), which we from now on will write without
explicitly stating the dependence of the many-electron wavefunction $|\Psi_n^{(0)}\rangle$ on the coordinates. In reality, this is of course not possible. However, we will ignore this fact until
we discuss practical methods for the calculation of molecular properties in Part III.
There, we will distinguish between two types of computational methods. In the first
type of methods one makes approximations to the perturbation theory expressions
that were derived assuming that Eq. (3.14) can be solved exactly, whereas in the second type of methods one starts from an approximate solution to Eq. (3.14) and derives
expressions for the molecular properties using perturbation theory with approximate
states. We will later also use the fact that the functions $\{|\Psi_n^{(0)}\rangle\}$ are eigenfunctions of
the Hermitian operator $\hat{H}^{(0)}$ and therefore form a complete orthonormal set. Furthermore, we restrict ourselves here to non-degenerate perturbation theory, which means
that the unperturbed state $|\Psi_0^{(0)}\rangle$ of the system cannot be degenerate.

Finally, we assume that the eigenfunction $|\Psi_0(\vec{\mathcal{F}})\rangle$ and eigenvalue $E_0(\vec{\mathcal{F}})$ of the
full Hamiltonian $\hat{H}$ are close to those of the unperturbed Hamiltonian $\hat{H}^{(0)}$, i.e. the
perturbation by $\vec{\mathcal{F}}$ is indeed small. We can then expand the **perturbed wavefunction**

---

[4] It is called the generalized Hellmann–Feynman theorem because Hellmann (1937) and Feynman
(1939) considered originally the changes in energy due to a change in the geometry.

and energy, $|\Psi_0(\vec{\mathcal{F}})\rangle$ and $E_0(\vec{\mathcal{F}})$, in a formal power series in λ around the exact solutions $E_0^{(0)}$ and $|\Psi_0^{(0)}\rangle$ of the unperturbed Hamiltonian $\hat{H}^{(0)}$

$$E_0(\vec{\mathcal{F}}) = E_0^{(0)} + \lambda E_0^{(1)}(\vec{\mathcal{F}}) + \lambda^2 E_0^{(2)}(\vec{\mathcal{F}}) + \cdots \tag{3.15}$$

$$|\Psi_0(\vec{\mathcal{F}})\rangle = |\Psi_0^{(0)}\rangle + \lambda|\Psi_0^{(1)}(\vec{\mathcal{F}})\rangle + \lambda^2|\Psi_0^{(2)}(\vec{\mathcal{F}})\rangle + \cdots \tag{3.16}$$

The energy $E_0^{(m)}(\vec{\mathcal{F}})$ and wavefunction $|\Psi_0^{(m)}(\vec{\mathcal{F}})\rangle$ are called the mth-order correction to the energy and wavefunction. Terms like $\hat{H}^{(1)}|\Psi_0^{(2)}(\vec{\mathcal{F}})\rangle$ or $E_0^{(2)}(\vec{\mathcal{F}})|\Psi_0^{(1)}(\vec{\mathcal{F}})\rangle$ are then third-order terms. When we assume that the perturbations are small, we actually mean that with increasing order m the corrections to the energy and wavefunction become systematically smaller.

In order to make the derivations mathematically easier without changing the final expressions we require the perturbed wavefunction $|\Psi_0(\vec{\mathcal{F}})\rangle$ to be normalized in the following way, called **intermediate normalization**

$$\begin{aligned} 1 &= \langle \Psi_0^{(0)} \mid \Psi_0(\vec{\mathcal{F}}) \rangle \\ &= \langle \Psi_0^{(0)} \mid \Psi_0^{(0)} \rangle + \lambda \langle \Psi_0^{(0)} \mid \Psi_0^{(1)}(\vec{\mathcal{F}}) \rangle + \lambda^2 \langle \Psi_0^{(0)} \mid \Psi_0^{(2)}(\vec{\mathcal{F}}) \rangle + \cdots \end{aligned} \tag{3.17}$$

As the unperturbed wavefunction $|\Psi_0^{(0)}\rangle$ is normalized ($\langle \Psi_0^{(0)} \mid \Psi_0^{(0)} \rangle = 1$), this leads to the following conditions on the higher-order ($n > 0$) corrections to the wavefunction

$$\langle \Psi_0^{(0)} \mid \Psi_0^{(n)}(\vec{\mathcal{F}}) \rangle = 0 \tag{3.18}$$

which we will use later. Although this is very convenient for the derivation of the higher-order corrections to the energy, one should realize that the wavefunction $|\Psi_0(\vec{\mathcal{F}})\rangle$ is not normalized, if one truncates the power series Eq. (3.16) at a finite order. Therefore, one has to renormalize the truncated wavefunction at each order. However, this is not necessary for the second-order correction to the energy that we will derive here.

We are ready now to insert the power-series expansions of the perturbed wavefunction and energy in the Schrödinger equation, Eq. (3.1),

$$\left(\hat{H}^{(0)} + \lambda \hat{H}^{(1)} + \lambda^2 \hat{H}^{(2)} \right) \left(|\Psi_0^{(0)}\rangle + \lambda|\Psi_0^{(1)}(\vec{\mathcal{F}})\rangle + \lambda^2|\Psi_0^{(2)}(\vec{\mathcal{F}})\rangle + \cdots \right) \tag{3.19}$$

$$= \left(E_0^{(0)} + \lambda E_0^{(1)}(\vec{\mathcal{F}}) + \lambda^2 E_0^{(2)}(\vec{\mathcal{F}}) + \cdots \right)$$

$$\times \left(|\Psi_0^{(0)}\rangle + \lambda|\Psi_0^{(1)}(\vec{\mathcal{F}})\rangle + \lambda^2|\Psi_0^{(2)}(\vec{\mathcal{F}})\rangle + \cdots \right)$$

The two sides of this equation are power series in λ. Therefore, the terms multiplied by the same powers of λ, i.e. terms of the same order, have to be equal on both sides in order for the whole equation to be fulfilled. We thus obtain a series of equations, where the zeroth-order equation is just the equation for the unperturbed Hamiltonian, Eq. (3.14) again. The first- and second-order equations are

$$\hat{H}^{(0)}|\Psi_0^{(1)}(\vec{\mathcal{F}})\rangle + \hat{H}^{(1)}|\Psi_0^{(0)}\rangle = E_0^{(0)}|\Psi_0^{(1)}(\vec{\mathcal{F}})\rangle + E_0^{(1)}(\vec{\mathcal{F}})|\Psi_0^{(0)}\rangle \tag{3.20}$$

Time-Independent Perturbation Theory

$$\hat{H}^{(0)}|\Psi_0^{(2)}(\vec{\mathcal{F}})\rangle + \hat{H}^{(1)}|\Psi_0^{(1)}(\vec{\mathcal{F}})\rangle + \hat{H}^{(2)}|\Psi_0^{(0)}\rangle$$
$$= E_0^{(0)}|\Psi_0^{(2)}(\vec{\mathcal{F}})\rangle + E_0^{(1)}(\vec{\mathcal{F}})|\Psi_0^{(1)}(\vec{\mathcal{F}})\rangle + E_0^{(2)}(\vec{\mathcal{F}})|\Psi_0^{(0)}\rangle \qquad (3.21)$$

and the general mth-order equation for $m > 2$ can be written as

$$\hat{H}^{(0)}|\Psi_0^{(m)}(\vec{\mathcal{F}})\rangle + \hat{H}^{(1)}|\Psi_0^{(m-1)}(\vec{\mathcal{F}})\rangle + \hat{H}^{(2)}|\Psi_0^{(m-2)}(\vec{\mathcal{F}})\rangle$$
$$= E_0^{(0)}|\Psi_0^{(m)}(\vec{\mathcal{F}})\rangle + \sum_{i=1}^{m-1} E_0^{(i)}(\vec{\mathcal{F}})|\Psi_0^{(m-i)}(\vec{\mathcal{F}})\rangle + E_0^{(m)}(\vec{\mathcal{F}})|\Psi_0^{(0)}\rangle \qquad (3.22)$$

These equations are inhomogeneous differential equations, which can sometimes be solved analytically for the first-, second- and higher-order corrections to the wavefunction. But normally they are solved by expanding the mth-order correction to the wavefunction $|\Psi_0^{(m)}(\vec{\mathcal{F}})\rangle$ in a complete basis of functions, which fulfill the same boundary conditions as the unknown function. The eigenfunctions $\{|\Psi_n^{(0)}\rangle\}$ of the unperturbed Hamiltonian $\hat{H}^{(0)}$ form a complete set and are therefore usually chosen as the basis set for the mth-order correction to the wavefunction

$$|\Psi_0^{(m)}(\vec{\mathcal{F}})\rangle = \sum_{n \neq 0} |\Psi_n^{(0)}\rangle\, C_{n0}^{(m)}(\vec{\mathcal{F}}) \qquad (3.23)$$

The expansion coefficients $C_{n0}^{(m)}(\vec{\mathcal{F}})$ are defined as the projection of the mth-order correction to the wavefunction against the corresponding basis functions $\langle \Psi_n^{(0)}|$ [see Exercise 3.1]

$$C_{n0}^{(m)}(\vec{\mathcal{F}}) = \langle \Psi_n^{(0)} | \Psi_0^{(m)}(\vec{\mathcal{F}})\rangle \qquad (3.24)$$

The term with $n = 0$ can be excluded from the summation in Eq. (3.23) because comparison of Eq. (3.24) with Eq. (3.18) shows that the corresponding coefficient $C_{00}^{(m)}(\vec{\mathcal{F}})$ vanishes as a direct consequence of intermediate normalization of the wavefunction. This is one of the reasons for using the intermediate normalization.

Exercise 3.1 Prove the relation for the coefficients of the mth-order correction to the wavefunction Eq. (3.24).

Inserting the expansion of the mth-order correction to the wavefunction, Eq. (3.23), into the mth-order equation, Eq. (3.22), gives

$$\hat{H}^{(0)} \sum_{k \neq 0} |\Psi_k^{(0)}\rangle\, C_{k0}^{(m)}(\vec{\mathcal{F}}) + \sum_{i=1}^{2} \hat{H}^{(i)} \sum_{k} |\Psi_k^{(0)}\rangle\, C_{k0}^{(m-i)}(\vec{\mathcal{F}})$$
$$= E_0^{(0)} \sum_{k \neq 0} |\Psi_k^{(0)}\rangle\, C_{k0}^{(m)}(\vec{\mathcal{F}}) + \sum_{i=1}^{m} E_0^{(i)}(\vec{\mathcal{F}}) \sum_{k} |\Psi_k^{(0)}\rangle\, C_{k0}^{(m-i)}(\vec{\mathcal{F}}) \qquad (3.25)$$

which holds for all orders m, if we define $C_{k0}^{(0)}(\vec{\mathcal{F}}) = \delta_{k0}$ and $C_{k0}^{(-1)}(\vec{\mathcal{F}}) = 0$. Expressions for the coefficients $C_{n0}^{(m)}(\vec{\mathcal{F}})$ can then be derived from this equation by projecting against the corresponding basis function $\langle \Psi_n^{(0)}|$ and isolation of $C_{n0}^{(m)}(\vec{\mathcal{F}})$. For the first-order coefficients this leads to [see Exercise 3.2]

$$C_{n0}^{(1)}(\vec{\mathcal{F}}) = \frac{\langle \Psi_n^{(0)}|\hat{H}^{(1)}|\Psi_0^{(0)}\rangle}{E_0^{(0)} - E_n^{(0)}} \qquad (3.26)$$

and the **first-order correction to the wavefunction** is therefore given as

$$|\Psi_0^{(1)}(\vec{\mathcal{F}})\rangle = \sum_{n \neq 0} |\Psi_n^{(0)}\rangle \frac{\langle \Psi_n^{(0)}|\hat{H}^{(1)}|\Psi_0^{(0)}\rangle}{E_0^{(0)} - E_n^{(0)}} \qquad (3.27)$$

For the coefficients of the mth-order correction to the wavefunction one obtains

$$C_{n0}^{(m)}(\vec{\mathcal{F}}) = \sum_{i=1}^{2} \sum_k \frac{\langle \Psi_n^{(0)}|\hat{H}^{(i)}|\Psi_k^{(0)}\rangle}{E_0^{(0)} - E_n^{(0)}} C_{k0}^{(m-i)}(\vec{\mathcal{F}}) - \sum_{i=1}^{m-1} E_0^{(i)}(\vec{\mathcal{F}}) \frac{C_{n0}^{(m-i)}(\vec{\mathcal{F}})}{E_0^{(0)} - E_n^{(0)}} \qquad (3.28)$$

Exercise 3.2 Derive the expression for the first-order coefficient $C_{n0}^{(1)}(\vec{\mathcal{F}})$, Eq. (3.26).

Exercise 3.3 Derive an expression for the second-order correction to the wavefunction.

When we project Eqs. (3.20) to (3.22) on the unperturbed ground state $\langle \Psi_0^{(0)}|$ and rearrange them, we can derive expressions for the first-, second- and higher-order ($m > 2$) **corrections to the energy** [see Exercise 3.4]

$$E_0^{(1)}(\vec{\mathcal{F}}) = \langle \Psi_0^{(0)}|\hat{H}^{(1)}|\Psi_0^{(0)}\rangle \qquad (3.29)$$

$$E_0^{(2)}(\vec{\mathcal{F}}) = \langle \Psi_0^{(0)}|\hat{H}^{(2)}|\Psi_0^{(0)}\rangle + \langle \Psi_0^{(0)}|\hat{H}^{(1)}|\Psi_0^{(1)}(\vec{\mathcal{F}})\rangle \qquad (3.30)$$

$$E_0^{(m)}(\vec{\mathcal{F}}) = \langle \Psi_0^{(0)}|\hat{H}^{(2)}|\Psi_0^{(m-2)}(\vec{\mathcal{F}})\rangle + \langle \Psi_0^{(0)}|\hat{H}^{(1)}|\Psi_0^{(m-1)}(\vec{\mathcal{F}})\rangle \qquad (3.31)$$

Exercise 3.4 Derive Eqs. (3.29), (3.30) and (3.31) for the first-, second- and mth-order corrections to the energy.

We should note that the first-order correction to the energy is independent of the changes in the wavefunction. In general, with the mth-order wavefunction one can calculate the energy up to order $2m + 1$

$$E_0^{(2m+1)}(\vec{\mathcal{F}}) = \langle \Psi_0^{(m)}(\vec{\mathcal{F}})|\hat{H}^{(2)}|\Psi_0^{(m-1)}(\vec{\mathcal{F}})\rangle + \langle \Psi_0^{(m-1)}(\vec{\mathcal{F}})|\hat{H}^{(2)}|\Psi_0^{(m)}(\vec{\mathcal{F}})\rangle$$
$$+ \langle \Psi_0^{(m)}(\vec{\mathcal{F}})|\hat{H}^{(1)}|\Psi_0^{(m)}(\vec{\mathcal{F}})\rangle$$
$$- \sum_{i,j=1}^{m} E_0^{(2m+1-i-j)}(\vec{\mathcal{F}}) \langle \Psi_0^{(i)}(\vec{\mathcal{F}})|\Psi_0^{(j)}(\vec{\mathcal{F}})\rangle \qquad (3.32)$$

which is the well-known **$2m+1$ rule** (Löwdin, 1965).

Exercise 3.5 Illustrate the $2m+1$ rule by deriving an expression for the third-order correction to the energy that includes only the first-order correction to the wavefunction.

Hint: Use the fact that $\langle \Psi_0^{(0)} | \hat{H}^{(1)} | \Psi_0^{(2)}(\vec{\mathcal{F}}) \rangle = \langle \Psi_0^{(2)}(\vec{\mathcal{F}}) | \hat{H}^{(1)} | \Psi_0^{(0)} \rangle^*$ and find an expression for $\langle \Psi_0^{(2)}(\vec{\mathcal{F}}) | \hat{H}^{(1)} | \Psi_0^{(0)} \rangle$ from Eq. (3.20).

On insertion of the first-order correction to the wavefunction in Eq. (3.30) we can then immediately write down the expression for the **second-order correction to the energy**

$$E_0^{(2)}(\vec{\mathcal{F}}) = \langle \Psi_0^{(0)} | \hat{H}^{(2)} | \Psi_0^{(0)} \rangle + \sum_{n \neq 0} \frac{\langle \Psi_0^{(0)} | \hat{H}^{(1)} | \Psi_n^{(0)} \rangle \langle \Psi_n^{(0)} | \hat{H}^{(1)} | \Psi_0^{(0)} \rangle}{E_0^{(0)} - E_n^{(0)}} \quad (3.33)$$

The second-order energy correction thus consists of two terms: a ground-state expectation value[5] over the second-order Hamiltonian $\hat{H}^{(2)}$ and a so-called sum-over-states term, which involves a summation over all excited states of the system and transition moments between the ground state and these excited states with the first-order Hamiltonian $\hat{H}^{(1)}$. Finally, we can insert the expressions for the first- and second-order perturbation Hamiltonians, Eq. (2.108),

$$E_0^{(2)}(\vec{\mathcal{F}}) = \sum_{\alpha,\beta,\cdots} \mathcal{F}_{\alpha\cdots} \left(\langle \Psi_0^{(0)} | \hat{O}_{\alpha\beta\cdots}^{\mathcal{F}\mathcal{F}} | \Psi_0^{(0)} \rangle + \sum_{n \neq 0} \frac{\langle \Psi_0^{(0)} | \hat{O}_{\alpha\cdots}^{\mathcal{F}} | \Psi_n^{(0)} \rangle \langle \Psi_n^{(0)} | \hat{O}_{\beta\cdots}^{\mathcal{F}} | \Psi_0^{(0)} \rangle}{E_0^{(0)} - E_n^{(0)}} \right) \mathcal{F}_{\beta\cdots}$$
(3.34)

which explicitly shows the quadratic dependence of $E_0^{(2)}(\vec{\mathcal{F}})$ on components of the field $\vec{\mathcal{F}}$.

3.3 Time-Independent Response Theory

Although time-independent perturbation theory is mostly used for the derivation of energy corrections, it is not restricted to this but can be applied to any physical observable P and its expectation value as defined in Eq. (2.8). In this section, we want to illustrate this for the case of an observable whose corresponding operator is obtained as a derivative of the Hamiltonian with respect to the component $\mathcal{F}_{\alpha\cdots}$ of a field $\vec{\mathcal{F}}$, i.e. the field-dependent operator $\hat{P}(\vec{\mathcal{F}})$ defined in Eqs. (3.8) to (3.10).

Considering now the time-independent expectation value of this operator in the presence of the field,

$$\langle \Psi_0(\vec{\mathcal{F}}) | \hat{P}(\vec{\mathcal{F}}) | \Psi_0(\vec{\mathcal{F}}) \rangle = \langle \Psi_0(\vec{\mathcal{F}}) | \hat{P} | \Psi_0(\vec{\mathcal{F}}) \rangle + \langle \Psi_0(\vec{\mathcal{F}}) | \hat{P}^{(1)}(\vec{\mathcal{F}}) | \Psi_0(\vec{\mathcal{F}}) \rangle \quad (3.35)$$

---

[5] Sometimes, the $\langle \Psi_0^{(0)} | \hat{H}^{(2)} | \Psi_0^{(0)} \rangle$ term is considered to be only first order because it does not include a summation over excited states. However, this is a misunderstanding and is not in agreement with the partitioning of the Hamiltonian in Eq. (2.101).

we can make use of the expansion of the field-dependent wavefunction $|\Psi_0(\vec{\mathcal{F}})\rangle$, Eq. (3.16), and obtain to first order

$$\langle\Psi_0(\vec{\mathcal{F}})|\hat{P}(\vec{\mathcal{F}})|\Psi_0(\vec{\mathcal{F}})\rangle = \langle\Psi_0^{(0)}|\hat{P}|\Psi_0^{(0)}\rangle + \langle\Psi_0^{(0)}|\hat{P}^{(1)}(\vec{\mathcal{F}})|\Psi_0^{(0)}\rangle \qquad (3.36)$$
$$+ \langle\Psi_0^{(0)}|\hat{P}|\Psi_0^{(1)}(\vec{\mathcal{F}})\rangle + \langle\Psi_0^{(1)}(\vec{\mathcal{F}})|\hat{P}|\Psi_0^{(0)}\rangle + \cdots$$

The last three terms are linear in the components of the field and represent the **linear response** of the expectation value $\langle\Psi_0^{(0)}|\hat{P}|\Psi_0^{(0)}\rangle$ to the field $\vec{\mathcal{F}}$. Therefore, we want to call this application of time-independent perturbation theory to the case of an expectation value in the presence of a field also **time-independent response theory**.

Inserting the expressions for the first-order wavefunction, Eq. (3.27), and for the operator $\hat{P}$, Eqs. (3.9) and (3.10), in the first-order contributions gives

$$\langle\Psi_0(\vec{\mathcal{F}})|\hat{P}(\vec{\mathcal{F}})|\Psi_0(\vec{\mathcal{F}})\rangle = \langle\Psi_0^{(0)}|\hat{P}|\Psi_0^{(0)}\rangle \qquad (3.37)$$
$$+ \sum_{\beta\ldots}\left(\langle\Psi_0^{(0)}|\hat{O}_{\alpha\beta\ldots}^{\mathcal{F}\mathcal{F}} + \hat{O}_{\beta\alpha\ldots}^{\mathcal{F}\mathcal{F}}|\Psi_0^{(0)}\rangle + \sum_{n\neq 0}\frac{\langle\Psi_0^{(0)}|\hat{O}_{\alpha\ldots}^{\mathcal{F}}|\Psi_n^{(0)}\rangle\langle\Psi_n^{(0)}|\hat{O}_{\beta\ldots}^{\mathcal{F}}|\Psi_0^{(0)}\rangle}{E_0^{(0)} - E_n^{(0)}}\right.$$
$$\left.+ \sum_{n\neq 0}\frac{\langle\Psi_0^{(0)}|\hat{O}_{\beta\ldots}^{\mathcal{F}}|\Psi_n^{(0)}\rangle\langle\Psi_n^{(0)}|\hat{O}_{\alpha\ldots}^{\mathcal{F}}|\Psi_0^{(0)}\rangle}{E_0^{(0)} - E_n^{(0)}}\right)\mathcal{F}_{\beta\ldots}$$

One should note that due to our choice of the operator $\hat{P}(\vec{\mathcal{F}})$ as first derivatives of the Hamiltonian we could have derived the terms linear in $\mathcal{F}_{\beta\ldots}$ also as the first derivative of $E_0^{(2)}(\vec{\mathcal{F}})$, Eq. (3.34), with respect to $\mathcal{F}_{\alpha\ldots}$. However, in Section 3.11 we are going to extend our treatment of response theory to the case of time-dependent fields, where the energy is not the eigenvalue of the Hamiltonian and where we can only work with expectation values.

3.4 Second Derivatives of the Energy

In this section, we want to derive expressions for the second derivatives of the energy with respect to two components $\mathcal{F}_{\alpha\ldots}$ and $\mathcal{F}_{\beta\ldots}$ of the general electromagnetic field without relying on perturbation theory. According to the Hellmann–Feynman theorem, Eq. (3.7), the second derivative of the energy is equal to the first derivative of the expectation value of the derivative of the Hamiltonian for a non-zero value of the field, $|\vec{\mathcal{F}}| \neq 0$, i.e.

$$\frac{d^2 E_0(\vec{\mathcal{F}})}{d\mathcal{F}_{\beta\ldots}d\mathcal{F}_{\alpha\ldots}} = \frac{d}{d\mathcal{F}_{\beta\ldots}}\langle\Psi_0(\vec{\mathcal{F}})|\frac{\partial\hat{H}(\vec{\mathcal{F}})}{\partial\mathcal{F}_{\alpha\ldots}}|\Psi_0(\vec{\mathcal{F}})\rangle$$
$$= \langle\frac{d\Psi_0(\vec{\mathcal{F}})}{d\mathcal{F}_{\beta\ldots}}|\frac{\partial\hat{H}(\vec{\mathcal{F}})}{\partial\mathcal{F}_{\alpha\ldots}}|\Psi_0(\vec{\mathcal{F}})\rangle + \langle\Psi_0(\vec{\mathcal{F}})|\frac{\partial\hat{H}(\vec{\mathcal{F}})}{\partial\mathcal{F}_{\alpha\ldots}}|\frac{d\Psi_0(\vec{\mathcal{F}})}{d\mathcal{F}_{\beta\ldots}}\rangle$$

$$+ \langle \Psi_0(\vec{\mathcal{F}}) | \frac{\partial^2 \hat{H}(\vec{\mathcal{F}})}{\partial \mathcal{F}_\beta ... \partial \mathcal{F}_\alpha ...} | \Psi_0(\vec{\mathcal{F}}) \rangle \quad (3.38)$$

Inserting the first, Eq. (3.8), and second derivative of the Hamiltonian,

$$\frac{\partial^2 \hat{H}(\vec{\mathcal{F}})}{\partial \mathcal{F}_\beta ... \partial \mathcal{F}_\alpha ...} = \hat{O}_{\alpha\beta...}^{\mathcal{F}\mathcal{F}} + \hat{O}_{\beta\alpha...}^{\mathcal{F}\mathcal{F}} \quad (3.39)$$

we obtain for the second derivative of the energy evaluated at zero field strength, $|\vec{\mathcal{F}}| = 0$,

$$\left. \frac{d^2 E_0(\vec{\mathcal{F}})}{d\mathcal{F}_\beta ... d\mathcal{F}_\alpha ...} \right|_{|\vec{\mathcal{F}}|=0} = \left. \frac{d}{d\mathcal{F}_\beta...} \langle \Psi_0(\vec{\mathcal{F}}) | \frac{\partial \hat{H}(\vec{\mathcal{F}})}{\partial \mathcal{F}_\alpha ...} | \Psi_0(\vec{\mathcal{F}}) \rangle \right|_{|\vec{\mathcal{F}}|=0}$$

$$= \left. \langle \frac{d\Psi_0(\vec{\mathcal{F}})}{d\mathcal{F}_\beta...} | \hat{O}_{\alpha...}^{\mathcal{F}} | \Psi_0(\vec{\mathcal{F}}) \rangle \right|_{|\vec{\mathcal{F}}|=0} + \left. \langle \Psi_0(\vec{\mathcal{F}}) | \hat{O}_{\alpha...}^{\mathcal{F}} | \frac{d\Psi_0(\vec{\mathcal{F}})}{d\mathcal{F}_\beta...} \rangle \right|_{|\vec{\mathcal{F}}|=0}$$

$$+ \langle \Psi_0^{(0)} | \hat{O}_{\alpha\beta...}^{\mathcal{F}\mathcal{F}} + \hat{O}_{\beta\alpha...}^{\mathcal{F}\mathcal{F}} | \Psi_0^{(0)} \rangle \quad (3.40)$$

In Part III we will apply these expressions directly, but here we want to combine them with perturbation theory as developed in Section 3.2. Inserting thus the perturbation theory expansion of the perturbed wavefunction $|\Psi_0(\vec{\mathcal{F}})\rangle$, Eq. (3.16), in the right-hand side of Eq. (3.40) we obtain for the second derivative of the energy

$$\left. \frac{d^2 E_0(\vec{\mathcal{F}})}{d\mathcal{F}_\beta ... d\mathcal{F}_\alpha ...} \right|_{|\vec{\mathcal{F}}|=0} = \left. \langle \Psi_0^{(0)} | \hat{O}_{\alpha...}^{\mathcal{F}} | \frac{d\Psi_0^{(1)}(\vec{\mathcal{F}})}{d\mathcal{F}_\beta...} \rangle \right|_{|\vec{\mathcal{F}}|=0} + \left. \langle \frac{d\Psi_0^{(1)}(\vec{\mathcal{F}})}{d\mathcal{F}_\beta...} \right|_{|\vec{\mathcal{F}}|=0} | \hat{O}_{\alpha...}^{\mathcal{F}} | \Psi_0^{(0)} \rangle \quad (3.41)$$

$$+ \langle \Psi_0^{(0)} | \hat{O}_{\alpha\beta...}^{\mathcal{F}\mathcal{F}} + \hat{O}_{\beta\alpha...}^{\mathcal{F}\mathcal{F}} | \Psi_0^{(0)} \rangle + \cdots$$

Inserting the expression for the first-order wavefunction, Eq. (3.27), would bring us back to the expression for the second-order energy correction, Eq. (3.34).

3.5 Density Matrices

The expectation value of a general one-electron but spin-free operator $\hat{O} = \sum_i \hat{o}(i)$ in the unperturbed ground state $|\Psi_0^{(0)}\rangle$ is given as

$$\langle \Psi_0^{(0)} | \hat{O} | \Psi_0^{(0)} \rangle \quad (3.42)$$

$$= \int \cdots \int \Psi_0^{(0)*}(\vec{x}_1, \vec{x}_2, \cdots, \vec{x}_N) \sum_i \hat{o}(i) \, \Psi_0^{(0)}(\vec{x}_1, \vec{x}_2, \cdots, \vec{x}_N) \, d\vec{x}_1 \cdots d\vec{x}_N$$

Electrons are indistinguishable and each of the terms in $\sum_i \hat{o}(i)$ will thus give the same result

$$\langle \Psi_0^{(0)} | \hat{O} | \Psi_0^{(0)} \rangle \quad (3.43)$$

$$= N \int \cdots \int \Psi_0^{(0)*}(\vec{x}_1, \vec{x}_2, \cdots, \vec{x}_N) \, \hat{o}(1) \, \Psi_0^{(0)}(\vec{x}_1, \vec{x}_2, \cdots, \vec{x}_N) \, d\vec{x}_1 \cdots d\vec{x}_N$$

40 Perturbation Theory

The operator $\hat{O}$ is spin free and we can therefore integrate over the spin

$$\langle \Psi_0^{(0)} | \hat{O} | \Psi_0^{(0)} \rangle \tag{3.44}$$

$$= N \int \cdots \int \Psi_0^{(0)*}(\vec{r}_1,\vec{r}_2,\cdots,\vec{r}_N)\, \hat{o}(1)\, \Psi_0^{(0)}(\vec{r}_1,\vec{r}_2,\cdots,\vec{r}_N)\, d\vec{r}_1 \cdots d\vec{r}_N$$

In order to make use of the definition of the reduced one-electron density matrix Eq. (2.22) we need to rearrange the kernel of the integral

$$\langle \Psi_0^{(0)} | \hat{O} | \Psi_0^{(0)} \rangle \tag{3.45}$$

$$= N \int \cdots \int_{\vec{r}\,'_1 = \vec{r}_1} \hat{o}(1)\, \Psi_0^{(0)}(\vec{r}_1,\vec{r}_2,\cdots,\vec{r}_N)\, \Psi_0^{(0)*}(\vec{r}\,'_1,\vec{r}_2,\cdots,\vec{r}_N)\, d\vec{r}_1 \cdots d\vec{r}_N$$

The subscript $\vec{r}\,'_1 = \vec{r}_1$ on the integral means that $\vec{r}\,'_1$ should be set equal to $\vec{r}_1$ after $\hat{o}(1)$ has acted on $\Psi_0^{(0)}(\vec{r}_1,\vec{r}_2,\cdots,\vec{r}_N)$ but before the integration is carried out. This procedure is necessary in the case that $\hat{o}(1)$ includes a differential operator that then should only act on $\Psi_0^{(0)}(\vec{r}_1,\vec{r}_2,\cdots,\vec{r}_N)$ and not on $\Psi_0^{(0)*}(\vec{r}\,'_1,\vec{r}_2,\cdots,\vec{r}_N)$. We can now use the definition of the reduced one-electron density matrix Eq. (2.22), which gives

$$\langle \Psi_0^{(0)} | \hat{O} | \Psi_0^{(0)} \rangle = \int_{\vec{r}\,'_1 = \vec{r}_1} \hat{o}(1)\, P(\vec{r}_1, \vec{r}\,'_1)\, d\vec{r}_1 \tag{3.46}$$

In the presence of the perturbing field $\vec{\mathcal{F}}$ also the ground-state electron density Eq. (2.20) and ground-state reduced one-electron density matrix Eq. (2.22) become field dependent

$$P(\vec{r}, \vec{\mathcal{F}}) = \langle \Psi_0(\{\vec{r}_i\}, \vec{\mathcal{F}}) | \hat{D}(\vec{r}) | \Psi_0(\{\vec{r}_i\}, \vec{\mathcal{F}}) \rangle \tag{3.47}$$

$$P(\vec{r}, \vec{r}\,', \vec{\mathcal{F}}) = N \int \cdots \int \Psi_0(\vec{r}, \vec{r}_2, \cdots, \vec{r}_N, \vec{\mathcal{F}})\, \Psi_0^*(\vec{r}\,', \vec{r}_2, \cdots, \vec{r}_N, \vec{\mathcal{F}})\, d\vec{r}_2 \cdots d\vec{r}_N \tag{3.48}$$

In continuation of the perturbation expansion of the wavefunction in Eq. (3.16) one can then also expand the perturbed electron density and perturbed reduced density matrix in a perturbation series

$$P(\vec{r}, \vec{\mathcal{F}}) = P(\vec{r}) + \sum_\alpha P_\alpha^{(1)}(\vec{r})\, \mathcal{F}_\alpha + \cdots \tag{3.49}$$

$$P(\vec{r}, \vec{r}\,', \vec{\mathcal{F}}) = P(\vec{r}, \vec{r}\,') + \sum_\alpha P_\alpha^{(1)}(\vec{r}, \vec{r}\,')\, \mathcal{F}_\alpha + \cdots \tag{3.50}$$

where the first-order corrections to the electron density and to the reduced one-electron density matrix are given as

$$P_\alpha^{(1)}(\vec{r}) = \langle \Psi_0^{(1)}(\{\vec{r}_i\}) | \hat{D}(\vec{r}) | \Psi_0^{(0)}(\{\vec{r}_i\}) \rangle + \langle \Psi_0^{(0)}(\{\vec{r}_i\}) | \hat{D}(\vec{r}) | \Psi_0^{(1)}(\{\vec{r}_i\}) \rangle \tag{3.51}$$

$$P_\alpha^{(1)}(\vec{r}, \vec{r}\,') = N \int \cdots \int \Psi_0^{(1)}(\vec{r}, \vec{r}_2, \cdots, \vec{r}_N)\, \Psi_0^{(0)*}(\vec{r}\,', \vec{r}_2, \cdots, \vec{r}_N)\, d\vec{r}_2 \cdots d\vec{r}_N$$

$$+ N \int \cdots \int \Psi_0^{(0)}(\vec{r}, \vec{r}_2, \cdots, \vec{r}_N)\, \Psi_0^{(1)*}(\vec{r}\,', \vec{r}_2, \cdots, \vec{r}_N)\, d\vec{r}_2 \cdots d\vec{r}_N \tag{3.52}$$

In extension of Eq. (3.46) we can then express the first-order correction to a field-dependent expectation value of the general but spin-free operator $\hat{O}$ with the first-order reduced density matrix $P_\alpha^{(1)}(\vec{r}, \vec{r}\,')$ as

$$\langle \Psi_0(\vec{\mathcal{F}})|\hat{O}|\Psi_0(\vec{\mathcal{F}})\rangle^{(1)} = \langle \Psi_0^{(0)}|\hat{O}|\Psi_0^{(1)}(\vec{\mathcal{F}})\rangle + \langle \Psi_0^{(1)}(\vec{\mathcal{F}})|\hat{O}|\Psi_0^{(0)}\rangle$$

$$= \sum_\alpha \mathcal{F}_\alpha \int_{\vec{r}\,'_1 = \vec{r}_1} \hat{o}(1)\, P_\alpha^{(1)}(\vec{r}_1, \vec{r}\,'_1)\, d\vec{r}_1 \qquad (3.53)$$

3.6 The Ehrenfest Theorem

In Section 3.3 we looked at the dependence of an expectation value on a perturbing field $\vec{\mathcal{F}}$ and expanded the expectation value in powers of this perturbation. In this section, we want to study now the time evolution of an expectation value of an arbitrary operator $\hat{P}$. Finally, in the section on time-dependent response theory, Section 3.11, we will combine both and study the effects of a time-dependent perturbation $\mathcal{F}_{\alpha\ldots}(t)$.

Let us study the time dependence of an expectation value by deriving an expression for the time derivative of an expectation value, i.e. an equation of motion for the expectation value of the operator $\hat{P}$

$$\frac{d}{dt}\langle \Psi_0^{(0)}(t)|\hat{P}|\Psi_0^{(0)}(t)\rangle \qquad (3.54)$$

$$= \langle \frac{\partial \Psi_0^{(0)}(t)}{\partial t}|\hat{P}|\Psi_0^{(0)}(t)\rangle + \langle \Psi_0^{(0)}(t)|\frac{\partial \hat{P}}{\partial t}|\Psi_0^{(0)}(t)\rangle + \langle \Psi_0^{(0)}(t)|\hat{P}|\frac{\partial \Psi_0^{(0)}(t)}{\partial t}\rangle$$

The time derivative of the wavefunction is given by the time-dependent electronic Schrödinger equation, Eq. (2.13),

$$\frac{\partial}{\partial t}|\Psi_0^{(0)}(t)\rangle = -\frac{i}{\hbar}\hat{H}|\Psi_0^{(0)}(t)\rangle \qquad (3.55)$$

and correspondingly for the complex conjugate of the wavefunction

$$\frac{\partial}{\partial t}\langle \Psi_0^{(0)}(t)| = \frac{i}{\hbar}\langle \Psi_0^{(0)}(t)|\hat{H} \qquad (3.56)$$

Inserted in Eq. (3.54) we obtain

$$\frac{d}{dt}\langle \Psi_0^{(0)}(t)|\hat{P}|\Psi_0^{(0)}(t)\rangle = -\frac{i}{\hbar}\langle \Psi_0^{(0)}(t)|[\hat{P},\hat{H}]|\Psi_0^{(0)}(t)\rangle + \langle \Psi_0^{(0)}(t)|\frac{\partial \hat{P}}{\partial t}|\Psi_0^{(0)}(t)\rangle$$

$$(3.57)$$

The second term vanishes, if the operator, $\hat{P}$, itself is independent of time, as in all cases we consider here. We arrive thus at the **Ehrenfest theorem**

$$\frac{d}{dt}\langle \Psi_0^{(0)}(t)|\hat{P}|\Psi_0^{(0)}(t)\rangle = -\frac{i}{\hbar}\langle \Psi_0^{(0)}(t)|[\hat{P},\hat{H}]|\Psi_0^{(0)}(t)\rangle \qquad (3.58)$$

One should note that the Ehrenfest theorem is derived solely by application of the time-dependent Schrödinger equation. It contains therefore the same information and is often applied as an alternative to the time-dependent Schrödinger equation, when the time evolution of a system is to be determined, as in Sections 11.2 and 11.3. In these cases the operator $\hat{P}$ would typically be one of the operators that govern the time dependence of the system (Olsen and Jørgensen, 1985; Christiansen et al., 1998b; Olsen et al., 2005).

3.7 The Off-Diagonal Hypervirial Theorem

A second very useful theorem can be derived from the Ehrenfest theorem, if one considers the wavefunctions of two different stationary states and the unperturbed Hamiltonian instead of a general time-dependent wavefunction (Chen, 1964)

$$i\hbar \frac{d}{dt}\langle \Psi_m^{(0)}(t)|\hat{P}|\Psi_n^{(0)}(t)\rangle = \langle \Psi_m^{(0)}(t)|[\hat{P},\hat{H}^{(0)}]|\Psi_n^{(0)}(t)\rangle \quad (3.59)$$

Being stationary states, Eq. (2.14), $|\Psi_m^{(0)}(t)\rangle = |\Psi_m^{(0)}\rangle e^{-\frac{i}{\hbar}E_m^{(0)}t}$ and $|\Psi_n^{(0)}(t)\rangle = |\Psi_n^{(0)}\rangle e^{-\frac{i}{\hbar}E_n^{(0)}t}$ are eigenfunctions of $\hat{H}^{(0)}$ and we can write

$$i\hbar \frac{d}{dt}\langle \Psi_m^{(0)}(t)|\hat{P}|\Psi_n^{(0)}(t)\rangle = (E_n^{(0)} - E_m^{(0)})\langle \Psi_m^{(0)}(t)|\hat{P}|\Psi_n^{(0)}(t)\rangle \quad (3.60)$$

The combination of Eqs. (3.59) and (3.60) gives for $m = n$ the **hypervirial theorem**

$$\langle \Psi_m^{(0)}|[\hat{P},\hat{H}^{(0)}]|\Psi_m^{(0)}\rangle = 0 \quad (3.61)$$

and for $m \neq n$ the **off-diagonal hypervirial theorem**

$$\langle \Psi_m^{(0)}|[\hat{P},\hat{H}^{(0)}]|\Psi_n^{(0)}\rangle = (E_n^{(0)} - E_m^{(0)})\langle \Psi_m^{(0)}|\hat{P}|\Psi_n^{(0)}\rangle \quad (3.62)$$

Several important relations can be derived from the off-diagonal hypervirial relation (Hansen and Bouman, 1979), if one chooses an operator $\hat{P}$ that does not commute with the Hamiltonian $\hat{H}^{(0)}$. The most important one is obtained for $\hat{P} = \hat{O}_\alpha^r$, which is a cartesian component of the **sum of the position operators** of all electrons defined as[6]

$$\hat{O}_\alpha^r = \sum_i^N \hat{r}_{i,\alpha} \quad (3.63)$$

Recalling the well-known commutator relation

$$[\hat{O}^r,\hat{H}^{(0)}] = \frac{i\hbar}{m_e}\hat{O}^p \quad (3.64)$$

where $\hat{O}^p$ is the **total canonical momentum operator** of the electrons, whose cartesian components are defined as

$$\hat{O}_\alpha^p = \sum_i^N \hat{p}_{i,\alpha} \quad (3.65)$$

---

[6] The definitions of all operators are also collected in Appendix A.

we obtain the off-diagonal hypervirial relation for transition moments of the electronic position and momentum operators

$$\left(E_n^{(0)} - E_m^{(0)}\right) \langle \Psi_m^{(0)} | \hat{\vec{O}}^r | \Psi_n^{(0)} \rangle = \frac{i\hbar}{m_e} \langle \Psi_m^{(0)} | \hat{\vec{O}}^p | \Psi_n^{(0)} \rangle \tag{3.66}$$

and the hypervirial relation for the canonical momentum operator

$$\langle \Psi_m^{(0)} | \hat{\vec{O}}^p | \Psi_m^{(0)} \rangle = 0 \tag{3.67}$$

which both of will be used several times in the following chapters.

3.8 The Interaction Picture

Before we can start with the discussion of time-dependent perturbation theory in the form of response theory, we need to introduce an alternative formulation of quantum mechanics, called the **interaction or Dirac representation**. In general, several representations of the wavefunctions or state vectors and of the operators of quantum mechanics are equivalent, i.e. valid, as long as the expectation values of operators $\langle \Psi_0 | \hat{O} | \Psi_0 \rangle$ or inner products of the wavefunctions $\langle \Psi_0 | \Psi_n \rangle$ are always the same. Measurable quantities and thus the physics are contained in the expectation values or inner products, whereas operators and wavefunctions are mathematical constructs used in a particular formulation of the theory. One example of this was already discussed in Section 2.9 on gauge transformations of the vector and scalar potentials. In the present section we want to look at a transformation that is related to the time dependence of the wavefunctions and operators.

The operators $\hat{H}$ and $\hat{O}$ and the wavefunctions $|\Psi_0(t)\rangle$ that we have used so far are said to be in the **Schrödinger picture**, where the wavefunctions carry the time dependence and the operators are time independent apart from the case of an explicit time dependence due to a time-dependent perturbation. The Schrödinger representation is the natural choice for time-independent systems.

The **interaction or Dirac representation** becomes, on the other hand, useful, if one deals with a system that is described by a time-dependent Hamiltonian such as

$$\hat{H}(t) = \hat{H}^{(0)} + \hat{H}^{(1)}(t) \tag{3.68}$$

The wavefunctions $|\Psi_0^I(t)\rangle$ and operators $\hat{O}_\alpha^I(t)$ in the **interaction picture** are defined as

$$|\Psi_0^I(t)\rangle = e^{\frac{i}{\hbar}\hat{H}^{(0)}t} |\Psi_0(t)\rangle \tag{3.69}$$

$$\hat{O}_\alpha^I(t) = e^{\frac{i}{\hbar}\hat{H}^{(0)}t} \hat{O}_\alpha \, e^{-\frac{i}{\hbar}\hat{H}^{(0)}t} \tag{3.70}$$

where one should recall that any exponential operator $e^{\hat{A}}$ is defined via its Taylor series expansion as

$$e^{\hat{A}} = 1 + \hat{A} + \frac{1}{2!}(\hat{A})^2 + \cdots \tag{3.71}$$

44   Perturbation Theory

The essence of the interaction picture can be illustrated when we assume for a moment that the time-dependent perturbation Hamiltonian $\hat{H}^{(1)}(t)$ vanishes. The time-dependent wavefunction $|\Psi_0(t)\rangle$ in the Schrödinger picture then becomes equal to the unperturbed but still time-dependent wavefunction

$$|\Psi_0^{(0)}(t)\rangle = e^{-\frac{i}{\hbar}E_0^{(0)}t}|\Psi_0^{(0)}\rangle = e^{-\frac{i}{\hbar}\hat{H}^{(0)}t}|\Psi_0^{(0)}\rangle \tag{3.72}$$

where the last equality holds, because $|\Psi_0^{(0)}\rangle$ is an eigenfunction of $\hat{H}^{(0)}$ with eigenvalue $E_0^{(0)}$, see Eq. (3.14). The effect of the transformation to the interaction picture is consequently, that the time dependence of the unperturbed wavefunction is removed

$$|\Psi_0^{(0),I}(t)\rangle = e^{\frac{i}{\hbar}\hat{H}^{(0)}t}\,|\Psi_0^{(0)}(t)\rangle = e^{\frac{i}{\hbar}\hat{H}^{(0)}t}e^{-\frac{i}{\hbar}\hat{H}^{(0)}t}|\Psi_0^{(0)}\rangle = |\Psi_0^{(0)}\rangle \tag{3.73}$$

As the time dependence of the unperturbed wavefunctions is simply a rotation in the complex plane, Eq. (2.6), we can say that in the interaction picture this rotation is frozen out or that we have switched to a **rotating frame** that rotates with the time dependence of the unperturbed wavefunctions. The time evolution of the perturbed wavefunction $|\Psi_0^I(t)\rangle$ is in the interaction picture thus governed by the perturbation Hamiltonian $\hat{H}^{(1)}(t)$ alone, as we will see in the following section. This will greatly simplify the derivation of time-dependent wavefunctions in the following section and is the motivation for introducing the interaction picture here.

3.9   Time-Dependent Perturbation Theory

When dealing with time-dependent fields one has to find solutions to the time-dependent electronic Schrödinger equation

$$i\hbar\frac{\partial}{\partial t}|\Psi_0(t,\vec{\mathcal{F}})\rangle = \left[\hat{H}^{(0)} + \hat{H}^{(1)}(t)\right]|\Psi_0(t,\vec{\mathcal{F}})\rangle \tag{3.74}$$

In the rest of this chapter we will not consider time-dependent magnetic perturbations and have therefore neglected the second-order perturbation Hamiltonian $\hat{H}^{(2)}(t)$. Generalizing Eq. (2.108) we write the time-dependent first-order perturbation Hamiltonian $\hat{H}^{(1)}(t)$ as

$$\hat{H}^{(1)}(t) = \sum_{\beta\cdots}\hat{O}^{\mathcal{F}}_{\beta\cdots}\mathcal{F}_{\beta\cdots}(t) \tag{3.75}$$

In the length gauge, Eq. (2.122), the operator $\hat{O}^{\mathcal{F}}_{\beta\cdots}$ could be the electric dipole or quadrupole operator, defined in Appendix A. It depends on coordinates and momenta of the electrons but it is independent of time, whereas we assume that the time-dependent field $\mathcal{F}_{\beta\cdots}(t)$ does not depend on any electronic variables. The subscript $\beta\cdots$ again denotes components of a tensor of appropriate rank. On the other hand, in the velocity gauge, Eq. (2.125), the operator $\hat{O}^{\mathcal{F}}_{\beta\cdots}$ is equal to the total canonical momentum operator $\hat{O}^p$, Eq. (3.65), and the time-dependent field in Eq. (3.75) is replaced by the time-dependent vector potential $\vec{\hat{A}}(t)$. However, in the following we will discuss only the length gauge.

For monochromatic linear polarized radiation in the **dipole approximation**[7] the time dependence of a component of the field vector can be expressed as

$$\mathcal{F}_{\beta\ldots}(t) = \mathcal{F}_{\beta\ldots}(\omega)\cos(\omega t) = \frac{1}{2}\mathcal{F}_{\beta\ldots}(\omega)\left(e^{i\omega t} + e^{-i\omega t}\right) \tag{3.76}$$

For a general pulse of coherent but polychromatic radiation this then becomes

$$\mathcal{F}_{\beta\ldots}(t) = \frac{1}{2}\int_0^\infty d\omega\, \mathcal{F}_{\beta\ldots}(\omega)\left(e^{i\omega t} + e^{-i\omega t}\right) = \frac{1}{2}\int_{-\infty}^\infty d\omega\, \mathcal{F}_{\beta\ldots}(\omega)\, e^{-i\omega t} \tag{3.77}$$

which we can recognize as the Fourier transform of $\mathcal{F}_{\beta\ldots}(t)$ into Fourier components $\frac{1}{2}\mathcal{F}_{\beta\ldots}(\omega)$. The perturbation Hamiltonian can then also be expressed in terms of its Fourier components[8]

$$\hat{H}^{(1)}(t) = \int_{-\infty}^\infty d\omega\, \hat{H}^{(1)}(\omega) = \frac{1}{2}\sum_{\beta\ldots}\hat{O}_{\beta\ldots}^\omega \int_{-\infty}^\infty d\omega\, \mathcal{F}_{\beta\ldots}(\omega)\, e^{-i\omega t} \tag{3.78}$$

$\mathcal{F}_{\beta\ldots}(\omega)$ are the frequency or Fourier components of the time-dependent field $\mathcal{F}_{\beta\ldots}(t)$, whereas the operator $\hat{O}_{\beta\ldots}^\omega$ depends only on coordinates and momenta of the electrons. It is actually not affected by the Fourier transformation and thus just the same operator as $\hat{O}_{\beta\ldots}^\mathcal{F}$. The superscript ω is a pure label, like the superscript $\mathcal{F}$, attached to it in order to associate it with the field oscillating with frequency ω and to remind us of the fact that we are in the frequency domain.

There exist many different but essentially equivalent approaches (Dirac, 1958; Langhoff et al., 1972; Zubarev, 1974; Olsen and Jørgensen, 1985; Pickup, 1992) for obtaining the time-dependent wavefunction. Here, we will use the **interaction** or **Dirac representation** of the time-dependent wavefunction

$$|\Psi_0^I(t,\vec{\mathcal{F}})\rangle = e^{\frac{i}{\hbar}\hat{H}^{(0)}t}|\Psi_0(t,\vec{\mathcal{F}})\rangle \tag{3.79}$$

introduced in the previous section, which is more convenient for the derivation of response functions in Section 3.11. We begin by taking the time derivative $i\hbar\frac{\partial}{\partial t}$ of $|\Psi_0^I(t,\vec{\mathcal{F}})\rangle$

$$i\hbar\frac{\partial}{\partial t}|\Psi_0^I(t,\vec{\mathcal{F}})\rangle = -\hat{H}^{(0)}e^{\frac{i}{\hbar}\hat{H}^{(0)}t}|\Psi_0(t,\vec{\mathcal{F}})\rangle + e^{\frac{i}{\hbar}\hat{H}^{(0)}t}\,i\hbar\frac{\partial}{\partial t}|\Psi_0(t,\vec{\mathcal{F}})\rangle \tag{3.80}$$

In the second term we can make use of the time-dependent Schrödinger equation, (3.74), which leads [see Exercise 3.6] to the **equation of motion** for $|\Psi_0^I(t,\vec{\mathcal{F}})\rangle$ in the interaction picture

$$i\hbar\frac{\partial}{\partial t}|\Psi_0^I(t,\vec{\mathcal{F}})\rangle = \hat{H}^{(1),I}(t)\,|\Psi_0^I(t,\vec{\mathcal{F}})\rangle \tag{3.81}$$

---

[7] Details of the dipole approximation are discussed in Section 7.1.
[8] In principle, the frequency ω in the exponential should be replaced by $\omega' \equiv \omega + i\eta$. The small positive infinitesimal η ensures then that $\hat{H}^{(1)}(t)$ is zero for $t = -\infty$ and that the perturbation builds up adiabatically from $t = -\infty$. However, this is omitted in the following for the sake of a less-complex notation. Similarly, the effects of switching on the static perturbations are ignored throughout this book.

46 Perturbation Theory

where $\hat{H}^{(1),I}(t)$ is the interaction representation, Eq. (3.70), of the perturbation Hamiltonian. We can see that the time dependence of the perturbed wavefunction, is, in the interaction picture, indeed governed by the perturbation Hamiltonian alone, as discussed in the last section.

Exercise 3.6 Fill in the missing steps in the derivation of the equation of motion, Eq. (3.81), for the time-dependent wavefunction in the interaction picture $|\Psi_0^I(t,\vec{\mathcal{F}})\rangle$.

The time-dependent wavefunction $|\Psi_0^I(t,\vec{\mathcal{F}})\rangle$ in the interaction picture is then obtained by integration

$$|\Psi_0^I(t,\vec{\mathcal{F}})\rangle - |\Psi_0^I(-\infty,\vec{\mathcal{F}})\rangle = \frac{1}{i\hbar}\int_{-\infty}^{t} dt'\, \hat{H}^{(1),I}(t')\, |\Psi_0^I(t',\vec{\mathcal{F}})\rangle \qquad (3.82)$$

Recalling the time dependence of the unperturbed Schrödinger wavefunctions, Eq. (3.72), and that the perturbed Schrödinger wavefunction $|\Psi_0(t,\vec{\mathcal{F}})\rangle$ is equal to the unperturbed time-dependent Schrödinger wavefunction in the limit of $t \to -\infty$ we can see that

$$|\Psi_0^I(-\infty,\vec{\mathcal{F}})\rangle = \lim_{t\to-\infty} e^{\frac{i}{\hbar}\hat{H}^{(0)}t}\, |\Psi_0(t,\vec{\mathcal{F}})\rangle = |\Psi_0^{(0)}\rangle \qquad (3.83)$$

Moving the unperturbed wavefunction to the right-hand side of Eq. (3.82), and premultiplying the resulting equation with $e^{-\frac{i}{\hbar}\hat{H}^{(0)}t}$ we obtain an equation for the time-dependent wavefunction $\Psi_0(t,\vec{\mathcal{F}})$ in the Schrödinger picture

$$|\Psi_0(t,\vec{\mathcal{F}})\rangle = e^{-\frac{i}{\hbar}\hat{H}^{(0)}t}\, |\Psi_0^{(0)}\rangle + \frac{1}{i\hbar}\int_{-\infty}^{t} dt'\, e^{-\frac{i}{\hbar}\hat{H}^{(0)}(t-t')}\, \hat{H}^{(1)}(t')\, |\Psi_0(t',\vec{\mathcal{F}})\rangle \qquad (3.84)$$

With this equation the Dirac or interaction representation has served its purpose and from now on we will work again with the wavefunctions in the Schrödinger picture.

Equation (3.84) is not really a solution to the differential equation, Eq. (3.81), but only the integral version of it. The unknown time-dependent wavefunction, $\Psi_0(t,\vec{\mathcal{F}})$, is expressed as the unperturbed wavefunction and a correction term that depends on the unknown wavefunction again. In analogy to the expansion of time-independent wavefunction in Eq. (3.16) we now expand the time-dependent wavefunction also in a perturbation series

$$|\Psi_0(t,\vec{\mathcal{F}})\rangle = |\Psi_0^{(0)}(t)\rangle + |\Psi_0^{(1)}(t,\vec{\mathcal{F}})\rangle + |\Psi_0^{(2)}(t,\vec{\mathcal{F}})\rangle + \cdots \qquad (3.85)$$

where $|\Psi_0^{(0)}(t)\rangle$ is the unperturbed but time-dependent wavefunction as, e.g., given in Eq. (3.72) and therefore the first term on the right-hand side of Eq. (3.84). Contrary to the time-independent case we do not need to introduce a formal ordering parameter λ here, because we are not going to split the time-dependent Schrödinger equation into separate equations for each order. The second term, i.e. the integral term on the right-hand side of Eq. (3.84), will then give rise to all higher-order corrections in the expansion of the time-dependent wavefunction. They can be obtained by iteratively solving the integral equation.

In the first iteration, the unknown time-dependent wavefunction $|\Psi_0(t',\vec{\mathcal{F}})\rangle$ in the integral in Eq. (3.84) is approximated by the zeroth-order term, leading to

$$|\Psi_0(t,\vec{\mathcal{F}})\rangle = e^{-\frac{i}{\hbar}\hat{H}^{(0)}t}|\Psi_0^{(0)}\rangle + e^{-\frac{i}{\hbar}\hat{H}^{(0)}t}\frac{1}{i\hbar}\int_{-\infty}^{t} dt' e^{\frac{i}{\hbar}\hat{H}^{(0)}t'}\hat{H}^{(1)}(t')e^{-\frac{i}{\hbar}\hat{H}^{(0)}t'}|\Psi_0^{(0)}\rangle + \cdots \quad (3.86)$$

We can then identify the first-order correction to the time-dependent wavefunction as

$$|\Psi_0^{(1)}(t,\vec{\mathcal{F}})\rangle = e^{-\frac{i}{\hbar}\hat{H}^{(0)}t}\frac{1}{i\hbar}\int_{-\infty}^{t} dt'\, \hat{H}^{(1),I}(t')\,|\Psi_0^{(0)}\rangle \quad (3.87)$$

where we have used the Dirac representation of operators, Eq. (3.70), for the sole purpose of a more compact equation; a trick that we will make use of more often in the rest of this chapter.

The second-order correction to the wavefunction is obtained, if we iterate once more on Eq. (3.84), which means that we let the unknown function $|\Psi_0(t',\vec{\mathcal{F}})\rangle$ in the integral in Eq. (3.84) be equal to $|\Psi_0^{(0)}(t')\rangle + |\Psi_0^{(1)}(t',\vec{\mathcal{F}})\rangle$. This gives rise to one term more

$$|\Psi_0^{(2)}(t,\vec{\mathcal{F}})\rangle = e^{-\frac{i}{\hbar}\hat{H}^{(0)}t}\left(\frac{1}{i\hbar}\right)^2 \int_{-\infty}^{t} dt' \int_{-\infty}^{t'} dt''\, \hat{H}^{(1),I}(t')\,\hat{H}^{(1),I}(t'')\,|\Psi_0^{(0)}\rangle \quad (3.88)$$

which we can identify as the second-order correction to the time-dependent wavefunction. In the same way one can derive also higher-order corrections.

3.10 Transition Probabilities and Rates

The compact integral expressions for the time-dependent wavefunction in Eq. (3.86) and in particular for the first-order correction in Eq. (3.87) will be employed in the derivation of response functions in the following section. However, for the interpretation of the time-dependent wavefunctions it is useful to expand them in the complete set of unperturbed wavefunctions Eq. (2.14) analogous to the perturbed wavefunctions of time-independent perturbation theory in Eq. (3.23)

$$|\Psi_0^{(1)}(t,\vec{\mathcal{F}})\rangle = \sum_{n\neq 0} e^{-\frac{i}{\hbar}\hat{H}^{(0)}t}|\Psi_n^{(0)}\rangle\, C_{n0}^{(1)}(t,\vec{\mathcal{F}}) \quad (3.89)$$

where the time-dependent first-order coefficients are defined as

$$C_{n0}^{(1)}(t,\vec{\mathcal{F}}) = \langle e^{-\frac{i}{\hbar}\hat{H}^{(0)}t}\Psi_n^{(0)}\,|\,\Psi_0^{(1)}(t,\vec{\mathcal{F}})\rangle \quad (3.90)$$

The coefficients $C_{n0}^{(1)}(t,\vec{\mathcal{F}})$ and their norm $|C_{n0}^{(1)}(t,\vec{\mathcal{F}})|^2$ can then be interpreted as probability amplitude and probability, respectively, for finding the system in the stationary state $|\Psi_n^{(0)}\rangle$ at time t. As the system was originally, i.e. before the perturbation was turned on, in the state $\Psi_0^{(0)}$ we can interpret $|C_{n0}^{(1)}(t,\vec{\mathcal{F}})|^2$ as the probability for a transition from state $\Psi_0^{(0)}$ to $\Psi_n^{(0)}$ and thus as the **transition probability**.

48 Perturbation Theory

Inserting the expression for the first-order wavefunction Eq. (3.87) we obtain for the first-order coefficient

$$C_{n0}^{(1)}(t,\vec{\mathcal{F}}) = \langle \Psi_n^{(0)} | e^{\frac{i}{\hbar}\hat{H}^{(0)}t} e^{-\frac{i}{\hbar}\hat{H}^{(0)}t} \frac{1}{i\hbar} \int_{-\infty}^{t} dt' \, \hat{H}^{(1),I}(t') | \Psi_0^{(0)} \rangle$$

$$= \frac{1}{i\hbar} \int_{-\infty}^{t} dt' \, \langle \Psi_n^{(0)} | e^{\frac{i}{\hbar}\hat{H}^{(0)}t'} \hat{H}^{(1)}(t') e^{-\frac{i}{\hbar}\hat{H}^{(0)}t'} | \Psi_0^{(0)} \rangle \quad (3.91)$$

Using the fact that the states $\Psi_0^{(0)}$ and $\Psi_n^{(0)}$ are eigenfunctions of the Hamiltonian $\hat{H}^{(0)}$ and defining a transition angular frequency ω_{n0} as

$$\omega_{n0} = \frac{E_n^{(0)} - E_0^{(0)}}{\hbar} \quad (3.92)$$

we can write the first-order probability amplitude as

$$C_{n0}^{(1)}(t,\vec{\mathcal{F}}) = \frac{1}{i\hbar} \int_{-\infty}^{t} dt' \, \langle \Psi_n^{(0)} | \hat{H}^{(1)}(t') | \Psi_0^{(0)} \rangle e^{i\omega_{n0}t'} \quad (3.93)$$

Inserting now the expressions for the perturbation Hamiltonian, Eq. (3.75) and Eq. (3.77) the probability amplitude becomes

$$C_{n0}^{(1)}(t,\vec{\mathcal{F}}) = \frac{1}{2i\hbar} \sum_{\beta\ldots} \langle \Psi_n^{(0)} | \hat{O}_{\beta\ldots}^{\omega} | \Psi_0^{(0)} \rangle \int_0^{\infty} d\omega \, \mathcal{F}_{\beta\ldots}(\omega)$$

$$\times \left(\int_{-\infty}^{t} dt' e^{i(\omega_{n0}-\omega)t'} + \int_{-\infty}^{t} dt' e^{i(\omega_{n0}+\omega)t'} \right) \quad (3.94)$$

where we have made use of the fact that the operator $\hat{O}_{\beta\ldots}^{\omega}$ does not depend on the frequency of the radiation. Integration over t' then gives

$$C_{n0}^{(1)}(t,\vec{\mathcal{F}}) = \frac{1}{2i\hbar} \sum_{\beta\ldots} \langle \Psi_n^{(0)} | \hat{O}_{\beta\ldots}^{\omega} | \Psi_0^{(0)} \rangle$$

$$\times \int_0^{\infty} d\omega \, \mathcal{F}_{\beta\ldots}(\omega) \left(\frac{e^{i(\omega_{n0}-\omega)t} - 1}{i(\omega_{n0}-\omega)} + \frac{e^{i(\omega_{n0}+\omega)t} - 1}{i(\omega_{n0}+\omega)} \right) \quad (3.95)$$

The first term in the second line is negligible unless the frequency of the radiation is close to the transition frequency, i.e. $\omega \simeq \omega_{n0}$, whereas in the second term the frequency needs to be $\omega \simeq -\omega_{n0}$. This implies that the first term corresponds to a transition from state $\Psi_0^{(0)}$ to state $\Psi_n^{(0)}$ and thus to the absorption of a photon of energy $\hbar\omega$, while the second term corresponds to the transition from $\Psi_n^{(0)}$ to $\Psi_0^{(0)}$ and therefore to the emission of a photon. Both conditions cannot be fulfilled at the same time and we can therefore consider the two processes separately.

The transition probability for absorption is thus given as

$$\left|C_{n0}^{(1)}(t,\vec{\mathcal{F}})\right|^2 = \frac{1}{2\hbar^2}\sum_{\beta\cdots}\left|\langle\Psi_n^{(0)}|\hat{O}_{\beta\cdots}^\omega|\Psi_0^{(0)}\rangle\right|^2 \int_0^\infty d\omega\, \mathcal{F}_{\beta\cdots}^2(\omega)\frac{2\sin\left(\frac{1}{2}(\omega_{n0}-\omega)t\right)}{(\omega_{n0}-\omega)^2} \quad (3.96)$$

As a function of the frequency ω the term $\frac{2\sin\left(\frac{1}{2}(\omega_{n0}-\omega)t\right)}{(\omega_{n0}-\omega)^2}$ has one sharp maximum for $\omega \simeq \omega_{n0}$, which is going to dominate the integral. We can therefore approximate the amplitude of the field $\mathcal{F}_{\beta\cdots}(\omega)$ by its value $\mathcal{F}_{\beta\cdots}(\omega_{n0})$ at this frequency, take it out of the integral, extend the lower integration limit to $-\infty$ and evaluate the remaining standard integral, leading to

$$\left|C_{n0}^{(1)}(t,\vec{\mathcal{F}})\right|^2 = \frac{\pi}{2\hbar^2}t\sum_{\beta\cdots}\left|\langle\Psi_n^{(0)}|\hat{O}_{\beta\cdots}^\omega|\Psi_0^{(0)}\rangle\right|^2 \mathcal{F}_{\beta\cdots}^2(\omega_{n0}) \quad (3.97)$$

which shows that the transition probability to first order in perturbation theory increases linear with time. Now, we can define finally a **transition rate**, $W_{n0}^{(1)}$, to first order as the time derivative of the transition probability

$$W_{n0}^{(1)} = \frac{d}{dt}\left|C_{n0}^{(1)}(t,\vec{\mathcal{F}})\right|^2 = \frac{\pi}{2\hbar^2}\sum_{\beta\cdots}\left|\langle\Psi_n^{(0)}|\hat{O}_{\beta\cdots}^\omega|\Psi_0^{(0)}\rangle\right|^2 \mathcal{F}_{\beta\cdots}^2(\omega_{n0}) \quad (3.98)$$

and understand why the matrix element $\langle\Psi_n^{(0)}|\hat{O}_{\beta\cdots}^\omega|\Psi_0^{(0)}\rangle$ is often called a **transition moment** $M_{n0,\beta\cdots}$.

3.11 Time-Dependent Response Theory

Having derived the first few terms in a perturbation expansion of the time-dependent, perturbed wavefunction $|\Psi_0(t,\vec{\mathcal{F}})\rangle$, Eq. (3.85), we are now ready to look at a perturbation expansion of an expectation value $\langle\Psi_0(t,\vec{\mathcal{F}})|\hat{P}|\Psi_0(t,\vec{\mathcal{F}})\rangle$ in the presence of a general, time-dependent field $\vec{\mathcal{F}}(t)$, where $\hat{P}$ could be any kind of operator. But again, we are here mostly interested in operators that are obtained as derivatives of the Hamiltonian with respect to a component of a field $\vec{\mathcal{F}}$, i.e. the operators $\hat{P}$ defined in Eqs. (3.8) and (3.9). The expansion will thus be a generalization of the expansion in Eq. (3.36) to the time-dependent case, but with the restriction that we will only consider field-independent operators $\hat{P}$ in the derivations in order to reduce the complexity of the equations.

We can now insert the expansion of the time-dependent wavefunction $|\Psi_0(t,\vec{\mathcal{F}})\rangle$, Eq. (3.85), and the first-order correction, Eq. (3.87), in the time-dependent expectation value

$$\langle\Psi_0(t,\vec{\mathcal{F}})|\hat{P}|\Psi_0(t,\vec{\mathcal{F}})\rangle$$
$$= \langle\Psi_0^{(0)}(t)|\hat{P}|\Psi_0^{(0)}(t)\rangle + \langle\Psi_0^{(0)}(t)|\hat{P}|\Psi_0^{(1)}(t,\vec{\mathcal{F}})\rangle + \langle\Psi_0^{(1)}(t,\vec{\mathcal{F}})|\hat{P}|\Psi_0^{(0)}(t)\rangle + \cdots$$

50 Perturbation Theory

$$= \langle \Psi_0^{(0)}(t) | \hat{P} | \Psi_0^{(0)}(t) \rangle + \frac{1}{i\hbar} \int_{-\infty}^{t} dt' \, \langle \Psi_0^{(0)}(t) | \hat{P} \, e^{-\frac{i}{\hbar} \hat{H}^{(0)} t} \hat{H}^{(1),I}(t') | \Psi_0^{(0)} \rangle$$

$$- \frac{1}{i\hbar} \int_{-\infty}^{t} dt' \, \langle \Psi_0^{(0)} | \hat{H}^{(1),I}(t') \, e^{\frac{i}{\hbar} \hat{H}^{(0)} t} \hat{P} | \Psi_0^{(0)}(t) \rangle + \cdots \quad (3.99)$$

where we have assumed that the time-dependent wavefunction is normalized. Using the definition of $|\Psi_0^{(0)}(t)\rangle$, Eq. (3.72), and introducing the commutator $[\hat{A}, \hat{B}] = \hat{A}\hat{B} - \hat{B}\hat{A}$ of two operators $\hat{A}$ and $\hat{B}$, we can write this more compactly as

$$\langle \Psi_0(t, \vec{\mathcal{F}}) | \hat{P} | \Psi_0(t, \vec{\mathcal{F}}) \rangle \quad (3.100)$$

$$= \langle \Psi_0^{(0)} | \hat{P} | \Psi_0^{(0)} \rangle + \frac{1}{i\hbar} \int_{-\infty}^{t} dt' \, \langle \Psi_0^{(0)} | [\hat{P}^I(t), \hat{H}^{(1),I}(t')] | \Psi_0^{(0)} \rangle + \cdots$$

The upper integration limit can be extended to ∞,[9] if we introduce the **Heaviside step function** $\Theta(t)$, which is equal to 1 for $t \geq 0$ and zero for $t < 0$, yielding

$$\langle \Psi_0(t, \vec{\mathcal{F}}) | \hat{P} | \Psi_0(t, \vec{\mathcal{F}}) \rangle \quad (3.101)$$

$$= \langle \Psi_0^{(0)} | \hat{P} | \Psi_0^{(0)} \rangle + \frac{1}{i\hbar} \int_{-\infty}^{\infty} dt' \, \Theta(t - t') \, \langle \Psi_0^{(0)} | [\hat{P}^I(t), \hat{H}^{(1),I}(t')] | \Psi_0^{(0)} \rangle + \cdots$$

Finally, we insert the definition of the first-order perturbation Hamiltonian, Eq. (3.75), and obtain

$$\langle \Psi_0(t, \vec{\mathcal{F}}) | \hat{P} | \Psi_0(t, \vec{\mathcal{F}}) \rangle = \langle \Psi_0^{(0)} | \hat{P} | \Psi_0^{(0)} \rangle \quad (3.102)$$

$$+ \frac{1}{i\hbar} \int_{-\infty}^{\infty} dt' \, \Theta(t - t') \, \langle \Psi_0^{(0)} | [\hat{P}^I(t), \sum_{\beta \ldots} \hat{O}_{\beta \ldots}^{\mathcal{F},I}(t') \mathcal{F}_{\beta \ldots}(t')] | \Psi_0^{(0)} \rangle + \cdots$$

or

$$\langle \Psi_0(t, \vec{\mathcal{F}}) | \hat{P} | \Psi_0(t, \vec{\mathcal{F}}) \rangle = \langle \Psi_0^{(0)} | \hat{P} | \Psi_0^{(0)} \rangle \quad (3.103)$$

$$+ \sum_{\beta \ldots} \int_{-\infty}^{\infty} dt' \, \mathcal{F}_{\beta \ldots}(t') \left\{ \Theta(t - t') \frac{1}{i\hbar} \langle \Psi_0^{(0)} | [\hat{P}^I(t), \hat{O}_{\beta \ldots}^{\mathcal{F},I}(t')] | \Psi_0^{(0)} \rangle \right\} + \cdots$$

where we have moved the perturbing time-dependent field $\mathcal{F}_{\beta \ldots}(t')$ out of the integral over the electronic coordinates. The second term is in complete analogy to the time-independent case in Eq. (3.36) again linear in the components of the now time-dependent field $\vec{\mathcal{F}}(t)$ and thus represents the **time-dependent linear response** of the expectation value $\langle \Psi_0^{(0)} | \hat{P} | \Psi_0^{(0)} \rangle$ to the time-dependent field.

The expression in "{}" in Eq. (3.103) is the coefficient of this linear term and expresses how susceptible the expectation value is to changes by the time-dependent field. It is therefore called the **linear response function in the time domain** (Olsen and Jørgensen, 1985) and is given its own symbol $\langle\langle \hat{P}^I(t) \, ; \, \hat{O}_{\beta \ldots}^{\mathcal{F},I}(t') \rangle\rangle$ defined as

---

[9] This is necessary for the Fourier transformation of this expression to the frequency domain later on.

$$\langle\langle\hat{P}^I(t); \hat{O}_{\beta\ldots}^{\mathcal{F},I}(t')\rangle\rangle = \Theta(t-t')\frac{1}{i\hbar}\langle\Psi_0^{(0)}|\left[\hat{P}^I(t),\hat{O}_{\beta\ldots}^{\mathcal{F},I}(t')\right]|\Psi_0^{(0)}\rangle \tag{3.104}$$

It is often also called the **polarization propagator in the time domain** (Zubarev, 1974; Linderberg and Öhrn, 1973; Jørgensen and Simons, 1981), while mathematically it is a **double-time Green's function**.

With the above definition of the linear response function or polarization propagator we can rewrite the expansion of the time-dependent, perturbed expectation value as

$$\langle\Psi_0(t,\vec{\mathcal{F}})|\hat{P}|\Psi_0(t,\vec{\mathcal{F}})\rangle = \langle\Psi_0^{(0)}|\hat{P}|\Psi_0^{(0)}\rangle + \int_{-\infty}^{\infty} dt' \sum_{\beta\ldots}\langle\langle\hat{P}^I(t); \hat{O}_{\beta\ldots}^{\mathcal{F},I}(t')\rangle\rangle \mathcal{F}_{\beta\ldots}(t')$$
$$+ \cdots \tag{3.105}$$

From the definition of the polarization propagator in Eq. (3.104) one could get the impression that it depends on two times, t and t'. However, this is not the case, the propagator depends only on the time difference $t - t'$ because

$$\langle\langle\hat{P}^I(t); \hat{O}_{\beta\ldots}^{\mathcal{F},I}(t')\rangle\rangle = \Theta(t-t')\frac{1}{i\hbar}\langle\Psi_0^{(0)}|\left[\hat{P}^I(t),\hat{O}_{\beta\ldots}^{\mathcal{F},I}(t')\right]|\Psi_0^{(0)}\rangle$$
$$= \Theta(t-t')\frac{1}{i\hbar}\langle\Psi_0^{(0)}|\left[e^{\frac{i}{\hbar}\hat{H}^{(0)}t}\hat{P}e^{-\frac{i}{\hbar}\hat{H}^{(0)}t},\hat{O}_{\beta\ldots}^{\mathcal{F},I}(t')\right]|\Psi_0^{(0)}\rangle$$
$$= \Theta(t-t')\frac{1}{i\hbar}\langle\Psi_0^{(0)}|\left[\hat{P},e^{-\frac{i}{\hbar}\hat{H}^{(0)}t}\hat{O}_{\beta\ldots}^{\mathcal{F},I}(t')e^{\frac{i}{\hbar}\hat{H}^{(0)}t}\right]|\Psi_0^{(0)}\rangle$$
$$= \Theta(t-t')\frac{1}{i\hbar}\langle\Psi_0^{(0)}|\left[\hat{P},\hat{O}_{\beta\ldots}^{\mathcal{F},I}(t'-t)\right]|\Psi_0^{(0)}\rangle \tag{3.106}$$

which follows from the fact that $|\Psi_0^{(0)}\rangle$ is an eigenfunction of $\hat{H}^{(0)}$ [see Exercise 3.7]. We can thus rewrite the definition of the polarization propagator

$$\langle\langle\hat{P}^I(t); \hat{O}_{\beta\ldots}^{\mathcal{F},I}(t')\rangle\rangle = \langle\langle\hat{P}; \hat{O}_{\beta\ldots}^{\mathcal{F},I}(t'-t)\rangle\rangle$$
$$= \Theta(t-t')\frac{1}{i\hbar}\langle\Psi_0^{(0)}|\left[\hat{P},\hat{O}_{\beta\ldots}^{\mathcal{F},I}(t'-t)\right]|\Psi_0^{(0)}\rangle \tag{3.107}$$

and change the variable from $t'-t$ to t in the following. In addition to this mathematical proof the physical interpretation of the polarization propagator also requires that it depends only on the time difference. A propagator is a quantity that propagates, i.e. transfers something in time or space from one point to another. The polarization propagator transfers the disturbance in the system created at the time t' by the perturbation $\hat{O}_{\beta\ldots}^{\mathcal{F},I}\mathcal{F}_{\beta\ldots}$ to the time t where it then leads to a change in the expectation value of the operator $\hat{P}$. Its value tells us then how much the property represented by the operator $\hat{P}$ is changed at time t by the action of the perturbation $\hat{O}_{\beta\ldots}^{\mathcal{F},I}\mathcal{F}_{\beta\ldots}$ since time t', which obviously has to be independent of the precise point on the time axis at which t' is.

Exercise 3.7 Fill in the missing steps in the derivation of Eq. (3.106), which proves that the polarization propagator depends only on the time difference $t - t'$.

52 Perturbation Theory

Hint: Expand the commutator and the exponential operator $e^{\frac{i}{\hbar}\hat{H}^{(0)}t}$ and use the fact that $|\Psi_0^{(0)}\rangle$ is an eigenfunction of $\hat{H}^{(0)}$.

We will later on in Chapter 7 see that the measurable molecular properties related to the response functions or polarization propagators are normally defined to be frequency dependent and not time dependent. We define therefore the Fourier transform of the time dependent polarization propagator as

$$\langle\langle \hat{P} ; \hat{O}_{\beta\ldots}^{\omega} \rangle\rangle_\omega = \int_{-\infty}^{\infty} \langle\langle \hat{P} ; \hat{O}_{\beta\ldots}^{\mathcal{F},I}(t) \rangle\rangle\, e^{-i\omega t}\, dt \qquad (3.108)$$

With this **frequency-dependent polarization propagator**, $\langle\langle \hat{P} ; \hat{O}_{\beta\ldots}^{\omega} \rangle\rangle_\omega$, and the Fourier components, $\mathcal{F}_{\beta\ldots}(\omega)$, of the time-dependent field from Eq. (3.77) we can rewrite the expansion of the time-dependent, perturbed expectation value, Eq. (3.105), as

$$\langle\Psi_0(t,\vec{\mathcal{F}})|\hat{P}|\Psi_0(t,\vec{\mathcal{F}})\rangle = \langle\Psi_0^{(0)}|\hat{P}|\Psi_0^{(0)}\rangle + \frac{1}{2}\int_{-\infty}^{\infty} d\omega\, e^{-i\omega t} \sum_{\beta\ldots} \langle\langle \hat{P} ; \hat{O}_{\beta\ldots}^{\omega} \rangle\rangle_\omega \mathcal{F}_{\beta\ldots}(\omega)$$

$$+\cdots \qquad (3.109)$$

However, Eq. (3.108) is only the definition of the Fourier transform, which then has to be applied to the expression for the time-dependent polarization propagator in Eq. (3.107) with $t'-t$ replaced by t. This leads us [see Exercise 3.8] to the **spectral representation** of the polarization propagator or linear response function

$$\langle\langle \hat{P} ; \hat{O}_{\beta\ldots}^{\omega} \rangle\rangle_\omega = \sum_{n\neq 0} \frac{\langle\Psi_0^{(0)}|\hat{P}|\Psi_n^{(0)}\rangle\langle\Psi_n^{(0)}|\hat{O}_{\beta\ldots}^{\omega}|\Psi_0^{(0)}\rangle}{\hbar\omega + E_0^{(0)} - E_n^{(0)}}$$

$$+ \sum_{n\neq 0} \frac{\langle\Psi_0^{(0)}|\hat{O}_{\beta\ldots}^{\omega}|\Psi_n^{(0)}\rangle\langle\Psi_n^{(0)}|\hat{P}|\Psi_0^{(0)}\rangle}{-\hbar\omega + E_0^{(0)} - E_n^{(0)}} \qquad (3.110)$$

Exercise 3.8 Derive the expression for the polarization propagator in the frequency domain, Eq. (3.110), from the expression in the time domain, Eq. (3.107).

Hint: Try to insert the resolution of the identity $\sum_n |\Psi_n^{(0)}\rangle\langle\Psi_n^{(0)}| = 1$ between $\hat{P}$ and $\hat{O}_{\beta\ldots}^{\mathcal{F},I}(t)$ in Eq. (3.107). Use, furthermore, that

$$\frac{1}{i}\int_{-\infty}^{\infty} e^{iat}\,\Theta(t)\,dt = \lim_{\eta\to 0^+}\int_{-\infty}^{\infty}\frac{\delta(a-x)}{x+i\eta}\,dx = \lim_{\eta\to 0^+}\frac{1}{a+i\eta} = \frac{1}{a}$$

From the spectral representation, Eq. (3.110), we can easily verify the general symmetry property of the polarization propagator

$$\langle\langle \hat{P} ; \hat{O}_{\beta\ldots}^{\omega} \rangle\rangle_\omega = \langle\langle \hat{O}_{\beta\ldots}^{\omega} ; \hat{P} \rangle\rangle_{-\omega} \qquad (3.111)$$

and the special cases for two real and hermitian operators or for two purely imaginary and hermitian operators

$$\langle\langle \hat{P} ; \hat{O}^\omega_{\beta...} \rangle\rangle_\omega = \langle\langle \hat{P} ; \hat{O}^\omega_{\beta...} \rangle\rangle_{-\omega} \quad \text{for } P \text{ and } O^\omega_{\beta...} \text{ both real or both imaginary} \tag{3.112}$$

and for one real hermitian and one imaginary hermitian operator

$$\langle\langle \hat{P} ; \hat{O}^\omega_{\beta...} \rangle\rangle_\omega = -\langle\langle \hat{P} ; \hat{O}^\omega_{\beta...} \rangle\rangle_{-\omega} \quad \text{for } P \text{ real and } O^\omega_{\beta...} \text{ imaginary hermitian} \tag{3.113}$$

However, the main application of the spectral resolution lies in the interpretation of the polarization propagator. We can see from Eq. (3.110), that the polarization propagator has a pole if the frequency of the perturbation ω matches one of the excitation energies $\pm(E_n^{(0)} - E_0^{(0)})/\hbar$ of the system. The polarization propagator contains thus information about the excited electronic states of a molecule and its spectra, as will be discussed in more detail in Section 7.4.

In the static case, $\omega = 0$, the spectral representation, Eq. (3.110), reduces to the sum-over-states contribution for the second derivative of a second-order energy correction, Eq. (3.34),

$$\langle\langle \hat{O}^\mathcal{F}_{\alpha...} ; \hat{O}^\mathcal{F}_{\beta...} \rangle\rangle_{\omega=0} = \sum_{n\neq 0} \frac{\langle \Psi_0^{(0)} | \hat{O}^\mathcal{F}_{\alpha...} | \Psi_n^{(0)} \rangle \langle \Psi_n^{(0)} | \hat{O}^\mathcal{F}_{\beta...} | \Psi_0^{(0)} \rangle}{E_0^{(0)} - E_n^{(0)}}$$

$$+ \sum_{n\neq 0} \frac{\langle \Psi_0^{(0)} | \hat{O}^\mathcal{F}_{\beta...} | \Psi_n^{(0)} \rangle \langle \Psi_n^{(0)} | \hat{O}^\mathcal{F}_{\alpha...} | \Psi_0^{(0)} \rangle}{E_0^{(0)} - E_n^{(0)}}$$

$$= \frac{\partial^2 E_0^{(2)}}{\partial \mathcal{F}_\alpha \partial \mathcal{F}_\beta ...} \tag{3.114}$$

where we have finally replaced the symbol for the operator $\hat{P}$ by $\hat{O}^\mathcal{F}_{\alpha...}$ according to its definition as the first derivative of the Hamiltonian with respect to the $\mathcal{F}_\alpha$ component of a field $\mathcal{F}$, i.e. Eq. (3.9), and have replaced $\hat{O}^\omega_{\beta...}$ by $\hat{O}^\mathcal{F}_{\beta...}$, because the perturbation is now time and frequency independent. In the second equality we have furthermore assumed that the field $\mathcal{F}$ gives rise to only a first-order perturbation Hamiltonian $\hat{H}^{(1)}$. This equivalence between static linear response functions and second-order energy derivatives implies that all properties that are obtained from this second-order energy correction can also be obtained from the zero-frequency limit of a polarization propagator or linear response function.

The comparison with Eqs. (3.34) and (3.37) also shows that the linear response function will only give the contributions to the second-order properties that depend on the first-order wavefunction and first-order perturbation Hamiltonian. The contributions that are expectation values of $\hat{H}^{(2)}$ must be obtained by choosing our operator $\hat{P}$ to be the perturbation-dependent operator $\hat{P}(\vec{\mathcal{F}})$ defined in Eqs. (3.8) to (3.10). The expansion of the expectation value, Eq. (3.109), then becomes

54 Perturbation Theory

$$\langle \Psi_0(t,\vec{\mathcal{F}})|\hat{P}(\vec{\mathcal{F}})|\Psi_0(t,\vec{\mathcal{F}})\rangle = \langle \Psi_0^{(0)}|\hat{P}|\Psi_0^{(0)}\rangle \qquad (3.115)$$

$$+ \frac{1}{2}\int_{-\infty}^{\infty} d\omega_1\, e^{-i\omega_1 t} \sum_{\beta\ldots} \left(\langle \Psi_0^{(0)}|\hat{O}_{\alpha\beta\ldots}^{\mathcal{F}\mathcal{F}} + \hat{O}_{\beta\alpha\ldots}^{\mathcal{F}\mathcal{F}}|\Psi_0^{(0)}\rangle + \langle\langle \hat{P}\,;\,\hat{O}_{\beta\ldots}^{\omega}\rangle\rangle_\omega \right) \mathcal{F}_{\beta\ldots}(\omega)$$

$$+\cdots$$

In the static case, $\omega = 0$, this is reduced to

$$\langle \Psi_0(\vec{\mathcal{F}})|\hat{P}(\vec{\mathcal{F}})|\Psi_0(\vec{\mathcal{F}})\rangle = \langle \Psi_0^{(0)}|\hat{P}|\Psi_0^{(0)}\rangle$$

$$+ \sum_{\beta\ldots} \left(\langle \Psi_0^{(0)}|\hat{O}_{\alpha\beta\ldots}^{\mathcal{F}\mathcal{F}} + \hat{O}_{\beta\alpha\ldots}^{\mathcal{F}\mathcal{F}}|\Psi_0^{(0)}\rangle + \langle\langle \hat{P}\,;\,\hat{O}_{\beta\ldots}^{\mathcal{F}}\rangle\rangle_{\omega=0} \right) \mathcal{F}_{\beta\ldots} + \cdots \qquad (3.116)$$

This equation, which is the perturbation theory expansion of an expectation value in the presence of a static field $\vec{\mathcal{F}}$ in static response functions or polarization propagators, is another way of writing Eq. (3.36). Therefore, we can identify the static response function as the first derivative of the first-order correction to a perturbed expectation value, i.e.

$$\langle\langle \hat{P}\,;\,\hat{O}_{\beta\ldots}^{\mathcal{F}}\rangle\rangle_{\omega=0} = \frac{\partial}{\partial \mathcal{F}_{\beta\ldots}} \left(\langle \Psi_0^{(0)}|\hat{P}|\Psi_0^{(1)}(\vec{\mathcal{F}})\rangle + \langle \Psi_0^{(1)}(\vec{\mathcal{F}})|\hat{P}|\Psi_0^{(0)}\rangle \right) \qquad (3.117)$$

or relate it according to Eq. (3.53) to the first-order reduced density matrix $P_\alpha^{(1)}(\vec{r},\vec{r}\,')$

$$\langle\langle \hat{P}\,;\,\hat{O}_{\beta\ldots}^{\mathcal{F}}\rangle\rangle_{\omega=0} = \int_{\vec{r}\,'_1=\vec{r}_1} \hat{p}(1)\, P_\beta^{(1)}(\vec{r}_1,\vec{r}\,'_1)\, d\vec{r}_1 \qquad (3.118)$$

In later chapters we will identify molecular properties as first derivatives of the energy with respect to different static fields $\vec{\mathcal{F}}$. There, we will return to this expansion in Eq. (3.116), when we are interested in the value of some of these molecular properties in the presence of the static field $\vec{\mathcal{F}}$ or another field. In particular, this expression will become important for the case of perturbations that give rise to a second-order perturbation Hamiltonian $\hat{H}^{(2)}$.

In the following, we want to go back to the expansion of the time-dependent expectation value in Eq. (3.99) and derive expressions for the next terms. Collecting all second-order terms in the expansion

$$\langle \Psi_0^{(1)}(t,\vec{\mathcal{F}})|\hat{P}|\Psi_0^{(1)}(t,\vec{\mathcal{F}})\rangle + \langle \Psi_0^{(2)}(t,\vec{\mathcal{F}})|\hat{P}|\Psi_0^{(0)}(t)\rangle + \langle \Psi_0^{(0)}(t)|\hat{P}|\Psi_0^{(2)}(t,\vec{\mathcal{F}})\rangle \qquad (3.119)$$

and inserting the first- and second-order corrections to the time-dependent wavefunction, Eqs. (3.87) and (3.88), the next and thus quadratic term, in the expansion of the time-dependent expectation value, Eq. (3.100), becomes [see Exercise 3.9]

$$\left(\frac{1}{i\hbar}\right)^2 \int_{-\infty}^{t} dt' \int_{-\infty}^{t'} dt''\, \langle \Psi_0^{(0)}|\left[\left[\hat{P}^I(t),\hat{H}^{(1),I}(t')\right],\hat{H}^{(1),I}(t'')\right]|\Psi_0^{(0)}\rangle \qquad (3.120)$$

or

$$\left(\frac{1}{i\hbar}\right)^2 \int_{-\infty}^{\infty} dt' \, \Theta(t-t') \int_{-\infty}^{\infty} dt'' \, \Theta(t'-t'') \, \langle \Psi_0^{(0)} | \left[\left[\hat{P}^I(t), \hat{H}^{(1),I}(t') \right], \hat{H}^{(1),I}(t'') \right] | \Psi_0^{(0)} \rangle \quad (3.121)$$

if we make use of the Heaviside step function again and extend the integration limits.

Exercise 3.9 Fill in the missing steps in the derivation of Eqs. (3.120) and (3.121).

The next step is then again to insert the first-order perturbation Hamiltonian, Eq. (3.75) leading to

$$\int_{-\infty}^{\infty} dt' \int_{-\infty}^{\infty} dt'' \sum_{\beta\cdots,\gamma\cdots} \mathcal{F}_{\beta\cdots}(t') \mathcal{F}_{\gamma\cdots}(t'') \quad (3.122)$$

$$\times \left\{ \Theta(t-t') \, \Theta(t'-t'') \left(\frac{1}{i\hbar}\right)^2 \langle \Psi_0^{(0)} | \left[\left[\hat{P}^I(t), \hat{O}_{\beta\cdots}^{\mathcal{F},I}(t') \right], \hat{O}_{\gamma\cdots}^{\mathcal{F},I}(t'') \right] | \Psi_0^{(0)} \rangle \right\}$$

The expression in "{}" in Eq. (3.122) is the coefficient of the term quadratic in the perturbing field $\mathcal{F}$ and is therefore called the **quadratic response function in the time domain** (Olsen and Jørgensen, 1985). Extending the notation for the linear response function it gets its own symbol $\langle\langle \hat{P}^I(t) \, ; \, \hat{O}_{\beta\cdots}^{\mathcal{F},I}(t'), \hat{O}_{\gamma\cdots}^{\mathcal{F},I}(t'') \rangle\rangle$ defined as

$$\langle\langle \hat{P}^I(t) \, ; \, \hat{O}_{\beta\cdots}^{\mathcal{F},I}(t'), \hat{O}_{\gamma\cdots}^{\mathcal{F},I}(t'') \rangle\rangle \quad (3.123)$$

$$= \Theta(t-t')\Theta(t'-t'') \left(\frac{1}{i\hbar}\right)^2 \langle \Psi_0^{(0)} | \left[\left[\hat{P}^I(t), \hat{O}_{\beta\cdots}^{\mathcal{F},I}(t') \right], \hat{O}_{\gamma\cdots}^{\mathcal{F},I}(t'') \right] | \Psi_0^{(0)} \rangle$$

This notation can be extended to higher-order response functions, by adding more operators after the ";", i.e. $\langle\langle \hat{A} \, ; \, \hat{B}, \hat{C}, \hat{D}, \cdots \rangle\rangle$.

In analogy to the linear response function, the quadratic response function does not depend on the three times, t, t' and t'' but rather on the two time differences $t_1 = t' - t$ and $t_2 = t'' - t$

$$\langle\langle \hat{P}^I(t) \, ; \, \hat{O}_{\beta\cdots}^{\mathcal{F},I}(t'), \hat{O}_{\gamma\cdots}^{\mathcal{F},I}(t'') \rangle\rangle = \langle\langle \hat{P} \, ; \, \hat{O}_{\beta\cdots}^{\mathcal{F},I}(t_1), \hat{O}_{\gamma\cdots}^{\mathcal{F},I}(t_2) \rangle\rangle \quad (3.124)$$

$$= \Theta(-t_1)\Theta(-t_2) \left(\frac{1}{i\hbar}\right)^2 \langle \Psi_0^{(0)} | \left[\left[\hat{P}, \hat{O}_{\beta\cdots}^{\mathcal{F},I}(t_1) \right], \hat{O}_{\gamma\cdots}^{\mathcal{F},I}(t_2) \right] | \Psi_0^{(0)} \rangle$$

Applying Fourier transformations, Eq. (3.108), to both times we obtain the **quadratic response function in the frequency domain**

$$\langle\langle \hat{P} ; \hat{O}^{\omega_1}_{\beta\ldots}, \hat{O}^{\omega_2}_{\gamma\ldots} \rangle\rangle_{\omega_1,\omega_2}$$

$$= \sum_{n\neq 0, m\neq 0} \left\{ \frac{\langle\Psi_0^{(0)}|\hat{P}|\Psi_n^{(0)}\rangle\langle\Psi_n^{(0)}|\overline{\hat{O}^{\omega_1}_{\beta\ldots}}|\Psi_m^{(0)}\rangle\langle\Psi_m^{(0)}|\hat{O}^{\omega_2}_{\gamma\ldots}|\Psi_0^{(0)}\rangle}{\left(E_n^{(0)} - E_0^{(0)} - \hbar(\omega_1+\omega_2)\right)\left(E_m^{(0)} - E_0^{(0)} - \hbar\omega_2\right)} \right.$$

$$+ \frac{\langle\Psi_0^{(0)}|\hat{P}|\Psi_n^{(0)}\rangle\langle\Psi_n^{(0)}|\overline{\hat{O}^{\omega_2}_{\beta\ldots}}|\Psi_m^{(0)}\rangle\langle\Psi_m^{(0)}|\hat{O}^{\omega_1}_{\gamma\ldots}|\Psi_0^{(0)}\rangle}{\left(E_n^{(0)} - E_0^{(0)} - \hbar(\omega_1+\omega_2)\right)\left(E_m^{(0)} - E_0^{(0)} - \hbar\omega_1\right)}$$

$$+ \frac{\langle\Psi_0^{(0)}|\hat{O}^{\omega_1}_{\gamma\ldots}|\Psi_n^{(0)}\rangle\langle\Psi_n^{(0)}|\overline{\hat{P}}|\Psi_m^{(0)}\rangle\langle\Psi_m^{(0)}|\hat{O}^{\omega_2}_{\beta\ldots}|\Psi_0^{(0)}\rangle}{\left(E_n^{(0)} - E_0^{(0)} + \hbar\omega_1\right)\left(E_m^{(0)} - E_0^{(0)} - \hbar\omega_2\right)} \quad (3.125)$$

$$+ \frac{\langle\Psi_0^{(0)}|\hat{O}^{\omega_2}_{\beta\ldots}|\Psi_n^{(0)}\rangle\langle\Psi_n^{(0)}|\overline{\hat{P}}|\Psi_m^{(0)}\rangle\langle\Psi_m^{(0)}|\hat{O}^{\omega_1}_{\gamma\ldots}|\Psi_0^{(0)}\rangle}{\left(E_n^{(0)} - E_0^{(0)} + \hbar\omega_2\right)\left(E_m^{(0)} - E_0^{(0)} - \hbar\omega_1\right)}$$

$$+ \frac{\langle\Psi_0^{(0)}|\hat{O}^{\omega_2}_{\beta\ldots}|\Psi_n^{(0)}\rangle\langle\Psi_n^{(0)}|\overline{\hat{O}^{\omega_1}_{\gamma\ldots}}|\Psi_m^{(0)}\rangle\langle\Psi_m^{(0)}|\hat{P}|\Psi_0^{(0)}\rangle}{\left(E_n^{(0)} - E_0^{(0)} + \hbar\omega_2\right)\left(E_m^{(0)} - E_0^{(0)} + \hbar(\omega_1+\omega_2)\right)}$$

$$+ \left. \frac{\langle\Psi_0^{(0)}|\hat{O}^{\omega_1}_{\beta\ldots}|\Psi_n^{(0)}\rangle\langle\Psi_n^{(0)}|\overline{\hat{O}^{\omega_2}_{\gamma\ldots}}|\Psi_m^{(0)}\rangle\langle\Psi_m^{(0)}|\hat{P}|\Psi_0^{(0)}\rangle}{\left(E_n^{(0)} - E_0^{(0)} + \hbar\omega_1\right)\left(E_m^{(0)} - E_0^{(0)} + \hbar(\omega_1+\omega_2)\right)} \right\}$$

where $\langle\Psi_n^{(0)}|\overline{\hat{P}}|\Psi_m^{(0)}\rangle$ denotes $\langle\Psi_n^{(0)}|\hat{P} - \langle\Psi_0^{(0)}|\hat{P}|\Psi_0^{(0)}\rangle|\Psi_m^{(0)}\rangle$, i.e. the matrix element of a fluctuation operator.

Finally, we can now write the expansion of the time-dependent expectation value, Eq. (3.99), in terms of linear and quadratic response functions in the time domain, Eqs. (3.107) and (3.124), as

$$\langle\Psi_0(t,\vec{\mathcal{F}})|\hat{P}|\Psi_0(t,\vec{\mathcal{F}})\rangle = \langle\Psi_0^{(0)}|\hat{P}|\Psi_0^{(0)}\rangle + \int_{-\infty}^{\infty} dt' \sum_{\beta\ldots} \langle\langle \hat{P} ; \hat{O}^{\mathcal{F},I}_{\beta\ldots}(t'-t) \rangle\rangle \mathcal{F}_{\beta\ldots}(t')$$

$$+ \frac{1}{2}\int_{-\infty}^{\infty} dt' \int_{-\infty}^{\infty} dt'' \sum_{\beta\gamma\ldots} \langle\langle \hat{P} ; \hat{O}^{\mathcal{F},I}_{\beta\ldots}(t'-t), \hat{O}^{\mathcal{F},I}_{\gamma\ldots}(t''-t) \rangle\rangle \mathcal{F}_{\beta\ldots}(t') \mathcal{F}_{\gamma\ldots}(t'') + \cdots$$

(3.126)

or in terms of linear and quadratic response functions in the frequency domain, Eqs. (3.110) and (3.125), as

$$\langle\Psi_0(t,\vec{\mathcal{F}})|\hat{P}|\Psi_0(t,\vec{\mathcal{F}})\rangle = \langle\Psi_0^{(0)}|\hat{P}|\Psi_0^{(0)}\rangle$$

$$+ \frac{1}{2}\int_{-\infty}^{\infty} d\omega_1 \, e^{-i\omega t} \sum_{\beta\ldots} \langle\langle \hat{P} ; \hat{O}^{\omega}_{\beta\ldots} \rangle\rangle_{\omega_1} \mathcal{F}_{\beta\ldots}(\omega)$$

$$+ \frac{1}{8} \int_{-\infty}^{\infty} d\omega_1 \, e^{-i\omega_1 t} \int_{-\infty}^{\infty} d\omega_2 \, e^{-i\omega_2 t} \sum_{\beta\cdots,\gamma\cdots} \langle\!\langle \hat{P} ; \hat{O}_{\beta\cdots}^{\omega_1}, \hat{O}_{\gamma\cdots}^{\omega_2} \rangle\!\rangle_{\omega_1,\omega_2} \mathcal{F}_{\beta\cdots}(\omega_1) \mathcal{F}_{\gamma\cdots}(\omega_2)$$
$$+ \cdots \quad . \tag{3.127}$$

3.12 Matrix Representation of the Propagator

In the last section we have derived an expression for the linear response function or polarization propagator in the frequency domain, Eq. (3.110). However, application of this expression requires that one knows all unperturbed excited states $\Psi_n^{(0)}$ of the system and their energies $E_n^{(0)}$ or the excitation energies $E_n^{(0)} - E_0^{(0)}$ and corresponding transition moments $\langle \Psi_0^{(0)} | \hat{O}^\omega | \Psi_n^{(0)} \rangle$. In the following, we want to derive now an alternative but equally exact matrix representation of the polarization propagator, where this is not required. Actually, by comparing this matrix representation with Eq. (3.110) will show us a way to obtain excitation energies and transition moments from the polarization propagator. Furthermore, this matrix expression will become the basis for approximate polarization propagator methods in Sections 10.3 and 10.4.

We start by taking the time derivative of the time-dependent linear response function, Eq. (3.107),

$$i\hbar \frac{d}{dt} \langle\!\langle \hat{P} ; \hat{O}_{\beta\cdots}^{\mathcal{F},I}(t) \rangle\!\rangle \tag{3.128}$$

$$= i\hbar \frac{d\Theta(-t)}{dt} \frac{1}{i\hbar} \langle \Psi_0^{(0)} | \left[\hat{P}, \hat{O}_{\beta\cdots}^{\mathcal{F},I}(t)\right] | \Psi_0^{(0)} \rangle + \Theta(-t) \frac{d}{dt} \langle \Psi_0^{(0)} | \left[\hat{P}, \hat{O}_{\beta\cdots}^{\mathcal{F},I}(t)\right] | \Psi_0^{(0)} \rangle$$

Next, we use the fact that the time derivative of an operator in the interaction picture is the commutator of this operator with the Hamiltonian [see Exercise 3.10], i.e.

$$\frac{d}{dt} \hat{O}_{\beta\cdots}^{\mathcal{F},I}(t) = \frac{1}{i\hbar} [\hat{O}_{\beta\cdots}^{\mathcal{F},I}(t), \hat{H}^{(0)}] \tag{3.129}$$

and that the Heaviside step function $\Theta(t)$ is the integral of the Dirac δ function

$$\Theta(t) = \int_{-\infty}^{t} \delta(t') \, dt' \tag{3.130}$$

leading to

$$i\hbar \frac{d}{dt} \langle\!\langle \hat{P} ; \hat{O}_{\beta\cdots}^{\mathcal{F},I}(t) \rangle\!\rangle \tag{3.131}$$

$$= -\delta(-t) \langle \Psi_0^{(0)} | \left[\hat{P}, \hat{O}_{\beta\cdots}^{\mathcal{F},I}(t)\right] | \Psi_0^{(0)} \rangle - \Theta(-t) \frac{1}{i\hbar} \langle \Psi_0^{(0)} | \left[\hat{P}, \left[\hat{H}^{(0)}, \hat{O}_{\beta\cdots}^{\mathcal{F},I}(t)\right]\right] | \Psi_0^{(0)} \rangle$$

Exercise 3.10 Prove that the time derivative of an operator in the interaction picture is given by the commutator of this operator with the Hamiltonian, Eq. (3.129).

58 Perturbation Theory

The Dirac δ function is symmetric and the second term on the right-hand side is again a linear response function, so that we can write

$$i\hbar \frac{d}{dt} \langle\!\langle \hat{P} ; \hat{O}_{\beta\ldots}^{\mathcal{F},I}(t) \rangle\!\rangle = -\delta(t)\langle \Psi_0^{(0)} | \left[\hat{P}, \hat{O}_{\beta\ldots}^{\mathcal{F},I}(t)\right] | \Psi_0^{(0)} \rangle$$
$$- \langle\!\langle \hat{P} ; [\hat{H}^{(0)}, \hat{O}_{\beta\ldots}^{\mathcal{F},I}(t)] \rangle\!\rangle \qquad (3.132)$$

Because of the Dirac δ function $\delta(t)$ the operator $\hat{O}_{\beta\ldots}^{\mathcal{F},I}(t)$ in the first term can be replaced by $\hat{O}_{\beta\ldots}^{\mathcal{F},I}(t=0) = \hat{O}_{\beta\ldots}$ and we thus obtain the **equation of motion for the polarization propagator** in the time domain

$$i\hbar \frac{d}{dt} \langle\!\langle \hat{P} ; \hat{O}_{\beta\ldots}^{\mathcal{F},I}(t) \rangle\!\rangle = -\delta(t)\langle \Psi_0^{(0)} | \left[\hat{P}, \hat{O}_{\beta\ldots}\right] | \Psi_0^{(0)} \rangle$$
$$- \langle\!\langle \hat{P} ; [\hat{H}^{(0)}, \hat{O}_{\beta\ldots}^{\mathcal{F},I}(t)] \rangle\!\rangle \qquad (3.133)$$

More interesting for us is, however, the **equation of motion** transformed to the frequency domain

$$\hbar\omega \langle\!\langle \hat{P} ; \hat{O}_{\beta\ldots}^{\omega} \rangle\!\rangle_\omega = \langle \Psi_0^{(0)} | [\hat{P}, \hat{O}_{\beta\ldots}^{\omega}] | \Psi_0^{(0)} \rangle + \langle\!\langle \hat{P} ; [\hat{H}^{(0)}, \hat{O}_{\beta\ldots}^{\omega}] \rangle\!\rangle_\omega \qquad (3.134)$$

which can be derived [see Exercise 3.11] by using the inverse Fourier transform of the polarization propagator, which is defined as

$$\langle\!\langle \hat{P} ; \hat{O}_{\beta\ldots}^{\mathcal{F},I}(t) \rangle\!\rangle = \frac{1}{2\pi} \int_{-\infty}^{\infty} d\omega \, \langle\!\langle \hat{P} ; \hat{O}_{\beta\ldots}^{\omega} \rangle\!\rangle_\omega e^{i\omega t} \qquad (3.135)$$

and another representation of the Dirac δ function

$$\delta(t) = \frac{1}{2\pi} \int_{-\infty}^{\infty} e^{i\omega t} d\omega \qquad (3.136)$$

Exercise 3.11 Derive the equation of motion in the frequency domain for the polarization propagator, Eq. (3.134), from the equation of motion in the time domain, Eq. (3.133).

The equation of motion in the frequency domain can also be derived directly from the spectral representation of the polarization propagator in the frequency domain, Eq. (3.110), if one realizes that $\frac{ab}{c+d} = \frac{ab}{c} - \frac{d}{c}\frac{ab}{c+d}$. Applying this trick to the first term in Eq. (3.110) gives

$$\frac{\langle \Psi_0^{(0)} | \hat{P} | \Psi_n^{(0)} \rangle \langle \Psi_n^{(0)} | \hat{O}_{\beta\ldots}^{\omega} | \Psi_0^{(0)} \rangle}{\hbar\omega + E_0^{(0)} - E_n^{(0)}} = \frac{\langle \Psi_0^{(0)} | \hat{P} | \Psi_n^{(0)} \rangle \langle \Psi_n^{(0)} | \hat{O}_{\beta\ldots}^{\omega} | \Psi_0^{(0)} \rangle}{\hbar\omega} \qquad (3.137)$$
$$- \frac{\langle \Psi_0^{(0)} | \hat{P} | \Psi_n^{(0)} \rangle \langle \Psi_n^{(0)} | \hat{O}_{\beta\ldots}^{\omega} | \Psi_0^{(0)} \rangle \left(E_0^{(0)} - E_n^{(0)}\right)}{\hbar\omega \left(\hbar\omega + E_0^{(0)} - E_n^{(0)}\right)}$$

Using the same trick also for the second term we can rewrite the polarization propagator in the frequency domain as

$$\langle\langle \hat{P} ; \hat{O}^\omega_{\beta...} \rangle\rangle_\omega = \frac{1}{\hbar\omega} \sum_n \left\{ \langle\Psi_0^{(0)}|\hat{P}|\Psi_n^{(0)}\rangle\langle\Psi_n^{(0)}|\hat{O}^\omega_{\beta...}|\Psi_0^{(0)}\rangle \right. \qquad (3.138)$$

$$\left. - \langle\Psi_0^{(0)}|\hat{O}^\omega_{\beta...}|\Psi_n^{(0)}\rangle\langle\Psi_n^{(0)}|\hat{P}|\Psi_0^{(0)}\rangle \right\}$$

$$- \frac{1}{\hbar\omega} \sum_{n\neq 0} \left\{ \frac{\langle\Psi_0^{(0)}|\hat{P}|\Psi_n^{(0)}\rangle\langle\Psi_n^{(0)}|\hat{O}^\omega_{\beta...}|\Psi_0^{(0)}\rangle \left(E_0^{(0)} - E_n^{(0)}\right)}{\hbar\omega + E_0^{(0)} - E_n^{(0)}} \right.$$

$$\left. - \frac{\langle\Psi_0^{(0)}|\hat{O}^\omega_{\beta...}|\Psi_n^{(0)}\rangle\langle\Psi_n^{(0)}|\hat{P}|\Psi_0^{(0)}\rangle \left(E_0^{(0)} - E_n^{(0)}\right)}{-\hbar\omega + E_0^{(0)} - E_n^{(0)}} \right\}$$

where we have extended the first sum to include also $n = 0$ because these terms vanish anyway. In the next step we use in the second sum the fact that the states $|\Psi_n^{(0)}\rangle$ are eigenfunctions of the Hamiltonian $\hat{H}^{(0)}$ with eigenvalue $E_n^{(0)}$, Eq. (3.14), and that they thus form a complete set. In the first sum we can make use of the resolution of the identity

$$1 = \sum_n |\Psi_n^{(0)}\rangle\langle\Psi_n^{(0)}| \qquad (3.139)$$

which leads to

$$\langle\langle \hat{P} ; \hat{O}^\omega_{\beta...} \rangle\rangle_\omega = \frac{1}{\hbar\omega}\langle\Psi_0^{(0)}|[\hat{P},\hat{O}^\omega_{\beta...}]|\Psi_0^{(0)}\rangle$$

$$+ \frac{1}{\hbar\omega} \sum_n \left\{ \frac{\langle\Psi_0^{(0)}|\hat{P}|\Psi_n^{(0)}\rangle\langle\Psi_n^{(0)}|[\hat{H}^{(0)},\hat{O}^\omega_{\beta...}]|\Psi_0^{(0)}\rangle}{\hbar\omega + E_0^{(0)} - E_n^{(0)}} \right.$$

$$\left. + \frac{\langle\Psi_0^{(0)}|[\hat{H}^{(0)},\hat{O}^\omega_{\beta...}]|\Psi_n^{(0)}\rangle\langle\Psi_n^{(0)}|\hat{P}|\Psi_0^{(0)}\rangle}{-\hbar\omega + E_0^{(0)} - E_n^{(0)}} \right\} \qquad (3.140)$$

where the second and third terms together are the spectral representation of the $\langle\langle \hat{P} ; [\hat{H}^{(0)},\hat{O}^\omega_{\beta...}] \rangle\rangle_\omega$ polarization propagator and we have therefore derived the equation of motion in the frequency domain, Eq. (3.134), directly from the spectral representation of the polarization propagator.

Before we continue in the derivation of a matrix representation of the polarization propagator, we want to mention that by taking the zero-frequency limit of the equation of motion in the frequency domain, we obtain the following relation between a polarization propagator and a ground-state expectation value

$$\langle\Psi_0^{(0)}|[\hat{P},\hat{O}^\omega_{\beta...}]|\Psi_0^{(0)}\rangle = \langle\langle \hat{P} ; [\hat{O}^\omega_{\beta...}, \hat{H}^{(0)}] \rangle\rangle_{\omega=0} \qquad (3.141)$$

which will become very useful in later chapters.

The equation of motion in the frequency domain, Eq. (3.134), tells us that a polarization propagator of two operators $\hat{P}$ and $\hat{O}^\omega_{\beta...}$ is equal to an expectation value

60 Perturbation Theory

of the commutator of these operators plus another, more complicated polarization propagator, in which the second operator $\hat{O}^\omega_{\beta\ldots}$ was replaced by the commutator of the unperturbed Hamiltonian with this operator. This might look like a step in the wrong direction. However, iterating on this equation leads us to the so-called **moment expansion of the polarization propagator** in terms of a series of expectation values of nested commutators

$$\langle\langle \hat{P}\,;\,\hat{O}^\omega_{\beta\ldots}\rangle\rangle_\omega = \left(\frac{1}{\hbar\omega}\right) \langle\Psi_0^{(0)}|[\hat{P},\hat{O}^\omega_{\beta\ldots}]|\Psi_0^{(0)}\rangle$$

$$+ \left(\frac{1}{\hbar\omega}\right)^2 \langle\Psi_0^{(0)}|\left[\hat{P},\left[\hat{H}^{(0)},\hat{O}^\omega_{\beta\ldots}\right]\right]|\Psi_0^{(0)}\rangle$$

$$+ \left(\frac{1}{\hbar\omega}\right)^3 \langle\Psi_0^{(0)}|\left[\hat{P},\left[\hat{H}^{(0)},\left[\hat{H}^{(0)},\hat{O}^\omega_{\beta\ldots}\right]\right]\right]|\Psi_0^{(0)}\rangle + \cdots \quad (3.142)$$

This series can be expressed in a more compact form by using the so-called **superoperator formalism** (Goscinski and Lukman, 1970). We introduce this formalism here, as we had introduced the interaction picture in Section 3.8, in order to facilitate our derivations. The final equations will, however, be written without any superoperators. The superoperator formalism is one level of abstraction higher than the Hilbert vector space of quantum mechanics. In the infinite-dimensional Hilbert space the vectors of the vector space are given as quantum mechanical wavefunctions and the transformations performed on the vectors in the vector space are given by the quantum mechanical operators. The binary product[10] defined in Hilbert space is the overlap integral $\langle\Psi_k|\Psi_l\rangle$ between two wavefunctions, Ψ_k and Ψ_l. In the superoperator formalism we now have an infinite-dimensional vector space, where the quantum mechanical operators, e.g. our operators $\hat{P}$ or $\hat{O}^\omega_{\beta\ldots}$, are the vectors in the vector space and the **superoperator binary product**, $(\hat{P}\,|\,\hat{O}^\omega_{\beta\ldots})$, is defined in the following way

$$(\hat{P}\,|\,\hat{O}^\omega_{\beta\ldots}) = \langle\Psi_0^{(0)}|[\hat{P}^\dagger,\hat{O}^\omega_{\beta\ldots}]|\Psi_0^{(0)}\rangle \quad (3.143)$$

The state $\Psi_0^{(0)}$ employed in the definition of the superoperator binary product is often called the **reference state** and need not be the ground state of the system. The transformations working on the vectors in this vector space of operators, i.e. the operators, are called superoperators and are here denoted with a wide hat as, e.g. in $\widehat{O}$. Commonly, only the **superoperator Hamiltonian** $\widehat{H}^{(0)}$ and the **superoperator identity operator** $\widehat{I}$ are used, which are defined as

$$\widehat{H}^{(0)}\hat{O}^\omega_{\beta\ldots} = [\hat{H}^{(0)},\hat{O}^\omega_{\beta\ldots}] \quad (3.144)$$

and

$$\widehat{I}\,\hat{O}^\omega_{\beta\ldots} = \hat{O}^\omega_{\beta\ldots}. \quad (3.145)$$

---

[10] The binary product in a general vector space is the generalization of the scalar or dot product of two vectors in the normal 3-dimensional vector space $\{R\}^3$.

The action of a superoperator $\hat{O}_1$ on the normal operator $\hat{O}_2$ is thus to build the commutator $[\hat{O}_1, \hat{O}_2]$ of the operator $\hat{O}_1$ (no longer as superoperator) with the normal operator.

Making use of this superoperator formalism, the moment expansion of the polarization propagator can be written as

$$\langle\langle \hat{P}\,;\, \hat{O}^\omega_{\beta...} \rangle\rangle_\omega = \left(\frac{1}{\hbar\omega}\right) \langle\Psi_0^{(0)}|[\hat{P}, \widehat{I}\,\hat{O}^\omega_{\beta...}]|\Psi_0^{(0)}\rangle$$

$$+ \left(\frac{1}{\hbar\omega}\right)^2 \langle\Psi_0^{(0)}|[\hat{P}, \widehat{H}^{(0)}\hat{O}^\omega_{\beta...}]|\Psi_0^{(0)}\rangle$$

$$+ \left(\frac{1}{\hbar\omega}\right)^3 \langle\Psi_0^{(0)}|[\hat{P}, (\widehat{H}^{(0)})^2\hat{O}^\omega_{\beta...}]|\Psi_0^{(0)}\rangle + \cdots \quad (3.146)$$

or

$$\langle\langle \hat{P}\,;\, \hat{O}^\omega_{\beta...} \rangle\rangle_\omega = \left(\frac{1}{\hbar\omega}\right) (\hat{P}^\dagger \mid \widehat{I}\,\hat{O}^\omega_{\alpha...}) + \left(\frac{1}{\hbar\omega}\right)^2 (\hat{P}^\dagger \mid \widehat{H}^{(0)}\hat{O}^\omega_{\beta...})$$

$$+ \left(\frac{1}{\hbar\omega}\right)^3 (\hat{P}^\dagger \mid (\widehat{H}^{(0)})^2\hat{O}^\omega_{\beta...}) + \cdots \quad (3.147)$$

This looks very much like the series expansion of $\frac{1}{1-x}$. We define therefore the **superoperator resolvent** by the series expansion

$$\left(\hbar\omega\widehat{I} - \widehat{H}^{(0)}\right)^{-1} = \left(\frac{1}{\hbar\omega}\right)\left\{\widehat{I} + \sum_{n=1}^{\infty}\left(\frac{\widehat{H}^{(0)}}{\hbar\omega}\right)^n\right\} \quad (3.148)$$

and use it in the moment expansion of the polarization propagator

$$\langle\langle \hat{P}\,;\, \hat{O}^\omega_{\beta...} \rangle\rangle_\omega = (\hat{P}^\dagger \mid (\hbar\omega\widehat{I} - \widehat{H}^{(0)})^{-1}\hat{O}^\omega_{\beta...}) \quad (3.149)$$

However, this is only a cosmetic change, because the superoperator resolvent is an inverse operator and is only defined through its series expansion in Eq. (3.148). The way in which we can proceed, is to find a matrix representation of the superoperator resolvent. In order to do that we need a complete set of basis vectors like in the normal Hilbert space. However, the vectors in the superoperator formalism are operators and we therefore need a complete set of operators. Such a set of operators $\{\hat{h}_n\}$ consists of a complete set of excitation and de-excitation operators with respect to the reference state $|\Psi_0^{(0)}\rangle$. This means that all other states $|\Psi_n^{(0)}\rangle$ of the system or all excited states of the system, if the reference state is the ground state, must be generated by operating on the reference state $|\Psi_0^{(0)}\rangle$, i.e.

$$\hat{h}_n|\Psi_0^{(0)}\rangle = |\Psi_n^{(0)}\rangle \quad (3.150)$$

With this complete set of excitation and de-excitation operators $\{\hat{h}_n\}$, arranged either as column vector $\hat{\mathbf{h}}$ or as row vector $\hat{\mathbf{h}}^T$, we can, like in other vector spaces, also find an expression for the superoperator identity operator $\widehat{I}$, which is called a resolution of the superoperator identity,

$$\widehat{I} = |\hat{\boldsymbol{h}}^T)(\hat{\boldsymbol{h}} \,|\, \hat{\boldsymbol{h}}^T)^{-1}(\hat{\boldsymbol{h}}\,| \quad (3.151)$$

The desired matrix representation of the superoperator resolvent is then obtained in two steps by the **inner projection technique** (Pickup and Goscinski, 1973), where the superoperator resolvent is projected in the space of the complete set of excitation and de-excitation operators $\{\hat{h}_n\}$. First, we insert the resolution of the superoperator identity twice in Eq. (3.149) leading to

$$\langle\!\langle \hat{P}\,;\, \hat{O}^\omega_{\beta\ldots}\rangle\!\rangle_\omega = (\hat{P}^\dagger\,|\,\hat{\boldsymbol{h}}^T)(\hat{\boldsymbol{h}}\,|\,\hat{\boldsymbol{h}}^T)^{-1}(\hat{\boldsymbol{h}}\,|\,(\hbar\omega\widehat{I} - \widehat{H}^{(0)})^{-1}\,|\,\hat{\boldsymbol{h}}^T)(\hat{\boldsymbol{h}}\,|\,\hat{\boldsymbol{h}}^T)^{-1}(\hat{\boldsymbol{h}}\,|\,\hat{O}^\omega_{\beta\ldots}) \quad (3.152)$$

In the second step we need to find an alternative expression for $(\hat{\boldsymbol{h}}\,|\,(\hbar\omega\widehat{I} - \widehat{H}^{(0)})^{-1}\,|\,\hat{\boldsymbol{h}}^T)$. To that purpose we start with the definition of the superoperator resolvent, i.e.

$$(\hbar\omega\widehat{I} - \widehat{H}^{(0)})^{-1}(\hbar\omega\widehat{I} - \widehat{H}^{(0)}) = \widehat{I} \quad (3.153)$$

Inserting this in the superoperator binary products between the complete set of operators arranged as a matrix, $(\hat{\boldsymbol{h}}\,|\,\hat{\boldsymbol{h}}^T)$,

$$(\hat{\boldsymbol{h}}\,|\,(\hbar\omega\widehat{I} - \widehat{H}^{(0)})^{-1}(\hbar\omega\widehat{I} - \widehat{H}^{(0)})\,|\,\hat{\boldsymbol{h}}^T) = (\hat{\boldsymbol{h}}\,|\,\hat{\boldsymbol{h}}^T) \quad (3.154)$$

and inserting the resolution of the identity, we obtain

$$(\hat{\boldsymbol{h}}\,|\,(\hbar\omega\widehat{I} - \widehat{H}^{(0)})^{-1}\,|\,\hat{\boldsymbol{h}}^T)(\hat{\boldsymbol{h}}\,|\,\hat{\boldsymbol{h}}^T)^{-1}(\hat{\boldsymbol{h}}\,|\,(\hbar\omega\widehat{I} - \widehat{H}^{(0)})\,|\,\hat{\boldsymbol{h}}^T) = (\hat{\boldsymbol{h}}\,|\,\hat{\boldsymbol{h}}^T) \quad (3.155)$$

Multiplying this equation from the right first by $(\hat{\boldsymbol{h}}\,|\,(\hbar\omega\widehat{I} - \widehat{H}^{(0)})\,|\,\hat{\boldsymbol{h}}^T)^{-1}$ and then by $(\hat{\boldsymbol{h}}\,|\,\hat{\boldsymbol{h}}^T)$ we arrive at the desired alternative expression

$$(\hat{\boldsymbol{h}}\,|\,(\hbar\omega\widehat{I} - \widehat{H}^{(0)})^{-1}\,|\,\hat{\boldsymbol{h}}^T) = (\hat{\boldsymbol{h}}\,|\,\hat{\boldsymbol{h}}^T)(\hat{\boldsymbol{h}}\,|\,(\hbar\omega\widehat{I} - \widehat{H}^{(0)})\,|\,\hat{\boldsymbol{h}}^T)^{-1}(\hat{\boldsymbol{h}}\,|\,\hat{\boldsymbol{h}}^T) \quad (3.156)$$

Using this relation in Eq. (3.152) leads us to an exact matrix representation of the polarization propagator in the superoperator formalism

$$\langle\!\langle \hat{P}\,;\, \hat{O}^\omega_{\beta\ldots}\rangle\!\rangle_\omega = (\hat{P}^\dagger\,|\,\hat{\boldsymbol{h}}^T)(\hat{\boldsymbol{h}}\,|\,(\hbar\omega\widehat{I} - \widehat{H}^{(0)})\,|\,\hat{\boldsymbol{h}}^T)^{-1}(\hat{\boldsymbol{h}}\,|\,\hat{O}^\omega_{\beta\ldots}) \quad (3.157)$$

This expression no longer contains the inverse of operators but the inverse of matrix representations of the operators. From comparison of Eq. (3.149) and Eq. (3.157) we may conclude that a matrix representation of the superoperator resolvent is given by

$$\left(\hbar\omega\widehat{I} - \widehat{H}^{(0)}\right)^{-1} = |\hat{\boldsymbol{h}}^T)(\hat{\boldsymbol{h}}\,|\,(\hbar\omega\widehat{I} - \widehat{H}^{(0)})\,|\,\hat{\boldsymbol{h}}^T)^{-1}(\hat{\boldsymbol{h}}\,| \quad (3.158)$$

Now we can insert the definitions of the superoperator binary product, Eq. (3.143), the superoperator Hamiltonian, Eq. (3.144) and the superoperator identity, Eq. (3.145), in Eq. (3.157) and arrive thereby at our final expression for a **matrix representation of the polarization propagator**

$$\langle\!\langle \hat{P}\,;\, \hat{O}^\omega_{\beta\ldots}\rangle\!\rangle_\omega = \boldsymbol{T}^T(\hat{P})\,(\hbar\omega\boldsymbol{S} - \boldsymbol{E})^{-1}\,\boldsymbol{T}(\hat{O}^\omega_{\beta\ldots}) \quad (3.159)$$

where $\boldsymbol{T}^T(\hat{P})$ and $\boldsymbol{T}(\hat{O}^\omega_{\beta\ldots})$ are, respectively, row and column vectors and are called **property gradient vectors**. They have the elements

$$T^T_i(\hat{P}) = \langle\Psi^{(0)}_0|\,[\hat{P},\hat{h}_i]\,|\Psi^{(0)}_0\rangle \quad (3.160)$$

and
$$T_j(\hat{O}^\omega_{\beta\ldots}) = \langle \Psi_0^{(0)} | [\hat{h}_j^\dagger, \hat{O}^\omega_{\beta\ldots}] | \Psi_0^{(0)} \rangle \tag{3.161}$$

The combination $(\hbar\omega S - E)$ is also called the **principal propagator** M and consists of the **electronic Hessian matrix** E with the elements

$$E_{ij} = \langle \Psi_0^{(0)} | [\hat{h}_i^\dagger, [\hat{H}^{(0)}, \hat{h}_j]] | \Psi_0^{(0)} \rangle \tag{3.162}$$

and the **overlap matrix** S with the elements

$$S_{ij} = \langle \Psi_0^{(0)} | [\hat{h}_i^\dagger, \hat{h}_j] | \Psi_0^{(0)} \rangle \tag{3.163}$$

As mentioned before, the complete set of operators $\{\hat{h}_n\}$ consists in general of excitation $\{{}^e\hat{h}_n\}$ and de-excitation $\{{}^d\hat{h}_n\}$ operators. The principal propagator matrix can therefore be divided into four blocks

$$M = \hbar\omega S - E = \hbar\omega \begin{pmatrix} {}^{ee}S & {}^{ed}S \\ {}^{de}S & {}^{dd}S \end{pmatrix} - \begin{pmatrix} {}^{ee}E & {}^{ed}E \\ {}^{de}E & {}^{dd}E \end{pmatrix} \tag{3.164}$$

where ${}^{ee}E$ and ${}^{ee}S$, arise from excitation and excitation operators, ${}^{ed}E$ and ${}^{ed}S$, arise from excitation and de-excitation operators, and so forth. Correspondingly, the property gradient vectors $T^T(\hat{P})$ and $T(\hat{O}^\omega_{\beta\ldots})$ can be divided into two contributions ${}^eT^T(\hat{P})$ and ${}^dT^T(\hat{P})$ or ${}^eT(\hat{O}^\omega_{\beta\ldots})$ and ${}^dT(\hat{O}^\omega_{\beta\ldots})$.

The matrix representation in Eq. (3.159) is exact as long as the set of operators $\{\hat{h}_n\}$ is complete and as long as $|\Psi_0^{(0)}\rangle$ is an eigenfunction of $\hat{H}^{(0)}$. Contrary to Eq. (3.110) the matrix representation Eq. (3.159) involves no excited states of the system, but only the ground state $|\Psi_0^{(0)}\rangle$ or in the general case the reference state. It is thus a much more convenient starting point for approximate polarization propagator methods, which will be discussed in more detail in Sections 10.3 and 10.4. Here, we can already mention that these approximate polarization propagator methods are obtained by making an approximation to the exact wavefunction $|\Psi_0^{(0)}\rangle$ and by truncating the otherwise infinite set of operators $\{\hat{h}_n\}$.

On the other hand, the spectral representation of the polarization propagator, Eq. (3.110), can be obtained from the matrix representation, Eq. (3.159), if one chooses the set of operators $\{\hat{h}_n\}$ to be the following set of operators

$$\{\hat{h}_n\} = \{{}^e\hat{h}_n, {}^d\hat{h}_n\} = \{|\Psi_n^{(0)}\rangle\langle\Psi_0^{(0)}|, |\Psi_0^{(0)}\rangle\langle\Psi_n^{(0)}|\} \tag{3.165}$$

which are a special type of projection operators and are called the **state transfer operators**, because they are made from the eigenstates $|\Psi_0^{(0)}\rangle$ and $|\Psi_n^{(0)}\rangle$ of the Hamiltonian $\hat{H}^{(0)}$. With these operators the two off-diagonal blocks ${}^{ed}E$ and ${}^{de}E$ of the electronic Hessian matrix and ${}^{ed}S$ and ${}^{de}S$ of the overlap matrix become zero [see Exercise 3.12] and the propagator in Eq. (3.159) can be written as

$$\langle\langle \hat{P}; \hat{O}^\omega_{\beta\ldots} \rangle\rangle_\omega = {}^eT^T(\hat{P}) \left(\hbar\omega\,{}^{ee}S - {}^{ee}E\right)^{-1} {}^eT(\hat{O}^\omega_{\beta\ldots})$$
$$+ {}^dT^T(\hat{P}) \left(\hbar\omega\,{}^{dd}S - {}^{dd}E\right)^{-1} {}^dT(\hat{O}^\omega_{\beta\ldots}) \tag{3.166}$$

64 Perturbation Theory

as the inverse of a block-diagonal matrix is a block-diagonal matrix whose diagonal blocks are the inverse of the diagonal blocks of the original matrix.

Noting, furthermore [see Exercise 3.13] that the state-transfer operators reduce the elements of the diagonal blocks of the electronic Hessian matrix to

$$^{ee}E_{nm} = \langle \Psi_0^{(0)} | {^e\hat{h}_n}^\dagger, [\hat{H}^{(0)}, {^e\hat{h}_m}] | \Psi_0^{(0)} \rangle = \delta_{nm} \left(E_n^{(0)} - E_0^{(0)} \right) \quad (3.167)$$

$$^{dd}E_{nm} = -\langle \Psi_0^{(0)} | [\hat{H}^{(0)}, {^d\hat{h}_m}], {^d\hat{h}_n}^\dagger | \Psi_0^{(0)} \rangle = \delta_{nm} \left(E_n^{(0)} - E_0^{(0)} \right) \quad (3.168)$$

and the elements of the diagonal blocks of the overlap matrix S to

$$^{ee}S_{nm} = \langle \Psi_0^{(0)} | {^e\hat{h}_n}^\dagger \, {^e\hat{h}_m} | \Psi_0^{(0)} \rangle = \delta_{nm} \quad (3.169)$$

$$^{dd}S_{nm} = -\langle \Psi_0^{(0)} | {^d\hat{h}_m} \, {^d\hat{h}_n}^\dagger | \Psi_0^{(0)} \rangle = -\delta_{nm} \quad (3.170)$$

and finally the elements of the property gradient vectors to

$$^{e}T_n^T(\hat{P}) = \langle \Psi_0^{(0)} | \hat{P} \, {^e\hat{h}_n} | \Psi_0^{(0)} \rangle = \langle \Psi_0^{(0)} | \hat{P} | \Psi_n^{(0)} \rangle \quad (3.171)$$

$$^{d}T_n^T(\hat{P}) = -\langle \Psi_0^{(0)} | {^d\hat{h}_n} \, \hat{P} | \Psi_0^{(0)} \rangle = -\langle \Psi_n^{(0)} | \hat{P} | \Psi_0^{(0)} \rangle \quad (3.172)$$

$$^{e}T_n(\hat{O}_{\beta...}^\omega) = \langle \Psi_0^{(0)} | {^e\hat{h}_n^\dagger} \, \hat{O}_{\beta...}^\omega | \Psi_0^{(0)} \rangle = \langle \Psi_n^{(0)} | \hat{O}_{\beta...}^\omega | \Psi_0^{(0)} \rangle \quad (3.173)$$

$$^{d}T_n(\hat{O}_{\beta...}^\omega) = -\langle \Psi_0^{(0)} | \hat{O}_{\beta...}^\omega \, {^d\hat{h}_n^\dagger} | \Psi_0^{(0)} \rangle = -\langle \Psi_0^{(0)} | \hat{O}_{\beta...}^\omega | \Psi_n^{(0)} \rangle \quad (3.174)$$

we can see that the ee block of the matrix representation in Eq. (3.166) leads to the first sum in the spectral representation of the polarization propagator, Eq. (3.110), while the dd block leads to the second sum in Eq. (3.110). This implies that by reversing this line of argument one could in principle derive directly and without resorting to the superoperator formalism the non-diagonal matrix representation of the polarization propagator, Eq. (3.159), from the spectral representation in Eq. (3.110).

Exercise 3.12 Show that the off-diagonal blocks ^{ed}E and ^{de}E of the electronic Hessian matrix and the off-diagonal blocks ^{ed}S and ^{de}S of the overlap matrix vanish if one chooses the state transfer operators, Eq. (3.165), as operators $\{\hat{h}_n\}$.

Exercise 3.13 Prove Eqs. (3.167) to (3.174).

3.13 Pseudo-Perturbation Theory

In the discussion of the spectral representation of the polarization propagator in Section 3.11 we have seen that the electronic vertical excitation energies of the system $(E_n^{(0)} - E_0^{(0)})$ are the poles of the polarization propagator. In the matrix representation Eq. (3.159) a polarization propagator has a pole, if the principal propagator matrix $(\boldsymbol{E} - \hbar\omega \boldsymbol{S})$ becomes singular. This leads to the homogeneous linear equations

$$(\boldsymbol{E} - \hbar\omega\boldsymbol{S})\,\boldsymbol{R} = 0 \tag{3.175}$$

and thus to the generalized eigenvalue problem

$$\boldsymbol{E}\boldsymbol{R}_n = \hbar\omega_n \boldsymbol{S}\boldsymbol{R}_n \tag{3.176}$$

where $\hbar\omega_n$ is an eigenvalue of the electronic Hessian matrix and therefore equal to a vertical excitation energy $(E_n^{(0)} - E_0^{(0)})$ and $\boldsymbol{R}_n$ is the corresponding eigenvector. In the case of an asymmetric Hessian matrix $\boldsymbol{E}$, as in the coupled cluster response theory described in Section 11.4, $\boldsymbol{R}_n$ is the right eigenvector, while a left eigenvector $\boldsymbol{L}_n$ is obtained as

$$\boldsymbol{L}_n \boldsymbol{E} = \boldsymbol{L}_n \boldsymbol{S} \hbar\omega_n \tag{3.177}$$

The eigenvectors are normally orthonormalized with the overlap matrix $\boldsymbol{S}$ as metric

$$\boldsymbol{L}_m \boldsymbol{S} \boldsymbol{R}_n = \delta_{mn} \tag{3.178}$$

For symmetric Hessian matrices the left and right eigenvectors are the same.

Often one knows, or can easily obtain, the eigenvalues and eigenvectors of an approximation to the original eigenvalue problem

$$\boldsymbol{E}^{(0)} \boldsymbol{R}_n^{(0)} = \hbar\omega_n^{(0)} \boldsymbol{S}^{(0)} \boldsymbol{R}_n^{(0)} \tag{3.179}$$

$$\boldsymbol{L}_n^{(0)} \boldsymbol{E}^{(0)} = \boldsymbol{L}_n^{(0)} \boldsymbol{S}^{(0)} \hbar\omega_n^{(0)} \tag{3.180}$$

In **pseudo-perturbation theory** (Christiansen et al., 1996) one builds on this fact and finds approximations to the eigenvalues and eigenvectors by applying the techniques of perturbation theory from Section 3.2 to this eigenvalue problem. The Hessian and overlap matrices are then partitioned into the zeroth-order parts $\boldsymbol{E}^{(0)}$ and $\boldsymbol{S}^{(0)}$ and a remainder that is treated as first and second order

$$\boldsymbol{E} = \boldsymbol{E}^{(0)} + \boldsymbol{E}^{(1)} + \boldsymbol{E}^{(2)} + \cdots \tag{3.181}$$

$$\boldsymbol{S} = \boldsymbol{S}^{(0)} + \boldsymbol{S}^{(1)} + \boldsymbol{S}^{(2)} + \cdots \tag{3.182}$$

The eigenvalues and eigenvectors of $\boldsymbol{E}$ are then also expanded in a perturbation series as

$$\omega_n = \omega_n^{(0)} + \omega_n^{(1)} + \omega_n^{(2)} + \cdots \tag{3.183}$$

$$\boldsymbol{R}_n = \boldsymbol{R}_n^{(0)} + \boldsymbol{R}_n^{(1)} + \boldsymbol{R}_n^{(2)} + \cdots \tag{3.184}$$

$$\boldsymbol{L}_n = \boldsymbol{L}_n^{(0)} + \boldsymbol{L}_n^{(1)} + \boldsymbol{L}_n^{(2)} + \cdots \tag{3.185}$$

$$\tag{3.186}$$

In the following it will be convenient to choose the first-order matrices $\boldsymbol{E}^{(1)}$ and $\boldsymbol{S}^{(1)}$ in such a way that their contribution to first order is zero, i.e.

$$\boldsymbol{L}_n^{(0)} \boldsymbol{E}^{(1)} \boldsymbol{R}_n^{(0)} = 0 \tag{3.187}$$

$$\boldsymbol{L}_n^{(0)} \boldsymbol{S}^{(1)} \boldsymbol{R}_n^{(0)} = 0 \tag{3.188}$$

66 Perturbation Theory

Inserting the expansions in the eigenvalue problem Eq. (3.176) and separating orders then gives first- and second-order equations

$$E^{(1)} R_n^{(0)} + E^{(0)} R_n^{(1)} = \left(\hbar\omega_n^{(1)} S^{(0)} + \hbar\omega_n^{(0)} S^{(1)}\right) R_n^{(0)} + \hbar\omega_n^{(0)} S^{(0)} R_n^{(1)} \quad (3.189)$$

$$E^{(2)} R_n^{(0)} + E^{(1)} R_n^{(1)} + E^{(0)} R_n^{(2)}$$
$$= \left(\hbar\omega_n^{(2)} S^{(0)} + \hbar\omega_n^{(0)} S^{(2)} + \hbar\omega_n^{(1)} S^{(1)}\right) R_n^{(0)}$$
$$+ \left(\hbar\omega_n^{(1)} S^{(0)} + \hbar\omega_n^{(0)} S^{(1)}\right) R_n^{(1)} + \hbar\omega_n^{(0)} S^{(0)} R_n^{(2)} \quad (3.190)$$

which projected against the zeroth-order left eigenvector leads to the following expressions for the first- and second-order corrections to the eigenvalues [see Exercise 3.14]

$$\hbar\omega_n^{(1)} = L_n^{(0)} \left(E^{(1)} - \hbar\omega_n^{(0)} S^{(1)}\right) R_n^{(0)} = 0 \quad (3.191)$$

$$\hbar\omega_n^{(2)} = L_n^{(0)} \left(E^{(2)} - \hbar\omega_n^{(0)} S^{(2)}\right) R_n^{(0)} + L_n^{(0)} \left(E^{(1)} - \hbar\omega_n^{(0)} S^{(1)}\right) R_n^{(1)} \quad (3.192)$$

Exercise 3.14 Derive the results for the first- and second-order corrections to the eigenvalues using Eqs. (3.187) and (3.188) and the zeroth-order eigenvalue problems Eq. (3.179).

Rearranging the first-order equation one obtains for the first-order right eigenvector

$$R_n^{(1)} = -\left(E^{(0)} - \hbar\omega_n^{(0)} S^{(0)}\right)^{-1} \left(E^{(1)} - \hbar\omega_n^{(0)} S^{(1)}\right) R_n^{(0)} \quad (3.193)$$

and thus finally for the second-order correction to the eigenvalue

$$\hbar\omega_n^{(2)} = L_n^{(0)} \left(E^{(2)} - \hbar\omega_n^{(0)} S^{(2)}\right) R_n^{(0)} \quad (3.194)$$
$$- L_n^{(0)} \left(E^{(1)} - \hbar\omega_n^{(0)} S^{(1)}\right) \left(E^{(0)} - \hbar\omega_n^{(0)} S^{(0)}\right)^{-1}$$
$$\times \left(E^{(1)} - \hbar\omega_n^{(0)} S^{(1)}\right) R_n^{(0)}$$

Comparison with Eq. (3.33) shows that the second-order correction in pseudo-perturbation theory has essentially the same structure as in regular time-independent perturbation theory.

3.14 Further Reading

Time-Independent Perturbation Theory

H. F. Hameka, *Advanced Quantum Chemistry: Theory of Interactions between Molecules and Electromagnetic Fields*, Addison-Wesley, Reading (1965): Chapter 4–2.

Ehrenfest Theorem

R. Shankar, *Principles of Quantum Mechanics*, Plenum, New York (1980): Chapter 6.

Time-Dependent Perturbation and Interaction Picture

R. Shankar, *Principles of Quantum Mechanics*, Plenum, New York (1980): Chapter 18.

R. Loudon, *The Quantum Theory of Light*, Oxford University Press, Oxford, 2nd edn, 1983: Chapter 5.14.

D. J. Tannor, *Introduction to Quantum Mechanics: A Time-Dependent Perspective*, University Science Books, Sausalito, 2007: Chapter 9.

F. Schwabl, *Advanced Quantum Mechanics*, Springer, Berlin (1999): Chapters 4.3 and 15.2.

Response Theory

D. N. Zubarev, *Nonequilibrium Statistical Thermodynamics*, Consultants Bureau, New York, 1974: Chapter III.

J. Linderberg and Y. Öhrn, *Propagators in Quantum Chemistry*, Academic Press, London, 1973.

P. Jørgensen and J. Simons, *Second Quantization-Based Methods in Quantum Chemistry*, Academic Press, New York, 1981: Chapter 6.

Part II

Definition of Properties

4
Electric Properties

In the first part of the book we have derived the Hamiltonian for the interaction of molecules with electromagnetic fields. Furthermore, we have employed time-independent perturbation theory or static response theory in order to obtain expressions for the corrections to the energy and wavefunction of a molecule due to the interaction with electromagnetic fields. We are thus well prepared for defining many different molecular properties in this and the following chapters and for deriving quantum mechanical expressions for them.

Some of the properties discussed here are well known from pre-quantum mechanical physics and chemistry. For these properties we will start with their classical definitions and then translate them to quantum mechanical expressions. The electric dipole moment and higher-order electric moments or the electric fields at the positions of the nuclei are typical examples. One can define an electric dipole moment for a collection of discrete point charges as well as for a continuous distribution of positive and negative charges with charge density $\rho(\vec{r})$ independent of whether the charge density follows the laws of classical or quantum mechanics. Similarly, one can define a magnetic dipole moment as soon as a current density $\vec{j}(\vec{r})$ is given.

In this chapter we will discuss electric properties and start with the electrostatic potential of the charges in a molecule, because it leads straightforwardly to a definition of electric moments. Afterwards, we will look at changes in the electric moments due to external electric fields and finally we will derive expressions for the electric field and field gradients due to the charges in a molecule.

4.1 Electric Multipole Expansion

The electric charges in a molecule, i.e. the charges of the nuclei at fixed positions and the charges of the distributed electrons, give rise to an electric field that we can represent by the associated **electrostatic potential** $\phi^\rho(\vec{R})$.[1] Other molecules in the neighbourhood of this molecule will experience and react to this field. Knowledge of the electrostatic potential $\phi^\rho(\vec{R})$ around a molecule is, therefore, important for, e.g., the study of long-range intermolecular interactions.

For a distribution of charges the electrostatic potential $\phi^\rho(\vec{R})$ at a point $\vec{R}$ is given as superposition of the potentials due to the individual charges. In the case of a

---

[1] The superscript ρ is supposed to indicate that the electrostatic potential is due to a charge distribution.

72   Electric Properties

continuous distribution of charges with **charge density** $\rho(\vec{r})$ the summation becomes an integration and the electrostatic potential then reads

$$\phi^\rho(\vec{R}) = \frac{1}{4\pi\epsilon_0} \int_{\vec{r}} \frac{\rho(\vec{r})}{|\vec{R}-\vec{r}|} d\vec{r} \qquad (4.1)$$

where $d\vec{r}$ stands for $dx\, dy\, dz$ throughout this chapter and $\int_{\vec{r}}$ denotes a triple integral $\int\int\int$ over the appropriate volume with volume element $d\vec{r}$. This expression is exact but often it is not particularly useful because an integration has to be performed for each observation point $\vec{R}$ and because complete knowledge of the charge distribution $\rho(\vec{r})$ is required. In the following, we will derive an alternative expression for this electrostatic potential that is only exact in the limit of an infinite series, but that neither requires a separate integration for every observation point $\vec{R}$ nor complete knowledge of the charge density $\rho(\vec{r})$.

We will start by expanding $\frac{1}{|\vec{R}-\vec{r}|}$ in a Taylor series around an origin $\vec{R}_O$ within the charge distribution

$$\frac{1}{|\vec{R}-\vec{r}|} = \frac{1}{|\vec{R}-\vec{R}_O|} + \sum_\alpha \left(\frac{\partial}{\partial r_\alpha} \frac{1}{|\vec{R}-\vec{r}|}\right)_{\vec{r}=\vec{R}_O} (r_\alpha - R_{O,\alpha}) \qquad (4.2)$$

$$+ \frac{1}{2} \sum_{\alpha\beta} \left(\frac{\partial^2}{\partial r_\alpha \partial r_\beta} \frac{1}{|\vec{R}-\vec{r}|}\right)_{\vec{r}=\vec{R}_O} (r_\alpha - R_{O,\alpha})(r_\beta - R_{O,\beta})$$

$$+ \ldots$$

where the derivatives have to be evaluated at the point $\vec{r} = \vec{R}_O$. The Greek subscripts α, β, etc. denote again vector or tensor components in the molecule-fixed cartesian coordinate system. A summation over a Greek subscript will here and in the following denote summation over all three cartesian components. For an observation point $\vec{R}$ far from the charge distribution this series should converge rapidly. Inserting this expansion in the expression for the electrostatic potential, Eq. (4.1), we obtain

$$\phi^\rho(\vec{R}) = \frac{1}{4\pi\epsilon_0} \left[\frac{1}{|\vec{R}-\vec{R}_O|} \int_{\vec{r}} \rho(\vec{r})\, d\vec{r} \right.$$

$$+ \sum_\alpha \left(\frac{\partial}{\partial r_\alpha} \frac{1}{|\vec{R}-\vec{r}|}\right)_{\vec{r}=\vec{R}_O} \int_{\vec{r}} \rho(\vec{r}) (r_\alpha - R_{O,\alpha})\, d\vec{r}$$

$$+ \frac{1}{2} \sum_{\alpha\beta} \left(\frac{\partial^2}{\partial r_\alpha \partial r_\beta} \frac{1}{|\vec{R}-\vec{r}|}\right)_{\vec{r}=\vec{R}_O} \int_{\vec{r}} \rho(\vec{r}) (r_\alpha - R_{O,\alpha})(r_\beta - R_{O,\beta})\, d\vec{r}$$

$$\left. + \ldots \right] \qquad (4.3)$$

Equation (4.3) includes a series of integrals over the charge density $\rho(\vec{r})$ multiplied by increasing powers of $(r_\alpha - R_{O,\alpha})$. These type of integrals $\int x^n f(x)\, dx$ are a well-known concept and are called the nth-order moments of the function $f(x)$. We define therefore the cartesian components of the **moments of the charge distribution** $\rho(\vec{r})$, also called the **electric moments**, as

$$q = \int_{\vec{r}} \rho(\vec{r}) \, d\vec{r} \tag{4.4}$$

$$\mu_\alpha(\vec{R}_O) = \int_{\vec{r}} (r_\alpha - R_{O,\alpha}) \, \rho(\vec{r}) \, d\vec{r} \tag{4.5}$$

$$Q_{\alpha\beta}(\vec{R}_O) = \int_{\vec{r}} (r_\alpha - R_{O,\alpha})(r_\beta - R_{O,\beta}) \, \rho(\vec{r}) \, d\vec{r} \tag{4.6}$$

The zeroth-order electric moment q is the total charge, μ_α is a cartesian component of the first-order electric moment, called the **electric dipole moment**, and $Q_{\alpha\beta}$ is a cartesian component of the **second-order electric moment tensor**. The definition of the components of the first electric moment, Eq. (4.5), is the generalization to a continuous charge distribution $\rho(\vec{r})$ and to an arbitrary origin $\vec{R}_O$ of the classical expression

$$\vec{\mu} = \sum_i q_i \vec{R}_i \tag{4.7}$$

for the electric dipole moment of a set of point charges $\{q_i\}$ with position vectors $\{\vec{R}_i\}$ relative to the origin of the coordinate system. For two point charges q_1 and $q_2 = -q_1$ the classical expression is reduced to the well-known expression $|\vec{\mu}| = q_1 d$, where $d = |\vec{R}_1 - \vec{R}_2|$ is the distance between the two point charges.

Frequently, a **traceless**[2] **quadrupole moment tensor** Θ is defined, which has only five independent elements given as

$$\Theta_{\alpha\beta}(\vec{R}_O) = \frac{1}{2} \int_{\vec{r}} \left[3(r_\alpha - R_{O,\alpha})(r_\beta - R_{O,\beta}) - \delta_{\alpha\beta} (\vec{r} - \vec{R}_O)^2 \right] \rho(\vec{r}) \, d\vec{r} \tag{4.8}$$

Since the sum of the diagonal elements is zero, one can obtain one diagonal element from the sum of the other two diagonal elements. The quadrupole moment tensor measures essentially the deviation of the charge distribution $\rho(\vec{r})$ from spherical symmetry. To avoid confusion it is customary to call the moment Q defined in Eq. (4.6) the **second electric moment** and Θ as defined in Eq. (4.8) the **electric quadrupole moment**.

With this definition of the electric moments we have already achieved our first goal, because we have removed or at least hidden the integration over the charge distribution in the Taylor expansion of the electrostatic potential

$$\phi^\rho(\vec{R}) = \frac{1}{4\pi\epsilon_0} \left[\frac{q}{|\vec{R} - \vec{R}_O|} + \sum_\alpha \left(\frac{\partial}{\partial r_\alpha} \frac{1}{|\vec{R} - \vec{r}|} \right)_{\vec{r} = \vec{R}_O} \mu_\alpha(\vec{R}_O) \right.$$

$$\left. + \frac{1}{2} \sum_{\alpha\beta} \left(\frac{\partial^2}{\partial r_\alpha \partial r_\beta} \frac{1}{|\vec{R} - \vec{r}|} \right)_{\vec{r} = \vec{R}_O} Q_{\alpha\beta}(\vec{R}_O) + \cdots \right] \tag{4.9}$$

---

[2] whose trace is zero: $\sum_\alpha \Theta_{\alpha\alpha} = 0$.

74 *Electric Properties*

Finally, we have to evaluate the derivatives in Eq. (4.9) [see Exercise 4.1] that then yields the **multipole expansion of the electrostatic potential**

$$\phi^\rho(\vec{R}) = \frac{1}{4\pi\epsilon_0} \frac{q}{|\vec{R}-\vec{R}_O|} + \frac{1}{4\pi\epsilon_0} \sum_\alpha \mu_\alpha(\vec{R}_O) \frac{R_\alpha - R_{O,\alpha}}{|\vec{R}-\vec{R}_O|^3} \qquad (4.10)$$

$$+ \frac{1}{8\pi\epsilon_0} \sum_{\alpha\beta} Q_{\alpha\beta}(\vec{R}_O) \frac{3(R_\alpha - R_{O,\alpha})(R_\beta - R_{O,\beta}) - \delta_{\alpha\beta}(\vec{R}-\vec{R}_O)^2}{|\vec{R}-\vec{R}_O|^5} + \ldots$$

Exercise 4.1 Derive the two derivatives

$$\left(\frac{\partial}{\partial r_\alpha} \frac{1}{|\vec{R}-\vec{r}|} \right)_{\vec{r}=\vec{R}_O}$$

$$\left(\frac{\partial^2}{\partial r_\alpha \partial r_\beta} \frac{1}{|\vec{R}-\vec{r}|} \right)_{\vec{r}=\vec{R}_O}$$

used in the derivation of Eq. (4.10).

Convergence of the multipole series for a particular observation point $\vec{R}$ depends on the precise form of the charge distribution $\rho(\vec{r})$ and on the distance $(\vec{R}-\vec{R}_O)$ between the observation point and the charge distribution. However, one can expect in general that the contribution from the higher moments in this series will become negligible as the distance between the observation point $\vec{R}$ and the origin $\vec{R}_O$ increases and the potential will then be described accurately by the charge and dipole moment terms alone.

Apart from motivating the definition of the electric moments, the importance of the multipole expansion in Eq. (4.10) lies in the fact that we can calculate the electrostatic potential $\phi^\rho(\vec{R})$ for any point $\vec{R}$ from the simple formula in Eq. (4.10) as soon as we know the electric multipole moments of the corresponding charge distribution $\rho(\vec{r})$ instead of evaluating the more complicated expression in Eq. (4.1) for each $\vec{R}$. A major application of electric multipole moments is thus the description and calculation of intermolecular forces (Buckingham, 1967).

An important feature of the electric multipole moments, as defined in Eqs. (4.4)–(4.6) and (4.8), is that the first non-vanishing moment of a charge distribution is independent of the choice of the origin $\vec{R}_O$. However, all the higher moments depend on this origin. Thus, the dipole moment of a neutral molecule or the quadrupole moment of a neutral and non-polar[3] molecule are both independent of the origin $\vec{R}_O$, whereas the dipole moment of an ion or the quadrupole moment of a neutral but polar[3] molecule will depend on the origin $\vec{R}_O$ [see Exercise 4.2].

Exercise 4.2 Show that the dipole moment of an ion and the quadrupole moment of a neutral but polar molecule depend on the origin $\vec{R}_O$.

---

[3] A molecule is polar when it has a dipole moment.

4.2 Potential Energy in an Electric Field

Electric multipole moments also play an important role in the description of interactions between molecules and external electric fields. The potential energy E of a distribution of charges $\rho(\vec{r})$ immersed in an external inhomogeneous static electric field is given as[4]

$$E(\vec{\mathcal{E}}, \boldsymbol{\mathcal{E}}) = \int_{\vec{r}} \rho(\vec{r})\, \phi^{\mathcal{E}}(\vec{r})\, d\vec{r}, \tag{4.11}$$

where $\phi^{\mathcal{E}}(\vec{r})$ is the scalar potential associated with the electric field as defined in Eq. (2.33). This is simply the generalization of $E = q\,\phi^{\mathcal{E}}$ for a single charge q to the case of a continuous charge distribution $\rho(\vec{r})$. Analogous to the expression for the electrostatic potential of a charge distribution in Eq. (4.1), this expression for the potential energy is exact, but evaluation of Eq. (4.11) requires that the charge density $\rho(\vec{r})$ and the electric potential $\phi^{\mathcal{E}}(\vec{r})$ are known for all values of $\vec{r}$. A more useful expression can be obtained again, if we expand the scalar potential in a Taylor series around $\vec{R}_O$

$$\phi^{\mathcal{E}}(\vec{r}) = \phi^{\mathcal{E}}(\vec{R}_O) + \sum_{\alpha}(r_\alpha - R_{O,\alpha})\,\frac{\partial \phi^{\mathcal{E}}(\vec{r})}{\partial r_\alpha}\bigg|_{\vec{r}=\vec{R}_O}$$

$$+ \frac{1}{2}\sum_{\alpha\beta}(r_\alpha - R_{O,\alpha})(r_\beta - R_{O,\beta})\,\frac{\partial^2 \phi^{\mathcal{E}}(\vec{r})}{\partial r_\alpha \partial r_\beta}\bigg|_{\vec{r}=\vec{R}_O} + \ldots \tag{4.12}$$

where the derivatives have to be evaluated again at $\vec{r}=\vec{R}_O$. The first derivatives of the scalar potential $\phi^{\mathcal{E}}$ are the components of the electric-field vector $\mathcal{E}_\alpha$

$$\mathcal{E}_\alpha(\vec{R}_O) = -\frac{\partial \phi^{\mathcal{E}}(\vec{r})}{\partial r_\alpha}\bigg|_{\vec{r}=\vec{R}_O} \tag{4.13}$$

and the second derivatives are the components of the electric-field gradient tensor $\mathcal{E}_{\alpha\beta}$

$$\mathcal{E}_{\alpha\beta}(\vec{R}_O) = -\frac{\partial^2 \phi^{\mathcal{E}}(\vec{r})}{\partial r_\alpha \partial r_\beta}\bigg|_{\vec{r}=\vec{R}_O} \tag{4.14}$$

etc. The expansion of the scalar potential is thus

$$\phi^{\mathcal{E}}(\vec{r}) = \phi^{\mathcal{E}}(\vec{R}_O) - \sum_{\alpha}(r_\alpha - R_{O,\alpha})\mathcal{E}_\alpha(\vec{R}_O)$$

$$- \frac{1}{2}\sum_{\alpha\beta}(r_\alpha - R_{O,\alpha})(r_\beta - R_{O,\beta})\mathcal{E}_{\alpha\beta}(\vec{R}_O) + \ldots \tag{4.15}$$

---

[4] $(\vec{\mathcal{E}}, \boldsymbol{\mathcal{E}})$ indicates that the energy depends on the electric-field vector $\vec{\mathcal{E}}$, the electric-field gradient tensor $\boldsymbol{\mathcal{E}}$ and possibly higher derivatives of the electric field.

On insertion of Eq. (4.15) in Eq. (4.11) one obtains for the potential energy

$$E(\vec{\mathcal{E}}, \mathcal{E}) = \phi^{\mathcal{E}}(\vec{R}_O) \int_{\vec{r}} \rho(\vec{r}) \, d\vec{r} - \sum_{\alpha} \mathcal{E}_\alpha(\vec{R}_O) \int_{\vec{r}} (r_\alpha - R_{O,\alpha}) \, \rho(\vec{r}) \, d\vec{r}$$

$$- \frac{1}{2} \sum_{\alpha\beta} \mathcal{E}_{\alpha\beta}(\vec{R}_O) \int_{\vec{r}} (r_\alpha - R_{O,\alpha})(r_\beta - R_{O,\beta}) \, \rho(\vec{r}) \, d\vec{r} + \ldots \quad (4.16)$$

The integrals over $\rho(\vec{r})$ are again the electric moments defined in Eqs. (4.4)–(4.6).

The energy $E(\vec{\mathcal{E}}, \mathcal{E})$ of the interaction between a charge distribution and a static but inhomogeneous electric field can therefore be expressed in terms of the electric moments of the charge distribution

$$E(\vec{\mathcal{E}}, \mathcal{E}) = q \, \phi^{\mathcal{E}}(\vec{R}_O) - \sum_{\alpha} \mu_\alpha(\vec{R}_O) \mathcal{E}_\alpha(\vec{R}_O) - \frac{1}{2} \sum_{\alpha\beta} Q_{\alpha\beta}(\vec{R}_O) \mathcal{E}_{\alpha\beta}(\vec{R}_O) + \ldots \quad (4.17)$$

similar to the multipole expansion of the electrostatic potential of a charge distribution in Eq. (4.10). Alternatively, using the quadrupole moment tensor Θ we can write for the interaction energy [see Exercise 4.3] and [see Exercise 4.4]

$$E(\vec{\mathcal{E}}, \mathcal{E}) = q \, \phi^{\mathcal{E}}(\vec{R}_O) - \sum_{\alpha} \mu_\alpha(\vec{R}_O) \mathcal{E}_\alpha(\vec{R}_O) - \frac{1}{3} \sum_{\alpha\beta} \Theta_{\alpha\beta}(\vec{R}_O) \mathcal{E}_{\alpha\beta}(\vec{R}_O) + \ldots \quad (4.18)$$

Exercise 4.3 Show that the contribution $\frac{1}{2} \sum_{\alpha\beta} Q_{\alpha\beta}(\vec{R}_O) \mathcal{E}_{\alpha\beta}(\vec{R}_O)$ of the second-order electric moment to the interaction energy in Eq. (4.17) is unchanged if an arbitrary constant C is added to the diagonal elements $Q_{\alpha\alpha}$ of the second-order electric moment.

Hint : Recall that we only use the scalar potential $\phi^{\mathcal{E}}(\vec{R}_O)$ far away from the charges that originally generated it and that therefore it satisfies Laplace's equation $\nabla^2 \phi^{\mathcal{E}}(\vec{R}_O) = 0$. This leads to an equation that can be used in the solution of this exercise.

Exercise 4.4 Show that Eqs. (4.17) and (4.18) are equivalent.

The electric multipole moments of a charge distribution, defined in Eqs. (4.4)–(4.6) or (4.8), can therefore not only be used to express the electrostatic potential ϕ^ρ created by this charge distribution in surroundings but also the interaction energy E of the same charge distribution with an external scalar potential $\phi^{\mathcal{E}}$. The electric moments therefore play an important role in the description of intermolecular interactions, where one molecule is considered to give rise to an electric potential that the other molecule feels. One part of the interaction energy is therefore expanded in the multipole moments of both molecules (Buckingham, 1967).

Furthermore, Eq. (4.18) shows that we can define the cartesian components of the dipole moment and quadrupole moment alternatively as derivatives[5] of the interaction energy with respect to the field strength $\mathcal{E}_\alpha$ or field gradient $\mathcal{E}_{\alpha\beta}$

---

[5] In Section 4.4 we will see that the derivatives have to be evaluated for zero field or field gradient strength, $\mathcal{E}_\alpha = 0$ or $\mathcal{E}_{\alpha\beta} = 0$, in order to obtain the permanent moments.

$$\mu_\alpha(\vec{R}_O) = -\frac{\partial E(\vec{\mathcal{E}})}{\partial \mathcal{E}_\alpha(\vec{R}_O)} \tag{4.19}$$

$$Q_{\alpha\beta}(\vec{R}_O) = -2\frac{\partial E(\vec{\mathcal{E}})}{\partial \mathcal{E}_{\alpha\beta}(\vec{R}_O)} \tag{4.20}$$

$$\Theta_{\alpha\beta}(\vec{R}_O) = -3\frac{\partial E(\vec{\mathcal{E}})}{\partial \mathcal{E}_{\alpha\beta}(\vec{R}_O)} \tag{4.21}$$

in addition to the definitions as integrals over the charge density given in Eqs. (4.5) and (4.8). These definitions will be used in the derivation of quantum mechanical expressions for the moments in the next section and in Part III.

4.3 Quantum Mechanical Expressions for Electric Moments

In the previous two sections we found two alternative sets of expressions for the electric moments from which we can start deriving quantum mechanical formulas. The first one, the original definitions in Eqs. (4.5)–(4.6) and (4.8), are integrals over the charge density $\rho(\vec{r})$, whereas in the second type of expressions, Eqs. (4.19), (4.20) and (4.21), the moments are defined as derivatives of the energy. Consequently, we could quantise the original definitions of the moments Eqs. (4.5), (4.6) and (4.8) or apply the derivatives Eqs. (4.19), (4.20) and (4.21) to the quantum-mechanical expression for the energy of a molecule in the presence of an electrostatic scalar potential $\phi^{\mathcal{E}}$. Actually, there is a third option that is based on the Hellmann–Feynman theorem, Eq. (3.11), which says that the derivative of the energy with respect to a parameter in the Hamiltonian can be calculated as the expectation value of an operator that is the corresponding derivative of the Hamiltonian.

Let us look first at the transition of the original definitions as integrals over the charge density, Eqs. (4.5), (4.6) and (4.8), to quantum mechanics that we will illustrate for the example of the electric dipole moment. In the Born–Oppenheimer approximation, Section 2.2, the electrons in a molecule form a continuous charge distribution $\rho^{el}(\vec{r})$, whereas the discrete nuclear charges are located at fixed points $\vec{R}_K$. The expression, Eq. (4.5) for the α-component of the electric dipole moment can therefore be rewritten as

$$\mu_\alpha(\vec{R}_O) = \int_{\vec{r}} (r_\alpha - R_{O,\alpha})\, \rho^{el}(\vec{r})\, d\vec{r} + \sum_K Z_K e\, (R_{K,\alpha} - R_{O,\alpha}) \tag{4.22}$$

The transition to quantum mechanics can now be made by inserting the quantum mechanical expression for the charge density of N electrons in the state $|\Psi_0^{(0)}\rangle$, Eq. (2.23).

$$\mu_\alpha(\vec{R}_O) = \int_{\vec{r}} (r_\alpha - R_{O,\alpha})\, (-e)\langle \Psi_0^{(0)} | \sum_i^N \delta(\vec{r}_i - \vec{r}) | \Psi_0^{(0)}\rangle\, d\vec{r}$$

$$+ \sum_K Z_K e\, (R_{K,\alpha} - R_{O,\alpha}) \tag{4.23}$$

78 Electric Properties

Recalling the properties of the Dirac δ function in three dimensions, Eq. (2.18), and the fact that the Dirac δ function is a symmetrical function of its argument

$$\delta(\vec{r} - \vec{a}) = \delta(\vec{a} - \vec{r}) \qquad (4.24)$$

we can evaluate the integral over $\vec{r}$ in Eq. (4.23). The **quantum mechanical expression** for the cartesian components of the **electric dipole moment** is therefore given as

$$\mu_\alpha(\vec{R}_O) = \langle \Psi_0^{(0)} | -e \sum_i^N (r_{i,\alpha} - R_{O,\alpha}) | \Psi_0^{(0)} \rangle + \sum_K Z_K e (R_{K,\alpha} - R_{O,\alpha}) \qquad (4.25)$$

In complete analogy we can derive quantum mechanical expressions for the components of the second electric moment tensor $Q_{\alpha\beta}$ and the quadrupole moment tensor $\Theta_{\alpha\beta}$

$$Q_{\alpha\beta}(\vec{R}_O) = \langle \Psi_0^{(0)} | -e \sum_i^N (r_{i,\alpha} - R_{O,\alpha})(r_{i,\beta} - R_{O,\beta}) | \Psi_0^{(0)} \rangle$$

$$+ \sum_K Z_K e (R_{K,\alpha} - R_{O,\alpha})(R_{K,\beta} - R_{O,\beta}) \qquad (4.26)$$

$$\Theta_{\alpha\beta}(\vec{R}_O) = \frac{1}{2} \langle \Psi_0^{(0)} | -e \sum_i^N 3(r_{i,\alpha} - R_{O,\alpha})(r_{i,\beta} - R_{O,\beta}) - \delta_{\alpha\beta} (\vec{r}_i - \vec{R}_O)^2 | \Psi_0^{(0)} \rangle$$

$$+ \frac{1}{2} \sum_K Z_K e \left[3(R_{K,\alpha} - R_{O,\alpha})(R_{K,\beta} - R_{O,\beta}) - \delta_{\alpha\beta} (\vec{R}_K - \vec{R}_O)^2 \right] \qquad (4.27)$$

In the second approach we will use the fact that the moments are defined as derivatives of the energy of a molecule in the presence of an inhomogeneous electric field, Eqs. (4.19), (4.20) and (4.21). In order to apply these definitions we need to find an expression for the energy of a molecule in the presence of an inhomogeneous electric field. Here, we are using perturbation theory as developed in Section 3.2. The first step is thus to define the perturbation Hamiltonian operators $\hat{H}^{(1)}$ and to derive explicit expressions for them in terms of components of the electric field $\mathcal{E}_\alpha(\vec{R}_O)$ and field gradient tensor $\mathcal{E}_{\alpha\beta}(\vec{R}_O)$. The electric field and field gradient enter the molecular Hamiltonian in the form of the scalar potential $\hat{\phi}^\mathcal{E}(\vec{r})$. From Eq. (4.15) we can see that the electrostatic potential at the position $\vec{r}$ of particle (electron or nucleus) for an electric field with non-zero gradient is given as

$$\hat{\phi}^\mathcal{E}(\vec{r}) = -(\vec{r} - \vec{R}_O) \cdot \vec{\mathcal{E}}(\vec{R}_O)$$

$$- \frac{1}{2} \sum_{\alpha\beta} \left[(r_\alpha - R_{O,\alpha})(r_\beta - R_{O,\beta}) - \frac{1}{3} \delta_{\alpha\beta} (\vec{r} - \vec{R}_O)^2 \right] \mathcal{E}_{\alpha\beta}(\vec{R}_O) \qquad (4.28)$$

where we have not included the constant term, $\phi^\mathcal{E}(\vec{R}_O)$, from the expansion of the electrostatic potential in Eq. (4.15), because it does not depend on any electronic coordinate. When we insert this in the general expression for the molecular Hamiltonian, Eq. (2.101), we can write the first-order perturbation Hamiltonian as

$$\hat{H}^{(1)} = -\sum_{\alpha}\left[\hat{O}_{\alpha}^{\mathcal{E}}(\vec{R}_O) + \hat{\Omega}_{\alpha}^{\mathcal{E}}(\vec{R}_O)\right]\mathcal{E}_{\alpha}(\vec{R}_O) - \sum_{\alpha\beta}\left[\hat{O}_{\alpha\beta}^{\nabla\mathcal{E}}(\vec{R}_O) + \hat{\Omega}_{\alpha\beta}^{\nabla\mathcal{E}}(\vec{R}_O)\right]\mathcal{E}_{\alpha\beta}(\vec{R}_O)$$
(4.29)

where the perturbation operators of the electrons are defined as[6]

$$\hat{O}_{\alpha}^{\mathcal{E}}(\vec{R}_O) = \sum_{i}^{N}\hat{o}_{i,\alpha}^{\mathcal{E}}(\vec{R}_O) = -e\sum_{i}^{N}(r_{i,\alpha} - R_{O,\alpha})$$

$$\equiv \hat{\mu}_{\alpha}(\vec{R}_O)$$
(4.30)

$$\hat{O}_{\alpha\beta}^{\nabla\mathcal{E}}(\vec{R}_O) = \sum_{i}^{N}\hat{o}_{i,\alpha\beta}^{\nabla\mathcal{E}}(\vec{R}_O) = -\frac{e}{2}\sum_{i}^{N}\left[(r_{i,\alpha} - R_{O,\alpha})(r_{i,\beta} - R_{O,\beta}) - \frac{1}{3}\delta_{\alpha\beta}(\vec{r}_i - \vec{R}_O)^2\right]$$

$$\equiv \frac{1}{3}\hat{\Theta}_{\alpha\beta}(\vec{R}_O)$$
(4.31)

The N-electron operators $\hat{\vec{\mu}}(\vec{R}_O)$ and $\hat{\Theta}(\vec{R}_O)$ will in the following often be called **the electric dipole operator** and **the electric quadrupole operator**, respectively. Although we are working within the Born–Oppenheimer approximation we have included the interaction of the electric field and field gradient with the nuclear charges in the molecular Hamiltonian in Eq. (2.101). This interaction then leads to nuclear contributions to the perturbation Hamiltonian operators. The operators $\hat{\Omega}_{\alpha}^{\mathcal{E}}$ and $\hat{\Omega}_{\alpha\beta}^{\nabla\mathcal{E}}$

$$\hat{\Omega}_{\alpha}^{\mathcal{E}}(\vec{R}_O) = \sum_{K}Z_K e(R_{K,\alpha} - R_{O,\alpha})$$
(4.32)

$$\hat{\Omega}_{\alpha\beta}^{\nabla\mathcal{E}}(\vec{R}_O) = \frac{1}{2}\sum_{K}Z_K e\left[(R_{K,\alpha} - R_{O,\alpha})(R_{K,\beta} - R_{O,\beta}) - \frac{1}{3}\delta_{\alpha\beta}(\vec{R}_K - \vec{R}_O)^2\right]$$
(4.33)

are such terms that give rise to the nuclear contributions to the electric dipole and quadrupole moments. Here and in the following, we will use $\hat{O}$ or $\hat{o}_i$ for operators relating to electrons, while $\hat{\Omega}$ will be nuclear operators.

Having defined the perturbation Hamiltonians and perturbation operators we can now derive expressions for the cartesian components of the electric moments as derivatives of the energy in the presence of perturbing fields according to Eqs. (4.19) and (4.21). We have now two possibilities: either we make use of the perturbation theory expansion of the energy, Eq. (3.15), or of the Hellmann-Feynman theorem. Let us start with perturbation theory. Because the moments are first derivatives we only need to consider the first-order energy correction, Eq. (3.29),

$$\mu_{\alpha}(\vec{R}_O) = -\frac{\partial E_0^{(1)}(\vec{\mathcal{E}})}{\partial \mathcal{E}_{\alpha}(\vec{R}_O)} = -\frac{\partial}{\partial \mathcal{E}_{\alpha}(\vec{R}_O)}\langle\Psi_0^{(0)}|\hat{H}^{(1)}|\Psi_0^{(0)}\rangle$$
(4.34)

---

[6] All perturbation operators derived in this and the following chapters are collected in Appendix A.

80 Electric Properties

$$\Theta_{\alpha\beta}(\vec{R}_O) = -3\frac{\partial E_0^{(1)}(\vec{\mathcal{E}})}{\partial \mathcal{E}_{\alpha\beta}(\vec{R}_O)} = -3\frac{\partial}{\partial \mathcal{E}_{\alpha\beta}(\vec{R}_O)}\langle \Psi_0^{(0)}|\hat{H}^{(1)}|\Psi_0^{(0)}\rangle \qquad (4.35)$$

Alternatively, we can employ the Hellmann–Feynman theorem, Eq. (3.11), and obtain the cartesian components of the electric moments directly as expectation values of the corresponding derivatives of the Hamiltonian

$$\mu_\alpha(\vec{R}_O) = \langle \Psi_0^{(0)}|-\frac{\partial \hat{H}}{\partial \mathcal{E}_\alpha(\vec{R}_O)}|\Psi_0^{(0)}\rangle \qquad (4.36)$$

$$\Theta_{\alpha\beta}(\vec{R}_O) = 3\,\langle \Psi_0^{(0)}|-\frac{\partial \hat{H}}{\partial \mathcal{E}_{\alpha\beta}(\vec{R}_O)}|\Psi_0^{(0)}\rangle \qquad (4.37)$$

However, the first derivatives of the first-order perturbation Hamiltonian in Eqs. (4.34) and (4.35) and of the full molecular electronic Hamiltonian are obviously the same

$$\frac{\partial \hat{H}}{\partial \mathcal{E}_\alpha(\vec{R}_O)} = \frac{\partial \hat{H}^{(1)}}{\partial \mathcal{E}_\alpha(\vec{R}_O)} = -\hat{\mu}_\alpha(\vec{R}_O) - \hat{\Omega}_\alpha^{\mathcal{E}}(\vec{R}_O) \qquad (4.38)$$

$$\frac{\partial \hat{H}}{\partial \mathcal{E}_{\alpha\beta}(\vec{R}_O)} = \frac{\partial \hat{H}^{(1)}}{\partial \mathcal{E}_{\alpha\beta}(\vec{R}_O)} = -\frac{1}{3}\hat{\Theta}_{\alpha\beta}(\vec{R}_O) - \hat{\Omega}_{\alpha\beta}^{\nabla\mathcal{E}}(\vec{R}_O) \qquad (4.39)$$

and the quantum mechanical expressions for the cartesian components of the total electric dipole and quadrupole moments are therefore given as

$$\mu_\alpha(\vec{R}_O) = \langle \Psi_0^{(0)}|\hat{\mu}_\alpha(\vec{R}_O)|\Psi_0^{(0)}\rangle + \hat{\Omega}_\alpha^{\mathcal{E}}(\vec{R}_O) \qquad (4.40)$$

$$\Theta_{\alpha\beta}(\vec{R}_O) = \langle \Psi_0^{(0)}|\hat{\Theta}_{\alpha\beta}(\vec{R}_O)|\Psi_0^{(0)}\rangle + 3\,\hat{\Omega}_{\alpha\beta}^{\nabla\mathcal{E}}(\vec{R}_O) \qquad (4.41)$$

These expression are of course identical to the ones derived previously in Eqs. (4.25) and (4.27), apart from the fact that they are written in terms of the moment operators here.

This proves that the different definitions are indeed equivalent as long as we know the exact solutions to the unperturbed Schrödinger equation, Eq. (3.14). However, this will no longer be the case when we work with approximate wavefunctions in Part III and therefore we have discussed all the alternative definitions in detail here.

4.4 Induced Electric Moments and Polarizabilities

So far, it has been assumed that the distribution of charges is fixed and is not influenced by the external electric field apart from a change in its energy. However, at least the electrons are moving and therefore the charge distribution will redistribute itself in the presence of the external electric field in such a way that the total energy is minimized. One says that the charge distribution will be polarized. As a result, the cartesian components of the electric moments of the charge distribution will change and their values will depend on the strength of the field

$$\mu_\alpha(\vec{\mathcal{E}}, \mathcal{E}) = \mu_\alpha + \mu_\alpha^{ind}(\vec{\mathcal{E}}, \mathcal{E}) \tag{4.42}$$

$$\Theta_{\alpha\beta}(\vec{\mathcal{E}}, \mathcal{E}) = \Theta_{\alpha\beta} + \Theta_{\alpha\beta}^{ind}(\vec{\mathcal{E}}, \mathcal{E}) \tag{4.43}$$

One can say that the field-dependent moments $\vec{\mu}^{ind}(\vec{\mathcal{E}}, \mathcal{E})$ and $\Theta^{ind}(\vec{\mathcal{E}}, \mathcal{E})$[7] are induced by the external field in addition to the field-independent, so-called permanent, moments $\vec{\mu}$, Θ, which we have introduced in the previous sections.

Traditionally, the dependence of the cartesian components of the electric moments, $\mu_\alpha(\vec{\mathcal{E}}, \mathcal{E})$ and $\Theta_{\alpha\beta}(\vec{\mathcal{E}}, \mathcal{E})$, on powers of the strength of an external field $\mathcal{E}_\alpha$ and field gradient $\mathcal{E}_{\alpha\beta}$ is expressed in the following way (Buckingham, 1967)

$$\mu_\alpha(\vec{\mathcal{E}}, \mathcal{E}) = \mu_\alpha + \sum_\beta \alpha_{\alpha\beta} \mathcal{E}_\beta + \frac{1}{2} \sum_{\beta\gamma} \beta_{\alpha\beta\gamma} \mathcal{E}_\beta \mathcal{E}_\gamma + \frac{1}{6} \sum_{\beta\gamma\delta} \gamma_{\alpha\beta\gamma\delta} \mathcal{E}_\beta \mathcal{E}_\gamma \mathcal{E}_\delta$$

$$+ \frac{1}{3} \sum_{\beta\gamma} A_{\alpha,\beta\gamma} \mathcal{E}_{\beta\gamma} + \frac{1}{3} \sum_{\beta\gamma\delta} B_{\alpha,\beta,\gamma\delta} \mathcal{E}_\beta \mathcal{E}_{\gamma\delta} + \cdots \tag{4.44}$$

$$\Theta_{\alpha\beta}(\vec{\mathcal{E}}, \mathcal{E}) = \Theta_{\alpha\beta} + \sum_\gamma A_{\gamma,\alpha\beta} \mathcal{E}_\gamma + \frac{1}{2} \sum_{\gamma\delta} B_{\gamma,\delta,\alpha\beta} \mathcal{E}_\gamma \mathcal{E}_\delta$$

$$+ \sum_{\gamma\delta} C_{\alpha\beta,\gamma\delta} \mathcal{E}_{\gamma\delta} + \cdots \tag{4.45}$$

where here and in the rest of this section the origin dependence "$(\vec{R}_O)$" of the moments is not written out explicitly as well, as $\mathcal{E}_\alpha$ and $\mathcal{E}_{\alpha\beta}$ are meant as abbreviations for $\mathcal{E}_\alpha(\vec{R}_O)$ and $\mathcal{E}_{\alpha\beta}(\vec{R}_O)$, respectively, in order to reduce the complexity of the notation. The expansion coefficients, $\alpha_{\alpha\beta}$, $\beta_{\alpha\beta\gamma}$, $\gamma_{\alpha\beta\gamma\delta}$, $A_{\gamma,\alpha\beta}$, $B_{\alpha,\beta,\gamma\delta}$ and $C_{\alpha\beta,\gamma\delta}$, on the other hand, are independent of the origin $\vec{R}_O$ that is shown in Section 4.5.

These expansions serve mainly as definitions of the polarizabilities and hyperpolarizabilities as proportionality constants in the correction terms to the permanent moments. The **dipole polarizability** α is a second-rank tensor with nine cartesian components $\alpha_{\alpha\beta}$, the **dipole–quadrupole polarizability** and **first dipole hyperpolarizability** are third-rank tensors with 27 cartesian components $A_{\alpha,\beta\gamma}$ and $\beta_{\alpha\beta\gamma}$, while the **quadrupole–quadrupole polarizability**, the **dipole–quadrupole hyperpolarizability** and the **second dipole hyperpolarizability** are fourth-rank tensors with 81 cartesian components $C_{\alpha\beta,\gamma\delta}$, $B_{\alpha,\beta,\gamma\delta}$ and $\gamma_{\alpha\beta\gamma\delta}$.[8]

Based on the expansions, one can express the polarizabilities and hyperpolarizabilities as first and higher derivatives of the field-dependent moments $\vec{\mu}(\vec{\mathcal{E}}, \mathcal{E})$ and $\Theta(\vec{\mathcal{E}}, \mathcal{E})$ with respect to the components of the electric field and field gradient[9]

---

[7] $(\vec{\mathcal{E}}, \mathcal{E})$ indicates that the moments depend on the electric-field vector $\vec{\mathcal{E}}$ and on the electric-field gradient tensor $\mathcal{E}$.

[8] The notation for the cartesian component indices of the various quadrupole (hyper) polarizabilities, e.g. $B_{\alpha,\beta,\gamma\delta}$, differs slightly from the one for the pure dipole (hyper)polarizabilities, e.g. $\beta_{\alpha\beta\gamma}$, in order to mark more clearly which are the quadrupole components.

[9] These definitions are also collected in the second and third columns of Table B.1 of appendix B.

$$\alpha_{\alpha\beta} = \left.\frac{\partial \mu_\alpha(\vec{\mathcal{E}})}{\partial \mathcal{E}_\beta}\right|_{|\vec{\mathcal{E}}|=0} \tag{4.46}$$

$$A_{\alpha,\gamma\delta} = 3\left.\frac{\partial \mu_\alpha(\boldsymbol{\mathcal{E}})}{\partial \mathcal{E}_{\gamma\delta}}\right|_{|\boldsymbol{\mathcal{E}}|=0} = \left.\frac{\partial \Theta_{\gamma\delta}(\vec{\mathcal{E}})}{\partial \mathcal{E}_\alpha}\right|_{|\vec{\mathcal{E}}|=0} \tag{4.47}$$

$$C_{\gamma\delta,\alpha\beta} = \left.\frac{\partial \Theta_{\gamma\delta}(\boldsymbol{\mathcal{E}})}{\partial \mathcal{E}_{\alpha\beta}}\right|_{|\boldsymbol{\mathcal{E}}|=0} \tag{4.48}$$

$$\beta_{\alpha\beta\gamma} = \left.\frac{\partial^2 \mu_\alpha(\vec{\mathcal{E}})}{\partial \mathcal{E}_\gamma \partial \mathcal{E}_\beta}\right|_{|\vec{\mathcal{E}}|=0} \tag{4.49}$$

$$B_{\alpha,\beta,\gamma\delta} = 3\left.\frac{\partial^2 \mu_\alpha(\vec{\mathcal{E}},\boldsymbol{\mathcal{E}})}{\partial \mathcal{E}_{\gamma\delta}\partial \mathcal{E}_\beta}\right|_{|\vec{\mathcal{E}}|=|\boldsymbol{\mathcal{E}}|=0} = \left.\frac{\partial^2 \Theta_{\gamma\delta}(\vec{\mathcal{E}},\boldsymbol{\mathcal{E}})}{\partial \mathcal{E}_\beta \partial \mathcal{E}_\alpha}\right|_{|\vec{\mathcal{E}}|=|\boldsymbol{\mathcal{E}}|=0} \tag{4.50}$$

$$\gamma_{\alpha\beta\gamma\delta} = \left.\frac{\partial^3 \mu_\alpha(\vec{\mathcal{E}})}{\partial \mathcal{E}_\delta \partial \mathcal{E}_\gamma \partial \mathcal{E}_\beta}\right|_{|\vec{\mathcal{E}}|=0} \tag{4.51}$$

The expansion of the electric moments in Eqs. (4.44) and (4.45) also explains why the polarizabilities are sometimes called the **linear response** of the moments to an electric field or field gradient, the first hyperpolarizabilities the **quadratic response** and the second-order hyperpolarizability the **cubic response**.

The isotropic or mean polarizabilities normally measured for molecules in the liquid or gas phase are defined as (Buckingham, 1967)

$$\overline{\alpha} = \frac{1}{3}\sum_\alpha \alpha_{\alpha\alpha} \tag{4.52}$$

$$\overline{B} = \frac{2}{15}\sum_{\alpha\beta} B_{\alpha,\beta,\alpha\beta} \tag{4.53}$$

$$\overline{C} = \frac{1}{5}\sum_{\alpha\beta} C_{\alpha\beta,\alpha\beta} \tag{4.54}$$

whereas the anisotropy of the dipole polarizability $\Delta\alpha$ is defined as

$$\Delta\alpha = \sqrt{\frac{\sum_{\alpha\beta}(3\alpha_{\alpha\beta}\alpha_{\alpha\beta} - \alpha_{\alpha\alpha}\alpha_{\beta\beta})}{2}} \tag{4.55}$$

Similarly, one defines two isotropic averages for the first hyperpolarizability

$$\beta_\| = \frac{1}{5}\sum_\alpha (\beta_{z\alpha\alpha} + \beta_{\alpha z\alpha} + \beta_{\alpha\alpha z}) \tag{4.56}$$

$$\beta_\perp = \frac{1}{5}\sum_\alpha (2\beta_{z\alpha\alpha} - 3\beta_{\alpha z\alpha} + 2\beta_{\alpha\alpha z}) \tag{4.57}$$

where the molecular z-axis is parallel to the electric dipole moment vector of the molecule, and also two isotropic averages for the second hyperpolarizability

$$\bar{\gamma} = \gamma_\parallel = \frac{1}{15} \sum_{\alpha\beta} (\gamma_{\alpha\alpha\beta\beta} + \gamma_{\alpha\beta\alpha\beta} + \gamma_{\alpha\beta\beta\alpha}) \tag{4.58}$$

$$\gamma_\perp = \frac{1}{15} \sum_{\alpha\beta} (2\gamma_{\alpha\beta\beta\alpha} - \gamma_{\alpha\alpha\beta\beta}) \tag{4.59}$$

Since we are concerned with a polarizable charge distribution, i.e. a charge distribution that will change when we turn on the electric field or field gradient, we have to be careful when deriving an expression for the energy in the presence of the external field and field gradient. It is not possible to obtain the correct expression for the energy E of the charge distribution by simply inserting the field and field-gradient-dependent moments, $\vec{\mu}(\vec{\mathcal{E}}, \mathcal{E})$ and $\Theta(\vec{\mathcal{E}}, \mathcal{E})$ from Eqs. (4.44) and (4.45), in the multipole expansion of the energy in Eq. (4.18), because the moments are now functions of the fields. Instead, we have to consider an infinitesimal change in energy dE due to an infinitesimal change in the fields and to integrate dE from zero field strength to $\vec{\mathcal{E}}$. We will illustrate this for the case of the dipole moment and its dependence on the electric field. From Eq. (4.19) we can see that

$$dE = -\sum_\alpha \mu_\alpha(\vec{\mathcal{E}}) \, d\mathcal{E}_\alpha = -\vec{\mu}(\vec{\mathcal{E}}) \cdot d\vec{\mathcal{E}} \tag{4.60}$$

The energy can now be obtained by integration on both sides, which gives

$$E(\vec{\mathcal{E}}) - E^{(0)} = -\int_0^{\vec{\mathcal{E}}} \vec{\mu}(\vec{\mathcal{E}}') \cdot d\vec{\mathcal{E}}' \tag{4.61}$$

The integral on the right-hand side of Eq. (4.61) is a line integral over a vector field $\vec{\mu}(\mathcal{E})$ in the space defined by the components of the electric field. Normally, a line integral is defined over coordinate space and looks like this $\int \vec{A}(\vec{r}) \cdot d\vec{r}$. However, if we identify the electric-field vector $\vec{\mathcal{E}}$ as the generalization of a position vector $\vec{r}$ we can see that the integral in Eq. (4.61) is indeed a line integral. A line integral is independent of the path when the vector function $\vec{A}(\vec{r})$ is the gradient of a scalar single-valued field with continuous derivatives. Recalling that the components of the dipole moment are the partial derivatives of the energy with respect to the components of the electric field, we can conclude that the integral in Eq. (4.61) is independent of the integration path and we can carry out the integration in three steps: from $\vec{\mathcal{E}}' = (0,0,0)$ to $(\mathcal{E}_x, 0, 0)$, from $(\mathcal{E}_x, 0, 0)$ to $(\mathcal{E}_x, \mathcal{E}_y, 0)$ and finally to $(\mathcal{E}_x, \mathcal{E}_y, \mathcal{E}_z)$. But before we can carry out the integration we have to insert the expansion of the dipole moment in the presence of an electric field, Eq. (4.44) and obtain

$$E(\vec{\mathcal{E}}) - E^{(0)} \tag{4.62}$$

$$= -\sum_\alpha \int_0^{\vec{\mathcal{E}}} \left(\mu_\alpha + \sum_\beta \alpha_{\alpha\beta} \mathcal{E}'_\beta + \frac{1}{2} \sum_{\beta\gamma} \beta_{\alpha\beta\gamma} \mathcal{E}'_\beta \mathcal{E}'_\gamma + \frac{1}{6} \sum_{\beta\gamma\delta} \gamma_{\alpha\beta\gamma\delta} \mathcal{E}'_\beta \mathcal{E}'_\gamma \mathcal{E}'_\delta + \ldots \right) d\mathcal{E}'_\alpha$$

84 Electric Properties

Integration along the path described above yields the following expression for the energy of a polarizable charge distribution

$$E(\vec{\mathcal{E}}) = E^{(0)} \tag{4.63}$$

$$-\sum_{\alpha}\mu_\alpha\mathcal{E}_\alpha - \frac{1}{2}\sum_{\alpha\beta}\alpha_{\alpha\beta}\mathcal{E}_\alpha\mathcal{E}_\beta - \frac{1}{6}\sum_{\alpha\beta\gamma}\beta_{\alpha\beta\gamma}\mathcal{E}_\alpha\mathcal{E}_\beta\mathcal{E}_\gamma - \frac{1}{24}\sum_{\alpha\beta\gamma\delta}\gamma_{\alpha\beta\gamma\delta}\mathcal{E}_\alpha\mathcal{E}_\beta\mathcal{E}_\gamma\mathcal{E}_\delta$$

Exercise 4.5 Verify the result of the line integral in Eq. (4.62) for the case of an electric field $(\mathcal{E}_x, \mathcal{E}_y, 0)$ and an expansion of the dipole moment in the polarizability and first hyperpolarizability only.

In the same way, the contribution of the quadrupole and mixed dipole–quadrupole polarizabilities to the energy can be obtained. The final expression for the energy of a polarizable charge distribution in the presence of an inhomogeneous electric field then reads (Buckingham, 1967)

$$E(\vec{\mathcal{E}}, \mathcal{E}) = E^{(0)} + q\,\phi^{\mathcal{E}} - \sum_\alpha \mu_\alpha \mathcal{E}_\alpha - \frac{1}{3}\sum_{\alpha\beta}\Theta_{\alpha\beta}\mathcal{E}_{\alpha\beta}$$

$$-\frac{1}{2}\sum_{\alpha\beta}\alpha_{\alpha\beta}\mathcal{E}_\alpha\mathcal{E}_\beta - \frac{1}{6}\sum_{\alpha\beta\gamma\delta}C_{\alpha\beta,\gamma\delta}\mathcal{E}_{\alpha\beta}\mathcal{E}_{\gamma\delta} - \frac{1}{3}\sum_{\alpha\beta\gamma}A_{\alpha,\beta\gamma}\mathcal{E}_\alpha\mathcal{E}_{\beta\gamma}$$

$$-\frac{1}{6}\sum_{\alpha\beta\gamma}\beta_{\alpha\beta\gamma}\mathcal{E}_\alpha\mathcal{E}_\beta\mathcal{E}_\gamma - \frac{1}{6}\sum_{\alpha\beta\gamma\delta}B_{\alpha,\beta,\gamma\delta}\mathcal{E}_\alpha\mathcal{E}_\beta\mathcal{E}_{\gamma\delta}$$

$$-\frac{1}{24}\sum_{\alpha\beta\gamma\delta}\gamma_{\alpha\beta\gamma\delta}\mathcal{E}_\alpha\mathcal{E}_\beta\mathcal{E}_\gamma\mathcal{E}_\delta - \ldots \,. \tag{4.64}$$

where again the origin dependence "$(\vec{R}_O)$" of the permanent moments and of the potential $\phi^{\mathcal{E}}$ was dropped and $\mathcal{E}_\alpha$ and $\mathcal{E}_{\alpha\beta}$ are used as abbreviations for $\mathcal{E}_\alpha(\vec{R}_O)$ and $\mathcal{E}_{\alpha\beta}(\vec{R}_O)$, respectively.

The importance of Eq. (4.64) lies in the fact that one can calculate from it the change in energy of a charge distribution due to an external electric field or field gradient of arbitrary strength. One only needs to know the various polarizabilities and hyperpolarizabilities. In the same way as the charge distribution in a molecule is influenced by an external electric field it is also modified by the electric field due to other molecules in the neighbourhood. Permanent electric moments of the surrounding molecules induce additional electric moments in molecules leading to another contribution to the intermolecular interaction energy. This so-called induction energy is determined by the polarizabilities of the molecules and detailed knowledge of the polarizabilities is therefore also important for the description of intermolecular forces (Buckingham, 1967).

Furthermore, Eq. (4.64) also allows us to define the various polarizabilities and hyperpolarizabilities as derivatives of the energy. The first derivatives of the energy with respect to the field and field gradient are the electric dipole moment and electric quadrupole moment, as shown in Eqs. (4.19) and (4.21). Evaluating them at zero electric field and field gradient yields the permanent moments. The second derivatives of the energy give the polarizabilities, the third derivatives give the first hyperpolarizabilities and so on[10]

$$\alpha_{\alpha\beta} = -\left.\frac{\partial^2 E(\vec{\mathcal{E}})}{\partial \mathcal{E}_\beta \partial \mathcal{E}_\alpha}\right|_{|\vec{\mathcal{E}}|=0} \tag{4.65}$$

$$A_{\alpha,\beta\gamma} = -3\left.\frac{\partial^2 E(\vec{\mathcal{E}}, \boldsymbol{\mathcal{E}})}{\partial \mathcal{E}_{\beta\gamma} \partial \mathcal{E}_\alpha}\right|_{|\vec{\mathcal{E}}|=|\boldsymbol{\mathcal{E}}|=0} \tag{4.66}$$

$$C_{\alpha\beta,\gamma\delta} = -3\left.\frac{\partial^2 E(\boldsymbol{\mathcal{E}})}{\partial \mathcal{E}_{\gamma\delta} \partial \mathcal{E}_{\alpha\beta}}\right|_{|\boldsymbol{\mathcal{E}}|=0} \tag{4.67}$$

$$\beta_{\alpha\beta\gamma} = -\left.\frac{\partial^3 E(\vec{\mathcal{E}})}{\partial \mathcal{E}_\gamma \partial \mathcal{E}_\beta \partial \mathcal{E}_\alpha}\right|_{|\vec{\mathcal{E}}|=0} \tag{4.68}$$

$$B_{\alpha,\beta,\gamma\delta} = -3\left.\frac{\partial^3 E(\vec{\mathcal{E}}, \boldsymbol{\mathcal{E}})}{\partial \mathcal{E}_{\gamma\delta} \partial \mathcal{E}_\beta \partial \mathcal{E}_\alpha}\right|_{|\vec{\mathcal{E}}|=|\boldsymbol{\mathcal{E}}|=0} \tag{4.69}$$

$$\gamma_{\alpha\beta\gamma\delta} = -\left.\frac{\partial^4 E(\vec{\mathcal{E}})}{\partial \mathcal{E}_\delta \partial \mathcal{E}_\gamma \partial \mathcal{E}_\beta \partial \mathcal{E}_\alpha}\right|_{|\vec{\mathcal{E}}|=0} \tag{4.70}$$

4.5 Quantum Mechanical Expressions for Polarizabilities

In the previous section we have defined the tensor components $\alpha_{\alpha\beta}$, $A_{\alpha,\beta\gamma}$ and $C_{\alpha\beta,\gamma\delta}$ of the electric dipole, dipole–quadrupole and quadrupole-quadrupole polarizability tensors as derivatives of the energy $E(\vec{\mathcal{E}}, \boldsymbol{\mathcal{E}})$ in the presence of a field and field gradient, Eqs. (4.65) to (4.67), or alternatively as derivatives of the perturbation dependent electric dipole $\vec{\mu}(\vec{\mathcal{E}}, \boldsymbol{\mathcal{E}})$ and quadrupole moment $\Theta(\vec{\mathcal{E}}, \boldsymbol{\mathcal{E}})$, Eqs. (4.46) to (4.48), see also Table B.1. Furthermore, we have seen in Sections 3.3 and 4.3 that the electronic contributions to the electric dipole and quadrupole moments can be expressed as expectation values of the electric dipole and quadrupole moment operators, $\hat{\vec{\mu}}(\vec{R}_O)$ and $\hat{\Theta}(\vec{R}_O)$ for the electrons, respectively. Both definitions can be used to derive quantum mechanical expressions for the polarizabilities.

Let us start with the first definition as derivatives of the energy, Eqs. (4.65) to (4.67). Again we will use the perturbation theory expression for the perturbed energy, Eq. (3.15), but differentiate it now twice with respect to the appropriate components of the field or field gradient. This leads us immediately to the second-order correction to the energy, because the first-order correction depends only linearly on the fields. We can therefore express the polarizabilities as

---

[10] These definitions are also collected in the last column of Table B.1 of Appendix B.

86 Electric Properties

$$\alpha_{\alpha\beta} = -\left.\frac{\partial^2 E_0^{(2)}(\vec{\mathcal{E}})}{\partial \mathcal{E}_\alpha \partial \mathcal{E}_\beta}\right|_{|\vec{\mathcal{E}}|=0}$$

$$= -\frac{\partial^2}{\partial \mathcal{E}_\alpha \partial \mathcal{E}_\beta} \sum_{n\neq 0} \left.\frac{\langle \Psi_0^{(0)} | \hat{H}^{(1)} | \Psi_n^{(0)}\rangle\langle \Psi_n^{(0)} | \hat{H}^{(1)} | \Psi_0^{(0)}\rangle}{E_0^{(0)} - E_n^{(0)}}\right|_{|\vec{\mathcal{E}}|=0} \quad (4.71)$$

$$A_{\alpha,\beta\gamma} = -3 \left.\frac{\partial^2 E_0^{(2)}(\vec{\mathcal{E}}, \boldsymbol{\mathcal{E}})}{\partial \mathcal{E}_\alpha \partial \mathcal{E}_{\beta\gamma}}\right|_{|\vec{\mathcal{E}}|=|\boldsymbol{\mathcal{E}}|=0}$$

$$= -3 \frac{\partial^2}{\partial \mathcal{E}_\alpha \partial \mathcal{E}_{\beta\gamma}} \sum_{n\neq 0} \left.\frac{\langle \Psi_0^{(0)} | \hat{H}^{(1)} | \Psi_n^{(0)}\rangle\langle \Psi_n^{(0)} | \hat{H}^{(1)} | \Psi_0^{(0)}\rangle}{E_0^{(0)} - E_n^{(0)}}\right|_{|\vec{\mathcal{E}}|=|\boldsymbol{\mathcal{E}}|=0} \quad (4.72)$$

$$C_{\alpha\beta,\gamma\delta} = -3 \left.\frac{\partial^2 E_0^{(2)}(\boldsymbol{\mathcal{E}})}{\partial \mathcal{E}_{\alpha\beta} \partial \mathcal{E}_{\gamma\delta}}\right|_{|\boldsymbol{\mathcal{E}}|=0}$$

$$= -3 \frac{\partial^2}{\partial \mathcal{E}_{\alpha\beta} \partial \mathcal{E}_{\gamma\delta}} \sum_{n\neq 0} \left.\frac{\langle \Psi_0^{(0)} | \hat{H}^{(1)} | \Psi_n^{(0)}\rangle\langle \Psi_n^{(0)} | \hat{H}^{(1)} | \Psi_0^{(0)}\rangle}{E_0^{(0)} - E_n^{(0)}}\right|_{|\boldsymbol{\mathcal{E}}|=0} \quad (4.73)$$

However, in Section 4.3 we have seen that the derivatives of the first-order perturbation Hamiltonian, $\hat{H}^{(1)}$, with respect to a component of the electric field $\mathcal{E}_\alpha$ and field gradient $\mathcal{E}_{\alpha\beta}$ are the cartesian components of the electric dipole and quadrupole moment operators, $\hat{\mu}_\alpha(\vec{R}_O) + \hat{\Omega}_\alpha^{\mathcal{E}}(\vec{R}_O)$ and $\hat{\Theta}_{\alpha\beta}(\vec{R}_O) + \hat{\Omega}_{\alpha\beta}^{\nabla\mathcal{E}}(\vec{R}_O)$, Eqs. (4.38) and (4.39). Using them we arrive at the final **sum-over-states expression** for the components of the electric dipole polarizability tensor

$$\alpha_{\alpha\beta} = -\sum_{n\neq 0} \left\{ \frac{\langle \Psi_0^{(0)} | \hat{\mu}_\alpha | \Psi_n^{(0)}\rangle\langle \Psi_n^{(0)} | \hat{\mu}_\beta | \Psi_0^{(0)}\rangle}{E_0^{(0)} - E_n^{(0)}} + \frac{\langle \Psi_0^{(0)} | \hat{\mu}_\beta | \Psi_n^{(0)}\rangle\langle \Psi_n^{(0)} | \hat{\mu}_\alpha | \Psi_0^{(0)}\rangle}{E_0^{(0)} - E_n^{(0)}} \right\} \quad (4.74)$$

the electric dipole–quadrupole polarizability tensor

$$A_{\alpha,\beta\gamma} = -\sum_{n\neq 0} \frac{\langle \Psi_0^{(0)} | \hat{\mu}_\alpha | \Psi_n^{(0)}\rangle\langle \Psi_n^{(0)} | \hat{\Theta}_{\beta\gamma} | \Psi_0^{(0)}\rangle}{E_0^{(0)} - E_n^{(0)}}$$

$$- \sum_{n\neq 0} \frac{\langle \Psi_0^{(0)} | \hat{\Theta}_{\beta\gamma} | \Psi_n^{(0)}\rangle\langle \Psi_n^{(0)} | \hat{\mu}_\alpha | \Psi_0^{(0)}\rangle}{E_0^{(0)} - E_n^{(0)}} \quad (4.75)$$

and the electric quadrupole–quadrupole polarizability tensor

$$C_{\alpha\beta,\gamma\delta} = -\frac{1}{3} \sum_{n\neq 0} \frac{\langle \Psi_0^{(0)} | \hat{\Theta}_{\alpha\beta} | \Psi_n^{(0)}\rangle\langle \Psi_n^{(0)} | \hat{\Theta}_{\gamma\delta} | \Psi_0^{(0)}\rangle}{E_0^{(0)} - E_n^{(0)}}$$

$$-\frac{1}{3}\sum_{n\neq 0}\frac{\langle\Psi_0^{(0)}|\hat{\Theta}_{\gamma\delta}|\Psi_n^{(0)}\rangle\langle\Psi_n^{(0)}|\hat{\Theta}_{\alpha\beta}|\Psi_0^{(0)}\rangle}{E_0^{(0)}-E_n^{(0)}} \tag{4.76}$$

One should note that there is no contribution from the nuclear operators $\hat{\Omega}_\alpha^{\mathcal{E}}(\vec{R}_O)$ and $\hat{\Omega}_{\alpha\beta}^{\nabla\mathcal{E}}(\vec{R}_O)$ and that the polarizabilities are independent of the origin $\vec{R}_O$. The explanation for that is that neither $\hat{\Omega}_\alpha^{\mathcal{E}}(\vec{R}_O)$ nor $\vec{R}_O$ act on the electronic wavefunctions and that the unperturbed states are orthogonal $\langle\Psi_0^{(0)}|\Psi_n^{(0)}\rangle = 0$. The transition moments over $\hat{\Omega}_\alpha^{\mathcal{E}}(\vec{R}_O)$ and $\hat{\Omega}_{\alpha\beta}^{\nabla\mathcal{E}}(\vec{R}_O)$ reduce therefore to

$$\langle\Psi_0^{(0)}|\hat{\Omega}_\alpha^{\mathcal{E}}(\vec{R}_O)|\Psi_n^{(0)}\rangle = \hat{\Omega}_\alpha^{\mathcal{E}}(\vec{R}_O)\langle\Psi_0^{(0)}|\Psi_n^{(0)}\rangle = 0 \tag{4.77}$$

and the origin dependence of the transition moment over the electronic dipole moment operator vanishes

$$\langle\Psi_0^{(0)}|\hat{\mu}_\alpha(\vec{R}_O)|\Psi_n^{(0)}\rangle = \langle\Psi_0^{(0)}|-e\sum_i^N r_{i,\alpha}|\Psi_n^{(0)}\rangle + \langle\Psi_0^{(0)}|eR_{O,\alpha}|\Psi_n^{(0)}\rangle$$

$$= \langle\Psi_0^{(0)}|-e\sum_i^N r_{i,\alpha}|\Psi_n^{(0)}\rangle + eR_{O,\alpha}\langle\Psi_0^{(0)}|\Psi_n^{(0)}\rangle$$

$$= \langle\Psi_0^{(0)}|-e\sum_i^N r_{i,\alpha}|\Psi_n^{(0)}\rangle \tag{4.78}$$

Since the polarizabilities can be derived as second derivatives of the energy, they are often also called **second-order properties**.

Calculating polarizabilities from the sum-over-states expressions requires knowledge of all excited states $\Psi_n^{(0)}$ and their energies $E_n^{(0)}$. Equations (4.74) to (4.76) are therefore mainly used in the interpretation of calculated or measured polarizabilities. The explicit knowledge of exited states can be avoided by employing the second approach, where we want to derive the polarizabilities as first derivatives of the corresponding field-dependent moments, Eqs. (4.46) to (4.48). In Section 4.3 it was shown that the permanent moments can be written as expectation values of the appropriate electric moment operators with unperturbed wavefunctions. Perturbation-dependent moments can be obtained as expectation values of the electric dipole moment operator with the perturbed wavefunction, Eq. (3.35), as was shown in section 3.3. We can therefore apply the general expression for the first derivative of a perturbation-dependent expectation value, Eq. (3.40), to the polarizabilities and obtain expressions involving first derivatives of the perturbed wavefunction for the tensor components of the dipole polarizability

$$\alpha_{\alpha\beta} = \left.\frac{\partial\mu_\alpha(\vec{\mathcal{E}})}{\partial\mathcal{E}_\beta}\right|_{|\vec{\mathcal{E}}|=0} = \left.\frac{\partial}{\partial\mathcal{E}_\beta}\langle\Psi_0(\vec{\mathcal{E}})|\hat{\mu}_\alpha|\Psi_0(\vec{\mathcal{E}})\rangle\right|_{|\vec{\mathcal{E}}|=0}$$

$$= \langle \Psi_0^{(0)} | \hat{\mu}_\alpha | \frac{\partial \Psi_0(\vec{\mathcal{E}})}{\partial \mathcal{E}_\beta} \rangle \bigg|_{|\vec{\mathcal{E}}|=0} + \langle \frac{\partial \Psi_0(\vec{\mathcal{E}})}{\partial \mathcal{E}_\beta} \bigg|_{|\vec{\mathcal{E}}|=0} | \hat{\mu}_\alpha | \Psi_0^{(0)} \rangle \qquad (4.79)$$

of the quadrupole polarizability

$$C_{\alpha\beta,\gamma\delta} = \frac{\partial \Theta_{\alpha\beta}(\mathcal{E})}{\partial \mathcal{E}_{\gamma\delta}}\bigg|_{|\mathcal{E}|=0} = \frac{\partial}{\partial \mathcal{E}_{\gamma\delta}} \langle \Psi_0(\mathcal{E}) | \hat{\Theta}_{\alpha\beta} | \Psi_0(\mathcal{E}) \rangle \bigg|_{|\mathcal{E}|=0}$$

$$= \langle \Psi_0^{(0)} | \hat{\Theta}_{\alpha\beta} | \frac{\partial \Psi_0(\mathcal{E})}{\partial \mathcal{E}_{\gamma\delta}} \rangle \bigg|_{|\mathcal{E}|=0} + \langle \frac{\partial \Psi_0(\mathcal{E})}{\partial \mathcal{E}_{\gamma\delta}} \bigg|_{|\mathcal{E}|=0} | \hat{\Theta}_{\alpha\beta} | \Psi_0^{(0)} \rangle \qquad (4.80)$$

and of the dipole–quadrupole polarizability

$$A_{\alpha,\beta\gamma} = \frac{\partial \Theta_{\beta\gamma}(\vec{\mathcal{E}})}{\partial \mathcal{E}_\alpha}\bigg|_{|\vec{\mathcal{E}}|=|\mathcal{E}|=0} = \frac{\partial}{\partial \mathcal{E}_\alpha} \langle \Psi_0(\vec{\mathcal{E}}) | \hat{\Theta}_{\beta\gamma} | \Psi_0(\vec{\mathcal{E}}) \rangle \bigg|_{|\vec{\mathcal{E}}|=|\mathcal{E}|=0}$$

$$= \langle \Psi_0^{(0)} | \hat{\Theta}_{\beta\gamma} | \frac{\partial \Psi_0(\vec{\mathcal{E}})}{\partial \mathcal{E}_\alpha} \rangle \bigg|_{|\vec{\mathcal{E}}|=|\mathcal{E}|=0} + \langle \frac{\partial \Psi_0(\vec{\mathcal{E}})}{\partial \mathcal{E}_\alpha} \bigg|_{|\vec{\mathcal{E}}|=|\mathcal{E}|=0} | \hat{\Theta}_{\beta\gamma} | \Psi_0^{(0)} \rangle \qquad (4.81)$$

Alternatively, the dipole–quadrupole polarizability can be obtained as derivative of the electric-field-gradient-dependent dipole moment

$$A_{\alpha,\beta\gamma} = 3 \frac{\partial \mu_\alpha(\mathcal{E})}{\partial \mathcal{E}_{\beta\gamma}}\bigg|_{|\vec{\mathcal{E}}|=|\mathcal{E}|=0} = 3 \frac{\partial}{\partial \mathcal{E}_{\beta\gamma}} \langle \Psi_0(\mathcal{E}) | \hat{\mu}_\alpha | \Psi_0(\mathcal{E}) \rangle \bigg|_{|\vec{\mathcal{E}}|=|\mathcal{E}|=0}$$

$$= 3 \langle \Psi_0^{(0)} | \hat{\mu}_\alpha | \frac{\partial \Psi_0(\mathcal{E})}{\partial \mathcal{E}_{\beta\gamma}} \rangle \bigg|_{|\vec{\mathcal{E}}|=|\mathcal{E}|=0} + 3 \langle \frac{\partial \Psi_0(\mathcal{E})}{\partial \mathcal{E}_{\beta\gamma}} \bigg|_{|\vec{\mathcal{E}}|=|\mathcal{E}|=0} | \hat{\mu}_\alpha | \Psi_0^{(0)} \rangle \qquad (4.82)$$

In Part III we will come back to these expressions and evaluate the derivatives of approximate wavefunctions. However, here we will use the response formalism as developed in Section 3.11. Using Eq. (3.116) we can express the derivatives of the perturbation dependent expectation value in terms of polarization propagators or linear response functions and thus obtain for the tensor components of the polarizabilities

$$\alpha_{\alpha\beta} = \frac{\partial}{\partial \mathcal{E}_\beta} \langle \Psi_0(\vec{\mathcal{E}}) | \hat{\mu}_\alpha | \Psi_0(\vec{\mathcal{E}}) \rangle \bigg|_{|\vec{\mathcal{E}}|=0} = -\langle\langle \hat{\mu}_\alpha ; \hat{\mu}_\beta \rangle\rangle_{\omega=0} \qquad (4.83)$$

$$A_{\alpha,\beta\gamma} = \frac{\partial}{\partial \mathcal{E}_\alpha} \langle \Psi_0(\vec{\mathcal{E}}) | \hat{\Theta}_{\beta\gamma} | \Psi_0(\vec{\mathcal{E}}) \rangle \bigg|_{|\vec{\mathcal{E}}|=|\mathcal{E}|=0} = 3 \frac{\partial}{\partial \mathcal{E}_{\beta\gamma}} \langle \Psi_0(\mathcal{E}) | \hat{\mu}_\alpha | \Psi_0(\mathcal{E}) \rangle \bigg|_{|\vec{\mathcal{E}}|=|\mathcal{E}|=0}$$

$$= -\langle\langle \hat{\mu}_\alpha ; \hat{\Theta}_{\beta\gamma} \rangle\rangle_{\omega=0} \qquad (4.84)$$

$$C_{\alpha\beta,\gamma\delta} = \frac{\partial}{\partial \mathcal{E}_{\gamma\delta}} \langle \Psi_0(\mathcal{E}) | \hat{\Theta}_{\alpha\beta} | \Psi_0(\mathcal{E}) \rangle \bigg|_{|\mathcal{E}|=0} = -\frac{1}{3} \langle\langle \hat{\Theta}_{\alpha\beta} ; \hat{\Theta}_{\gamma\delta} \rangle\rangle_{\omega=0} \qquad (4.85)$$

Inserting the perturbation theory expansion of the perturbed wavefunction up to first order, Eq. (3.27), in Eqs. (4.79) to (4.82) or the expression for the static response function Eq. (3.114) in equations (4.83) to (4.85) leads us back to the sum-over-states expressions given in Eqs. (4.74)–(4.76), of course. The significance of the

4.6 Molecular Electric Fields and Field Gradients

In addition to the electric moments and polarizabilities, which have been considered up to now, the electric fields arising from a distribution of charges are also important for describing various molecular properties.

We have discussed several times already that knowledge of the electrostatic potential $\phi^\rho(\vec{R})$ due to a distribution of charges, as given in Eq. (4.1), is important for the study of intermolecular interactions. The first and second derivatives of this molecular electrostatic potential are, according to Eqs. (4.13) and (4.14), the molecular electric field $\vec{\mathcal{E}}^\rho(\vec{R})$ and the molecular electric field gradient tensor $\mathcal{E}^\rho(\vec{R})$ at a point $\vec{R}$ due to the molecular charge distribution $\rho(\vec{r})$. [see Exercise 4.6] Their cartesian components of the field and field gradient are thus given as

$$\mathcal{E}^\rho_\alpha(\vec{R}) = \frac{1}{4\pi\epsilon_0} \int_{\vec{r}} \rho(\vec{r}) \frac{R_\alpha - r_\alpha}{|\vec{R} - \vec{r}|^3}\, d\vec{r} \tag{4.86}$$

$$\mathcal{E}^\rho_{\alpha\beta}(\vec{R}) = \frac{1}{4\pi\epsilon_0} \int_{\vec{r}} \rho(\vec{r}) \left[\frac{\delta_{\alpha\beta}}{|\vec{R} - \vec{r}|^3} - 3\frac{(R_\alpha - r_\alpha)(R_\beta - r_\beta)}{|\vec{R} - \vec{r}|^5} \right] d\vec{r} \tag{4.87}$$

Exercise 4.6 Derive the expression (4.87) for the molecular electric-field gradient.

The molecular electric field gives rise to a force $\vec{F}$ acting on the charges in the charge distribution, where the contribution to the electric field from the charge in question has to be excluded, of course. For a charge distribution in equilibrium this force should obviously be zero. For example, the force acting on a nucleus K in a molecule would then be

$$\vec{F}_K = Z_K e\, \vec{\mathcal{E}}^\rho(\vec{R}_K) \tag{4.88}$$

and this force will be zero in the Born-Oppenheimer approximation, if the molecule is in its equilibrium geometry. In actual calculations using one of the approximate methods that are discussed in Part III, however, this will only be the case, if the equilibrium geometry of the molecule was determined with the same method as used in the calculation of this force.

Although the fields and field gradients are well defined for any point in space, it is not possible to measure them at an arbitrary point within the charge distribution. Fields can be probed by dipole moments and field gradients by quadrupole moments (see e.g. Eq. (4.18)). In order to measure the field at an arbitrary point one would have to bring a dipole moment there, which is of course not possible within a molecule. Only via the interaction with the nuclei in a molecule is it therefore possible to get information about some of these field quantities and only at the positions of the nuclei.

But nuclei do not have electric dipole moments and the molecular electric field can thus not be investigated in this way.

However, nuclei with a spin quantum number $I^K \geq 1$ possess an electric quadrupole moment Θ_K and one can study the molecular electric field gradient at the positions of the nuclei, $\mathcal{E}^\rho_{\alpha\beta}(\vec{R}_K)$, via the interaction with the nuclear electric quadrupole moment defined via an effective spin Hamiltonian

$$\hat{H}^{spin} = \frac{1}{2\hbar} \sum_K \frac{\Theta_K}{I_K(2I_K - 1)} \sum_{\alpha\beta} \hat{I}^K_\alpha \mathcal{E}^\rho_{\alpha\beta}(\vec{R}_K) \hat{I}^K_\beta \tag{4.89}$$

The product of the nuclear quadrupole moment with the electric-field-gradient tensor divided by $\hbar$, i.e. $\Theta_K \mathcal{E}^\rho(\vec{R}_K)/\hbar$, is called the **nuclear quadrupole coupling tensor**, while $\Theta_K \mathcal{E}^\rho_{zz}(\vec{R}_K)/\hbar$ is called the **nuclear quadrupole coupling constant**. The latter can be obtained from the hyperfine structure of rotational spectra, the quadrupole splitting of the lines in a Mössbauer spectrum or the linewidth of the lines in the NMR spectrum of a molecule containing the quadrupole nucleus K. As a traceless second-rank tensor the electric-field-gradient tensor $\mathcal{E}^\rho(\vec{R}_K)$ can be diagonalized. The three eigenvalues of the electric-field-gradient tensor and thus the components of $\mathcal{E}^\rho(\vec{R}_K)$ in its own principal axis coordinate system are called $\mathcal{E}^\rho_{aa}(\vec{R}_K)$, $\mathcal{E}^\rho_{bb}(\vec{R}_K)$ and $\mathcal{E}^\rho_{cc}(\vec{R}_K)$, where

$$|\mathcal{E}^\rho_{cc}(\vec{R}_K)| \geq |\mathcal{E}^\rho_{bb}(\vec{R}_K)| \geq |\mathcal{E}^\rho_{aa}(\vec{R}_K)| \tag{4.90}$$

From these eigenvalues one defines also an asymmetry parameter η_K as

$$\eta_K = \frac{\mathcal{E}^\rho_{aa}(\vec{R}_K) - \mathcal{E}^\rho_{bb}(\vec{R}_K)}{\mathcal{E}^\rho_{cc}(\vec{R}_K)} \tag{4.91}$$

which has values between zero and one. Finally, as a traceless tensor the eigenvalues are not independent of each other but fulfill the condition that

$$\mathcal{E}^\rho_{aa}(\vec{R}_K) + \mathcal{E}^\rho_{bb}(\vec{R}_K) + \mathcal{E}^\rho_{cc}(\vec{R}_K) = 0 \tag{4.92}$$

The classical expression for the electric field and field gradient due to a charge density $\rho(\vec{r})$, Eqs. (4.86) and (4.87), are analogous to the definition of the electric moments as integrals over the charge density in Eqs. (4.5) and (4.6). The quantum mechanical expressions for the tensor components of the electric field at an arbitrary observation point $\vec{R}$, $\mathcal{E}^\rho_\alpha(\vec{R})$, and of the field gradient at the position of a nucleus K, $\mathcal{E}^\rho_{\alpha\beta}(\vec{R}_K)$, can therefore be obtained in the same way as for the electric moments in Section 4.3. We only have to replace the integration over the classical charge density $\rho(\vec{r})$ with an expectation value for the electrons and with a summation over all nuclei in the case of the field and with a summation over all other nuclei, $L \neq K$, in the case of the field gradient

$$\mathcal{E}^\rho_\alpha(\vec{R}) = \frac{e}{4\pi\epsilon_0} \langle \Psi_0^{(0)} | \sum_i^N \frac{r_{i,\alpha} - R_\alpha}{|\vec{r}_i - \vec{R}|^3} | \Psi_0^{(0)} \rangle - \sum_K \frac{Z_K e}{4\pi\epsilon_0} \frac{R_{K,\alpha} - R_\alpha}{|\vec{R}_K - \vec{R}|^3} \tag{4.93}$$

$$\mathcal{E}_{\alpha\beta}^{\rho}(\vec{R}_K) = \frac{e}{4\pi\epsilon_0} \langle \Psi_0^{(0)} | \sum_i^N \left[3\frac{(r_{i,\alpha} - R_{K,\alpha})(r_{i,\beta} - R_{K,\beta})}{|\vec{r}_i - \vec{R}_K|^5} - \frac{\delta_{\alpha\beta}}{|\vec{r}_i - \vec{R}_K|^3} \right] | \Psi_0^{(0)} \rangle$$

$$- \frac{1}{4\pi\epsilon_0} \sum_{L \neq K} Z_L e \left[3\frac{(R_{L,\alpha} - R_{K,\alpha})(R_{L,\beta} - R_{K,\beta})}{|\vec{R}_L - \vec{R}_K|^5} - \frac{\delta_{\alpha\beta}}{|\vec{R}_L - \vec{R}_K|^3} \right] \quad (4.94)$$

Defining **electric-field operators**, $\hat{O}^\mu$ and $\hat{\Omega}^\mu$,

$$\hat{O}_\alpha^\mu(\vec{R}) = \sum_i^N \hat{o}_{i,\alpha}^\mu(\vec{R}) = \frac{e}{4\pi\epsilon_0} \sum_i^N \frac{r_{i,\alpha} - R_\alpha}{|\vec{r}_i - \vec{R}|^3} \quad (4.95)$$

$$\hat{\Omega}_\alpha^\mu(\vec{R}) = -\sum_K \frac{Z_K e}{4\pi\epsilon_0} \frac{R_{K,\alpha} - R_\alpha}{|\vec{R}_K - \vec{R}|^3} \quad (4.96)$$

and **electric-field-gradient operators**, $\hat{O}^\Theta$ and $\hat{\Omega}^\Theta$,

$$\hat{O}_{\alpha\beta}^\Theta(\vec{R}) = \sum_i^N \hat{o}_{i,\alpha\beta}^\Theta(\vec{R})$$

$$= \frac{e}{4\pi\epsilon_0} \sum_i^N \left[3\frac{(r_{i,\alpha} - R_{K,\alpha})(r_{i,\beta} - R_{K,\beta})}{|\vec{r}_i - \vec{R}_K|^5} - \frac{\delta_{\alpha\beta}}{|\vec{r}_i - \vec{R}_K|^3} \right] \quad (4.97)$$

$$\hat{\Omega}_{\alpha\beta}^\Theta(\vec{R}) = -\frac{1}{4\pi\epsilon_0} \sum_{L \neq K} Z_L e \left[3\frac{(R_{L,\alpha} - R_{K,\alpha})(R_{L,\beta} - R_{K,\beta})}{|\vec{R}_L - \vec{R}_K|^5} - \frac{\delta_{\alpha\beta}}{|\vec{R}_L - \vec{R}_K|^3} \right] \quad (4.98)$$

we can write the quantum mechanical expression for the electric-field and electric-field gradient at an arbitrary point $\vec{R}$ more simply as

$$\mathcal{E}_\alpha^\rho(\vec{R}) = \langle \Psi_0^{(0)} | \hat{O}_\alpha^\mu(\vec{R}) | \Psi_0^{(0)} \rangle + \hat{\Omega}_\alpha^\mu(\vec{R}) \quad (4.99)$$

$$\mathcal{E}_{\alpha\beta}^\rho(\vec{R}_K) = \langle \Psi_0^{(0)} | \hat{O}_{\alpha\beta}^\Theta(\vec{R}) | \Psi_0^{(0)} \rangle + \hat{\Omega}_{\alpha\beta}^\Theta(\vec{R}) \quad (4.100)$$

4.7 Further Reading

Electric Multipole Expansion and Interaction with Electric Fields

A. Hinchliffe and R. W. Munn, *Molecular Electromagnetism*, John Wiley and Sons Ltd, Chichester (1985): Chapters 1.6, 8.1 and 8.2.

A. Hinchliffe, *Ab Initio Determination of Molecular Properties*, Adam Hilger, Bristol (1987): Chapter 5.4.

D. W. Davies, *The Theory of the Electric and Magnetic Properties of Molecules*, John Wiley and Sons, London (1967): Appendix II.

V. D. Barger and M. G. Olsson, *Classical Electricity and Magnetism*, Allyn and Bacon, Boston (1987): Chapters 1–7, 2–11 and 4–1.

Polarizability

A. Hinchliffe and R. W. Munn, *Molecular Electromagnetism*, John Wiley and Sons Ltd, Chichester (1985): Chapters 8.4 and 8.5.

P. Atkins and R. Friedman, *Molecular Quantum Mechanics*, 4th edn. Oxford University Press, Oxford (2005): Chapters 12.1 and 12.2.

5
Magnetic Properties

In the previous chapter we have defined several electric properties as derivatives of the energy of a charge distribution $\rho(\vec{r})$ in the presence of an electric field or field gradient. Some of these properties were alternatively also defined as derivatives of the electric moments. Furthermore, quantum mechanical expressions were derived for all these properties using perturbation theory or static response theory as outlined in Chapter 3.

In the present chapter we will now define analogous magnetic properties and derive quantum mechanical expression for them in the same ways as in the electric case. However, there are some important differences. First, there will be more types of properties to be studied, because in addition to an external magnetic field we are also interested in the interaction with nuclear magnetic dipole moments. Secondly, magnetic properties exhibit a greater complexity than electric properties. This shows up already in the fact that the current density $\vec{j}(\vec{r})$, which takes over the role of the charge density $\rho(\vec{r})$ in this chapter, is a vector field and not a scalar field as $\rho(\vec{r})$. Furthermore the potential that will represent the fields or magnetic moments is the vector potential, $\vec{A}(\vec{r})$, i.e. again a vector field instead of the scalar potential of the electric case. Finally, the problem of gauge transformations as discussed in Section 2.9, is important for magnetic properties.

5.1 Magnetic Multipole Expansion

A dynamic system of charges with charge density $\rho(\vec{r})$ gives rise to a **current density** $\vec{j}(\vec{r})$

$$\vec{j}(\vec{r}) = \rho(\vec{r})\,\vec{v}(\vec{r}) \tag{5.1}$$

where $\vec{v}(\vec{r})$ is the velocity distribution. The **vector potential** $\vec{A}^j(\vec{R})$ due to this current density is given as

$$\vec{A}^j(\vec{R}) = \frac{\mu_0}{4\pi} \int_{\vec{r}} \frac{\vec{j}(\vec{r})}{|\vec{R}-\vec{r}|}\, d\vec{r} \tag{5.2}$$

94 Magnetic Properties

This expression is completely analogous to the scalar potential of a charge distribution $\rho(\vec{r})$, Eq. (4.1).[1] Using again the Taylor expansion of $\frac{1}{|\vec{R}-\vec{r}|}$ around an origin[2] $\vec{R}_{GO}$ within the charge distribution, Eq. (4.2), we can write a cartesian component of the vector potential as

$$A_\alpha^j(\vec{R}) = \frac{\mu_0}{4\pi} \frac{1}{|\vec{R} - \vec{R}_{GO}|} \int_{\vec{r}} j_\alpha(\vec{r}) \, d\vec{r}$$

$$+ \frac{\mu_0}{4\pi} \sum_\beta \frac{R_\beta - R_{GO,\beta}}{|\vec{R} - \vec{R}_{GO}|^3} \int_{\vec{r}} j_\alpha(\vec{r}) \, (r_\beta - R_{GO,\beta}) \, d\vec{r} + \ldots \qquad (5.3)$$

We can simplify this expression, when the charge density $\rho(\vec{r})$ and current density $\vec{j}(\vec{r})$ are independent of time, i.e. when we are dealing with a steady current that implies that

$$\vec{\nabla} \cdot \vec{j} = 0 \qquad (5.4)$$

For such a steady current distribution one can derive from the divergence theorem of vector calculus an expression [see Exercise 5.1]

$$\int_{\vec{r}} \left[\vec{\nabla} f(\vec{r}) \right] \cdot \vec{j}(\vec{r}) \, d\vec{r} = 0 \qquad (5.5)$$

which is valid for an arbitrary scalar function $f(\vec{r})$. With this expression we can show now that the monopole term in the Taylor expansion of the vector potential, Eq. (5.3) vanishes, because choosing $f = r_\alpha$ in Eq. (5.5) gives precisely the first term in Eq. (5.3) [see Exercise 5.2].

Exercise 5.1 Derive Eq. (5.5) from the divergence theorem for a bounding surface S' that completely encloses the current distribution

$$\int_{\vec{r}} \vec{\nabla} \cdot \left[f(\vec{r}) \vec{j}(\vec{r}) \right] d\vec{r} = \oint f(\vec{r}) \vec{j}(\vec{r}) \cdot d\vec{S}' = 0$$

Next, we will not consider terms higher than the second, i.e. the dipole term. This is in accordance with the **electric quadrupole approximation**, which is based on the fact that the effects of electric quadrupole and magnetic dipole terms are of the same order of magnitude and smaller than the electric dipole terms. They should therefore be treated together, as we do here, while second-order magnetic moments should be treated together with third electric moments, which we have not discussed either. The remaining (dipole) term

---

[1] This reflects the fundamental symmetry between space and time coordinates in special relativity and thus also in electromagnetism. The three space coordinates and time are collected in a so-called four-vector in special relativity. In the same way is the charge density $\rho(\vec{r})$ the fourth, i.e. time, component of the current-charge-density four-vector and the scalar potential $\phi(\vec{r}, t)$ is the fourth, i.e. time, component of the vector-scalar-potential four-vector.

[2] Contrary to the electric multipole expansion we want to denote the arbitrary origin with $\vec{R}_{GO}$ and call it the gauge origin as defined in Section 2.9.

Magnetic Multipole Expansion

$$A^j_\alpha(\vec{R}) = \frac{\mu_0}{4\pi} \sum_\beta \frac{R_\beta - R_{GO,\beta}}{|\vec{R} - \vec{R}_{GO}|^3} \int_{\vec{r}} (r_\beta - R_{GO,\beta})\, j_\alpha(\vec{r})\, d\vec{r} \tag{5.6}$$

can also be simplified if we write it as the sum of its symmetric and antisymmetric part

$$A^j_\alpha(\vec{R}) = \frac{\mu_0}{4\pi} \sum_\beta \frac{R_\beta - R_{GO,\beta}}{|\vec{R} - \vec{R}_{GO}|^3} \frac{1}{2} \left\{ \int_{\vec{r}} [(r_\beta - R_{GO,\beta})\, j_\alpha(\vec{r}) + (r_\alpha - R_{GO,\alpha})\, j_\beta(\vec{r})]\, d\vec{r} \right.$$

$$\left. + \int_{\vec{r}} [(r_\beta - R_{GO,\beta})\, j_\alpha(\vec{r}) - (r_\alpha - R_{GO,\alpha})\, j_\beta(\vec{r})]\, d\vec{r} \right\} \tag{5.7}$$

Using now $f = (r_\alpha - R_{GO,\alpha})(r_\beta - R_{GO,\beta})$ in Eq. (5.5) shows that the symmetric part of the dipole term vanishes [see Exercise 5.2] and we obtain finally

$$\vec{A}^j(\vec{R}) = \frac{\mu_0}{4\pi} \frac{1}{2} \int_{\vec{r}} \left[(\vec{r} - \vec{R}_{GO}) \times \vec{j}(\vec{r}) \right] \times \frac{(\vec{R} - \vec{R}_{GO})}{|\vec{R} - \vec{R}_{GO}|^3}\, d\vec{r} + \ldots \tag{5.8}$$

or

$$\vec{A}^j(\vec{R}) = \frac{\mu_0}{4\pi}\, \vec{m} \times \frac{(\vec{R} - \vec{R}_{GO})}{|\vec{R} - \vec{R}_{GO}|^3} + \ldots \tag{5.9}$$

for the vector potential, where the first-order magnetic moment $\vec{m}$, the **magnetic dipole moment**, is defined as

$$\vec{m} = \frac{1}{2} \int_{\vec{r}} (\vec{r} - \vec{R}_{GO}) \times \vec{j}(\vec{r})\, d\vec{r} = \frac{1}{2} \int_{\vec{r}} \rho(\vec{r})(\vec{r} - \vec{R}_{GO}) \times \vec{v}(\vec{r})\, d\vec{r} \tag{5.10}$$

Exercise 5.2 Prove that

$$\int_{\vec{r}} j_\alpha(\vec{r})\, d\vec{r} = 0$$

and

$$\int_{\vec{r}} [(r_\beta - R_{GO,\beta})\, j_\alpha(\vec{r}) + (r_\alpha - R_{GO,\alpha})\, j_\beta(\vec{r})]\, d\vec{r} = 0$$

starting from Eq. (5.5)

The absence of a zeroth-order moment in Eq. (5.9) reflects the fact that magnetic monopole moments do not exist. The magnetic dipole moments can thus be shown to be independent of the gauge origin $\vec{R}_{GO}$ [see Exercise 5.3] as a direct consequence of the absence of magnetic monopole moments.

Exercise 5.3 Show that the magnetic dipole moment, Eq. (5.10), is independent of the gauge origin $\vec{R}_{GO}$.

Higher magnetic moments are rarely encountered (Buckingham and Stiles, 1972) and are not considered here as stated earlier. Neither do magnetic dipole moments play such an important role as their electric counterparts since closed-shell molecules do not posses a permanent magnetic moment. Open-shell molecules can have non-zero permanent magnetic moments, which explains their **paramagnetism**. In the following we will look at first-order properties of open shell molecules in Sections 5.3 and 5.6, while we will only consider second-order or response properties of closed-shell molecules. On the other hand, nuclei with non zero spin have a magnetic moment that gives rise to many interesting interactions with the electrons such as NMR spectra and couplings in ESR spectra (see Sections 5.6 and 5.7). The expression for the vector potential, Eq. (5.9), will therefore be mainly used in the following for the vector potential of a nuclear magnetic moment.

5.2 Potential Energy in a Magnetic Induction

The potential energy of a distribution of charges immersed in an external homogeneous magnetic induction $\vec{\mathcal{B}}$ can be expressed in terms of the magnetic moments analogously to the electric field case in Section 4.2. In general, the potential energy E of a current distribution in the presence of an external magnetic induction is given by

$$E(\vec{\mathcal{B}}) = -\int_{\vec{r}} \vec{j}(\vec{r}) \cdot \vec{A}^{\mathcal{B}}(\vec{r}) \, d\vec{r} \tag{5.11}$$

where $\vec{A}^{\mathcal{B}}(\vec{r})$ is the vector potential associated with the magnetic induction $\vec{\mathcal{B}}$,

$$\vec{\mathcal{B}}(\vec{r}) = \vec{\nabla} \times \vec{A}^{\mathcal{B}}(\vec{r}) \tag{5.12}$$

which is the time-independent version of Eq. (2.34). A simpler expression for the potential energy can again be obtained now by expanding a component of the vector potential $A_\alpha^{\mathcal{B}}(\vec{r})$ in a Taylor series around the gauge origin $\vec{R}_{GO}$ (Lazzeretti, 1989)

$$A_\alpha^{\mathcal{B}}(\vec{r}) = A_\alpha^{\mathcal{B}}(\vec{R}_{GO}) + \sum_\beta (r_\beta - R_{GO,\beta}) \left(\frac{\partial A_\alpha^{\mathcal{B}}(\vec{r})}{\partial r_\beta}\right)_{\vec{r}=\vec{R}_{GO}} + \ldots \tag{5.13}$$

which leads to

$$E(\vec{\mathcal{B}}) = -\sum_\alpha A_\alpha^{\mathcal{B}}(\vec{R}_{GO}) \int_{\vec{r}} j_\alpha(\vec{r}) \, d\vec{r} - \sum_{\alpha\beta} \left(\frac{\partial A_\alpha^{\mathcal{B}}(\vec{r})}{\partial r_\beta}\right)_{\vec{r}=\vec{R}_{GO}} \int_{\vec{r}} (r_\beta - R_{GO,\beta}) j_\alpha(\vec{r}) \, d\vec{r}$$
$$+ \ldots \tag{5.14}$$

The integral in the first term was shown to vanish in the last section and the second term can again be rewritten (Eq. (5.5)) in terms of its antisymmetric part, such that

$$E(\vec{\mathcal{B}}) = -\sum_{\alpha\beta} \left(\frac{\partial A_\alpha^{\mathcal{B}}(\vec{r})}{\partial r_\beta}\right)_{\vec{r}=\vec{R}_{GO}} \frac{1}{2} \int_{\vec{r}} [(r_\beta - R_{GO,\beta}) j_\alpha(\vec{r}) - (r_\alpha - R_{GO,\alpha}) j_\beta(\vec{r})] \, d\vec{r}$$
$$+ \ldots \tag{5.15}$$

or in vector notation

$$E(\vec{\mathcal{B}}) = -\frac{1}{2}\int_{\vec{r}} \left[(\vec{r} - \vec{R}_{GO}) \times \vec{j}(\vec{r})\right] \cdot \left[\vec{\nabla} \times \vec{A}^{\mathcal{B}}(\vec{r})\right]_{\vec{r}=\vec{R}_{GO}} d\vec{r} + \ldots \quad (5.16)$$

Using the definition of the magnetic dipole moment given in Eq. (5.10) and the definition of the vector potential given in Eq. (5.12) the expansion of the energy can be written as

$$E(\vec{\mathcal{B}}) = -\vec{m} \cdot \vec{\mathcal{B}}(\vec{R}_{GO}) + \ldots \quad (5.17)$$

From this equation it can be seen that as an alternative to Eq. (5.10) the magnetic dipole moment can also be defined as the derivative of the potential energy with respect to the field induction $\mathcal{B}_\alpha$

$$m_\alpha = -\frac{\partial E(\vec{\mathcal{B}})}{\partial \mathcal{B}_\alpha(\vec{R}_{GO})} \quad (5.18)$$

5.3 Quantum Mechanical Expression for the Magnetic Moment

Quantum mechanical expressions for the permanent magnetic dipole moment can be derived in exactly the same way as the corresponding formulas for the electric dipole moment, in Section 4.3. We will therefore skip most of the derivations here and only discuss the final equations. There is, however, one interesting difference.

But first, we need to derive explicit expressions for the first-order perturbation Hamiltonian operator $\hat{H}^{(1)}$ for the case of a static and homogeneous magnetic induction $\vec{\mathcal{B}}$. The corresponding vector potential at the position of electron i can be obtained from the general expression in Eq. (2.121)

$$\hat{\vec{A}}^{\mathcal{B}}(\vec{r}_i) = \frac{1}{2}\vec{\mathcal{B}} \times (\vec{r}_i - \vec{R}_{GO}) \quad (5.19)$$

Inserting this vector potential in the general expression for the molecular Hamiltonian, Eq. (2.101), we can write the first-order perturbation Hamiltonian as [see Exercise 5.4]

$$\hat{H}^{(1)} = -\sum_\alpha \mathcal{B}_\alpha \left[\hat{O}_\alpha^{l\mathcal{B}}(\vec{R}_{GO}) + \hat{O}_\alpha^{s\mathcal{B}}\right] \quad (5.20)$$

where the perturbation operators are defined as[3]

$$\hat{O}_\alpha^{l\mathcal{B}}(\vec{R}_{GO}) = \sum_i^N \hat{o}_{i,\alpha}^{l\mathcal{B}}(\vec{R}_{GO})$$

$$= -\frac{e}{2m_e}\hat{L}_\alpha(\vec{R}_{GO}) = -\frac{e}{2m_e}\sum_i^N \hat{l}_{i,\alpha}(\vec{R}_{GO}) = -\frac{e}{2m_e}\sum_i^N \left[(\vec{r}_i - \vec{R}_{GO}) \times \hat{\vec{p}}_i\right]_\alpha$$

$$\equiv \hat{m}_\alpha^l(\vec{R}_{GO}) \quad (5.21)$$

---

[3] All perturbation operators derived in this chapter are also collected in Appendix A.

and

$$\hat{O}_\alpha^{sB} = \sum_i^N \hat{o}_{i,\alpha}^{sB} = -\frac{g_e e}{2m_e}\hat{S}_\alpha = -\frac{g_e e}{2m_e}\sum_i^N \hat{s}_{i,\alpha} \quad (5.22)$$

Exercise 5.4 Derive the first-order perturbation Hamiltonian for a homogeneous external magnetic induction, $\vec{B}$, by inserting the vector potential, Eq. (5.19), in the general expression of the molecular Hamiltonian, Eq. (2.101), retaining the first-order term.

The N-electron operator $\hat{\vec{m}}^l(\vec{R}_{GO})$ is called the **orbital magnetic dipole operator**, $\hat{\vec{l}}_i(\vec{R}_{GO})$ is the **orbital angular momentum operator** of electron i with respect to the gauge origin $\vec{R}_{GO}$ and $\hat{\vec{s}}_i$ is the **spin angular momentum operator** of electron i. The total orbital and spin angular moment operators of all N electrons are denoted by $\hat{\vec{L}}$ and $\hat{\vec{S}}$

$$\hat{\vec{L}} = \sum_i^N \hat{\vec{l}}_i \quad (5.23)$$

$$\hat{\vec{S}} = \sum_i^N \hat{\vec{s}}_i \quad (5.24)$$

The perturbation operators are again the first derivatives of the molecular electronic Hamiltonian

$$\left.\frac{\partial \hat{H}}{\partial B_\alpha}\right|_{|\vec{B}|=0} = \frac{\partial \hat{H}^{(1)}}{\partial B_\alpha} = -\hat{O}_\alpha^{lB} - \hat{O}_\alpha^{sB} \quad (5.25)$$

The derivation of a component of the **magnetic dipole moment** as first derivative of the perturbed energy, Eq. (5.18), via perturbation theory or the Hellmann–Feynman theorem then leads to the following expectation value

$$\begin{aligned} m_\alpha &= \langle \Psi_0^{(0)} | \hat{O}_\alpha^{lB} + \hat{O}_\alpha^{sB} | \Psi_0^{(0)} \rangle \\ &= \langle \Psi_0^{(0)} | \hat{m}_\alpha^l(\vec{R}_{GO}) | \Psi_0^{(0)} \rangle - \frac{g_e e}{2m_e}\langle \Psi_0^{(0)} | \hat{S}_\alpha | \Psi_0^{(0)} \rangle \\ &= -\frac{e}{2m_e}\langle \Psi_0^{(0)} | \hat{L}_\alpha(\vec{R}_{GO}) + g_e \hat{S}_\alpha | \Psi_0^{(0)} \rangle \end{aligned} \quad (5.26)$$

In the case of the electric dipole moment we could derive the quantum mechanical expression, Eq. (4.25), simply by replacing the electron density by the quantum mechanical expression for it. However, the analogous derivation of the quantum mechanical expression for the magnetic dipole moment starting from the classical definition of the magnetic dipole moment as an integral over the current density, Eq. (5.10), and replacing the classical current density by the quantum mechanical expectation value from Eq. (2.31), will lead only to the expectation value of the orbital magnetic dipole moment operator. The contribution from the electron spin cannot be obtained in this way, because spin has no classical analogue. This illustrates the danger in simply translating classical expressions to quantum mechanics and the advantage of defining properties as derivatives of the energy.

At the end of Section 5.1 it was briefly mentioned that permanent magnetic moments are not as important as their electric counterparts, because they vanish for closed-shell molecules. We want to prove this statement now starting from Eq. (5.26). The orbital part of the magnetic dipole moment is essentially an expectation value of the angular momentum operator $\hat{\vec{L}}$. The angular momentum operator is a hermitian operator, which implies that

$$\langle \Psi | \hat{\vec{L}} | \Psi \rangle = \langle \Psi | \hat{\vec{L}} | \Psi \rangle^* \tag{5.27}$$

For real wavefunctions Ψ the right-hand side becomes

$$\langle \Psi | \hat{\vec{L}} | \Psi \rangle = \langle \Psi | \hat{\vec{L}}^* | \Psi \rangle \tag{5.28}$$

or

$$\langle \Psi | \hat{\vec{L}} | \Psi \rangle = -\langle \Psi | \hat{\vec{L}} | \Psi \rangle \tag{5.29}$$

when we recall that $\hat{\vec{L}}$ is a purely imaginary operator. This means that the expectation value of $\hat{\vec{L}}$ is zero for real wavefunctions. One says that the **orbital angular momentum is quenched** for a molecule described by real wavefunctions. For molecules in orbitally non-degenerate states we can always choose the wavefunctions to be real and therefore they do not have a permanent orbital magnetic moment. Furthermore, when the state is also a singlet state, the expectation value $\langle \Psi_0^{(0)} | \hat{S}^2 | \Psi_0^{(0)} \rangle$ vanishes and the molecule has neither spin nor orbital permanent magnetic moment.

Among open-shell molecules only linear molecules with an odd number of electrons have permanent orbital magnetic moments. First, linear molecules can have orbitally degenerate states according to the Jahn–Teller theorem, secondly for even number of electrons the electrons are paired with respect to the eigenvalues of $\hat{l}_z$, which means that their total contribution to the angular momentum is zero again. On the other hand, open-shell molecules with an odd number of electrons, called radicals, have spin magnetic moments. Molecules with an even number of unpaired electrons in the ground state are rare, because it requires an orbitally degenerate state. O_2 with its triplet ground state is thus the prime example.

In the case of the electric moments we also had to include contributions from the nuclear charges. For the magnetic moments the situation is different. Nuclei with non-zero spin have of course a nuclear magnetic moment. However, due to the inverse dependence of the magnetic moment on the mass, see e.g. Eq. (5.26), the nuclear spin magnetic dipole moments are at least three orders of magnitude smaller than the electronic spin magnetic moments. Therefore, we do not include them here but will consider them in Sections 5.4 and 5.7 as perturbations of the electronic structure. Nuclear magnetic moments analogous to the electronic angular magnetic moment exist, but they require an angular motion of the nuclei and thus a rotation of the nuclear framework of the molecule. Since we are working in the Born–Oppenheimer approximation the nuclei are kept fixed in space and we do not have this contribution. However, we will consider this contribution and other couplings with molecular rotation in Chapter 6.

5.4 Induced Magnetic Moment, Magnetizability, and Nuclear Magnetic Shielding

In the presence of an external magnetic induction $\vec{\mathcal{B}}$ the energy of the distribution of moving charges changes according to Eq. (5.17). A polarizable distribution of charges will adjust itself in order to minimize the energy. This leads to a change in the current density, $\vec{j}(\vec{r})$, and in the moments of the current density, $\vec{m}$, such that an additional current density $\vec{j}^{ind}(\vec{r})$ and magnetic moment $\vec{m}^{ind}$ are induced. An important source of magnetic induction in molecules, apart from an external magnetic field, is the magnetic dipole moment $\vec{m}^K$ of a nucleus K in the molecule. The electronic magnetic dipole moment $\vec{m}(\vec{\mathcal{B}}, \{\vec{m}^K\})$ in the presence of an external magnetic induction $\vec{\mathcal{B}}$ and M nuclear magnetic moments, $\{\vec{m}^K\}$,[4] can again be expanded in a Taylor series as (Lazzeretti, 1989)

$$m_\alpha(\vec{\mathcal{B}}, \{\vec{m}^K\}) = m_\alpha + m_\alpha^{ind}(\vec{\mathcal{B}}, \{\vec{m}^K\})$$

$$m_\alpha(\vec{\mathcal{B}}, \{\vec{m}^K\}) = m_\alpha + \sum_\beta \xi_{\alpha\beta} \mathcal{B}_\beta - \sum_K \sum_\beta \sigma^K_{\beta\alpha} m^K_\beta + \ldots \quad (5.30)$$

where $\xi_{\alpha\beta}$ and $\sigma^K_{\beta\alpha}$ are cartesian components of the dipole **magnetizability** and **nuclear magnetic shielding tensor**, respectively, while the mean or isotropic magnetizability is thus defined as

$$\bar{\xi} = \frac{1}{3} \sum_\alpha \xi_{\alpha\alpha} \quad (5.31)$$

The magnetizability is the magnetic analogue to the polarizability and can be considered as the **linear response** of the molecular magnetic dipole moment to an external magnetic induction. The nuclear magnetic shielding tensor is similarly the **linear response** of the molecular magnetic dipole moment to the magnetic dipole moment of a nucleus K. It is the molecular property behind the **chemical shift** measured in nuclear magnetic resonance (NMR) spectroscopy and will be discussed in more detail in Section 5.7. Here, we should note already that the nuclear magnetic shielding tensor is not symmetric and that we choose to associate the first index, i.e. the row index, with the nuclear magnetic moment $\vec{m}^K$, while the second, i.e. the column index, will be associated with an external magnetic field in the following.

From the expansion in Eq. (5.30) we can define the components of the magnetizability and nuclear magnetic shielding tensor of nucleus K as first derivatives of the perturbed electronic magnetic dipole moment with respect to a component of the magnetic induction or the nuclear magnetic dipole moment of nucleus K[5]

---

[4] We will use the notation $\{\vec{m}^K\}$, when referring to all the nuclear moments in a molecule collectively.
[5] These definitions are collected in the second column of Table B.2 of Appendix B.

$$\xi_{\alpha\beta} = \left.\frac{\partial m_\alpha(\vec{B})}{\partial \mathcal{B}_\beta}\right|_{|\vec{B}|=0} \tag{5.32}$$

$$\sigma^K_{\beta\alpha} = -\left.\frac{\partial m_\alpha(\vec{m}^K)}{\partial m^K_\beta}\right|_{|\vec{m}^K|=0} \tag{5.33}$$

In order to derive an expression for the energy of a polarizable charge and current distribution we have to proceed as in the electric case. We consider first only the changes due to the magnetic induction $\vec{B}$. The infinitesimal change in the energy dE due to an infinitesimal change in the magnetic induction is according to Eq. (5.18) then

$$dE = -\sum_\alpha m_\alpha(\vec{B})\, d\mathcal{B}_\alpha = -\vec{m}(\vec{B}) \cdot d\vec{B} \tag{5.34}$$

The energy is again obtained by integration on both sides

$$E(\vec{B}) - E^{(0)} = -\int_0^{\vec{B}} \vec{m}(\vec{B}') \cdot d\vec{B}' \tag{5.35}$$

Inserting now the expansion of the electronic magnetic dipole moment from Eq. (5.30) we can write

$$E(\vec{B}) - E^{(0)} = -\sum_\alpha \int_0^{\vec{B}} \left(m_\alpha + \sum_\beta \xi_{\alpha\beta}\mathcal{B}'_\beta + \ldots \right) d\mathcal{B}'_\alpha \tag{5.36}$$

Evaluating the line integral as described for the electric case in Section 4.4 then yields

$$E(\vec{B}) = E^{(0)} - \sum_\alpha m_\alpha \mathcal{B}_\alpha - \frac{1}{2}\sum_{\alpha\beta} \xi_{\alpha\beta}\mathcal{B}_\alpha \mathcal{B}_\beta + \ldots \tag{5.37}$$

In the same way, the contribution of the nuclear magnetic shielding tensor to the energy can be obtained. The final expression for the energy of a polarizable charge and current distribution in the presence of a magnetic induction $\vec{B}$ and M nuclear magnetic moments, $\{\vec{m}^K\}$ then reads (Lazzeretti, 1989)

$$E(\vec{B}, \{\vec{m}^K\}) = E^{(0)} - \sum_\alpha m_\alpha \mathcal{B}_\alpha - \frac{1}{2}\sum_{\alpha\beta} \xi_{\alpha\beta}\mathcal{B}_\alpha \mathcal{B}_\beta + \sum_K \sum_{\alpha\beta} \sigma^K_{\alpha\beta} m^K_\alpha \mathcal{B}_\beta + \ldots \tag{5.38}$$

This expression includes only the contributions from the charge and current distribution. For the total energy of the system one would have to add the energy contributions $-\sum_K \vec{m}^K \cdot \vec{B}$ of the M nuclear magnetic moments $\vec{m}^K$ in the external magnetic induction $\vec{B}$. Equation (5.38) allows us to define some of the magnetic properties as derivatives of the energy. The first derivative is the magnetic dipole moment, Eq. (5.18). The permanent magnetic dipole moment $\vec{m}$ is obtained, if the derivative is evaluated at zero magnetic field. The magnetizability as well as the nuclear magnetic shielding tensor of a nucleus K, which we had already defined as first derivatives of the electronic magnetic dipole moment in the presence of an external magnetic induction

102 Magnetic Properties

or nuclear magnetic moments, Eqs. (5.32) and (5.33), can therefore also be defined[6] as second derivatives of the energy,

$$\xi_{\alpha\beta} = - \left.\frac{\partial^2 E(\vec{B})}{\partial B_\alpha \partial B_\beta}\right|_{|\vec{B}|=0} \tag{5.39}$$

$$\sigma^K_{\alpha\beta} = \left.\frac{\partial^2 E(\vec{B}, \vec{m}^K)}{\partial B_\beta \partial m^K_\alpha}\right|_{|\vec{B}|=|\vec{m}^K|=0} \tag{5.40}$$

similar to the polarizability tensor. They are therefore normally also called **second-order properties**.

5.5 Quantum Mechanical Expression for the Magnetizability

In the previous section we have defined the cartesian components of the **magnetizability** tensor $\xi_{\alpha\beta}$ as second derivatives of the energy $E(\vec{B})$ in the presence of a magnetic induction $\vec{B}$, Eq. (5.39), or alternatively as first derivatives of the magnetic-field-dependent electronic magnetic dipole moment $m_\alpha(\vec{B})$, Eq. (5.32). Both definitions can be used to derive quantum mechanical expressions for the magnetizability.

Let us start with the first definition as derivative of the energy. Again, we will use the perturbation theory expression for the perturbed energy, Eq. (3.15), but differentiate it now twice with respect to the appropriate components of the magnetic induction. This leads us immediately to the second-order correction to the energy, because the first-order correction depends only linearly on the fields. We can therefore express the magnetizability as

$$\begin{aligned}
\xi_{\alpha\beta} =& - \left.\frac{\partial^2 E_0^{(2)}(\vec{B})}{\partial B_\alpha \partial B_\beta}\right|_{|\vec{B}|=0} \\
=& - \left.\frac{\partial^2}{\partial B_\alpha \partial B_\beta} \langle \Psi_0^{(0)} | \hat{H}^{(2)} | \Psi_0^{(0)} \rangle \right|_{|\vec{B}|=0} \\
& - \left.\frac{\partial^2}{\partial B_\alpha \partial B_\beta} \sum_{n \neq 0} \frac{\langle \Psi_0^{(0)} | \hat{H}^{(1)} | \Psi_n^{(0)} \rangle \langle \Psi_n^{(0)} | \hat{H}^{(1)} | \Psi_0^{(0)} \rangle}{E_0^{(0)} - E_n^{(0)}} \right|_{|\vec{B}|=0}
\end{aligned} \tag{5.41}$$

One should recall that the magnetic perturbations enter the molecular Hamiltonian in Eq. (2.101) in the form of the vector potential $\vec{\hat{A}}$ and that $\vec{\hat{A}}$ contributes both to $\hat{H}^{(1)}$ and to $\hat{H}^{(2)}$. Therefore, we have to use the full expression for the second-order energy correction including the $\langle \Psi_0^{(0)} | \hat{H}^{(2)} | \Psi_0^{(0)} \rangle$ term for the magnetizability and in general all magnetic properties contrary to the electric analogue in Section 4.5.

In Section 5.3 we have obtained already the appropriate expression for $\hat{H}^{(1)}$ in Eq. (5.20). In the same way we can now derive the second-order perturbation Hamiltonian $\hat{H}^{(2)}$ for the case of an external magnetic induction as [see Exercise 5.5]

---

[6] These definitions are collected in the last column of Table B.2 of Appendix B.

$$\hat{H}^{(2)} = \sum_{\alpha\beta} \hat{O}^{BB}_{\alpha\beta}(\vec{R}_{GO})\, \mathcal{B}_\alpha\, \mathcal{B}_\beta \tag{5.42}$$

where the perturbation operator is defined as[7]

$$\hat{O}^{BB}_{\alpha\beta}(\vec{R}_{GO}) = \sum_i^N \hat{o}^{BB}_{i,\alpha\beta}(\vec{R}_{GO})$$

$$= \frac{e^2}{8 m_e} \sum_i \left[(\vec{r}_i - \vec{R}_{GO})^2 \delta_{\alpha\beta} - (r_{i,\alpha} - R_{GO,\alpha})(r_{i,\beta} - R_{GO,\beta}) \right] \tag{5.43}$$

Exercise 5.5 Derive the second-order perturbation Hamiltonian for a homogeneous external magnetic induction, $\vec{\mathcal{B}}$, by inserting the vector potential, Eq. (5.19), in the general expression of the molecular Hamiltonian, Eq. (2.101), retaining the second-order term.

We are now ready to evaluate the derivatives in Eq. (5.41) and obtain the **sum-over-states expression** for the components of the magnetizability tensor

$$\xi_{\alpha\beta} = -\langle \Psi_0^{(0)} | \hat{O}^{BB}_{\alpha\beta}(\vec{R}_{GO}) + \hat{O}^{BB}_{\beta\alpha}(\vec{R}_{GO}) | \Psi_0^{(0)} \rangle$$

$$- \sum_{n\neq 0} \frac{\langle \Psi_0^{(0)} | \hat{m}^l_\alpha(\vec{R}_{GO}) | \Psi_n^{(0)} \rangle \langle \Psi_n^{(0)} | \hat{m}^l_\beta(\vec{R}_{GO}) | \Psi_0^{(0)} \rangle}{E_0^{(0)} - E_n^{(0)}}$$

$$- \sum_{n\neq 0} \frac{\langle \Psi_0^{(0)} | \hat{m}^l_\beta(\vec{R}_{GO}) | \Psi_n^{(0)} \rangle \langle \Psi_n^{(0)} | \hat{m}^l_\alpha(\vec{R}_{GO}) | \Psi_0^{(0)} \rangle}{E_0^{(0)} - E_n^{(0)}} \tag{5.44}$$

One should note that there is no contribution from the electron spin $\vec{S}$ to the magnetizability, because transition matrix elements like $\langle \Psi_0^{(0)} | \hat{S}_\alpha | \Psi_n^{(0)} \rangle$ are zero due to the orthogonality of the unperturbed states.

In the second approach we want to derive the magnetizability as first derivative of the magnetic moment in the presence of a magnetic field, i.e. of a magnetic-field-dependent magnetic moment, according to Eq. (5.32). In Section 5.3 it was shown that the permanent magnetic moment can be obtained as the expectation value of the operator for the magnetic moment with the unperturbed wavefunction. For the magnetic-field-dependent magnetic moment we need then both a perturbation-dependent magnetic dipole moment operator, $\hat{m}^l(\vec{R}_{GO}, \vec{\mathcal{B}})$ as well as the perturbed wavefunction, $\Psi_0(\vec{\mathcal{B}})$. Contrary to the electric case in Section 4.5 the normal magnetic dipole moment operator, Eq. (5.21), differs from the magnetic dipole moment operator in the presence of a magnetic induction $\vec{\mathcal{B}}$. The latter is obtained by applying, Eq. (3.8), to the Hamiltonian, Eq. (2.101), with $H^{(1)}$ and $H^{(2)}$ as given in Eqs. (5.20) and (5.42).

$$\frac{\partial \hat{H}}{\partial \mathcal{B}_\alpha} = \hat{m}^l(\vec{R}_{GO}, \vec{\mathcal{B}}) = \hat{m}^l_\alpha(\vec{R}_{GO}) - \sum_\beta \left[\hat{O}^{BB}_{\alpha\beta}(\vec{R}_{GO}) + \hat{O}^{BB}_{\beta\alpha}(\vec{R}_{GO}) \right] \mathcal{B}_\beta \tag{5.45}$$

---

[7] All perturbation operators derived in this chapter are collected in Appendix A.

104 *Magnetic Properties*

Here, we have neglected the contribution from the spin of the electrons, i.e. $\hat{O}_\alpha^{sB}$, because it will not contribute to the magnetizability. Applying the general expression for the first derivative of a perturbation-dependent expectation value, Eq. (3.40), to the case of magnetic dipole moment in the presence of an external magnetic induction, we obtain expressions involving first derivatives of the perturbed wavefunction for the tensor components of the dipole magnetizability

$$\xi_{\alpha\beta} = \frac{\partial m_\alpha(\vec{\mathcal{B}})}{\partial \mathcal{B}_\beta}\bigg|_{|\vec{\mathcal{B}}|=0} = \frac{\partial}{\partial \mathcal{B}_\beta}\langle \Psi_0(\vec{\mathcal{B}})|\hat{m}_\alpha^l(\vec{R}_{GO},\vec{\mathcal{B}})|\Psi_0(\vec{\mathcal{B}})\rangle\bigg|_{|\vec{\mathcal{B}}|=0}$$

$$= \langle \frac{\partial \Psi_0(\vec{\mathcal{B}})}{\partial \mathcal{B}_\beta}\bigg|_{|\vec{\mathcal{B}}|=0} |\hat{m}_\alpha^l(\vec{R}_{GO})|\Psi_0^{(0)}\rangle + \langle \Psi_0^{(0)}|\hat{m}_\alpha^l(\vec{R}_{GO})| \frac{\partial \Psi_0(\vec{\mathcal{B}})}{\partial \mathcal{B}_\beta}\bigg|_{|\vec{\mathcal{B}}|=0}\rangle \quad (5.46)$$

$$- \langle \Psi_0^{(0)}|\left[\hat{O}_{\alpha\beta}^{BB}(\vec{R}_{GO}) + \hat{O}_{\beta\alpha}^{BB}(\vec{R}_{GO})\right]|\Psi_0^{(0)}\rangle$$

In Part III we will come back to these expressions and evaluate the derivatives of approximate wavefunctions. However, here we will use the response formalism as developed in Section 3.11. Using Eq. (3.116) we can express the derivatives of the perturbation-dependent expectation value in terms of polarization propagators or linear response functions and thus obtain for the tensor components of the magnetizability

$$\xi_{\alpha\beta} = \frac{\partial}{\partial \mathcal{B}_\beta}\langle \Psi_0(\vec{\mathcal{B}})|\hat{m}_\alpha^l(\vec{R}_{GO},\vec{\mathcal{B}})|\Psi_0(\vec{\mathcal{B}})\rangle\bigg|_{|\vec{\mathcal{B}}|=0} \quad (5.47)$$

$$= -\langle \Psi_0^{(0)}|\left[\hat{O}_{\alpha\beta}^{BB}(\vec{R}_{GO}) + \hat{O}_{\beta\alpha}^{BB}(\vec{R}_{GO})\right]|\Psi_0^{(0)}\rangle - \langle\langle \hat{m}_\alpha^l(\vec{R}_{GO}) ; \hat{m}_\beta^l(\vec{R}_{GO}) \rangle\rangle_{\omega=0}$$

Inserting the perturbation theory expansion of the perturbed wavefunction up to first order, Eq. (3.27), in Eq. (5.46) or the expression for the static response function Eq. (3.114) in Eq. (5.47) leads us back to the sum-over-states expressions given in Eq. (5.44), of course. The significance of the other expressions for the magnetizability will become clear in Part III where we will see that for approximate wavefunctions they can give different results than the normal sum-over-states expression in Eq. (5.44).

All the expressions for the magnetizability derived in this section show that there are two different contributions to the magnetizability and in general to many of the second-order magnetic properties.

$$\xi_{\alpha\beta} = \xi_{\alpha\beta}^{dia}(\vec{R}_{GO}) + \xi_{\alpha\beta}^{para}(\vec{R}_{GO}) \quad (5.48)$$

The first contribution is an expectation value of the second-order perturbation operator with the unperturbed wavefunction $\Psi_0^{(0)}$ of the system. It is negative and is called the **diamagnetic contribution** $\xi_{\alpha\beta}^{dia}$. The second contribution involves either the first derivative of the perturbed wavefunction, the first-order correction to the wavefunction, a sum over all the other unperturbed states $\{\Psi_n^{(0)}\}$ or a linear response function like in the case of the polarizability. This contribution is positive, because the energy difference in the denominator of the sum-over-states expression, Eq. (5.44) is negative, and is called the **paramagnetic contribution** $\xi^{para}_{\alpha\beta}$. Only the paramagnetic contribution is immediately expressible as polarization propagator, Eq. (5.47). However, also the diamagnetic contribution, i.e. the term involving a ground-state expectation value, can be reformulated as sum-over-states or polarization propagator expression, as will be shown for all magnetic properties in Section 5.9. For most closed-shell molecules the diamagnetic term is larger than the paramagnetic term and the molecule is said to be diamagnetic. This means that a probe of this molecule when brought into an inhomogeneous magnetic field will avoid the areas of higher magnetic induction, because the energy increases there, as we can see from Eq. (5.38). In a few closed-shell molecules the paramagnetic term is larger than the diamagnetic term, which means that the molecule is paramagnetic.[8] A probe of these molecules will be pulled into regions of higher magnetic induction because the energy is lower there, Eq. (5.38). Such a molecule will therefore behave like a molecule with a permanent magnetic moment, which explains why $\xi^{para}_{\alpha\beta}(\vec{R}_{GO})$ is called the paramagnetic contribution. This induced paramagnetism should be distinguished from paramagnetism due to a permanent magnetic dipole moment. For a bulk sample the latter paramagnetism becomes temperature dependent due to the Boltzmann averaging of the orientation of the permanent magnetic dipole moments, whereas the induced paramagnetism due to a positive magnetizability is temperature independent. This explains why it is sometimes also called temperature-independent paramagnetism (TIP).

In Section 5.10 we will discuss that **the dia- and paramagnetic contributions both depend quadratically on the gauge origin** $\vec{R}_{GO}$ but their sum, the magnetizability, is independent. This cancellation holds for exact unperturbed states $\{\Psi_n^{(0)}\}$ and for certain approximate methods such as the random phase approximation or coupled Hartree–Fock and multiconfigurational self-consistent field response methods, which will be discussed in Sections 10.3, 10.4, 11.1 and 11.2. The gauge-origin dependence of the dia- and paramagnetic contributions has the consequence that the separation in a dia- and paramagnetic contribution is arbitrary and that no physical meaning should be assigned to the two terms individually. Actually, alternative expressions have been derived for the diamagnetic contribution that are no longer expectation values but involve also a sum over all states $\{\Psi_n^{(0)}\}$ (Geertsen, 1989; Sauer et al., 1994a; Lazzeretti et al., 1994). These expressions are discussed and derived in Section 5.9.

5.6 Molecular Magnetic Fields and ESR Parameters

The current density $\vec{j}(\vec{r})$ of a system of moving charges will also give rise to a magnetic field or magnetic induction that we want to call the **permanent molecular magnetic induction** $\vec{B}^j(\vec{R})$ and add a superscript "j" in order to distinguish it from the external magnetic induction. An expression for it can be obtained by application of Eq. (5.12) to the expression (5.2) for the vector potential $\vec{A}^j(\vec{R})$ of a current distribution [see Exercise 5.6]

---

[8] For some examples see, e.g., Sauer et al. (1993).

106 Magnetic Properties

$$\vec{B}^j(\vec{R}) = \vec{\nabla} \times \vec{A}^j(\vec{R})$$

$$= -\frac{\mu_0}{4\pi} \int_{\vec{r}} \frac{(\vec{R}-\vec{r}) \times \vec{j}(\vec{r})}{|\vec{R}-\vec{r}|^3} \, d\vec{r}$$

$$= -\frac{\mu_0}{4\pi} \int_{\vec{r}} \rho(\vec{r}) \frac{(\vec{R}-\vec{r}) \times \vec{v}(\vec{r})}{|\vec{R}-\vec{r}|^3} \, d\vec{r} \qquad (5.49)$$

Exercise 5.6 Derive the expression (5.49) for the molecular magnetic induction.

Like the molecular electric field it is of course defined for any point in space, but within the charge distribution only the value at the position of the nuclei can be probed experimentally again. The interaction of such a nuclear magnetic dipole moment $\vec{m}^K$ with the molecular magnetic induction, $\vec{B}^j$, gives rise to a change in the energy of the distribution of charges

$$E(\vec{m}^K) = -\sum_\alpha m_\alpha^K \mathcal{B}_\alpha^j(\vec{R}_K) \qquad (5.50)$$

Consequently, the molecular magnetic induction at the position of some nucleus K can be defined as the derivative of the energy of the distribution of charges,

$$\mathcal{B}_\alpha^j(\vec{R}_K) = -\frac{\partial E(\vec{m}^K)}{\partial m_\alpha^K} \qquad (5.51)$$

with respect to its nuclear magnetic moment $\vec{m}^K$.[9]

In Section 5.3 on the magnetic moments it was shown that there can be two contributions to the magnetic moment: one from the orbital angular momentum of the electrons and one from the electronic spins, the latter being the more frequent case. The same applies therefore also to the permanent molecular magnetic induction $\vec{B}^j(\vec{R})$. Experimentally interesting is the spin molecular magnetic induction $\vec{B}^{j,s}(\vec{R}_K)$ of open-shell molecules at the position of a nucleus K with non-zero spin $\hat{I}_K$ and associated nuclear magnetic moment $\vec{m}^K$

$$\vec{m}^K = \frac{g_K \mu_N}{\hbar} \vec{I}^K \qquad (5.52)$$

where g_K and $\mu_N = \frac{e\hbar}{2m_p}$ are the nuclear g-factor of nucleus K and the nuclear magneton, respectively, and m_p is the proton mass. According to Eq. (5.50) this interaction of the nuclear spin and the spin of the electrons via their magnetic moments leads to a change in the energy of the electrons. The latter shows up as splittings of the lines in the **electron spin resonance (ESR)** spectra of radicals. The splittings are directly related to the **hyperfine coupling tensor** $a_{\alpha\beta}^K$, which is defined via an effective spin Hamiltonian

$$\hat{H}^{ESR} = \frac{2\pi}{\hbar} \sum_{\alpha\beta} \hat{I}_\alpha^K a_{\alpha\beta}^K \hat{S}_\beta \qquad (5.53)$$

---

[9] This definition is also listed in the first column of Table B.2 of Appendix B.

which acts only on electronic and nuclear spin states and is normally used in the analysis of ESR spectra. An expectation value of this operator for a particular spin state then gives the energy of this state. This energy is the same as Eq. (5.50) and an expression for the hyperfine tensor $a^K_{\alpha\beta}$ can therefore be extracted by comparing Eqs. (5.50) and (5.53)

$$a^K_{\alpha\beta} = -\frac{g_K \mu_N}{2\pi} \frac{\mathcal{B}^{j,s}_\alpha(\vec{R}_K)}{\langle \Psi^{(0)}_0 | \hat{S}_\beta | \Psi^{(0)}_0 \rangle} \tag{5.54}$$

where $\vec{\mathcal{B}}^{j,s}(\vec{R}_K)$ is the contribution of the electronic spin to the permanent molecular magnetic induction.

Like all the electric and magnetic moments and all other first derivatives of the energy, the molecular magnetic induction can be calculated as an expectation value. And as in the case of the magnetic moment in Section 5.3 the spin contribution is not obtained by simply translating the classical expression to quantum mechanics. We therefore will express it directly as an expectation value of the derivative of the first-order Hamiltonian $\hat{H}^{(1)}$ with respect to the components of the nuclear magnetic moment m^K_α of nucleus K. The "external" perturbation is therefore now a nuclear magnetic moment $\vec{m}^K$ and we need to insert its vector potential $\vec{A}^K(\vec{r}_i)$ in the molecular Hamiltonian, Eq. (2.101). The vector potential of a general magnetic moment was already derived in Section 5.1. Applying Eq. (5.9) we thus obtain for the value of the vector potential of $\vec{m}^K$ at the position of electron i

$$\vec{A}^K(\vec{r}_i) = \frac{\mu_0}{4\pi} \vec{m}^K \times \frac{(\vec{r}_i - \vec{R}_K)}{|\vec{r}_i - \vec{R}_K|^3} \tag{5.55}$$

Inserted in the molecular Hamiltonian, Eq. (2.101), the first-order perturbation Hamiltonian becomes [see Exercise 5.7]

$$\hat{H}^{(1)} = -\sum_\alpha \left(\hat{O}^{l\,m^K}_\alpha + \hat{O}^{s\,m^K}_\alpha \right) m^K_\alpha \tag{5.56}$$

$$= -\frac{g_K \mu_N}{\hbar} \sum_\alpha \left(\hat{O}^{l\,m^K}_\alpha + \hat{O}^{s\,m^K}_\alpha \right) I^K_\alpha \tag{5.57}$$

where the perturbation operators[10] are the **orbital paramagnetic** (OP) or **paramagnetic nuclear spin-electron orbit operator** (PSO)

$$\hat{O}^{l\,m^K}_\alpha = \sum_i^N \hat{o}^{l\,m^K}_{i,\alpha}$$

$$= -\frac{e}{m_e} \frac{\mu_0}{4\pi} \sum_i^N \frac{\hat{l}_{i,\alpha}(\vec{R}_K)}{|\vec{r}_i - \vec{R}_K|^3} = -\frac{e}{m_e} \frac{\mu_0}{4\pi} \sum_i^N \left(\frac{\vec{r}_i - \vec{R}_K}{|\vec{r}_i - \vec{R}_K|^3} \times \hat{\vec{p}}_i \right)_\alpha$$

$$\equiv \hat{O}^{OP}_{K,\alpha} \tag{5.58}$$

---

[10] All perturbation operators derived in this chapter are collected in Appendix A.

108 *Magnetic Properties*

and the sum of the **Fermi contact** and **spin-dipolar** operators

$$\hat{O}_\alpha^{sm^K} = \sum_i \hat{o}_{i,\alpha}^{sm^K} \equiv \hat{O}_{K,\alpha}^{FC} + \hat{O}_{K,\alpha}^{SD} \qquad (5.59)$$

where

$$\hat{O}_{K,\alpha}^{FC} = -\frac{g_e e \mu_0}{3 m_e} \sum_i^N \delta(\vec{r}_i - \vec{R}_K) \hat{s}_{i,\alpha} \qquad (5.60)$$

$$\hat{O}_{K,\alpha}^{SD} = -\frac{g_e e}{2 m_e} \frac{\mu_0}{4\pi} \sum_i^N \left\{ \frac{3 \left[\hat{\vec{s}}_i \cdot (\vec{r}_i - \vec{R}_K) \right] (r_{i,\alpha} - R_{K,\alpha})}{|\vec{r}_i - \vec{R}_K|^5} - \frac{\hat{s}_{i,\alpha}}{|\vec{r}_i - \vec{R}_K|^3} \right\} \qquad (5.61)$$

The operator $\hat{\vec{l}}_i(\vec{R}_K) = \left(\vec{r}_i - \vec{R}_K \right) \times \hat{\vec{p}}_i$, in Eq. (5.58), is the **angular momentum operator** of electron i again but now with respect to the position of the nucleus K.

Exercise 5.7 Derive the first-order perturbation Hamiltonian, Eqs. (5.56) to (5.61), for the vector potential of a nuclear magnetic moment, $\vec{m}^K$, by inserting the vector potential, Eq. (5.55), in the general expression of the molecular Hamiltonian, Eq. (2.101), retaining the first-order term.

The permanent molecular magnetic induction at the position of nucleus K can therefore be calculated as the following expectation value

$$\begin{aligned} B_\alpha^j(\vec{R}_K) &= B_\alpha^{j,l}(\vec{R}_K) + B_\alpha^{j,s}(\vec{R}_K) \\ &= \langle \Psi_0^{(0)} | \hat{O}_{K,\alpha}^{OP} | \Psi_0^{(0)} \rangle + \langle \Psi_0^{(0)} | \hat{O}_{K,\alpha}^{FC} + \hat{O}_{K,\alpha}^{SD} | \Psi_0^{(0)} \rangle \end{aligned} \qquad (5.62)$$

and the components $a_{\alpha\beta}^K$ of the ESR hyperfine coupling tensor, defined in Eq. (5.54), are thus given as

$$\begin{aligned} a_{\alpha\beta}^K &= -\frac{g_K \mu_N}{2\pi} \frac{\langle \Psi_0^{(0)} | \hat{O}_{K,\alpha}^{FC} + \hat{O}_{K,\alpha}^{SD} | \Psi_0^{(0)} \rangle}{\langle \Psi_0^{(0)} | \hat{S}_\beta | \Psi_0^{(0)} \rangle} \\ &= \frac{\mu_0}{6\pi m_e} \frac{e g_e g_K \mu_N}{\langle \Psi_0^{(0)} | \hat{S}_\beta | \Psi_0^{(0)} \rangle} \langle \Psi_0^{(0)} | \sum_i^N \delta(\vec{r}_i - \vec{R}_K) \hat{s}_{i,\alpha} | \Psi_0^{(0)} \rangle \\ &+ \frac{\mu_0}{16\pi^2 m_e} \frac{3 e g_e g_K \mu_N}{\langle \Psi_0^{(0)} | \hat{S}_\beta | \Psi_0^{(0)} \rangle} \langle \Psi_0^{(0)} | \sum_i^N \frac{\left[\hat{\vec{s}}_i \cdot (\vec{r}_i - \vec{R}_K) \right] (r_{i,\alpha} - R_{K,\alpha})}{|\vec{r}_i - \vec{R}_K|^5} | \Psi_0^{(0)} \rangle \\ &- \frac{\mu_0}{16\pi^2 m_e} \frac{e g_e g_K \mu_N}{\langle \Psi_0^{(0)} | \hat{S}_\beta | \Psi_0^{(0)} \rangle} \langle \Psi_0^{(0)} | \sum_i^N \frac{\hat{s}_{i,\alpha}}{|\vec{r}_i - \vec{R}_K|^3} | \Psi_0^{(0)} \rangle \end{aligned} \qquad (5.63)$$

There are two contributions – one from the Fermi contact and one from the spin-dipolar operator. The latter dipolar contribution is anisotropic and is averaged to zero for molecules in the gas or liquid phase, whereas the Fermi-contact contribution is isotropic and is therefore the only contribution to the ESR hyperfine coupling constants of gas- or liquid-phase molecules.

5.7 Induced Magnetic Fields and NMR Parameters

In Section 5.4 it was discussed that the interaction of a charge distribution with an external magnetic induction, $\vec{\mathcal{B}}$, or with a nuclear magnetic moment, $\vec{m}^L$, leads to an induced current density $\vec{j}^{ind}(\vec{r})$. According to Eq. (5.49) this will also give rise to an **induced molecular magnetic induction** $\vec{\mathcal{B}}^{j,ind}(\vec{R})$. In analogy to the induced magnetic moment we want to expand the induced molecular magnetic induction at an arbitrary point $\vec{R}$, $\vec{\mathcal{B}}^{j,ind}(\vec{R})$, in a series in the external magnetic induction $\vec{\mathcal{B}}$ and the magnetic moments $\{\vec{m}^L\}$ of nuclei L

$$\mathcal{B}^j_\alpha(\vec{R},\vec{\mathcal{B}},\{\vec{m}^L\}) = \mathcal{B}^j_\alpha(\vec{R}) + \mathcal{B}^{j,ind}_\alpha(\vec{R},\vec{\mathcal{B}},\{\vec{m}^L\})$$

$$\mathcal{B}^j_\alpha(\vec{R},\vec{\mathcal{B}},\{\vec{m}^L\}) = \mathcal{B}^j_\alpha(\vec{R}) - \sum_\beta \sigma_{\alpha\beta}(\vec{R})\,\mathcal{B}_\beta - \sum_L \sum_\beta K^L_{\alpha\beta}(\vec{R})\,m^L_\beta + \ldots \quad (5.64)$$

where $\sigma_{\alpha\beta}(\vec{R})$ is a cartesian component the **magnetic shielding tensor field** (Jensen and Hansen, 1999) and we define $K^L_{\alpha\beta}(\vec{R})$ as a cartesian component of the **reduced indirect nuclear spin-spin coupling tensor field**. They can therefore be defined[11] as first derivatives of the molecular magnetic induction $\vec{\mathcal{B}}^j(\vec{R})$ in the presence of an external magnetic induction $\vec{\mathcal{B}}$ and the magnetic moment of a nucleus L

$$\sigma_{\alpha\beta}(\vec{R}) = -\left.\frac{\partial \mathcal{B}^j_\alpha(\vec{R},\vec{\mathcal{B}})}{\partial \mathcal{B}_\beta}\right|_{|\vec{\mathcal{B}}|=0} \quad (5.65)$$

$$K^L_{\alpha\beta}(\vec{R}) = -\left.\frac{\partial \mathcal{B}^j_\alpha(\vec{R},\vec{m}^L)}{\partial m^L_\beta}\right|_{|\vec{m}^L|=0} \quad (5.66)$$

The product $\boldsymbol{\sigma}(\vec{R})\vec{\mathcal{B}}$ is the contribution to the induced electronic magnetic induction at point $\vec{R}$ coming from changes in the motion of the electrons due to an external magnetic induction, while the product $\boldsymbol{K}^L(\vec{R})\,\vec{m}^L$ is the additional electronic magnetic induction induced by the magnetic moment $\vec{m}^L$ and thus by the nuclear spin $\vec{I}^L$.

The magnetic shielding tensor field $\boldsymbol{\sigma}(\vec{R})$ is a generalization of the magnetic shielding tensor of NMR spectroscopy, introduced already in Section 5.4, to an arbitrary point $\vec{R}$ in space. This implies that the **nuclear magnetic shielding tensor** of a nucleus K is given as $\boldsymbol{\sigma}^K = \boldsymbol{\sigma}(\vec{R}_K)$. The **nuclear magnetic shielding constant** σ^K, which one is normally concerned with in liquid- or gas-phase NMR, is the trace of the nuclear magnetic shielding tensor

---

[11] These definitions are also included in Table B.2 of Appendix B.

110 *Magnetic Properties*

$$\sigma^K = \frac{1}{3} \sum_{\alpha\alpha} \sigma^K_{\alpha\alpha} \tag{5.67}$$

Another property, conceptually closely related to the magnetic shielding tensor field, is the nucleus independent chemical shift (NICS) (Schleyer et al., 1996), which is the generalization of the chemical shift (*vide infra*) to an arbitrary point in space. The reduced indirect nuclear spin-spin coupling tensor field $\boldsymbol{K}^L(\vec{R})$, finally, is the generalization of the **reduced indirect nuclear spin-spin coupling tensor** $\boldsymbol{K}^{KL}$ of nuclei K and L in such a way that the latter is the value of the reduced indirect nuclear spin-spin coupling tensor field at the position of the nuclear magnetic moment $\vec{m}^K$, i.e. $\boldsymbol{K}^{KL} = \boldsymbol{K}^L(\vec{R}_K)$. The reduced indirect nuclear spin-spin coupling constant K^{KL} between two nuclei K and L is again defined as the trace of the corresponding tensor

$$K^{KL} = \frac{1}{3} \sum_{\alpha\alpha} K^{KL}_{\alpha\alpha} \tag{5.68}$$

A nuclear magnetic moment $\vec{m}^K$ in a molecule thus experiences in the presence of an external magnetic induction $\vec{\mathcal{B}}$ not the pure external magnetic induction but a **local magnetic induction** $\vec{\mathcal{B}}^{loc}(\vec{R}_K)$, which is the sum of the external magnetic induction $\vec{\mathcal{B}}$ and the molecular magnetic induction $\vec{\mathcal{B}}^j$. In the following, we will consider only molecules without a permanent magnetic moment and therefore without a permanent molecular magnetic induction, $\vec{\mathcal{B}}^j = 0$, i.e. closed-shell molecules. The local magnetic induction at the position of nucleus K in a closed-shell molecule can then be written as

$$\mathcal{B}^{loc}_\alpha(\vec{R}_K) = \mathcal{B}_\alpha + \mathcal{B}^{j,ind}_\alpha(\vec{R}_K, \vec{\mathcal{B}}, \vec{m}^L) = \mathcal{B}_\alpha - \sum_\beta \sigma^K_{\alpha\beta} \mathcal{B}_\beta - \sum_{L \neq K} \sum_\beta K^{KL}_{\alpha\beta} m^L_\beta + \ldots$$

$$= \sum_\beta \left(\delta_{\alpha\beta} - \sigma^K_{\alpha\beta}\right) \mathcal{B}_\beta - \sum_{L \neq K} \sum_\beta K^{KL}_{\alpha\beta} m^L_\beta + \ldots \tag{5.69}$$

where $\delta_{\alpha\beta}$ is the Kronecker δ. We can recognize the two terms linear in the external magnetic induction as the well-known expression of NMR spectroscopy for the local field at nucleus K. This proves that our identification of the property $\sigma_{\alpha\beta}(\vec{R}_K)$ from Eq. (5.64) as the nuclear magnetic shielding tensor of nucleus K was indeed correct.

However, the shielding constant cannot be obtained from an NMR spectrum.[12] The parameter actually measured is the **chemical shift** δ, which is the relative difference of the Larmor or resonance frequency ν^K of nucleus K and the Larmor frequency $\nu^{K,ref}$ of the same nucleus in a reference molecule

$$\delta = \frac{\nu^K - \nu^{K,ref}}{\nu^{K,ref}} \times 10^6 \text{ in ppm} \tag{5.70}$$

The **Larmor** or **resonance frequency** is the frequency of an allowed transition between two nuclear spin states of nucleus K. The energy of a nuclear spin state in

---

[12] In Section 6.6 it will be shown that the paramagnetic contribution of the shielding tensor is related to the spin rotation constant, which can be measured in vibration-rotation spectra.

the presence of a local magnetic induction $\vec{\mathcal{B}}^{loc,K}$ is simply the interaction energy of the corresponding nuclear magnetic moment with the field, Eq. (5.17), i.e.

$$E_{m_{IK}} = -m_z^K \, \mathcal{B}_z^{loc}(\vec{R}_K) \qquad (5.71)$$

where we have used the standard notation that the direction of the magnetic field defines the z-axis. Inserting the relation between the nuclear magnetic moment and nuclear spin, Eq. (5.52), and using the fact that the z-component of the nuclear spin is quantized,

$$I_z = m_{IK} \hbar \qquad (5.72)$$

we can write the resonance frequency of an allowed transition ($\Delta m_{IK} = \pm 1$) as

$$\nu^K = \frac{|E_{m_{IK}+1} - E_{m_{IK}}|}{h} = \frac{g_K \mu_N}{h} \mathcal{B}_z^{loc}(\vec{R}_K) = \frac{g_K \mu_N}{h} \left(1 - \sigma^K\right) \mathcal{B}_z \qquad (5.73)$$

for a molecule in the gas or liquid phase.[13] The chemical shift can thus alternatively be written in terms of the shielding constants as

$$\delta = \frac{\sigma^{K,ref} - \sigma^K}{1 - \sigma^{K,ref}} \times 10^6 \approx \left(\sigma^{K,ref} - \sigma^K\right) \times 10^6 \qquad (5.74)$$

where $\sigma^{K,ref}$ is the nuclear magnetic shielding of the same type of nucleus in a reference substance added to the experimental sample. In order to distinguish the nuclear magnetic shielding σ^K from the chemical shift δ it is often also called the **absolute nuclear magnetic shielding**. From Eq. (5.74) we can see that a negative (positive) chemical shift implies that the nucleus is more (less) shielded than in the reference molecule.

The reduced indirect nuclear spin-spin coupling tensor $\mathbf{K}^{KL}$ is proportional to the normal **indirect nuclear spin-spin coupling tensor** $\mathbf{J}^{KL}$ measured in NMR spectra (Mills et al., 1993)

$$K_{\alpha\beta}^{KL} = \frac{h}{\mu_N^2 g_K g_L} J_{\alpha\beta}^{KL} \qquad (5.75)$$

where g_K and g_L are again the nuclear g-factors of the two nuclei K and L. The **indirect nuclear spin-spin coupling constant** J^{KL} finally is the trace of the corresponding tensor. The value of the reduced coupling constant K^{KL} is independent of the nuclear g-factors and thus independent of the particular isotope of a nucleus contrary to the measured coupling constant J^{KL}. Reduced spin-spin coupling constants K^{KL} involving e.g. the ^{1}H and ^{2}H isotopes of hydrogen are equal, whereas the normal coupling constants J^{KL} differ by the ratio of the two nuclear g-factors. Furthermore, reduced coupling constants between different pairs of nuclei differ only because of the different electronic environment and not also because of the differences in nuclear g-factors, which makes the comparison of reduced coupling constants often more meaningful.

---

[13] Obviously, for a molecule with more than one nucleus with spin there could be an additional contribution to the local field from the spin-spin couplings. However, this is not included in the definition of the chemical shift and is therefore omitted here.

On the other hand, it is the spin-spin coupling constant J^{KL}, which is traditionally defined through the **effective spin Hamiltonian** used in the analysis of NMR spectra

$$\hat{H}^{NMR} = -\sum_K \frac{g_K \mu_N}{\hbar} \sum_{\alpha\beta} \hat{I}^K_\alpha \left(\delta_{\alpha\beta} - \sigma^K_{\alpha\beta}\right) \mathcal{B}_\beta + \frac{2\pi}{\hbar} \sum_{K \neq L} \sum_{\alpha\beta} \hat{I}^K_\alpha \left(J^{KL}_{\alpha\beta} + D^{KL}_{\alpha\beta}\right) \hat{I}^L_\beta$$

(5.76)

Similar to the spin Hamiltonian of ESR spectroscopy, Eq. (5.53), $\hat{H}^{NMR}$ in Eq. (5.76) acts on nuclear spin states. $D^{KL}_{\alpha\beta}$ is a component of the **direct** through space **dipolar nuclear spin-spin coupling tensor** that, however, averages to zero in the gas and liquid phase.

All the correction terms to the pure nuclear spin–external magnetic induction term, $-\frac{g_K \mu_N}{\hbar} \vec{I}^K \cdot \vec{\mathcal{B}}$, involve interactions with the electrons and can therefore also be obtained from corrections to the electronic energy of the molecule. Namely, the interaction of the permanent and induced molecular magnetic induction, Eq. (5.64), with the magnetic moment $\vec{m}^K$ of the nuclei produces according to Eq. (5.50), $E = -\sum_\alpha m^K_\alpha \mathcal{B}^j_\alpha(\vec{R}_K)$, additional contributions to the electronic energy in Eq. (5.37). Following the same procedure as in Sections 4.4 and 5.4 one obtains

$$\Delta E(\vec{\mathcal{B}}, \{\vec{m}^K\}) = -\sum_K \sum_\alpha m^K_\alpha \mathcal{B}^j_\alpha + \sum_K \sum_{\alpha\beta} \sigma^K_{\alpha\beta} m^K_\alpha \mathcal{B}_\beta + \sum_{KL} \sum_{\alpha\beta} K^{KL}_{\alpha\beta} m^K_\alpha m^L_\beta + \ldots$$

$$= \sum_K \frac{g_K \mu_N}{\hbar} \left(-\sum_\alpha \hat{I}^K_\alpha \mathcal{B}^j_\alpha + \sum_{\alpha\beta} \sigma^K_{\alpha\beta} \hat{I}^K_\alpha \mathcal{B}_\beta\right) \quad (5.77)$$

$$+ \frac{2\pi}{\hbar} \sum_{KL} \sum_{\alpha\beta} J^{KL}_{\alpha\beta} \hat{I}^K_\alpha \hat{I}^L_\beta + \ldots$$

Comparison with the spin Hamiltonian Eq. (5.76) shows that the identification of the property $K^L_{\alpha\beta}(\vec{R}_K)$ in Eq. (5.64) as the reduced indirect nuclear spin-spin coupling tensor of nuclei K and L was correct and that the cartesian components of it can be defined as second derivative of the energy with respect to the components of the two nuclear magnetic moments $\vec{m}^K$ and $\vec{m}^L$

$$K^{KL} = \left.\frac{\partial^2 E(\vec{m}^K, \vec{m}^L)}{\partial m^K_\alpha \partial m^L_\beta}\right|_{|\vec{m}^K|=|\vec{m}^L|=0} \quad (5.78)$$

Furthermore, a comparison of Eq. (5.77) with Eq. (5.38) proves that $\sigma^K_{\alpha\beta}$ in Eq. (5.30) was indeed a component of the nuclear magnetic shielding tensor.

5.8 Quantum Mechanical Expression for the NMR Parameters

In this section, we are going to derive quantum mechanical expressions for the elements of the nuclear magnetic shielding and reduced indirect nuclear spin-spin coupling tensors, $\sigma^K_{\alpha\beta}$ and $K^{KL}_{\alpha\beta}$, of closed-shell molecules. According to Eqs. (5.40) and (5.78) they can be defined as second derivatives of the energy in the presence of an external magnetic induction $\vec{\mathcal{B}}$ and a nuclear magnetic moment $\vec{m}^K$ or two nuclear magnetic moments $\vec{m}^K$ and $\vec{m}^L$. Like always, we will use the perturbation theory expression for the perturbed energy, Eq. (3.15), but differentiate it now once with respect to a component of the nuclear magnetic moment m_α^K and once with respect to a component of the magnetic induction $\mathcal{B}_\beta$ in the case of the nuclear magnetic shielding tensor or once more with respect to a component of another nuclear magnetic moment m_β^L in the case of the reduced indirect nuclear spin-spin coupling tensor. We can therefore express the nuclear magnetic shielding tensor as

$$\sigma_{\alpha\beta}^K = \frac{\partial^2 E_0^{(2)}(\vec{\mathcal{B}}, \vec{m}^K)}{\partial m_\alpha^K \partial \mathcal{B}_\beta}\bigg|_{|\vec{\mathcal{B}}|=|\vec{m}^K|=0}$$

$$= \frac{\partial^2}{\partial m_\alpha^K \partial \mathcal{B}_\beta} \langle \Psi_0^{(0)} | \hat{H}^{(2)} | \Psi_0^{(0)} \rangle \bigg|_{|\vec{\mathcal{B}}|=|\vec{m}^K|=0}$$

$$+ \frac{\partial^2}{\partial m_\alpha^K \partial \mathcal{B}_\beta} \sum_{n \neq 0} \frac{\langle \Psi_0^{(0)} | \hat{H}^{(1)} | \Psi_n^{(0)} \rangle \langle \Psi_n^{(0)} | \hat{H}^{(1)} | \Psi_0^{(0)} \rangle}{E_0^{(0)} - E_n^{(0)}} \bigg|_{|\vec{\mathcal{B}}|=|\vec{m}^K|=0} \quad (5.79)$$

and the reduced indirect nuclear spin-spin coupling tensor as

$$K_{\alpha\beta}^{KL} = \frac{\partial^2 E_0^{(2)}(\vec{m}^K, \vec{m}^L)}{\partial m_\alpha^K \partial m_\beta^L}\bigg|_{|\vec{m}^K|=|\vec{m}^L|=0}$$

$$= \frac{\partial^2}{\partial m_\alpha^K \partial m_\beta^L} \langle \Psi_0^{(0)} | \hat{H}^{(2)} | \Psi_0^{(0)} \rangle \bigg|_{|\vec{m}^K|=|\vec{m}^L|=0}$$

$$+ \frac{\partial^2}{\partial m_\alpha^K \partial m_\beta^L} \sum_{n \neq 0} \frac{\langle \Psi_0^{(0)} | \hat{H}^{(1)} | \Psi_n^{(0)} \rangle \langle \Psi_n^{(0)} | \hat{H}^{(1)} | \Psi_0^{(0)} \rangle}{E_0^{(0)} - E_n^{(0)}} \bigg|_{|\vec{m}^K|=|\vec{m}^L|=0} \quad (5.80)$$

Again, as in the case of the magnetizability we have to use the full expression for the second-order energy correction including the diamagnetic $\langle \Psi_0^{(0)} | \hat{H}^{(2)} | \Psi_0^{(0)} \rangle$ terms. The first-order perturbation Hamiltonians for the interaction with an external magnetic induction and with a nuclear magnetic moment were already presented in Eqs. (5.20) and (5.56). They contain the orbital paramagnetic $\hat{O}_\alpha^{OP}$, Eq. (5.58), the Fermi contact $\hat{O}_\alpha^{FC}$, Eq. (5.60), and the spin-dipolar operator $\hat{O}_\alpha^{SD}$ Eq. (5.61).

But we also need the second-order perturbation Hamiltonians $\hat{H}^{(2)}$ that are bilinear in the external magnetic induction and a nuclear magnetic moment in the case of the shielding tensor and bilinear in two nuclear magnetic moments in the case of the coupling tensor. Inserting, therefore, the sum of the vector potential for an external field $\hat{\vec{A}}^\mathcal{B}(\vec{r}_i) = \frac{1}{2}\vec{\mathcal{B}} \times (\vec{r}_i - \vec{R}_{GO})$, Eq. (5.19), and for a nuclear magnetic moment $\vec{A}^K(\vec{r}_i) = \frac{\mu_0}{4\pi} \vec{m}^K \times \frac{(\vec{r}_i - \vec{R}_K)}{|\vec{r}_i - \vec{R}_K|^3}$, Eq. (5.55) in the general expression for the molecular Hamiltonian, Eq. (2.101), we can identify the two second-order perturbation Hamiltonian operators as [see Exercise 5.8]

Magnetic Properties

$$\hat{H}^{(2)} = \sum_K \sum_{\alpha\beta} \hat{O}_{\alpha\beta}^{m^K \mathcal{B}} \, m_\alpha^K \mathcal{B}_\beta \tag{5.81}$$

and

$$\hat{H}^{(2)} = \sum_{KL} \sum_{\alpha\beta} \hat{O}_{\alpha\beta}^{m^K m^L} \, m_\alpha^K m_\beta^L \tag{5.82}$$

where the perturbation operators are defined as[14]

$$\hat{O}_{\alpha\beta}^{m^K \mathcal{B}}(\vec{R}_{GO}) = \sum_i \hat{o}_{i,\alpha\beta}^{m^K \mathcal{B}}(\vec{R}_{GO}) \tag{5.83}$$

$$= \frac{e^2}{2m_e} \frac{\mu_0}{4\pi} \sum_i \left[(\vec{r}_i - \vec{R}_{GO}) \cdot \frac{(\vec{r}_i - \vec{R}_K)}{|\vec{r}_i - \vec{R}_K|^3} \delta_{\alpha\beta} - (r_{i,\alpha} - R_{GO,\alpha}) \frac{(r_{i,\beta} - R_{K,\beta})}{|\vec{r}_i - \vec{R}_K|^3} \right]$$

and

$$\hat{O}_{\alpha\beta}^{m^K m^L} = \sum_i \hat{o}_{i,\alpha\beta}^{m^K m^L} \tag{5.84}$$

$$= \frac{e^2}{2m_e} \left(\frac{\mu_0}{4\pi}\right)^2 \sum_i \left[\frac{(\vec{r}_i - \vec{R}_L)}{|\vec{r}_i - \vec{R}_L|^3} \cdot \frac{(\vec{r}_i - \vec{R}_K)}{|\vec{r}_i - \vec{R}_K|^3} \delta_{\alpha\beta} - \frac{(r_{i,\alpha} - R_{L,\alpha})}{|\vec{r}_i - \vec{R}_L|^3} \frac{(r_{i,\beta} - R_{K,\beta})}{|\vec{r}_i - \vec{R}_K|^3} \right]$$

The latter is often called the **orbital diamagnetic (OD)** or **diamagnetic nuclear spin-electron orbit operator (DSO)**, whereas the first could be called the diamagnetic shielding operator.

Exercise 5.8 Derive the second-order perturbation Hamiltonian, Eqs. (5.81) to (5.84), for the vector potential of a nuclear magnetic moment, $\vec{m}^K$ and an external magnetic induction, Eq. (5.19), by inserting the two vector potentials in the general expression of the molecular Hamiltonian, Eq. (2.101), retaining the second-order terms.

Taking the derivatives we arrive at the sum-over-states expressions for the nuclear magnetic shielding tensor

$$\sigma_{\alpha\beta}^K = \langle \Psi_0^{(0)} | \hat{O}_{\alpha\beta}^{m^K \mathcal{B}} | \Psi_0^{(0)} \rangle + \sum_{n \neq 0} \frac{\langle \Psi_0^{(0)} | \hat{O}_{K,\alpha}^{OP} | \Psi_n^{(0)} \rangle \langle \Psi_n^{(0)} | \hat{m}_\beta^l(\vec{R}_{GO}) | \Psi_0^{(0)} \rangle}{E_0^{(0)} - E_n^{(0)}}$$

$$+ \sum_{n \neq 0} \frac{\langle \Psi_0^{(0)} | \hat{m}_\beta^l(\vec{R}_{GO}) | \Psi_n^{(0)} \rangle \langle \Psi_n^{(0)} | \hat{O}_{K,\alpha}^{OP} | \Psi_0^{(0)} \rangle}{E_0^{(0)} - E_n^{(0)}} \tag{5.85}$$

$$= \sigma_{\alpha\beta}^{K,dia}(\vec{R}_{GO}) + \sigma_{\alpha\beta}^{K,para}(\vec{R}_{GO})$$

There are two contributions as in the case of the magnetizability that are called diamagnetic and paramagnetic term, in analogy to the magnetizability in Eq. (5.44). However, the two contributions have the opposite sign as their magnetizability counterparts. Equation (5.85) indicates that the two contributions also depend on the gauge origin but their sum, the shielding tensor, is independent. The separation in a

---

[14] All perturbation operators derived in this chapter are collected in Appendix A.

dia- and paramagnetic contribution is therefore again arbitrary and the diamagnetic contribution can also be rewritten in terms of a sum over all states (Geertsen, 1991; Smith et al., 1992; Sauer et al., 1994a,b; Lazzeretti et al., 1994) like the paramagnetic term as discussed in Section 5.9. A negative shielding constant implies that the nucleus is de-shielded and the local magnetic induction at the nucleus is larger than the external field.

Application of the second derivatives in Eq. (5.80) gives quantum mechanical expressions for the reduced indirect nuclear spin-spin coupling tensor

$$K_{\alpha\beta}^{KL} = K_{\alpha\beta}^{KL,OD} + K_{\alpha\beta}^{KL,OP} + K_{\alpha\beta}^{KL,FC} + K_{\alpha\beta}^{KL,SD} + K_{\alpha\beta}^{KL,FC/SD} \qquad (5.86)$$

where

$$K_{\alpha\beta}^{KL,OD} = 2\langle \Psi_0^{(0)} | \hat{O}_{\alpha\beta}^{m^K m^L} | \Psi_0^{(0)} \rangle \qquad (5.87)$$

$$K_{\alpha\beta}^{KL,OP} = \sum_{n\neq 0} \frac{\langle \Psi_0^{(0)} | \hat{O}_{K,\alpha}^{OP} | \Psi_n^{(0)} \rangle \langle \Psi_n^{(0)} | \hat{O}_{L,\beta}^{OP} | \Psi_0^{(0)} \rangle}{E_0^{(0)} - E_n^{(0)}}$$

$$+ \sum_{n\neq 0} \frac{\langle \Psi_0^{(0)} | \hat{O}_{L,\beta}^{OP} | \Psi_n^{(0)} \rangle \langle \Psi_n^{(0)} | \hat{O}_{K,\alpha}^{OP} | \Psi_0^{(0)} \rangle}{E_0^{(0)} - E_n^{(0)}} \qquad (5.88)$$

$$K_{\alpha\alpha}^{KL,FC} = \sum_{n\neq 0} \frac{\langle \Psi_0^{(0)} | \hat{O}_{K,\alpha}^{FC} | \Psi_n^{(0)} \rangle \langle \Psi_n^{(0)} | \hat{O}_{L,\alpha}^{FC} | \Psi_0^{(0)} \rangle}{E_0^{(0)} - E_n^{(0)}}$$

$$+ \sum_{n\neq 0} \frac{\langle \Psi_0^{(0)} | \hat{O}_{L,\alpha}^{FC} | \Psi_n^{(0)} \rangle \langle \Psi_n^{(0)} | \hat{O}_{K,\alpha}^{FC} | \Psi_0^{(0)} \rangle}{E_0^{(0)} - E_n^{(0)}} \qquad (5.89)$$

$$K_{\alpha\beta}^{KL,SD} = \sum_{n\neq 0} \frac{\langle \Psi_0^{(0)} | \hat{O}_{K,\alpha}^{SD} | \Psi_n^{(0)} \rangle \langle \Psi_n^{(0)} | \hat{O}_{L,\beta}^{SD} | \Psi_0^{(0)} \rangle}{E_0^{(0)} - E_n^{(0)}}$$

$$+ \sum_{n\neq 0} \frac{\langle \Psi_0^{(0)} | \hat{O}_{L,\beta}^{SD} | \Psi_n^{(0)} \rangle \langle \Psi_n^{(0)} | \hat{O}_{K,\alpha}^{SD} | \Psi_0^{(0)} \rangle}{E_0^{(0)} - E_n^{(0)}} \qquad (5.90)$$

$$K_{\alpha\beta}^{KL,FC/SD} = \sum_{n\neq 0} \frac{\langle \Psi_0^{(0)} | \hat{O}_{K,\alpha}^{FC} | \Psi_n^{(0)} \rangle \langle \Psi_n^{(0)} | \hat{O}_{L,\beta}^{SD} | \Psi_0^{(0)} \rangle}{E_0^{(0)} - E_n^{(0)}}$$

$$+ \sum_{n\neq 0} \frac{\langle \Psi_0^{(0)} | \hat{O}_{K,\alpha}^{SD} | \Psi_n^{(0)} \rangle \langle \Psi_n^{(0)} | \hat{O}_{L,\beta}^{FC} | \Psi_0^{(0)} \rangle}{E_0^{(0)} - E_n^{(0)}}$$

$$+ \sum_{n\neq 0} \frac{\langle \Psi_0^{(0)} | \hat{O}_{L,\beta}^{SD} | \Psi_n^{(0)} \rangle \langle \Psi_n^{(0)} | \hat{O}_{K,\alpha}^{FC} | \Psi_0^{(0)} \rangle}{E_0^{(0)} - E_n^{(0)}}$$

$$+ \sum_{n\neq 0} \frac{\langle \Psi_0^{(0)} | \hat{O}_{L,\beta}^{FC} | \Psi_n^{(0)} \rangle \langle \Psi_n^{(0)} | \hat{O}_{K,\alpha}^{SD} | \Psi_0^{(0)} \rangle}{E_0^{(0)} - E_n^{(0)}} \qquad (5.91)$$

116 *Magnetic Properties*

It consists of five contributions, where, however, the last contribution, the Fermi contact – spin dipolar cross-term is purely anisotropic and does not contribute to the trace of the tensor and thus to the coupling constant. The Fermi-contact term on the other hand is isotropic, i.e. the off-diagonal elements of the tensor all vanish. Furthermore the diagonal elements are all the same. Like the diamagnetic contributions to the magnetizability and nuclear magnetic shielding tensor, the orbital diamagnetic contribution to the indirect nuclear spin-spin coupling constant is normally written as an expectation value. However, it can be expressed alternatively as a sum over all states (Sauer, 1993) (see Section 5.9) like the diamagnetic contributions to the shielding and magnetizability tensor.

The Fermi contact, spin dipolar and their cross-term contain operators that include the electron spin operator $\hat{S}$. Application of these operators on a singlet reference state $|\Psi_0^{(0)}\rangle$ will give a linear combination of triplet states. The states $|\Psi_n^{(0)}\rangle$ thus have to be triplet states in order for the transition moments to be non-zero. For a singlet reference state the Fermi contact, spin dipolar and Fermi contact–spin dipolar cross-term therefore involve a sum over triplet states and triplet excitation energies $E_0^{(0)} - E_n^{(0)}$. The orbital paramagnetic operator, on the other hand, is spin free and the summation is therefore over states of the same spin symmetry as $|\Psi_0^{(0)}\rangle$.

The nuclear magnetic shielding and indirect nuclear spin-spin coupling tensor can also be obtained as derivatives of the molecular magnetic induction $\mathcal{B}_\alpha^j(\vec{R}_K, \vec{B}, \vec{m}^L)$ at the position of nucleus K with respect to either the external magnetic induction $\vec{B}$ or the nuclear magnetic moment of nucleus L, Eqs. (5.65) and (5.66). Furthermore the nuclear shielding tensor was also defined as the derivative of the molecular magnetic moment with respect to the nuclear magnetic moment of nucleus K, Eq. (5.33). Quantum mechanical expressions can thus be obtained as first derivatives of the corresponding perturbation-dependent first-order properties according to Eq. (3.41). The **operator for the molecular magnetic moment in the presence of nuclear magnetic moments** can be derived by Eq. (3.8)

$$\hat{m}_\alpha(\vec{R}_{GO}, \{\vec{m}^K\}) = \hat{O}_\alpha^{lB}(\vec{R}_{GO}) + \hat{O}_\alpha^{sB} - \sum_K \sum_\beta \hat{O}_{\beta\alpha}^{m^K B}(\vec{R}_{GO}) \, m_\beta^K$$

$$= \hat{m}_\alpha^l(\vec{R}_{GO}) + \hat{O}_\alpha^{sB} - \sum_K \sum_\beta \hat{O}_{\beta\alpha}^{m^K B}(\vec{R}_{GO}) \, m_\beta^K \quad (5.92)$$

The **operator for the molecular magnetic induction at nucleus K in the presence of an external magnetic induction $\vec{B}$ and nuclear magnetic moments** $\{\vec{m}^L\}$ is correspondingly given as

$$\mathcal{B}_\alpha^j(\vec{R}_K, \vec{B}, \{\vec{m}^L\}) = \hat{O}_\alpha^{lm^K} + \hat{O}_\alpha^{sm^K} - \sum_\beta \hat{O}_{\alpha\beta}^{m^K B}(\vec{R}_{GO}) \, B_\beta - 2 \sum_{L \neq K} \sum_\beta \hat{O}_{\alpha\beta}^{m^K m^L} m_\beta^L$$

$$= \hat{O}_{K,\alpha}^{OP} + \hat{O}_{K,\alpha}^{FC} + \hat{O}_{K,\alpha}^{SD} - \sum_\beta \hat{O}_{\alpha\beta}^{m^K B}(\vec{R}_{GO}) \, B_\beta$$

$$- 2 \sum_{L \neq K} \sum_\beta \hat{O}_{\alpha\beta}^{m^K m^L} m_\beta^L \quad (5.93)$$

Applying then the general expression for the first derivative of a perturbation-dependent expectation value, Eq. (3.41), to a component of the molecular magnetic induction $\mathcal{B}_\alpha^j(\vec{R}_K, \vec{\mathcal{B}})$ at the position of nucleus K in the presence of an external magnetic induction we obtain expressions for the tensor components of the nuclear magnetic shielding tensor of nucleus K involving first derivatives of the perturbed wavefunction

$$\sigma_{\alpha\beta}^K = -\left.\frac{\partial \mathcal{B}_\alpha^j(\vec{R}_K, \vec{\mathcal{B}})}{\partial \mathcal{B}_\beta}\right|_{|\vec{\mathcal{B}}|=0} = -\left.\frac{\partial}{\partial \mathcal{B}_\beta}\langle \Psi_0(\vec{\mathcal{B}})|\hat{\mathcal{B}}_\alpha^j(\vec{R}_K, \vec{\mathcal{B}})|\Psi_0(\vec{\mathcal{B}})\rangle\right|_{|\vec{\mathcal{B}}|=0} \quad (5.94)$$

$$= -\langle\Psi_0^{(0)}|\hat{O}_{K,\alpha}^{OP}|\left.\frac{\partial \Psi_0(\vec{\mathcal{B}})}{\partial \mathcal{B}_\beta}\right|_{|\vec{\mathcal{B}}|=0}\rangle - \langle\left.\frac{\partial \Psi_0(\vec{\mathcal{B}})}{\partial \mathcal{B}_\beta}\right|_{|\vec{\mathcal{B}}|=0}|\hat{O}_{K,\alpha}^{OP}|\Psi_0^{(0)}\rangle$$

$$+ \langle\Psi_0^{(0)}|\hat{O}_{\alpha\beta}^{m^K \mathcal{B}}(\vec{R}_{GO})|\Psi_0^{(0)}\rangle$$

Alternatively, we can take the first derivative of a component of the molecular magnetic dipole moment $m_\alpha(\vec{m}^K)$ in the presence of a magnetic nucleus K

$$\sigma_{\beta\alpha}^K = -\left.\frac{\partial m_\alpha(\vec{m}^K)}{\partial m_\beta^K}\right|_{|\vec{m}^K|=0} = -\left.\frac{\partial}{\partial m_\beta^K}\langle \Psi_0(\vec{m}^K)|\hat{m}_\alpha(\vec{R}_{GO}, \vec{m}^K)|\Psi_0(\vec{m}^K)\rangle\right|_{|\vec{m}^K|=0} \quad (5.95)$$

$$= -\langle\Psi_0^{(0)}|\hat{m}_\alpha^l(\vec{R}_{GO})|\left.\frac{\partial \Psi_0(\vec{m}^K)}{\partial m_\beta^K}\right|_{|\vec{m}^K|=0}\rangle - \langle\left.\frac{\partial \Psi_0(\vec{m}^K)}{\partial m_\beta^K}\right|_{|\vec{m}^K|=0}|\hat{m}_\alpha^l(\vec{R}_{GO})|\Psi_0^{(0)}\rangle$$

$$+ \langle\Psi_0^{(0)}|\hat{O}_{\beta\alpha}^{m^K \mathcal{B}}(\vec{R}_{GO})|\Psi_0^{(0)}\rangle$$

In both expressions we have assumed that we are looking at closed-shell molecules only and have therefore not included the contributions from $\hat{O}^{sB}$ and $\hat{O}^{sm^K}$.

Finally, taking the first derivative of the molecular magnetic induction $\mathcal{B}_\alpha^j(\vec{R}_K, \vec{m}^L)$ at the position of nucleus K in the presence of a magnetic nucleus L we obtain expressions for the reduced indirect nuclear spin-spin coupling constant of nuclei K and L

$$K_{\alpha\beta}^{KL} = -\left.\frac{\partial \mathcal{B}_\alpha^j(\vec{R}_K, \vec{m}^L)}{\partial m_\beta^L}\right|_{|\vec{m}^L|=0} = -\left.\frac{\partial}{\partial m_\beta^L}\langle \Psi_0(\vec{m}^L)|\hat{\mathcal{B}}_\alpha^j(\vec{R}_K, \vec{m}^L)|\Psi_0(\vec{m}^L)\rangle\right|_{|\vec{m}^L|=0} \quad (5.96)$$

$$= -\langle\Psi_0^{(0)}|\hat{O}_{K,\alpha}^{OP} + \hat{O}_{K,\alpha}^{FC} + \hat{O}_{K,\alpha}^{SD}|\left.\frac{\partial \Psi_0(\vec{m}^L)}{\partial m_\beta^L}\right|_{|\vec{m}^L|=0}\rangle$$

$$- \langle\left.\frac{\partial \Psi_0(\vec{m}^L)}{\partial m_\beta^L}\right|_{|\vec{m}^L|=0}|\hat{O}_{K,\alpha}^{OP} + \hat{O}_{K,\alpha}^{FC} + \hat{O}_{K,\alpha}^{SD}|\Psi_0^{(0)}\rangle + 2\langle\Psi_0^{(0)}|\hat{O}_{\alpha\beta}^{m^K m^L}|\Psi_0^{(0)}\rangle$$

In Part III we will come back to these expressions and evaluate the derivatives of approximate wavefunctions. However, here we will use the response formalism

118 Magnetic Properties

as developed in Section 3.11. Using Eq. (3.116) we can express the derivatives of the perturbation-dependent expectation value in terms of polarization propagators or linear response functions and thus obtain for the tensor components of the nuclear magnetic shielding tensor of nucleus K

$$\sigma_{\alpha\beta}^{K} = -\frac{\partial}{\partial \mathcal{B}_{\beta}} \langle \Psi_0(\vec{\mathcal{B}}) | \hat{\mathcal{B}}_{\alpha}^{j}(\vec{R}_K, \vec{\mathcal{B}}) | \Psi_0(\vec{\mathcal{B}}) \rangle \bigg|_{|\vec{\mathcal{B}}|=0}$$

$$= -\frac{\partial}{\partial m_{\alpha}^{K}} \langle \Psi_0(\vec{m}^K) | \hat{m}_{\beta}(\vec{R}_{GO}, \vec{m}^K) | \Psi_0(\vec{m}^K) \rangle \bigg|_{|\vec{m}^K|=0}$$

$$= \langle \Psi_0^{(0)} | \hat{O}_{\alpha\beta}^{m^K \mathcal{B}}(\vec{R}_{GO}) | \Psi_0^{(0)} \rangle + \langle\langle \hat{O}_{K,\alpha}^{OP} ; \hat{m}_{\beta}^{l}(\vec{R}_{GO}) \rangle\rangle_{\omega=0} \quad (5.97)$$

and of the reduced indirect nuclear spin-spin coupling constant of nuclei K and L

$$K_{\alpha\beta}^{KL} = -\frac{\partial}{\partial m_{\beta}^{L}} \langle \Psi_0(\vec{m}^L) | \hat{\mathcal{B}}_{\alpha}^{j}(\vec{R}_K, \vec{m}^L) | \Psi_0(\vec{m}^L) \rangle \bigg|_{|\vec{m}^L|=0} \quad (5.98)$$

$$= 2\langle \Psi_0^{(0)} | \hat{O}_{\alpha\beta}^{m^K m^L} | \Psi_0^{(0)} \rangle + \langle\langle \hat{O}_{K,\alpha}^{OP} ; \hat{O}_{L,\beta}^{OP} \rangle\rangle_{\omega=0}$$

$$+ \langle\langle \hat{O}_{K,\alpha}^{FC} + \hat{O}_{K,\alpha}^{SD} ; \hat{O}_{L,\beta}^{FC} + \hat{O}_{L,\beta}^{SD} \rangle\rangle_{\omega=0}$$

Analogous to the magnetizability, we can see that the diamagnetic contributions to the nuclear magnetic shielding and the reduced indirect nuclear spin-spin coupling tensors are normally not expressed as linear response functions. However, these diamagnetic contributions can also be reformulated as a sum-over-states or polarization propagator expression, as will be shown in Section 5.9.

Inserting the perturbation theory expansion of the perturbed wavefunction up to first order, Eq. (3.27), or the expression for the static response function Eq. (3.114) yields the same expressions for the nuclear magnetic shielding tensor and reduced indirect nuclear spin-spin coupling tensor as derived above.

5.9 Sum-over-States Expression for Diamagnetic Terms

The three second-order magnetic properties, the magnetizability, nuclear magnetic shielding and reduced indirect nuclear spin-spin coupling tensor, all consist of two contributions: a linear response or sum-over-states term with two first-order perturbation Hamiltonians, which is often called the paramagnetic term, and a ground-state expectation value of a second-order perturbation Hamiltonian, which is called the diamagnetic term. This asymmetry has profound consequences for the gauge-origin dependence of the magnetizability and shielding tensor, as will be discussed in Section 5.10. It is therefore desirable to find the same type of expression for both terms. Unfortunately, it seems impossible to find ground-state-expectation-value expressions for the paramagnetic terms, which would be highly desirable, because ground-state expectation values are computationally much simpler. On the other hand, it is possible to rewrite the diamagnetic contributions to the magnetizability $\xi_{\alpha\beta}^{dia}(\vec{R}_{GO})$, nuclear magnetic shielding $\sigma_{\alpha\beta}^{K,dia}(\vec{R}_{GO})$ and reduced indirect nuclear spin-spin coupling tensors $K_{\alpha\beta}^{KL,OD}$ (Geertsen, 1989, 1991; Smith et al., 1992; Sauer, 1993; Sauer et al., 1994a,b; Lazzeretti et al., 1994) as a sum-over-states or linear response functions that will be denoted with a superscript "Δ"

$$\xi_{\alpha\beta}^{\Delta}(\vec{R}_{GO}) = -\frac{1}{2m_e}\sum_{n\neq 0}\frac{\langle\Psi_0^{(0)}|\left[\hat{\vec{\mu}}(\vec{R}_{GO})\times\hat{\vec{m}}^l(\vec{R}_{GO})\right]_\alpha|\Psi_n^{(0)}\rangle\langle\Psi_n^{(0)}|\hat{O}_\beta^p|\Psi_0^{(0)}\rangle}{E_0^{(0)} - E_n^{(0)}}$$

$$-\frac{1}{2m_e}\sum_{n\neq 0}\frac{\langle\Psi_0^{(0)}|\hat{O}_\beta^p|\Psi_n^{(0)}\rangle\langle\Psi_n^{(0)}|\left[\hat{\vec{\mu}}(\vec{R}_{GO})\times\hat{\vec{m}}^l(\vec{R}_{GO})\right]_\alpha|\Psi_0^{(0)}\rangle}{E_0^{(0)} - E_n^{(0)}}$$

$$= -\frac{1}{2m_e}\langle\langle\left[\hat{\vec{\mu}}(\vec{R}_{GO})\times\hat{\vec{m}}^l(\vec{R}_{GO})\right]_\alpha ; \hat{O}_\beta^p\rangle\rangle_{\omega=0} \quad (5.99)$$

$$\sigma_{\alpha\beta}^{K,\Delta}(\vec{R}_{GO}) = \frac{1}{m_e c^2}\sum_{n\neq 0}\frac{\langle\Psi_0^{(0)}|\left[\hat{\vec{O}}^\mu(\vec{R}_K)\times\hat{\vec{m}}^l(\vec{R}_{GO})\right]_\alpha|\Psi_n^{(0)}\rangle\langle\Psi_n^{(0)}|\hat{O}_\beta^p|\Psi_0^{(0)}\rangle}{E_0^{(0)} - E_n^{(0)}}$$

$$+\frac{1}{m_e c^2}\sum_{n\neq 0}\frac{\langle\Psi_0^{(0)}|\hat{O}_\beta^p|\Psi_n^{(0)}\rangle\langle\Psi_n^{(0)}|\left[\hat{\vec{O}}^\mu(\vec{R}_K)\times\hat{\vec{m}}^l(\vec{R}_{GO})\right]_\alpha|\Psi_0^{(0)}\rangle}{E_0^{(0)} - E_n^{(0)}}$$

$$= \frac{1}{m_e c^2}\langle\langle\left[\hat{\vec{O}}^\mu(\vec{R}_K)\times\hat{\vec{m}}^l(\vec{R}_{GO})\right]_\alpha ; \hat{O}_\beta^p\rangle\rangle_{\omega=0} \quad (5.100)$$

$$K_{\alpha\beta}^{KL,\Delta} = \frac{1}{m_e c^2}\sum_{n\neq 0}\frac{\langle\Psi_0^{(0)}|\left[\hat{\vec{O}}^\mu(\vec{R}_K)\times\hat{\vec{O}}_L^{OP}\right]_\alpha|\Psi_n^{(0)}\rangle\langle\Psi_n^{(0)}|\hat{O}_\beta^p|\Psi_0^{(0)}\rangle}{E_0^{(0)} - E_n^{(0)}}$$

$$+\frac{1}{m_e c^2}\sum_{n\neq 0}\frac{\langle\Psi_0^{(0)}|\hat{O}_\beta^p|\Psi_n^{(0)}\rangle\langle\Psi_n^{(0)}|\left[\hat{\vec{O}}^\mu(\vec{R}_K)\times\hat{\vec{O}}_L^{OP}\right]_\alpha|\Psi_0^{(0)}\rangle}{E_0^{(0)} - E_n^{(0)}}$$

$$= \frac{1}{m_e c^2}\langle\langle\left[\hat{\vec{O}}^\mu(\vec{R}_K)\times\hat{\vec{O}}_L^{OP}\right]_\alpha ; \hat{O}_\beta^p\rangle\rangle_{\omega=0} \quad (5.101)$$

where $\hat{\vec{\mu}}$, $\hat{\vec{m}}^l$ and $\hat{\vec{O}}^\mu$ are the electric and magnetic dipole, Eqs. (4.30) and (5.21), and electric-field operators, Eq. (4.95), while $\hat{\vec{O}}^p$ is the total canonical momentum operator of the electrons, whose cartesian components are defined in Eq. (3.65).

In order to prove this for all three properties we start by rewriting the operators of the diamagnetic terms as

$$\hat{O}^{dia} = f\left(\hat{\vec{O}}_1\cdot\hat{\vec{O}}_2\, \boldsymbol{I}_3 - \hat{\vec{O}}_1\otimes\hat{\vec{O}}_2\right) \quad (5.102)$$

where $\boldsymbol{I}_3$ is the 3×3 unit matrix and the vector products $\hat{\vec{O}}_1\cdot\hat{\vec{O}}_2$ and $\hat{\vec{O}}_1\otimes\hat{\vec{O}}_2 = \hat{\vec{O}}_1\hat{\vec{O}}_2^T$ are the inner and outer or dyadic product of two vectors, which give a scalar and a

120 Magnetic Properties

Table 5.1 Operators and factors of the diamagnetic contributions to the magnetizability, nuclear magnetic shielding and reduced indirect nuclear spin-spin coupling tensors

	f	$\hat{O}_1$	$\hat{O}_2$
ξ^{dia}	$-\dfrac{e^2}{4m_e}$	$\hat{\vec{\mu}}(\vec{R}_{GO})$	$\hat{\vec{\mu}}(\vec{R}_{GO})$
$\sigma^{K,dia}$	$-\dfrac{1}{2m_e c^2}$	$\hat{\vec{\mu}}(\vec{R}_{GO})$	$\hat{O}^\mu(\vec{R}_K)$
$K^{KL,dia}$	$\dfrac{1}{2m_e c^4}$	$\hat{O}^\mu(\vec{R}_L)$	$\hat{O}^\mu(\vec{R}_K)$

3×3 matrix with elements $\left(\hat{\vec{O}}_1 \otimes \hat{\vec{O}}_2\right)_{\alpha\beta} = \hat{O}_{1,\alpha}\hat{O}_{2,\beta}$, respectively. The constants f and the operators $\hat{\vec{O}}_1$ and $\hat{\vec{O}}_2$ are given in Table 5.1.

The operators in Eq. (5.102) can be written as the following commutator [see Exercise 5.9]

$$\hat{\vec{O}}_1 \cdot \hat{\vec{O}}_2 \, I_3 - \hat{\vec{O}}_1 \otimes \hat{\vec{O}}_2 = -\frac{i}{\hbar}\left[\hat{\vec{O}}_2 \times \left(\hat{\vec{O}}_1 \times \hat{\vec{O}}^p\right), (\hat{\vec{O}}^r)^T\right] \qquad (5.103)$$

where $\hat{\vec{O}}^r$ is the sum of the position operators of all electrons as defined in Eq. (3.63).

Exercise 5.9 Prove that the operators for the diamagnetic contributions to the magnetizability, nuclear magnetic shielding and reduced indirect nuclear spin-spin coupling tensors can be written as the commutator given in Eq. (5.103).

Hint: Start by showing that $\left[\hat{\vec{O}}_1, (\hat{\vec{O}}^r)^T\right] = \left[\hat{\vec{O}}_2, (\hat{\vec{O}}^r)^T\right] = \mathbf{0}$

The diamagnetic contributions can therefore be written as

$$\begin{aligned}\langle\Psi_0^{(0)}|\hat{O}^{dia}|\Psi_0^{(0)}\rangle &= \frac{f}{i\hbar}\langle\Psi_0^{(0)}|\left[\hat{\vec{O}}_2 \times \left(\hat{\vec{O}}_1 \times \hat{\vec{O}}^p\right), (\hat{\vec{O}}^r)^T\right]|\Psi_0^{(0)}\rangle \\ &= \frac{f}{i\hbar}\left\{\langle\Psi_0^{(0)}|\left[\hat{\vec{O}}_2 \times \left(\hat{\vec{O}}_1 \times \hat{\vec{O}}^p\right)\right](\hat{\vec{O}}^r)^T|\Psi_0^{(0)}\rangle \right. \\ &\quad \left. -\langle\Psi_0^{(0)}|\hat{\vec{O}}^r\left[\hat{\vec{O}}_2 \times \left(\hat{\vec{O}}_1 \times \hat{\vec{O}}^p\right)\right]^T|\Psi_0^{(0)}\rangle\right\} \end{aligned} \qquad (5.104)$$

We can then insert the resolution of the identity, $\sum_n |\Psi_n^{(0)}\rangle\langle\Psi_n^{(0)}| = 1$, between the operators $\hat{\vec{O}}_2 \times \left(\hat{\vec{O}}_1 \times \hat{\vec{O}}^p\right)$ and $\hat{\vec{O}}^r$, leading to

$$\langle\Psi_0^{(0)}|\hat{O}^{dia}|\Psi_0^{(0)}\rangle = \frac{f}{i\hbar}\sum_n \left\{\langle\Psi_0^{(0)}|\hat{\vec{O}}_2\times\left(\hat{\vec{O}}_1\times\hat{\vec{O}}^p\right)|\Psi_n^{(0)}\rangle\langle\Psi_n^{(0)}|(\hat{\vec{O}}^r)^T|\Psi_0^{(0)}\rangle \right. \quad (5.105)$$

$$\left. -\langle\Psi_0^{(0)}|\hat{\vec{O}}^r|\Psi_n^{(0)}\rangle\langle\Psi_n^{(0)}|\left[\hat{\vec{O}}_2\times\left(\hat{\vec{O}}_1\times\hat{\vec{O}}^p\right)\right]^T|\Psi_0^{(0)}\rangle\right\}$$

Finally, using the off-diagonal hypervirial relation, Eq. (3.66) we arrive at the desired sum-over-states expression in Eqs. (5.99)–(5.101)

$$\langle\Psi_0^{(0)}|\hat{O}^{dia}|\Psi_0^{(0)}\rangle = \frac{f}{m_e}\sum_{n\neq 0}\frac{\langle\Psi_0^{(0)}|\hat{\vec{O}}_2\times\left(\hat{\vec{O}}_1\times\hat{\vec{O}}^p\right)|\Psi_n^{(0)}\rangle\langle\Psi_n^{(0)}|(\hat{\vec{O}}^p)^T|\Psi_0^{(0)}\rangle}{E_0^{(0)}-E_n^{(0)}}$$

$$+\frac{f}{m_e}\sum_{n\neq 0}\frac{\langle\Psi_0^{(0)}|\hat{\vec{O}}^p|\Psi_n^{(0)}\rangle\langle\Psi_n^{(0)}|\left[\hat{\vec{O}}_2\times\left(\hat{\vec{O}}_1\times\hat{\vec{O}}^p\right)\right]^T|\Psi_0^{(0)}\rangle}{E_0^{(0)}-E_n^{(0)}}$$

(5.106)

where $n=0$ was removed from the summation, because $\langle\Psi_0^{(0)}|\hat{\vec{O}}^p|\Psi_0^{(0)}\rangle = \mathbf{0}$ according to the hypervirial theorem Eq. (3.67).

5.10 The Gauge-Origin Problem

In Section 2.9, we have discussed the fact that the vector potential is not uniquely defined, i.e. adding the gradient of the so-called gauge function $\chi(\vec{r},t)$ to it will leave the magnetic induction unchanged but will change the Hamiltonian, see Eq. (2.115). Gauge invariance for the Schrödinger equation can, however, be obtained by similarly gauge transforming the wavefunction, Eq. (2.117). The physics is thus not changed by such a gauge transformation, which implies that all quantum mechanical expressions for properties should also be invariant under these gauge transformations.

A particularly important gauge function is $\chi(\vec{r}_i) = -\frac{1}{2}\vec{B}\times\vec{R}_{GO}\cdot\vec{r}_i$, Eq. (2.119), which leads to a dependence of the vector potential on an arbitrary gauge origin $\vec{R}_{GO}$. In the derivation of the magnetic properties in the previous sections we had already used a vector potential of this form. As a consequence, the dia- and paramagnetic contributions to the magnetizability tensor, Eq. (5.44), depend quadratically on this gauge origin, whereas the two contributions to the nuclear magnetic shielding tensor, Eq. (5.85), only linearly. For each property the sum of both contributions, however, must be independent of any kind of gauge transformation and thus also of this gauge origin $\vec{R}_{GO}$.

It can be shown [see Exercise 5.10] that this is indeed the case, if we assume that the wavefunctions are exact solutions of the unperturbed Schrödinger equation Eq. (3.14).

Exercise 5.10 The expressions for the diamagnetic and paramagnetic contributions to the magnetizability and nuclear magnetic shielding tensor in equations (5.44) and (5.85) depend

122 Magnetic Properties

on the gauge origin $\vec{R}_{GO}$ of the vector potential $\vec{A}(\vec{r}_i)$, Eq. (2.121), for the magnetic induction. Show that the sum of the diamagnetic and paramagnetic contributions, however, is independent of $\vec{R}_{GO}$.

Hint: Replace the gauge origin $\vec{R}_{GO}$ in Eq. (5.44) and Eq. (5.85) by $\vec{R}_{GO} + \vec{D}$, isolate the terms that depend on the arbitrary change $\vec{D}$ in the gauge origin and show that these terms cancel. Furthermore you might want to use the hypervirial relation Eq. (3.66) and the fact that the set of excited states $\Psi_n^{(0)}$ is complete, i.e.

$$\sum_n |\Psi_n^{(0)}\rangle\langle\Psi_n^{(0)}| = 1$$

However, for the latter discussion of approximate methods in Part III it is useful to derive explicit expressions for the gauge dependence of the trace of both properties. For a change in the gauge origin from $\vec{R}_{GO}$ to $\vec{R}_{GO} + \vec{D}$ one obtains [see Exercise 5.11]

$$\xi(\vec{R}_{GO} + \vec{D}) = \xi(\vec{R}_{GO}) + \vec{C}_1^\xi(\vec{R}_{GO})\vec{D} + \vec{D}\,\boldsymbol{C}_2^\xi\,\vec{D} \tag{5.107}$$

$$\sigma(\vec{R}_{GO} + \vec{D}) = \sigma(\vec{R}_{GO}) + \vec{C}_1^\sigma\,\vec{D} \tag{5.108}$$

where the gauge-origin dependence tensors are given as

$$\begin{aligned}
C_{1,\alpha}^\xi(\vec{R}_{GO}) = &-\frac{e}{3m_e}\sum_{\beta\gamma}\epsilon_{\alpha\beta\gamma}\Bigg(\sum_{n\neq 0}\frac{\langle\Psi_0^{(0)}|\hat{O}_\beta^p|\Psi_n^{(0)}\rangle\langle\Psi_n^{(0)}|\hat{m}_\gamma^l(\vec{R}_{GO})|\Psi_0^{(0)}\rangle}{E_0^{(0)}-E_n^{(0)}} \\
&-\sum_{n\neq 0}\frac{\langle\Psi_0^{(0)}|\hat{m}_\beta^l(\vec{R}_{GO})|\Psi_n^{(0)}\rangle\langle\Psi_n^{(0)}|\hat{O}_\gamma^p|\Psi_0^{(0)}\rangle}{E_0^{(0)}-E_n^{(0)}}\Bigg) \\
&-\frac{e}{3m_e}\langle\Psi_0^{(0)}|\hat{\mu}_\alpha(\vec{R}_{GO})|\Psi_0^{(0)}\rangle
\end{aligned} \tag{5.109}$$

$$\begin{aligned}
C_{2,\alpha\beta}^\xi = &-\frac{e^2}{12m_e^2}\Bigg(\sum_\gamma \delta_{\alpha\beta}\sum_{n\neq 0}\frac{\langle\Psi_0^{(0)}|\hat{O}_\gamma^p|\Psi_n^{(0)}\rangle\langle\Psi_n^{(0)}|\hat{O}_\gamma^p|\Psi_0^{(0)}\rangle}{E_0^{(0)}-E_n^{(0)}} \\
&-\sum_{n\neq 0}\frac{\langle\Psi_0^{(0)}|\hat{O}_\alpha^p|\Psi_n^{(0)}\rangle\langle\Psi_n^{(0)}|\hat{O}_\beta^p|\Psi_0^{(0)}\rangle}{E_0^{(0)}-E_n^{(0)}}\Bigg) \\
&-\frac{e^2}{6m_e}N\delta_{\alpha\beta}
\end{aligned} \tag{5.110}$$

$$C_{1,\alpha}^{\sigma} = -\frac{e}{6m_e} \sum_{\beta\gamma} \epsilon_{\alpha\beta\gamma} \left(\sum_{n\neq 0} \frac{\langle \Psi_0^{(0)} | \hat{O}_{\beta}^p | \Psi_n^{(0)} \rangle \langle \Psi_n^{(0)} | \hat{O}_{K,\gamma}^{OP} | \Psi_0^{(0)} \rangle}{E_0^{(0)} - E_n^{(0)}} \right.$$

$$\left. - \sum_{n\neq 0} \frac{\langle \Psi_0^{(0)} | \hat{O}_{K,\beta}^{OP} | \Psi_n^{(0)} \rangle \langle \Psi_n^{(0)} | \hat{O}_{\gamma}^p | \Psi_0^{(0)} \rangle}{E_0^{(0)} - E_n^{(0)}} \right)$$

$$- \frac{e}{3m_e c^2} \langle \Psi_0^{(0)} | \hat{O}_{\alpha}^{\mu}(\vec{R}_K) | \Psi_0^{(0)} \rangle \quad (5.111)$$

where $\epsilon_{\alpha\beta\gamma}$ is the Levi-Civita symbol (Mills et al., 1993), which is defined as

$$\epsilon_{\alpha\beta\gamma} = \begin{cases} +1 & \text{if } \alpha\beta\gamma \text{ is an even permutation of } x,y,z \\ -1 & \text{if } \alpha\beta\gamma \text{ is an odd permutation of } x,y,z \\ 0 & \text{otherwise} \end{cases} \quad (5.112)$$

Alternatively, we can write them more compactly in terms of polarization propagators as

$$C_{1,\alpha}^{\xi}(\vec{R}_{GO}) = -\frac{e}{3m_e} \langle\langle \hat{O}_{\beta}^p ; \hat{m}_{\gamma}^l(\vec{R}_{GO}) \rangle\rangle_{\omega=0} - \frac{e}{3m_e} \langle \Psi_0^{(0)} | \hat{\mu}_{\alpha}(\vec{R}_{GO}) | \Psi_0^{(0)} \rangle \quad (5.113)$$

$$C_{2,\alpha\beta}^{\xi} = -\frac{e^2}{24m_e^2} \left(\sum_{\gamma} \delta_{\alpha\beta} \langle\langle \hat{O}_{\gamma}^p ; \hat{O}_{\gamma}^p \rangle\rangle_{\omega=0} - \langle\langle \hat{O}_{\alpha}^p ; \hat{O}_{\beta}^p \rangle\rangle_{\omega=0} \right) - \frac{e^2}{6m_e} N\delta_{\alpha\beta}$$

$$C_{1,\alpha}^{\sigma} = -\frac{e}{6m_e} \sum_{\gamma} \epsilon_{\alpha\beta\gamma} \langle\langle \hat{O}_{\beta}^p ; \hat{O}_{K,\gamma}^{OP} \rangle\rangle_{\omega=0} - \frac{e}{3m_e c^2} \langle \Psi_0^{(0)} | \hat{O}_{\alpha}^{\mu}(\vec{R}_K) | \Psi_0^{(0)} \rangle$$

$$(5.114)$$

Exercise 5.11 Derive the expressions for gauge-origin dependence tensors of the trace of the magnetizability and nuclear magnetic shielding tensors in Eqs. (5.109) to (5.111).

Hint: Replace the gauge-origin $\vec{R}_{GO}$ in Eqs. (5.44) and (5.85) by $\vec{R}_{GO} + \vec{D}$, form the trace of the tensors and collect the terms that depend linearly or quadratically on $\vec{D}$.

In the case of exact unperturbed wavefunctions the gauge-dependence tensors are all zero, of course, but in the case of approximate methods this is not necessarily the case and they can then serve as a measure of the gauge-origin dependence of a particular result for the magnetizability or nuclear magnetic shielding.

Finally, it is actually possible for the case of the nuclear magnetic shielding tensor to derive an alternative expression for the diamagnetic contribution, such that the gauge-origin dependence disappears exactly from the diamagnetic and paramagnetic contribution. These expressions will then give gauge-origin-independent results for the nuclear magnetic shielding tensor also in the case of the approximate methods in part III. Several attempts have been made to achieve this (Geertsen, 1991; Smith et al., 1992; Sauer et al., 1994b; Lazzeretti et al., 1994). They are all based on the reformulation of the diamagnetic contribution as a sum over all states, as discussed in

124 Magnetic Properties

Section 5.9. The most useful one, because it gives gauge-origin-independent expressions for the whole shielding tensor is the **continuous transformation of origin of the current density**, whereby the **diamagnetic** contribution to the current density is set to **zero** (CTOCD-DZ) approach by Lazzeretti and coworkers (Lazzeretti et al., 1994). The CTOCD-DZ diamagnetic contribution is given as

$$\sigma_{\alpha\beta}^{K,\Delta} = \sum_{\gamma\delta} \epsilon_{\beta\gamma\delta} \left(\sum_{n\neq 0} \frac{\langle \Psi_0^{(0)} | \hat{O}_\gamma^p | \Psi_n^{(0)} \rangle \langle \Psi_n^{(0)} | \hat{O}_{K,\delta\alpha}^{CTOCD-DZ} | \Psi_0^{(0)} \rangle}{E_0^{(0)} - E_n^{(0)}} \right.$$

$$\left. + \sum_{n\neq 0} \frac{\langle \Psi_0^{(0)} | \hat{O}_{K,\delta\alpha}^{CTOCD-DZ} | \Psi_n^{(0)} \rangle \langle \Psi_n^{(0)} | \hat{O}_\gamma^p | \Psi_0^{(0)} \rangle}{E_0^{(0)} - E_n^{(0)}} \right) \quad (5.115)$$

$$= \sum_{\gamma\delta} \epsilon_{\beta\gamma\delta} \langle\langle \hat{O}_\gamma^p ; \hat{O}_{K,\delta\alpha}^{CTOCD-DZ} \rangle\rangle_{\omega=0} \quad (5.116)$$

where

$$\hat{O}_{K,\delta\alpha}^{CTOCD-DZ} = \frac{1}{4m_e} \left[\hat{\mu}_\delta(R_{GO}) \hat{O}_{K,\alpha}^{OP} + \hat{O}_{K,\alpha}^{OP} \hat{\mu}_\delta(R_{GO}) \right] \quad (5.117)$$

Exercise 5.12 Show that the sum of the paramagnetic contribution to the nuclear magnetic shielding tensor Eq. (5.85) and the CTOCD-DZ diamagnetic contribution Eq. (5.115) is independent of the gauge origin $\vec{R}_{GO}$ without making use of the hypervirial relation, Eq. (3.66), or the resolution of the identity, $\sum_n |\Psi_n^{(0)}\rangle\langle\Psi_n^{(0)}| = 1$.

5.11 Further Reading

Magnetic Multipole Expansion and Interaction with Magnetic Fields

D. W. Davies, *The Theory of the Electric and Magnetic Properties of Molecules*, John Wiley and Sons, London (1967): Appendix II.

V. D. Barger and M. G. Olsson, *Classical Electricity and Magnetism*, Allyn and Bacon, Boston (1987): Chapters 5–1, 6–4 and 7–1.

Magnetizability

W. H. Flygare, *Molecular Structure and Dynamics*, Prentice-Hall, Englewood Cliffs (1978): Chapter 6.8.

NMR Parameters

J. D. Memory, *Quantum Theory of Magnetic Resonance Parameters*, McGraw-Hill, New York (1968).

I. Ando and G. A. Webb, *Theory of NMR Parameters*, Academic Press, London (1983).

P. Lazzeretti, *Electric and Magnetic Properties of Molecules*, in *Handbook of Molecular Physics and Quantum Chemistry* ed. by S. Wilson, John Wiley & Sons, Chichester (2003): Volumen 3, Part 1, Chapter 3.

M. Kaupp, M. Bühl and V. G. Malkin, (ed.) *Calculation of NMR and EPR Parameters Theory and Applications*, Wiley-VCH, Weinheim (2004).

P. Atkins and R. Friedman, *Molecular Quantum Mechanics*, 4th edn. Oxford University Press, Oxford (2005): Chapters 13.11–13.13 and 13.17.

W. H. Flygare, *Molecular Structure and Dynamics*, Prentice-Hall, Englewood Cliffs (1978): Chapter 6.9.

EPR Parameters

M. Kaupp, M. Bühl and V. G. Malkin, (ed.) *Calculation of NMR and EPR Parameters Theory and Applications*, Wiley-VCH, Weinheim (2004).

P. Atkins and R. Friedman, *Molecular Quantum Mechanics*, 4th edn. Oxford University Press, Oxford (2005): Chapters 13.14–13.16.

6
Properties Related to Nuclear Motion

So far, we have restricted ourselves to the situation that the nuclei are fixed in space, i.e. we have considered molecular properties or contributions to the molecular properties that can be obtained from the electronic Schrödinger equation, Eq. (2.10), alone. In this chapter, and the following chapter we will finally lift this restriction and allow the nuclei to move again. In this chapter, we will look at properties that arise or at least have contributions due to a breakdown of the Born–Oppenheimer approximation. This means that in order to derive quantum mechanical expressions for these experimentally observable properties we have to take into account the coupling of nuclear and electronic motion, i.e. some of the terms that are neglected in the Born–Oppenheimer approximation.

6.1 Molecular Rotation as Source for Magnetic Moments

When a molecule rotates around its centre of nuclear masses R_{CM}, there are rotating charges, which give rise to an additional current density $\vec{j}^J(\vec{r})$. This current density can be expressed in terms of the angular momentum of the rotation $\vec{J}$ and the moment of inertia tensor $\boldsymbol{I}$ as

$$\vec{j}^J(\vec{r}) = \rho(\vec{r})\,\vec{v}(\vec{r}) = \rho(\vec{r})\,\vec{\omega} \times (\vec{r} - \vec{R}_{CM})$$
$$= \rho(\vec{r})(\boldsymbol{I}^{-1}\,\vec{J}) \times (\vec{r} - \vec{R}_{CM}) \qquad (6.1)$$

where $\vec{\omega} = \boldsymbol{I}^{-1}\,\vec{J}$ is the angular velocity of the rotating charges. According to Eqs. (5.10) this rotational current density will lead to a magnetic moment $\vec{m}^J$, called the **rotational magnetic moment**,

$$\vec{m}^{J,rig}(\vec{J}) = \frac{1}{2}\int_{\vec{r}} \rho(\vec{r})(\vec{r} - \vec{R}_{CM}) \times \left[(\boldsymbol{I}^{-1}\,\vec{J}) \times (\vec{r} - \vec{R}_{CM})\right]\,d\vec{r} \qquad (6.2)$$

In this expression, we have assumed that the charges rotate rigidly and therefore added the superscript "rig". This means in particular that the electrons move rigidly with the nuclear frame and do not lag behind. However, in a real molecule the electronic charge distribution will not rotate rigidly, but will be influenced by the fact that the nuclei rotate with angular momentum $\vec{J}$. This coupling between nuclear rotational motion and the motion of the electrons

Molecular Rotation as Source for Magnetic Moments

$$\hat{H}^{(1)} = \hat{\vec{L}}\, \mathbf{I}^{-1}\, \hat{\vec{J}} \tag{6.3}$$

is neglected in the Born–Oppenheimer approximation. The changes in the rotational magnetic moment and induction are thus a manifestation of the **breakdown of the Born–Oppenheimer approximation**.

There are several ways of deriving these contributions (Wick, 1933b, 1933a, 1948; Eshbach and Strandberg, 1952; Flygare and Benson, 1971). Here we will follow the original derivation (Wick, 1933b, 1933a) which makes use of the fact that going to the rotating frame of the nuclei leads, according to Larmor's theorem, (Rabi et al., 1954) to an apparent magnetic induction $\vec{\mathcal{B}}^J$ acting on the electrons

$$\vec{\mathcal{B}}^J = -\frac{2m_e}{e}\, \mathbf{I}^{-1}\, \vec{J} \tag{6.4}$$

for which we then can define a vector potential as

$$\vec{A}^J = -\frac{m_e}{e}\, \mathbf{I}^{-1}\, \vec{J} \times (\vec{r} - \vec{R}_{CM}) \tag{6.5}$$

In complete analogy to Eq. (5.30) this magnetic induction leads to an induced magnetic moment $\vec{m}^{J,ind}$ in addition to the magnetic moment $\vec{m}^{J,rig}$ of the rigidly rotating charge distribution

$$m_\alpha^J(\vec{J}) = m_\alpha^{J,rig}(\vec{J}) + m_\alpha^{J,ind}(\vec{J}) \tag{6.6}$$

Contrary to the definition of the properties in the previous chapters here one combines traditionally the rigid and induced or Born–Oppenheimer breakdown contribution in one property and defines only one property, the **rotational g tensor** g_J, as the proportionality tensor between the rotational magnetic moment $\vec{m}^J(\vec{J})$ and the rotational angular momentum $\vec{J}$

$$m_\alpha^J(\vec{J}) = \frac{\mu_N}{\hbar} \sum_\beta g_{J,\alpha\beta}\, J_\beta = \frac{\mu_N}{\hbar} \sum_\beta (g_{J,\alpha\beta}^{rig} + g_{J,\alpha\beta}^{ind})\, J_\beta \tag{6.7}$$

where $\mu_N = \frac{e\hbar}{2m_p}$ is the nuclear magneton again.

From the expansion in Eq. (6.7) we can define the components of the rotational g tensor as first derivatives[1] of the rotational magnetic moment with respect to a component of the rotational angular momentum $\vec{J}$ of the nuclei

$$g_{J,\alpha\beta} = \frac{\hbar}{\mu_N} \frac{\partial m_\alpha^J(\vec{J})}{\partial J_\beta}\bigg|_{|\vec{J}|=0} \tag{6.8}$$

The rotational magnetic moment $\vec{m}^J(\vec{J})$ can interact with an external magnetic induction $\vec{\mathcal{B}}$ like any other magnetic moment, Eq. (5.17). The resulting change in the energy is then

$$\Delta E(\vec{\mathcal{B}}, \vec{J}) = -\sum_\alpha \mathcal{B}_\alpha\, m_\alpha^J(\vec{J}) = -\frac{\mu_N}{\hbar} \sum_{\alpha\beta} \mathcal{B}_\alpha\, g_{J,\alpha\beta}\, J_\beta \tag{6.9}$$

---

[1] This definition is also given in the second column of Table B.3 of Appendix B.

128 *Properties Related to Nuclear Motion*

which allows us to define the components of the rotational g tensor as second derivatives of the corresponding interaction energies with respect to a component of the rotational angular momentum $\vec{J}$ of the nuclei and a component of the external magnetic induction[2]

$$g_{J,\alpha\beta} = -\frac{\hbar}{\mu_N} \left.\frac{\partial^2 E(\vec{\mathcal{B}},\vec{J})}{\partial \mathcal{B}_\alpha \partial J_\beta}\right|_{|\vec{J}|=|\vec{\mathcal{B}}|=0} \tag{6.10}$$

Although the rotational g tensor can be defined as second derivatives of the energy similar to the magnetizability, there exists a fundamental difference. The energy in Eq. (5.39) is the electronic energy, i.e. the solution to the electronic Schrödinger equation in the presence of scalar and vector potentials, Eq. (3.1), while the energy in Eq. (6.10) is in principle the solution to the full time-independent Schrödinger equation, Eq. (2.5), including the nuclear kinetic energy terms. Consequently, only the induced contributions can be obtained as derivatives of the electronic energy in the presence of the Born–Oppenheimer breakdown operator given in Eq. (6.3).

Nowadays, the diagonal components of the rotational g tensor in the principal axes coordinate system of the molecule[3] can be measured in many ways. The two original methods are both based on Eq. (6.9). Already in 1933 rotational magnetic moments and thus the rotational g tensor were measured by deflection of molecular beams in inhomogenous magnetic fields (Frisch and Stern, 1933; Estermann and Stern, 1933). Alternatively, one can study the changes in the rotational energies due to an external magnetic field, the so-called rotational Zeeman effect, which is normally expressed in terms of an effective rotational Hamiltonian as (Eshbach and Strandberg, 1952)

$$\hat{H}^{rot}(\vec{\mathcal{B}}) = \frac{\hat{J}_x^2}{2I_{xx}} + \frac{\hat{J}_y^2}{2I_{yy}} + \frac{\hat{J}_z^2}{2I_{zz}} - \frac{\mu_N}{\hbar}\vec{\mathcal{B}}\mathbf{g}_J\hat{\vec{J}} \tag{6.11}$$

and allows for the experimental determination of the rotational g tensor (Flygare and Benson, 1971; Flygare, 1974; Sutter and Flygare, 1976).

6.2 Quantum Mechanical Expression for the Rotational g Tensor

The derivation of the quantum mechanical expression for the rotational g tensor requires the derivation of quantum mechanical expressions for the rigid and induced contribution to the rotational magnetic moment. An expression for the first is most easily derived in analogy to the electric dipole moment in Section 4.3 by translating the classical expression, Eq. (6.2), to quantum mechanics. Before doing so, however, we want to make use of Lagrange's formula for a vector triple product [see Exercise 6.1]

$$\vec{A} \times \left(\vec{B} \times \vec{C}\right) = \vec{B}\left(\vec{A}\cdot\vec{C}\right) - \left(\vec{A}\cdot\vec{B}\right)\vec{C} \tag{6.12}$$

---

[2] This definition is also given in the last column of Table B.3 of Appendix B.
[3] In the principal axes coordinate system the moment-of-inertia tensor of the molecule is diagonal, but not necessarily the rotational g tensor.

and rewrite the classical expression for the rigid contribution to the rotational magnetic moment

$$\vec{m}^{J,rig}(\vec{J}) \tag{6.13}$$
$$= \frac{1}{2}\int_{\vec{r}} \rho(\vec{r}) \left[(\vec{r}-\vec{R}_{CM})\cdot(\vec{r}-\vec{R}_{CM})\,\mathbf{I}_3 - (\vec{r}-\vec{R}_{CM})\otimes(\vec{r}-\vec{R}_{CM})\right](\mathbf{I}^{-1}\,\vec{J})\,d\vec{r}$$

Exercise 6.1 Prove relation (6.12).

Comparison with Eq. (6.7) allows us to identify a component of the rigid contribution to the rotational g tensor as

$$g^{rig}_{J,\alpha\beta} = \frac{m_p}{e}\int_{\vec{r}} \rho(\vec{r})\left[(\vec{r}-\vec{R}_{CM})^2\delta_{\alpha\beta} - (r_\alpha - R_{CM,\alpha})(r_\beta - R_{CM,\beta})\right]\frac{1}{I_{\beta\beta}}\,d\vec{r} \tag{6.14}$$

In a molecule the charge distribution $\rho(\vec{r})$ consists of the discrete nuclear charges located at the points $\vec{R}_K$ and the continuous charge distribution $\rho^{el}(\vec{r})$ of the electrons. A quantum mechanical expression for the latter can be obtained again from Eq. (2.23). The rigid contribution to the rotational g tensor

$$g^{rig}_{J,\alpha\beta} = g^{nuc}_{J,\alpha\beta} + g^{rig,el}_{J,\alpha\beta} \tag{6.15}$$

thus has a nuclear

$$g^{nuc}_{J,\alpha\beta} = m_p \sum_K Z_K \left[(\vec{R}_K - \vec{R}_{CM})^2\delta_{\alpha\beta} - (R_{K,\alpha} - R_{CM,\alpha})(R_{K,\beta} - R_{CM,\beta})\right]\frac{1}{I_{\beta\beta}} \tag{6.16}$$

and an electronic contribution

$$g^{rig,el}_{J,\alpha\beta} = -\frac{2m_p}{m_e}\langle\Psi_0^{(0)}|\hat{O}^{JJ}_{\alpha\beta}(\vec{R}_{CM})|\Psi_0^{(0)}\rangle\frac{1}{I_{\beta\beta}} \tag{6.17}$$

where the electronic perturbation operator $\hat{O}^{JJ}_{\alpha\beta}(\vec{R}_{CM})$ is defined as

$$\hat{O}^{JJ}_{\alpha\beta}(\vec{R}_{CM}) = \sum_i^N \hat{o}^{JJ}_{i,\alpha\beta}(\vec{R}_{CM}) \tag{6.18}$$

$$= \frac{m_e}{2}\sum_i^N \left[(\vec{r}_i - \vec{R}_{CM})^2\delta_{\alpha\beta} - (r_{i,\alpha} - R_{CM,\alpha})(r_{i,\beta} - R_{CM,\beta})\right]$$

The derivation of the induced contribution, on the other hand, is very similar to the derivation for the magnetizability. We could start from the definition of the rotational g tensor as first derivative of the rotational magnetic moment, Eq. (6.8), which would then be the induced contribution to it, and use the response theory formalism of Section 3.11. Using Eq. (3.116) we could express the derivatives of the induced rotational magnetic moment in terms of a polarization propagator and ground-state expectation value. Here we will, however, make use of the definition as second

130 Properties Related to Nuclear Motion

derivative of the energy, Eq. (6.10). Applying this to the electronic energy in the presence of an external magnetic induction $\vec{B}$ and the internal Born–Oppenheimer breakdown perturbation in Eq. (6.3) we will obtain directly the induced contribution to the rotational g tensor.

But first, we have to derive the first- and second-order perturbation Hamiltonian operators [see Exercise 6.2]. In addition to the orbital magnetic dipole operator, Eq. (5.21), we also obtain a first-order operator from the vector potential due to the coupling with the rotation, Eq. (6.5) and second-order operator that is bilinear in the external magnetic induction and the coupling with the rotation

$$\hat{H}^{(1)} = \sum_{\alpha} \hat{O}_{\alpha}^{lJ}(\vec{R}_{CM})(\mathbf{I}^{-1}\,\vec{J})_{\alpha} \tag{6.19}$$

$$\hat{H}^{(2)} = \sum_{\alpha\beta} \hat{O}_{\alpha\beta}^{BJ}(\vec{R}_{CM},\vec{R}_{GO})\,B_{\alpha}(\mathbf{I}^{-1}\,\vec{J})_{\beta} \tag{6.20}$$

where the first- and second-order perturbation operators are given as

$$\hat{O}_{\alpha}^{lJ}(\vec{R}_{CM}) = \sum_{i}^{N} \hat{o}_{i,\alpha}^{lJ}(\vec{R}_{CM}) = -\sum_{i}^{N} \hat{l}_{i,\alpha}(\vec{R}_{CM}) \tag{6.21}$$

$$= \frac{2m_e}{e}\,\hat{m}_{\alpha}^{l}(\vec{R}_{CM})$$

and

$$\hat{O}_{\alpha\beta}^{BJ}(\vec{R}_{CM},\vec{R}_{GO}) = \sum_{i}^{N} \hat{o}_{i,\alpha\beta}^{BJ}(\vec{R}_{CM},\vec{R}_{GO}) \tag{6.22}$$

$$= -\frac{e}{2}\sum_{i}^{N}\left[(\vec{r}_i - \vec{R}_{CM})\cdot(\vec{r}_i - \vec{R}_{GO})\delta_{\alpha\beta} - (r_{i,\alpha} - R_{CM,\alpha})(r_{i,\beta} - R_{GO,\beta})\right]$$

One should note that both perturbation operators are similar to the corresponding operators for the magnetizability tensor, but not equal. Apart from constant factors, they differ such that for the rotational g tensor the first-order operator as well as one of the factors in the second-order operator are defined with respect to the nuclear centre of masses $\vec{R}_{CM}$ and not with respect to the arbitrary gauge origin $\vec{R}_{GO}$.

Exercise 6.2 Derive the second-order perturbation Hamiltonian for the induced contribution to the rotational g tensor, Eq. (6.22), by inserting the vector potentials, (5.19) and (6.5), in the general expression of the molecular Hamiltonian, Eq. (2.101), retaining the bilinear second-order term.

Using again perturbation theory for the perturbed energy we can obtain the second derivative of the electronic energy directly from the second-order correction to the energy

$$g^{ind}_{J,\alpha\beta} = -\frac{\hbar}{\mu_N} \left.\frac{\partial^2 E_0^{(2)}(\vec{\mathcal{B}},\vec{J})}{\partial \mathcal{B}_\alpha \partial J_\beta}\right|_{|\vec{J}|=|\vec{\mathcal{B}}|=0}$$

$$= m_p \frac{1}{I_{\beta\beta}} \left[\frac{4m_e}{e^2} \sum_{n\neq 0} \left(\frac{\langle \Psi_0^{(0)}|\hat{m}_\alpha^l(\vec{R}_{GO})|\Psi_n^{(0)}\rangle \langle \Psi_n^{(0)}|\hat{m}_\beta^l(\vec{R}_{CM})|\Psi_0^{(0)}\rangle}{E_0^{(0)} - E_n^{(0)}} \right.\right.$$

$$\left.+ \frac{\langle \Psi_0^{(0)}|\hat{m}_\beta^l(\vec{R}_{CM})|\Psi_n^{(0)}\rangle \langle \Psi_n^{(0)}|\hat{m}_\alpha^l(\vec{R}_{GO})|\Psi_0^{(0)}\rangle}{E_0^{(0)} - E_n^{(0)}} \right)$$

$$\left.- \frac{2}{e} \langle \Psi_0^{(0)}|\hat{O}^{\mathcal{B}J}_{\alpha\beta}(\vec{R}_{CM},\vec{R}_{GO})|\Psi_0^{(0)}\rangle \right] \quad (6.23)$$

The induced contribution consists therefore of a paramagnetic or sum-over-states contribution and a diamagnetic or ground-state expectation value term. Combining these with the contribution from the rigid charges, Eq. (6.15), yields

$$g_{J,\alpha\beta} = g^{para}_{J,\alpha\beta} + g^{dia}_{J,\alpha\beta} + g^{nuc}_{J,\alpha\beta}$$

$$= \frac{m_p}{I_{\beta\beta}} \frac{4m_e}{e^2} \sum_{n\neq 0} \left(\frac{\langle \Psi_0^{(0)}|\hat{m}_\alpha^l(\vec{R}_{GO})|\Psi_n^{(0)}\rangle \langle \Psi_n^{(0)}|\hat{m}_\beta^l(\vec{R}_{CM})|\Psi_0^{(0)}\rangle}{E_0^{(0)} - E_n^{(0)}} \right.$$

$$\left.+ \frac{\langle \Psi_0^{(0)}|\hat{m}_\beta^l(\vec{R}_{CM})|\Psi_n^{(0)}\rangle \langle \Psi_n^{(0)}|\hat{m}_\alpha^l(\vec{R}_{GO})|\Psi_0^{(0)}\rangle}{E_0^{(0)} - E_n^{(0)}} \right)$$

$$- \frac{m_p}{I_{\beta\beta}} \langle \Psi_0^{(0)}| \sum_i \left[(\vec{R}_{GO} - \vec{R}_{CM}) \cdot (\vec{r}_i - \vec{R}_{CM})\delta_{\alpha\beta} \right.$$

$$\left. - (r_{i,\alpha} - R_{CM,\alpha})(R_{GO,\beta} - R_{CM,\beta}) \right] |\Psi_0^{(0)}\rangle$$

$$+ \frac{m_p}{I_{\beta\beta}} \sum_K Z_K \left[(\vec{R}_K - \vec{R}_{CM})^2 \delta_{\alpha\beta} - (R_{K,\alpha} - R_{CM,\alpha})(R_{K,\beta} - R_{CM,\beta}) \right] \quad (6.24)$$

This would be the final expression for the rotational g tensor consisting of three terms: a paramagnetic term, a new diamagnetic-like term and a nuclear contribution. The first term can then also be expressed in terms of linear response functions according to Eq. (3.114), leading to

$$g^{para}_{J,\alpha\beta} = \frac{4m_p m_e}{e^2 I_{\beta\beta}} \langle\langle \hat{m}_\alpha^l(\vec{R}_{GO}) ; \hat{m}_\beta^l(\vec{R}_{CM}) \rangle\rangle_{\omega=0} \quad (6.25)$$

However in the spirit of section 5.9 the new diamagnetic contribution can also be written as a sum over all states [see Exercise 6.3]

132   Properties Related to Nuclear Motion

$$\langle \Psi_0^{(0)} | \sum_i \left[(\vec{r}_i - \vec{R}_{CM})(\vec{R}_{GO} - \vec{R}_{CM})\delta_{\alpha\beta} - (r_{i,\alpha} - R_{CM,\alpha})(R_{GO,\beta} - R_{CM,\beta}) \right] | \Psi_0^{(0)} \rangle$$

$$= -\frac{1}{m_e} \sum_{n\neq 0} \left(\frac{\langle \Psi_0^{(0)} | \sum_i [(\vec{R}_{GO} - \vec{R}_{CM}) \times \vec{p}_i]_\alpha | \Psi_n^{(0)} \rangle \langle \Psi_n^{(0)} | \sum_i l_{i,\beta}(\vec{R}_{CM}) | \Psi_0^{(0)} \rangle}{E_0^{(0)} - E_n^{(0)}} \right.$$

$$\left. + \frac{\langle \Psi_0^{(0)} | \sum_i l_{i,\beta}(\vec{R}_{CM}) | \Psi_n^{(0)} \rangle \langle \Psi_n^{(0)} | \sum_i [(\vec{R}_{GO} - \vec{R}_{CM}) \times \vec{p}_i]_\alpha | \Psi_0^{(0)} \rangle}{E_0^{(0)} - E_n^{(0)}} \right) \quad (6.26)$$

Exercise 6.3 Show that the diamagnetic contribution to the rotational g tensor can indeed be reformulated as a sum-over-states, Eq. (6.26).

Hint: Start by showing that for a constant vector $\vec{C}$ it holds that

$$\hat{\vec{\mu}}(\vec{R}_O) \cdot \vec{C} \, \boldsymbol{I}_3 - \hat{\vec{\mu}}(\vec{R}_O) \otimes \vec{C} = \frac{1}{i\hbar} \left[\vec{C} \times \hat{O}^r, \left(\hat{\vec{\mu}}(\vec{R}_O) \times \hat{O}^p \right)^T \right]$$

and then continue as in Section 5.9.

The effect of this reformulated diamagnetic contribution is that it replaces the dependence of one of the orbital angular momentum operators on the gauge origin R_{GO} in the sum-over-states term by a dependence on the centre of nuclear masses R_{CM}. An alternative expression for the rotational g tensor is thus

$$g_{J,\alpha\beta} = g_{J,\alpha\beta}^{nuc} + g_{J,\alpha\beta}^{el}$$

$$= \frac{m_p}{I_{\beta\beta}} \sum_K Z_K \left[(\vec{R}_K - \vec{R}_{CM})^2 \delta_{\alpha\beta} - (R_{K,\alpha} - R_{CM,\alpha})(R_{K,\beta} - R_{CM,\beta}) \right]$$

$$+ \frac{m_p}{m_e I_{\beta\beta}} \sum_{n\neq 0} \left(\frac{\langle \Psi_0^{(0)} | \sum_i l_{i,\alpha}(\vec{R}_{CM}) | \Psi_n^{(0)} \rangle \langle \Psi_n^{(0)} | \sum_i l_{i,\beta}(\vec{R}_{CM}) | \Psi_0^{(0)} \rangle}{E_0^{(0)} - E_n^{(0)}} \right.$$

$$\left. + \frac{\langle \Psi_0^{(0)} | \sum_i l_{i,\beta}(\vec{R}_{CM}) | \Psi_n^{(0)} \rangle \langle \Psi_n^{(0)} | \sum_i l_{i,\alpha}(\vec{R}_{CM}) | \Psi_0^{(0)} \rangle}{E_0^{(0)} - E_n^{(0)}} \right) \quad (6.27)$$

where we can see that it consists of a nuclear and an electronic sum-over-states contribution only. The latter term can then also be expressed in terms of linear response functions according to Eq. (3.114)

$$g_{J,\alpha\beta}^{el} = \frac{4 m_p m_e}{e^2 I_{\beta\beta}} \langle\!\langle \hat{m}_\alpha^l(\vec{R}_{CM}) ; \hat{m}_\beta^l(\vec{R}_{CM}) \rangle\!\rangle_{\omega=0} \quad (6.28)$$

The equivalence between the expressions in Eqs. (6.24) and (6.27) is based on the reformulation of the diamagnetic contribution as a sum-over-states or linear response function. As discussed in Section 5.9 this reformulation is exact, if we are dealing with the exact eigenstates of the unperturbed Hamiltonian. However, in approximate calculations this does not always hold and different values might be obtained from Eqs. (6.24) and (6.27).

On comparison of the electronic contribution with the expression for the paramagnetic contribution to the magnetizability in Eq. (5.44) we can see that they are proportional

$$g^{el}_{J,\alpha\beta} = -\frac{4m_p m_e}{e^2 I_{\beta\beta}} \xi^{para}_{\alpha\beta}(\vec{R}_{CM}) \qquad (6.29)$$

if one chooses the centre of nuclear masses, R_{CM}, as the gauge origin for the magnetizability. This relation has frequently been used for the experimental determination of the paramagnetic contribution to the magnetizability from measured rotational g tensors (Flygare and Benson, 1971; Flygare, 1974; Sutter and Flygare, 1976), as the nuclear contribution, $g^n_{J,\alpha\beta}$, can easily be calculated from the nuclear coordinates. Combined with a calculated diamagnetic contribution to the magnetizability one can thus obtain semi-experimental values of the magnetizability

$$\xi_{\alpha\beta} = \xi^{dia}_{\alpha\beta}(\vec{R}_{CM}) - \frac{e^2 I_{\beta\beta}}{4m_p m_e} \left(g_{J,\alpha\beta} - g^{nuc}_{J,\alpha\beta}\right) \qquad (6.30)$$

However, one should keep in mind that this relation only holds for a given fixed nuclear geometry, while the measured rotational g tensor is for a particular vibrational state. Direct application of it will therefore neglect the possibly large contributions from vibrational motion of the nuclei (see e.g. Lutnæs et al. (2009)), as discussed in Chapter 8. For an accurate determination of the magnetizability it is therefore necessary to correct explicitly for the vibrational corrections included in the measured rotational g tensor.

6.3 Rotational g Tensor and Electric Dipole Moment

In the previous section we have seen that the rotational g tensor is related to the paramagnetic contribution to the magnetizability, Eq. (6.29). Here, we will explore a relation between the rotational g tensor and the electric dipole moment. We will see that the latter is related to the difference between the rotational g tensor of two isotopologues of the same molecule, i.e. two molecules that differ only in the isotopes of one or more nuclei. We consider therefore a component of the rotational g tensor, $g'_{J,\alpha\beta}$, of one isotopologue with moment of inertia tensor $\boldsymbol{I}'$ and centre of nuclear masses, $\vec{R}'_{CM}$, which is shifted by a vector $\vec{D} = \vec{R}'_{CM} - \vec{R}_{CM}$ from the centre of nuclear masses of the second isotopologue with moment of inertia tensor $\boldsymbol{I}$

$$g'_{J,\alpha\beta} I'_{\beta\beta} = m_p \sum_K Z_K \left[(\vec{R}_K - \vec{R}'_{CM})^2 \delta_{\alpha\beta} - (R_{K,\alpha} - \vec{R}'_{CM,\alpha})(R_{K,\beta} - \vec{R}'_{CM,\beta})\right]$$
$$+ \frac{m_p}{m_e} \langle\langle \sum_i [(\vec{r}_i - \vec{R}'_{CM}) \times \vec{p}_i]_\alpha ; \sum_i [(\vec{r}_i - \vec{R}'_{CM}) \times \vec{p}_i]_\beta \rangle\rangle_{\omega=0} \qquad (6.31)$$

Using the relation between the position vectors of the two centres of mass we can rewrite this as

$$g'_{J,\alpha\beta} \, I'_{\beta\beta} = g_{J,\alpha\beta} \, I_{\beta\beta}$$

$$- m_p \sum_K Z_K \left[2\vec{D} \cdot (\vec{R}_K - \vec{R}_{CM})\delta_{\alpha\beta} - D_\alpha(R_{K,\beta} - R_{CM,\beta}) - D_\beta(R_{K,\alpha} - R_{CM,\alpha}) \right]$$

$$+ m_p \sum_K Z_K \left(\vec{D} \cdot \vec{D} \, \delta_{\alpha\beta} - D_\alpha D_\beta \right)$$

$$- \frac{m_p}{m_e} \sum_{\gamma\delta} D_\gamma \left(\epsilon_{\alpha\gamma\delta} \langle\langle \hat{O}^p_\delta \, ; \, \sum_i l_{i,\beta}(\vec{R}_{CM}) \rangle\rangle_{\omega=0} + \epsilon_{\beta\gamma\delta} \langle\langle \sum_i l_{i,\alpha}(\vec{R}_{CM}) \, ; \, \hat{O}^p_\delta \rangle\rangle_{\omega=0} \right)$$

$$+ \frac{m_p}{m_e} \sum_{\gamma\delta\zeta\eta} \epsilon_{\alpha\gamma\delta} \, \epsilon_{\beta\zeta\eta} D_\gamma D_\zeta \langle\langle \hat{O}^p_\delta \, ; \, \hat{O}^p_\eta \rangle\rangle_{\omega=0} \tag{6.32}$$

where we have made use of the Levi-Civita symbol, Eq. (5.112), again.

We are going to rewrite the three linear response functions now as ground-state expectation values similar to the derivations in Section 5.9. However, here we will not proceed via the sum-over-states expressions for the response function, but want to illustrate an alternative approach via the equation of motion of the polarization propagator for zero frequencies, Eq. (3.141). Recalling that $\hat{O}^p$ is the canonical conjugate momentum operator to $\hat{O}^r$, i.e. Eq. (3.64) we can make use of Eq. (3.141) and replace the three response functions by ground-state expectation values

$$g'_{J,\alpha\beta} \, I'_{\beta\beta} = g_{J,\alpha\beta} \, I_{\beta\beta}$$

$$- m_p \sum_K Z_K \left[2\vec{D} \cdot (\vec{R}_K - \vec{R}_{CM})\delta_{\alpha\beta} - D_\alpha(R_{K,\beta} - R_{CM,\beta}) - D_\beta(R_{K,\alpha} - R_{CM,\alpha}) \right]$$

$$+ \frac{m_p}{\imath\hbar} \sum_{\gamma\delta} D_\gamma \left(\epsilon_{\alpha\gamma\delta} \langle \Psi_0^{(0)} | [\hat{O}^r_\delta, \sum_i l_{i,\beta}(\vec{R}_{CM})] | \Psi_0^{(0)} \rangle \right.$$

$$\left. - \epsilon_{\beta\gamma\delta} \langle \Psi_0^{(0)} | [\sum_i l_{i,\alpha}(\vec{R}_{CM}), \hat{O}^r_\delta] | \Psi_0^{(0)} \rangle \right) \tag{6.33}$$

$$+ m_p \sum_K Z_K \left(\vec{D} \cdot \vec{D} \, \delta_{\alpha\beta} - D_\alpha D_\beta \right) + \frac{m_p}{\imath\hbar} \sum_{\gamma\delta\zeta\eta} \epsilon_{\alpha\gamma\delta} \, \epsilon_{\beta\zeta\eta} D_\gamma D_\zeta \langle \Psi_0^{(0)} | [\hat{O}^p_\delta, \hat{O}^r_\eta] | \Psi_0^{(0)} \rangle$$

Evaluating the commutators we obtain for the expectation values

$$\langle \Psi_0^{(0)} | [\hat{O}^r_\delta, \hat{O}^p_\eta] | \Psi_0^{(0)} \rangle = \imath\hbar N \delta_{\delta\eta} \tag{6.34}$$

$$\langle \Psi_0^{(0)} | [\hat{O}^r_\delta, \sum_i l_{i,\beta}(\vec{R}_{CM})] | \Psi_0^{(0)} \rangle = \imath\hbar \epsilon_{\delta\beta\zeta} \langle \Psi_0^{(0)} | \sum_i (r_{i,\zeta} - R_{CM,\zeta}) | \Psi_0^{(0)} \rangle \tag{6.35}$$

where N is the total number of electrons, and thus for the relation between the rotational g tensors

$$g'_{J,\alpha\beta} I'_{\beta\beta} = g_{J,\alpha\beta} I_{\beta\beta}$$

$$- m_p \sum_K Z_K \left[2\vec{D} \cdot (\vec{R}_K - \vec{R}_{CM})\delta_{\alpha\beta} - D_\alpha(R_{K,\beta} - R_{CM,\beta}) - D_\beta(R_{K,\alpha} - R_{CM,\alpha}) \right]$$

$$+ 2m_p \vec{D} \cdot \langle \Psi_0^{(0)} | \sum_i (\vec{r}_i - \vec{R}_{CM}) | \Psi_0^{(0)} \rangle \delta_{\alpha\beta}$$

$$- m_p D_\alpha \langle \Psi_0^{(0)} | \sum_i (r_{i,\beta} - R_{CM,\beta}) | \Psi_0^{(0)} \rangle - m_p D_\beta \langle \Psi_0^{(0)} | \sum_i (r_{i,\alpha} - R_{CM,\alpha}) | \Psi_0^{(0)} \rangle$$

$$+ m_p \sum_K Z_K \left(\vec{D} \cdot \vec{D}\, \delta_{\alpha\beta} - D_\alpha D_\beta \right) - m_p N \left(\vec{D} \cdot \vec{D}\, \delta_{\alpha\beta} - D_\alpha D_\beta \right) \quad (6.36)$$

Recalling the definition of the electric dipole moment, Eq. (4.25), introducing the total charge q of the molecule

$$q = e \sum_K Z_K - eN \quad (6.37)$$

and using the definition of $\vec{D} = \vec{R}'_{CM} - \vec{R}_{CM}$ we arrive finally at the relation between the rotational g tensor of two isotopologues of a molecule and its dipole moment

$$g'_{J,\alpha\beta} I'_{\beta\beta} = g_{J,\alpha\beta} I_{\beta\beta} \quad (6.38)$$

$$- 2\frac{m_p}{e}(\vec{R}'_{CM} - \vec{R}_{CM}) \cdot \vec{\mu}(\vec{R}_{CM})\, \delta_{\alpha\beta}$$

$$+ \frac{m_p}{e}(R'_{CM,\alpha} - R_{CM,\alpha})\mu_\beta(\vec{R}_{CM}) + \frac{m_p}{e}(R'_{CM,\beta} - R_{CM,\beta})\mu_\alpha(\vec{R}_{CM})$$

$$+ \frac{m_p}{e} q \left[(\vec{R}'_{CM} - \vec{R}_{CM})^2 \delta_{\alpha\beta} - (R'_{CM,\alpha} - R_{CM,\alpha})(R'_{CM,\beta} - R_{CM,\beta}) \right]$$

This relation allows us to determine experimentally the electric dipole moment by simply measuring the rotational g tensor of two isotopologues (Rosenblum et al., 1958). However, one should keep in mind that the expressions for the dipole moment as well as for the rotational g tensor, which were used in the derivation, are for a particular nuclear geometry. The effects of vibrational motion of the nuclei, as discussed in Chapter 8, are thus not included, while experimentally measured rotational g tensors are always for a particular vibrational state. Equation (6.38) can thus only be applied to measured rotational g tensors, if the changes due to the vibrational motion of the nuclei are negligible or explicitly corrected for prior to application of Eq. (6.38).

6.4 Rotational g Tensor and Electric Quadrupole Moment

The rotational g tensor can also be related to the electric quadrupole moment tensor Θ. We will in the following derive this relation for the xx diagonal component, which according to Eq. (4.27) is given as

$$\Theta_{xx}(\vec{R}_{CM}) = \sum_K Z_K e \left\{ (\vec{R}_{K,z} - \vec{R}_{CM,z})^2 - \frac{1}{2}[(\vec{R}_{K,x} - \vec{R}_{CM,x})^2 + (\vec{R}_{K,y} - \vec{R}_{CM,y})^2] \right\}$$
$$+ \frac{1}{e} \langle \Psi_0^{(0)} | \frac{1}{2} \left[\hat{\mu}_x^2(\vec{R}_{CM}) + \hat{\mu}_y^2(\vec{R}_{CM}) \right] - \hat{\mu}_z^2(\vec{R}_{CM}) | \Psi_0^{(0)} \rangle \quad (6.39)$$

where we have placed the otherwise arbitrary origin of the coordinate system on the centre of nuclear masses, $\vec{R}_{CM}$ and made use of the definition of the electric dipole moment operator, Eq. (4.30). Comparison with Eqs. (5.44) and (6.27) shows that the nuclear contribution to the quadrupole moment tensor is expressible in terms of components of the nuclear contribution to the rotational g tensor, $g_{J,\alpha\beta}^{nuc}$, and the electronic contribution in terms of components of the diamagnetic contribution to the magnetizability tensor, $\xi_{\alpha\beta}^{dia}(\vec{R}_{CM})$, as

$$\Theta_{xx}(\vec{R}_{CM}) = \frac{e}{m_p} \left[\frac{1}{2} \left(I_{xx} g_{xx,J}^{nuc} + I_{yy} g_{yy,J}^{nuc} \right) - I_{zz} g_{zz,J}^{nuc} \right]$$
$$- \frac{4m_e}{e} \left\{ \frac{1}{2} \left[\xi_{xx}^{dia}(\vec{R}_{CM}) + \xi_{yy}^{dia}(\vec{R}_{CM}) \right] - \xi_{zz}^{dia}(\vec{R}_{CM}) \right\} \quad (6.40)$$

Recalling the relation between the magnetizability and the rotational g tensor, Eq. (6.30), we can replace the diamagnetic and nuclear contributions by the total magnetizability and rotational g tensor and obtain finally

$$\Theta_{xx}(\vec{R}_{CM}) = \frac{e}{m_p} \left[\frac{1}{2} \left(I_{xx} g_{xx,J} + I_{yy} g_{yy,J} \right) - I_{zz} g_{zz,J} \right]$$
$$- \frac{4m_e}{e} \left[\frac{1}{2} (\xi_{xx} + \xi_{yy}) - \xi_{zz} \right] \quad (6.41)$$

Corresponding relations for the other diagonal components can be obtained by cyclic permutations of the coordinate triple xyz to yzx or zxy. Since the quadrupole moment of a polar molecule depends on the origin of the coordinate system it is important to remember that the centre of nuclear masses is automatically chosen as the origin in Eq. (6.41). Similar to the previously derived relations between the rotational g tensor and other molecular properties, Eqs. (6.30) and (6.38), one has to take care of vibrational corrections to the properties in Eq. (6.41), when it is applied to measured rotational g tensors and magnetizabilities.

6.5 Molecular Rotation as Source for Magnetic Fields

We will now look at the effect of molecular rotation on the magnetic field at the position $\hat{R}_K$ of a magnetic nucleus K. According to Eq. (5.49) the current density of the rigidly rotating charges will give rise to an additional contribution, a **rotational magnetic induction** $\vec{B}^{j,J}(\vec{R})$

$$\vec{B}^{j,J,rig}(\vec{R}_K, \vec{J}) = \frac{\mu_0}{4\pi} \int_{\vec{r}} \rho(\vec{r}) \frac{(\vec{r} - \vec{R}_K) \times \left[(\mathbf{I}^{-1} \vec{J}) \times (\vec{r} - \vec{R}_K) \right]}{|\vec{r} - \vec{R}_K|^3} d\vec{r} \quad (6.42)$$

where the position vectors are now with respect to the nucleus of interest. Again, we have assumed that the charges rotate rigidly and therefore added the superscript "rig". However, the coupling between the rotational motion of the nuclei and the motion of the electrons, Eq. (6.3), will again lead to an additional induced contribution, $\vec{\mathcal{B}}^{j,J,ind}$, so that the total rotational magnetic induction at the position of nucleus K can be written as

$$\mathcal{B}^{j,J}_\alpha(\vec{R}_K,\vec{J}) = \mathcal{B}^{j,J,rig}_\alpha(\vec{R}_K,\vec{J}) + \mathcal{B}^{j,J,ind}_\alpha(\vec{R}_K,\vec{J}) \tag{6.43}$$

Analogous to the rotational g tensor one defines a **spin rotation tensor** C^K as the proportionality tensor between the rotational magnetic induction $\vec{\mathcal{B}}^{j,J}(\vec{R}_K,\vec{J})$ and the rotational angular momentum $\vec{J}$ (Gunther-Mohr et al., 1954; Flygare, 1964)

$$\mathcal{B}^{j,J}_\alpha(\vec{R}_K,\vec{J}) = \frac{2\pi}{\mu_N g_K} \sum_\beta C^K_{\alpha\beta} J_\beta = \frac{2\pi}{\mu_N g_K} \sum_\beta (C^{K,rig}_{\alpha\beta} + C^{K,ind}_{\alpha\beta}) J_\beta \tag{6.44}$$

where $\mu_N = \frac{e\hbar}{2m_p}$ and g_K are the nuclear magneton and the nuclear g-factor of nucleus K. The induced contribution to the spin-rotation tensor is again a consequence of the breakdown of the Born–Oppenheimer approximation. The individual cartesian components of the spin-rotation tensor can thus be defined as first derivatives[4] of the rotational magnetic induction at the position of nucleus K with respect to a component of the rotational angular momentum $\vec{J}$ of the nuclei

$$C^K_{\alpha\beta} = \frac{\mu_N g_K}{2\pi} \frac{\partial \mathcal{B}^{j,J}_\alpha(\vec{R}_K;\vec{J})}{\partial J_\beta}\bigg|_{|\vec{J}|=0} \tag{6.45}$$

The rotational magnetic induction, $\vec{\mathcal{B}}^{j,J}(\vec{R}_K,\vec{J})$, at the position of a nucleus K can be probed by the magnetic moment of this nucleus, $\vec{m}^K$ like any other magnetic field within a molecule. The change in energy is then given by Eq. (5.50)

$$\Delta E(\vec{m}^K,\vec{J}) = -\sum_\alpha m^K_\alpha \mathcal{B}^{j,J}_\alpha(\vec{R}_K;\vec{J}) = -\frac{2\pi}{\mu_N g_K} \sum_{\alpha\beta} m^K_\alpha C^K_{\alpha\beta} J_\beta \tag{6.46}$$

which allows us to define the components of the spin-rotation tensor alternatively as second derivatives of this interaction energy with respect to a component of the rotational angular momentum $\vec{J}$ of the nuclei and a component of the nuclear magnetic moment $\vec{m}^K$.[5]

$$C^K_{\alpha\beta} = -\frac{\mu_N g_K}{2\pi} \frac{\partial^2 E(\vec{\mathcal{B}},\vec{m}^K)}{\partial m^K_\alpha \partial J_\beta}\bigg|_{|\vec{J}|=|\vec{m}^K|=0} \tag{6.47}$$

where we should keep in mind that it is only the induced, Born-Oppenheimer breakdown contribution, which can be obtained in this way as derivative of the perturbed electronic energy.

---

[4] This definition is also given in the fourth column of Table B.3 of Appendix B.
[5] This definition is also given in the last column of Table B.3 of Appendix B.

Equation (6.46) explains also the name spin-rotation tensor for $\mathbf{C}^K$, because it is the coupling tensor for the coupling of the rotational angular moment of the molecule $\vec{J}$ with the spin $\vec{I}^K$ of the nuclei $\{K\}$, which gives rise to an additional contribution to the rotational energy. This is normally expressed in terms of an effective rotational Hamiltonian as

$$\hat{H}^{rot} = \frac{\hat{J}_x^2}{2I_{xx}} + \frac{\hat{J}_y^2}{2I_{yy}} + \frac{\hat{J}_z^2}{2I_{zz}} - \frac{2\pi}{\hbar} \sum_K \hat{\vec{I}}^K \mathbf{C}^K \hat{\vec{J}} \tag{6.48}$$

6.6 Quantum Mechanical Expression for the Spin Rotation Tensor

The derivation of the quantum mechanical expressions for the spin rotation tensor is completely analogous to the one for the rotational g tensor. We will therefore just discuss the final expressions. The rigid contribution again consists of a nuclear and an electronic term

$$C_{\alpha\beta}^{K,rig} = \frac{\mu_N g_K e}{2\pi I_{\beta\beta}} \frac{\mu_0}{4\pi} \left\{ \sum_{L \neq K} Z_L \left[(\vec{R}_L - \vec{R}_K) \frac{(\vec{R}_L - \vec{R}_K)}{|\vec{R}_L - \vec{R}_K|^3} \delta_{\alpha\beta} \right.\right.$$
$$\left. - (R_{L,\alpha} - R_{K,\alpha}) \frac{(R_{L,\beta} - R_{K,\beta})}{|\vec{R}_L - \vec{R}_K|^3} \right] \tag{6.49}$$
$$\left. - \langle \Psi_0^{(0)} | \sum_i^N \left[(\vec{r}_i - \vec{R}_K) \frac{(\vec{r}_i - \vec{R}_K)}{|\vec{r}_i - \vec{R}_K|^3} \delta_{\alpha\beta} - (r_{i,\alpha} - R_{K,\alpha}) \frac{(r_{i,\beta} - R_{K,\beta})}{|\vec{r}_i - \vec{R}_K|^3} \right] | \Psi_0^{(0)} \rangle \right\}$$

For the derivation of the induced contribution we will start from the definition as second derivative of the electronic energy, Eq. (6.47) in the presence of a nuclear magnetic moment and the molecular rotation. The corresponding vector potentials, Eqs. (5.55) and (6.5), lead to two first-order perturbation Hamiltonians, Eqs. (5.56) and (6.19), and a new second-order-order-perturbation Hamiltonian [see Exercise 6.4]

$$\hat{H}^{(2)} = \sum_{\alpha\beta} \hat{O}_{\alpha\beta}^{m^K J}(\vec{R}_{CM}, \vec{R}_K)\, m_\alpha^K \,(\mathbf{I}^{-1}\, \vec{J})_\beta \tag{6.50}$$

where the perturbation operator is given as

$$\hat{O}_{\alpha\beta}^{m^K J}(\vec{R}_{CM}, \vec{R}_K) = \sum_i^N \hat{o}_{i,\alpha\beta}^{m^K J}(\vec{R}_{CM}, \vec{R}_K) \tag{6.51}$$

$$= -\frac{e\mu_0}{4\pi} \sum_i \left[(\vec{r}_i - \vec{R}_{CM}) \cdot \frac{(\vec{r}_i - \vec{R}_K)}{|\vec{r}_i - \vec{R}_K|^3} \delta_{\alpha\beta} - (r_{i,\alpha} - R_{CM,\alpha}) \frac{(r_{i,\beta} - R_{K,\beta})}{|\vec{r}_i - \vec{R}_K|^3} \right]$$

As for the rotational g tensor this operator is very similar to the one for the diamagnetic contribution to the nuclear magnetic shielding tensor but with one of the electronic position vectors defined with respect to the nuclear centre of masses $\vec{R}_{CM}$ and not with respect to the arbitrary gauge origin $\vec{R}_{GO}$.

Exercise 6.4 Derive the second-order perturbation Hamiltonian for the induced contribution to the spin rotation tensor, Eq. (6.51), by inserting the vector potentials, (5.55) and (6.5), in the general expression of the molecular Hamiltonian, Eq. (2.101), retaining the bilinear second-order term.

The second-order perturbation theory expression for a component of the spin rotation tensor then becomes

$$C_{\alpha\beta}^{K,ind} = -\frac{\mu_N g_K}{2\pi} \frac{\partial^2 E(\vec{B}, \vec{m}^K)}{\partial m_\alpha^K \partial J_\beta}\bigg|_{|\vec{J}|=|\vec{m}^K|=0}$$

$$= -\frac{\mu_N g_K}{2\pi I_{\beta\beta}} \left[\sum_{n\neq 0} \left(\frac{\langle \Psi_0^{(0)}|\hat{O}_{K,\alpha}^{OP}|\Psi_n^{(0)}\rangle \langle \Psi_n^{(0)}|\hat{L}_\beta(\vec{R}_{CM})|\Psi_0^{(0)}\rangle}{E_0^{(0)} - E_n^{(0)}} \right.\right.$$

$$\left.+ \frac{\langle \Psi_0^{(0)}|\hat{L}_\beta(\vec{R}_{CM})|\Psi_n^{(0)}\rangle \langle \Psi_n^{(0)}|\hat{O}_{K,\alpha}^{OP}|\Psi_0^{(0)}\rangle}{E_0^{(0)} - E_n^{(0)}} \right)$$

$$+ \langle \Psi_0^{(0)}|\hat{O}_{\alpha\beta}^{m^K J}(\vec{R}_{CM}, \vec{R}_K)|\Psi_0^{(0)}\rangle \Bigg] \quad (6.52)$$

The induced contribution consists again of a paramagnetic or sum-over-states contribution and a diamagnetic or ground-state expectation value term, which can be combined with the contribution from the rigid charges, Eq. (6.49),

$$C_{\alpha\beta}^K = C_{\alpha\beta}^{K,para} + C_{\alpha\beta}^{K,dia} + C_{\alpha\beta}^{K,nuc} \quad (6.53)$$

$$= -\frac{\mu_N g_K}{2\pi I_{\beta\beta}} \frac{e\mu_0}{4\pi} \left\{ \frac{4\pi}{e\mu_0} \sum_{n\neq 0} \frac{\langle \Psi_0^{(0)}|\hat{O}_{K,\alpha}^{OP}|\Psi_n^{(0)}\rangle \langle \Psi_n^{(0)}|\sum_i \left[(\vec{r}_i - \vec{R}_{CM}) \times \vec{p}_i\right]_\beta|\Psi_0^{(0)}\rangle}{E_0^{(0)} - E_n^{(0)}} \right.$$

$$+ \frac{4\pi}{e\mu_0} \sum_{n\neq 0} \frac{\langle \Psi_0^{(0)}|\sum_i \left[(\vec{r}_i - \vec{R}_{CM}) \times \vec{p}_i\right]_\beta|\Psi_n^{(0)}\rangle \langle \Psi_n^{(0)}|\hat{O}_{K,\alpha}^{OP}|\Psi_0^{(0)}\rangle}{E_0^{(0)} - E_n^{(0)}}$$

$$- \langle \Psi_0^{(0)}|\sum_i^N \left[(\vec{R}_K - \vec{R}_{CM})\frac{(\vec{r}_i - \vec{R}_K)}{|\vec{r}_i - \vec{R}_K|^3}\delta_{\alpha\beta} - (R_{K,\alpha} - R_{CM,\alpha})\frac{(r_{i,\beta} - R_{K,\beta})}{|\vec{r}_i - \vec{R}_K|^3}\right]|\Psi_0^{(0)}\rangle$$

$$- \sum_{L\neq K} Z_L \left[(\vec{R}_L - \vec{R}_K)\frac{(\vec{R}_L - \vec{R}_K)}{|\vec{R}_L - \vec{R}_K|^3}\delta_{\alpha\beta} - (R_{L,\alpha} - R_{K,\alpha})\frac{(R_{L,\beta} - R_{K,\beta})}{|\vec{R}_L - \vec{R}_K|^3}\right] \right\}$$

These are again a paramagnetic term, a new diamagnetic-like term and a nuclear contribution, where the first can be expressed as a linear response function according to Eq. (3.114)

140 Properties Related to Nuclear Motion

$$C_{\alpha\beta}^{K,para} = \frac{\mu_N g_K}{2\pi I_{\beta\beta}} \frac{2m_e}{e} \langle\langle \hat{O}_{K,\alpha}^{OP} ; \hat{m}_\beta^l(\vec{R}_{CM}) \rangle\rangle_{\omega=0} \qquad (6.54)$$

On the other hand, the new diamagnetic contribution can again be written as a sum over all states [see Exercise 6.5] following the derivations in Section 5.9

$$\langle \Psi_0^{(0)} | \sum_i^N \left[(\vec{R}_K - \vec{R}_{CM}) \frac{(\vec{r}_i - \vec{R}_K)}{|\vec{r}_i - \vec{R}_K|^3} \delta_{\alpha\beta} - (R_{K,\alpha} - R_{CM,\alpha}) \frac{(r_{i,\beta} - R_{K,\beta})}{|\vec{r}_i - \vec{R}_K|^3} \right] | \Psi_0^{(0)} \rangle$$

$$= \frac{4\pi}{e\mu_0} \sum_{n \neq 0} \left\{ \frac{\langle \Psi_0^{(0)} | \hat{O}_{K,\alpha}^{OP} | \Psi_n^{(0)} \rangle \langle \Psi_n^{(0)} | \sum_i \left[(\vec{R}_K - \vec{R}_{CM}) \times \vec{p}_i\right]_\beta | \Psi_0^{(0)} \rangle}{E_0^{(0)} - E_n^{(0)}} \right.$$

$$\left. + \frac{\langle \Psi_0^{(0)} | \sum_i \left[(\vec{R}_K - \vec{R}_{CM}) \times \vec{p}_i\right]_\beta | \Psi_n^{(0)} \rangle \langle \Psi_n^{(0)} | \hat{O}_{K,\alpha}^{OP} | \Psi_0^{(0)} \rangle}{E_0^{(0)} - E_n^{(0)}} \right\} \qquad (6.55)$$

Exercise 6.5 Show that the diamagnetic contribution to the spin rotation tensor can indeed be reformulated as a sum-over-states, Eq. (6.55).

Hint: Start by showing that for a constant vector $\vec{D}$ it holds that

$$\vec{D} \cdot \hat{\vec{O}}^\mu(\vec{R}_K) \, I_3 - \vec{D} \otimes \hat{\vec{O}}^\mu(\vec{R}_K) = \frac{\imath}{\hbar} \left[\hat{\vec{O}}^\mu(\vec{R}_K) \times \hat{\vec{O}}^p, \left(\vec{D} \times \hat{\vec{O}}^r\right)^T \right]$$

and then continue as in Section 5.9.

Combined with the paramagnetic term in Eq. (6.53) this reformulated diamagnetic contribution replaces in the paramagnetic term the dependence of the orbital angular momentum operator on the nuclear centre of mass R_{CM} by a dependence on the position of nucleus K. An alternative expression for the spin-rotation tensor is thus

$$C_{\alpha\beta}^K = C_{\alpha\beta}^{K,nuc} + C_{\alpha\beta}^{K,el}$$

$$= \frac{\mu_N g_K}{2\pi I_{\beta\beta}} \frac{e\mu_0}{4\pi} \sum_{L \neq K} Z_L \left[(\vec{R}_L - \vec{R}_K) \frac{(\vec{R}_L - \vec{R}_K)}{|\vec{R}_L - \vec{R}_K|^3} \delta_{\alpha\beta} \right.$$

$$\left. -(R_{L,\alpha} - R_{K,\alpha}) \frac{(R_{L,\beta} - R_{K,\beta})}{|\vec{R}_L - \vec{R}_K|^3} \right]$$

$$- \frac{\mu_N g_K}{2\pi I_{\beta\beta}} \sum_{n \neq 0} \left[\frac{\langle \Psi_0^{(0)} | \hat{O}_{K,\alpha}^{OP} | \Psi_n^{(0)} \rangle \langle \Psi_n^{(0)} | \hat{L}_\beta(\vec{R}_K) | \Psi_0^{(0)} \rangle}{E_0^{(0)} - E_n^{(0)}} \right.$$

$$\left. + \frac{\langle \Psi_0^{(0)} | \hat{L}_\beta(\vec{R}_K) | \Psi_n^{(0)} \rangle \langle \Psi_n^{(0)} | \hat{O}_{K,\alpha}^{OP} | \Psi_0^{(0)} \rangle}{E_0^{(0)} - E_n^{(0)}} \right] \qquad (6.56)$$

which consists of a nuclear and an electronic sum-over-states contribution only. The latter term can then also be expressed in terms of linear response functions according to Eq. (3.114)

$$C_{\alpha\beta}^{K,el} = \frac{\mu_N g_K}{2\pi I_{\beta\beta}} \frac{2m_e}{e} \langle\langle \hat{O}_{K,\alpha}^{OP} ; \hat{m}_\beta^l(\vec{R}_K) \rangle\rangle_{\omega=0} \qquad (6.57)$$

The two alternative expressions for the spin rotation tensor in equations (6.53) and (6.56) are equivalent, if we are dealing with the exact eigenstates of the unperturbed Hamiltonian. However, in approximate calculations this does not always hold and different values might be obtained from Eqs. (6.53) and (6.56).

Comparing Eq. (6.57) with the expression for the paramagnetic contribution to the nuclear magnetic shielding tensor in Eq. (5.97) we can see that the latter is proportional to the electronic contribution to the spin-rotation tensor, if one chooses the position of nucleus K as the gauge origin for the nuclear magnetic shielding tensor

$$C_{\alpha\beta}^{K,el} = \frac{m_e}{m_p} \frac{g_K \hbar}{2\pi I_{\beta\beta}} \sigma_{\alpha\beta}^{K,para}(\vec{R}_K) \qquad (6.58)$$

This relation is of great importance for NMR spectroscopy because it allows us to determine the absolute shielding tensor $\boldsymbol{\sigma}^K$ by a combination of the measured spin rotation tensor with its nuclear contribution, which can easily be calculated from the nuclear coordinates, and a calculated diamagnetic contribution

$$\sigma_{\alpha\beta} = \sigma_{\alpha\beta}^{dia}(\vec{R}_K) + \frac{m_e}{m_p} \frac{g_K \hbar}{2\pi I_{\beta\beta}} \left(C_{\alpha\beta}^K - C_{\alpha\beta}^{K,nuc} \right) \qquad (6.59)$$

This is the only possibility to determine experimental, or rather semi-experimental, absolute shielding constants, as one can only obtain differences in the shielding constants, i.e. chemical shifts, from NMR spectra as discussed in Section 5.7. However, one has to be careful in applying this relation similar to the relation between the rotational g tensor and the magnetizability, Eq. (6.30). First, one has to take care of the vibrational corrections in the measured spin rotation tensors and secondly NMR spectra are normally measured in the liquid phase so that solvent effects would have to be considered as well. Nevertheless, it has been used to establish absolute shielding scales for several light nuclei (Flygare, 1964; Hindermann and Cornwell, 1968; Jameson et al., 1980; Vaara et al., 1998; Puzzarini et al., 2009).

6.7 Non-Adiabatic Rotational and Vibrational Reduced Masses

In the previous sections we have studied Born–Oppenheimer-breakdown corrections to two molecular properties, the rotational g tensor and the nuclear spin-rotation constant, i.e. the effect of the coupling between nuclear and electronic motion on the electronic energies. In this and the following sections we will now turn our attention to the effect of this coupling on the motion of the nuclei and will discuss Born–Oppenheimer-breakdown corrections to the rotational and vibrational energies. For the sake of a simpler presentation we will illustrate it for a diatomic molecule AB, where there is only one vibrational mode that involves changes in the internuclear distance $R = |\vec{R}_A - \vec{R}_B|$ and that we have placed along the z-axis. Corrections to the Born-Oppenheimer expressions for vibration-rotation energies of diatomic molecules have been derived several times (Herman and Asgharian, 1966; Watson, 1973; Bunker and Moss, 1977; Watson, 1980; Herman and Ogilvie, 1998). Here we will follow mainly the derivation by Bunker and Moss (Bunker and Moss, 1977). After separation of the translation of the whole molecule and transformation to nuclear centre of mass coordinates one can write the field-free Hamiltonian for the electronic ground state of symmetry $^1\Sigma^+$ of a diatomic molecule as

$$\hat{H}^{(0)}_{nuc,e} = \hat{H}^{(0)} + \hat{H}' \tag{6.60}$$

where the zeroth-order electronic Hamiltonian, $\hat{H}^{(0)}$ is the Born–Oppenheimer molecular field-free electronic Hamiltonian defined in Eq. (2.9) but now with all position vectors defined relative to the centre of nuclear masses R_{CM}. The remaining three terms in the Hamiltonian,

$$\hat{H}' = -\frac{\hbar^2}{2\mu_n} \frac{\partial^2}{\partial R^2} + \frac{1}{2\mu_n R^2} \left[\hat{\vec{J}} - \hat{\vec{L}}(\vec{R}_{CM})\right]^2 + \frac{1}{2(m_A + m_B)} \sum_{i,j} \hat{\vec{p}}_i \cdot \hat{\vec{p}}_j \tag{6.61}$$

are the kinetic-energy operator for the vibrational motion of the two nuclei of masses m_A and m_B, the kinetic-energy operator for the rotation of the nuclei about their centre of mass and third a mass polarization term, where the nuclear reduced mass μ_n is defined as

$$\mu_n = \frac{m_A m_B}{m_A + m_B} \tag{6.62}$$

The angular momentum operator for rotation of the whole molecule about the molecular centre of mass is denoted $\hat{\vec{J}}$, and $\hat{\vec{L}}(\vec{R}_{CM})$ is the operator for total angular momentum of the electrons, Eq. (5.23), but now with respect to the nuclear centre of mass.

The eigenfunctions of the Hamiltonian in Eq. (6.60), the molecular wavefunctions $\Phi^{(0)}_{k,v,J}(R,\theta,\phi,\{\vec{r}_i\})$, are functions of both the electronic coordinates $\vec{r}_i$, the internuclear distance R and the two rotation angles θ and ϕ. Approximations to them are normally obtained by an approximate separation of nuclear and electronic coordinates. In the first step one solves the electronic Schrödinger equation, Eq. (2.10),

$$\hat{H}^{(0)} |\Psi^{(0)}_k(\{\vec{r}_i\}; R)\rangle = E^{(0)}_k(R) |\Psi^{(0)}_k(\{\vec{r}_i\}; R)\rangle \tag{6.63}$$

which yields a complete set of electronic wavefunctions $\{\Psi^{(0)}_k(\{\vec{r}_i\}; R)\}$. The molecular wavefunctions can then be expanded in this complete set

$$\Phi^{(0)}_{k,v,J}(R,\theta,\phi,\{\vec{r}_i\}) = \sum_k \Psi^{(0)}_k(\{\vec{r}_i\}; R) \, \Theta^{k,(0)}_{v,J}(R,\theta,\phi) \tag{6.64}$$

Similar to ESR and NMR spectroscopy, Sections 5.6 and 5.7, vibration-rotation spectra are interpreted in terms of an effective Hamiltonian for vibration-rotational motion of the nuclei. This Hamiltonian is in principle obtained by taking the expectation value of the molecular Hamiltonian, Eq. (6.60), over the corresponding electronic state $|\Psi_0^{(0)}(\{\vec{r}_i\}; R)\rangle$. However, the situation is complicated by the fact that the electronic energies depend on the internuclear distance R, which is the coordinate of the vibrational motion. Therefore, the Hamiltonian is first subjected to a unitary transformation (Bunker and Moss, 1977)

$$\hat{\tilde{H}} = e^{-\imath\lambda\hat{S}}\hat{H}e^{\imath\lambda\hat{S}} = \hat{\tilde{H}}^{(0)} + \lambda\hat{\tilde{H}}^{(1)} + \lambda^2\hat{\tilde{H}}^{(2)} + \cdots \quad (6.65)$$

in which the hermitian operator $\hat{S}$ is chosen such that the transformed Hamiltonian $\hat{\tilde{H}}$ does not couple different electronic wave functions through first order [see Exercise 6.6], i.e.

$$\langle\Psi_0^{(0)}(\{\vec{r}_i\}; R)|\hat{\tilde{H}}^{(1)}|\Psi_n^{(0)}(\{\vec{r}_i\}; R)\rangle = 0 \quad (6.66)$$

Exercise 6.6 Derive the expression for the first Hamiltonian $\hat{\tilde{H}}^{(1)}$ following the discussion in Bunker and Moss (Bunker and Moss, 1977).

In zeroth order one then obtains the Born–Oppenheimer nuclear Hamiltonian, Eq. (2.12), whereas going to second order gives an effective vibration-rotational Hamiltonian for the electronic ground state (Watson, 1973; Bunker and Moss, 1977; Watson, 1980; Herman and Ogilvie, 1998)

$$\hat{H}^{eff} = -\frac{\hbar^2}{2}\frac{\partial}{\partial R}\frac{1}{\mu_n}[1+\beta(R)]\frac{\partial}{\partial R} + \frac{1}{2\mu_n R^2}[1+\alpha(R)]\hat{\vec{J}}^2$$
$$+ E_0^{(0)}(R) + E^{ad}(R) + E^{nad}(R) \quad (6.67)$$

which includes four additional contributions: an adiabatic, $E^{ad}(R)$, and a non-adiabatic, $E^{nad}(R)$, contribution to the potential energy and non-adiabatic correction terms $\beta(R)$ and $\alpha(R)$ for the nuclear reduced masses in the kinetic energy operators for vibration and rotation. Adiabatic in this context means that this term is an expectation value over the wavefunction of the considered electronic state, here the ground state,

$$E^{ad}(R) = -\frac{\hbar^2}{2\mu_n}\langle\Psi_0^{(0)}(\{\vec{r}_i\}; R)|\frac{\partial^2}{\partial R^2}|\Psi_0^{(0)}(\{\vec{r}_i\}; R)\rangle$$
$$+ \frac{1}{2\mu_n R^2}\langle\Psi_0^{(0)}(\{\vec{r}_i\}; R)|L_x^2 + L_y^2|\Psi_0^{(0)}(\{\vec{r}_i\}; R)\rangle$$
$$+ \frac{1}{2(m_A + m_B)}\langle\Psi_0^{(0)}(\{\vec{r}_i\}; R)|\sum_{i,j}\hat{\vec{p}}_i\cdot\hat{\vec{p}}_j|\Psi_0^{(0)}(\{\vec{r}_i\}; R)\rangle \quad (6.68)$$

and thus comes from the first contribution to the second-order energy correction, Eq. (3.33). The effect of these corrections is that the potential energy of the nuclei does not only depend on the internuclear distance R but also on the relative momenta of the nuclei. Non-adiabatic, on the other hand, implies that they are obtained from the second contribution to the second-order energy-correction, Eq. (3.33), and involve a sum over other electronic, i.e. excited states. Both energy-correction terms, $E^{ad}(R)$ and $E^{nad}(R)$, are not related to any molecular properties and will therefore not be discussed further.

The corrections to the vibrational and rotational reduced nuclear masses, on the other hand, can be related to molecular electromagnetic properties. They also involve sum over excited states and are therefore non-adiabatic corrections

$$\beta(R) = -\frac{2}{\mu_n} \sum_{n \neq 0} \frac{\left(\langle \Psi_0^{(0)}(\{\vec{r}_i\}; R) | -i\hbar \frac{\partial}{\partial R} | \Psi_n^{(0)}(\{\vec{r}_i\}; R)\rangle\right)^2}{E_0^{(0)}(R) - E_n^{(0)}(R)} \quad (6.69)$$

$$\alpha(R) = \frac{2}{\mu_n R^2} \sum_{n \neq 0} \frac{\left|\langle \Psi_0^{(0)}(\{\vec{r}_i\}; R) | \hat{L}_\perp(\vec{R}_{CM}) | \Psi_n^{(0)}(\{\vec{r}_i\}; R)\rangle\right|^2}{E_0^{(0)}(R) - E_n^{(0)}(R)} \quad (6.70)$$

where we have assumed that the molecule is aligned along the z-axis and therefore $\hat{L}_x = \hat{L}_y$, which we thus denote here and in the rest of this section as $\hat{L}_\perp$. Both terms are second order in the coupling between the nuclear and the electronic motion. The rotational correction, $\alpha(R)$, is due to the coupling between the nuclear and electronic angular momentum, Eq. (6.3), whereas the vibrational correction, $\beta(R)$ arises due to a similar coupling between the linear momenta. They are thus a consequence of the breakdown of the Born–Oppenheimer approximation and are therefore often also called **Born–Oppenheimer breakdown (BOB) parameters**. Physically, they represent the contribution of the electrons to the reduced masses or the lagging behind of the electrons and one can therefore define effective reduced masses for vibration and for rotation as

$$\mu_{v,n}^{eff} = \frac{\mu_n}{1 + \beta(R)} \quad (6.71)$$

$$\mu_{J,n}^{eff} = \frac{\mu_n}{1 + \alpha(R)} \quad (6.72)$$

Recalling that $-i\hbar \frac{\partial}{\partial R}$ is a hermitian operator, i.e.

$$\langle \Psi_0^{(0)}(\{\vec{r}_i\}; R) | -i\hbar \frac{\partial \Psi_n^{(0)}(\{\vec{r}_i\}; R)}{\partial R} \rangle = \langle \Psi_n^{(0)}(\{\vec{r}_i\}; R) | -i\hbar \frac{\partial \Psi_0^{(0)}(\{\vec{r}_i\}; R)}{\partial R} \rangle^*$$

$$= -\langle \Psi_n^{(0)}(\{\vec{r}_i\}; R) | -i\hbar \frac{\partial \Psi_0^{(0)}(\{\vec{r}_i\}; R)}{\partial R} \rangle \quad (6.73)$$

where it was also used that the wavefunctions are real, we can rewrite the non-adiabatic vibrational correction alternatively as

$$\beta(R) = -\frac{2\hbar^2}{\mu_n} \sum_{n \neq 0} \frac{\langle \Psi_0^{(0)}(\{\vec{r}_i\}; R) | \frac{\partial}{\partial R} | \Psi_n^{(0)}(\{\vec{r}_i\}; R)\rangle \langle \Psi_n^{(0)}(\{\vec{r}_i\}; R) | \frac{\partial}{\partial R} | \Psi_0^{(0)}(\{\vec{r}_i\}; R)\rangle}{E_0^{(0)}(R) - E_n^{(0)}(R)}$$

$$(6.74)$$

In contrast to previous chapters we have until now in this section shown explicitly that the electronic wavefunctions depend not only on the position vectors of the electrons $\{\vec{r}_i\}$ but parametrically also on the internuclear distance R. In the following, we will not indicate the obvious dependence on the electronic coordinates for the sake of more compact formulas, but continue with showing the dependence of the electronic wavefunctions and Born–Oppenheimer energies on the internuclear distance.

Like all sum-over-states expressions we can also express the non-adiabatic corrections to the reduced masses as linear response functions

$$\beta(R) = -\frac{\hbar^2}{\mu_n} \langle\langle \frac{\partial}{\partial R} ; \frac{\partial}{\partial R} \rangle\rangle_{\omega=0} \tag{6.75}$$

$$\alpha(R) = \frac{1}{\mu_n R^2} \langle\langle \hat{L}_\perp(\vec{R}_{CM}) ; \hat{L}_\perp(\vec{R}_{CM}) \rangle\rangle_{\omega=0} \tag{6.76}$$

Comparison with Eq. (6.27) shows that the non-adiabatic correction to the rotational reduced mass is proportional to the electronic contribution to the rotational g factor $g^{el}_{J,n}(R)$

$$\alpha(R) = \frac{m_e}{m_p} g^{el}_{J,n}(R) \tag{6.77}$$

Here and in the following, we refer to the xx- or yy-component of the rotational g tensor of a diatomic molecule aligned along the z-axis as the rotational g factor $g_J(R)$, i.e.

$$g_J(R) \equiv g_{J,xx}(R) = g_{J,yy}(R) \tag{6.78}$$

Furthermore, we have added a subscript n in order to indicate that the moment of inertia tensor as well as the centre of mass is in terms of the nuclear masses. This relation might at first sight be surprising, like the relations between the rotational g tensor and the magnetizability, Eq. (6.29), or between the spin rotation tensor and the nuclear magnetic shielding, Eq. (6.58), because $\alpha(R)$ is quadratic in the coupling between rotation of the nuclei and electronic motion, Eq. (6.3), whereas the rotational g tensor is bilinear in the coupling with rotation and the interaction with an external magnetic induction. However, rotation and apparent magnetic fields are interrelated, as discussed already in Section 6.1. A corresponding magnetic property for the non-adiabatic vibrational correction $\beta(R)$ does not exist, as molecules do not acquire a magnetic moment during their vibrations. Nevertheless, Herman and coworkers (Herman and Asgharian, 1966; Herman and Ogilvie, 1998) defined a corresponding **vibrational g factor**, whose electronic contribution $g^{el}_{v,n}(R)$ is then proportional to the non-adiabatic vibrational correction

$$\beta(R) = \frac{m_e}{m_p} g^{el}_{v,n}(R) \tag{6.79}$$

The effective Hamiltonian for vibration-rotational motion of nuclei, Eq. (6.67), thus contains effective reduced masses

$$\mu^{eff}_{v/J,n} = \frac{\mu_n}{1 + \frac{m_e}{m_p} g^{el}_{v/J,n}(R)} \tag{6.80}$$

in which $g^{el}_{v/J,n}(R)$ is the electronic contribution to either a rotational or vibrational g factor and μ_n is the reduced mass in terms of nuclear masses. However, sometimes it is more convenient to express the effective Hamiltonian, Eq. (6.67), in terms of atomic masses M_A and M_B. Such a change of masses can be achieved approximately, if one assumes that the difference between an atomic and a nuclear mass is equal to the total mass of the electrons of the given atom, i.e. $Z_A m_e$ or $Z_B m_e$, where Z_A and Z_B are the atomic numbers of the two atoms. The inverse of an atomic reduced mass μ can then be expressed in terms of nuclear masses and atomic numbers as

$$\frac{1}{\mu} = \frac{m_A + Z_A m_e + m_B + Z_B m_e}{(m_A + Z_A m_e)(m_B + Z_B m_e)} = \frac{m_A + Z_A m_e + m_B + Z_B m_e}{m_A m_B \left[1 + m_e \left(\frac{Z_A}{m_A} + \frac{Z_B}{m_B}\right)\right] + m_e^2 Z_A Z_B} \quad (6.81)$$

Neglecting the term proportional to m_e^2 and expanding the remaining denominator as

$$\left[1 + m_e \left(\frac{Z_A}{m_A} + \frac{Z_B}{m_B}\right)\right]^{-1} \approx 1 - m_e \left(\frac{Z_A}{m_A} + \frac{Z_B}{m_B}\right) \quad (6.82)$$

leads to

$$\frac{1}{\mu} \approx \frac{m_A + m_B}{m_A m_B}\left[1 - m_e \frac{Z_A m_B^2 + Z_B m_A^2}{(m_A + m_B) m_A m_B}\right] - m_e^2 \frac{Z_A + Z_B}{m_A m_B}\left(\frac{Z_A}{m_A} + \frac{Z_B}{m_B}\right) \quad (6.83)$$

Neglecting once more the term proportional to m_e^2 one obtains an approximate relation between the atomic and nuclear reduced masses

$$\frac{1}{\mu} \approx \frac{1}{\mu_n}\left[1 - m_e \frac{Z_A m_B^2 + Z_B m_A^2}{(m_A + m_B) m_A m_B}\right] \quad (6.84)$$

The error made by this approximation is of order $10^{-7} u$ or smaller (Bak et al., 2005). Interestingly, this correction happens to be proportional to the nuclear contribution to the rotational g factor of a diatomic molecule, Eq. (6.16), expressed in terms of nuclear masses,

$$g_n^{nuc} = m_p \frac{Z_A m_B^2 + Z_B m_A^2}{(m_A + m_B) m_A m_B} \quad (6.85)$$

which is independent of internuclear distance. The change from nuclear to atomic masses thus introduces a term that is equal to the nuclear contribution to the rotational g factor of diatomic molecules.

Defining a "total vibrational g factor" of a diatomic molecule (Herman and Asgharian, 1966; Herman and Ogilvie, 1998)

$$g_{v,n} = g^{el}_{v,n} + g_n^{nuc} \quad (6.86)$$

where the subscript n indicates again the use of nuclear masses, we can approximate the effective reduced masses in Eq. (6.80) by the atomic reduced mass μ and a correction from the rotational or vibrational g factor as

$$\frac{1}{\mu^{eff}_{v/J,n}} \approx \frac{1}{\mu} + \frac{1}{\mu_n}\frac{m_e}{m_p} g_{v/J,n} \quad (6.87)$$

However, the correction terms still depend on the nuclear masses. In addition to the obvious factor $\frac{1}{\mu_n}$ the electronic contributions to the g factors, Eqs. (6.69) and (6.70), include a second factor $\frac{1}{\mu_n}$. Furthermore, the electronic contribution to the rotational g factor, Eq. (6.70), depends also on the masses, because the angular momentum operator $\hat{L}(\vec{R}_{CM})$ is defined with respect to the centre of nuclear masses. But we ignore this dependence here, although this dependence on masses allows the determination of the electric dipolar moment from the rotational g factors of isotopic variants as discussed in Section 6.3. The mass dependence of the nuclear contribution to the g factor, Eq. (6.85), on the other hand, is more complicated. However, the ratio $\frac{g_n^{nuc}}{g^{nuc}}$ is almost equal to $\frac{\mu}{\mu_n}$, which implies that we can write

$$\frac{1}{\mu_{v/J,n}^{eff}} \approx \frac{1}{\mu}\left[1+\left(\frac{\mu}{\mu_n}\right)^2\frac{m_e}{m_p}g_{v/J}\right] \tag{6.88}$$

or

$$\frac{1}{\mu_{v/J,n}^{eff}} \approx \frac{1}{\mu}\left(1+\frac{m_e}{m_p}g_{v/J}\right) \tag{6.89}$$

if one accepts that $\left(\frac{\mu}{\mu_n}\right)^2 \approx 1$ to the accuracy normally required here (Bak et al., 2005). The effective vibration-rotational Hamiltonian for the electronic ground state of a diatomic molecule, using atomic masses, can then finally be written as

$$\hat{H}^{eff} = -\frac{\hbar^2}{2}\frac{\partial}{\partial R}\frac{1}{\mu}\left[1+\frac{m_e}{m_p}g_v(R)\right]\frac{\partial}{\partial R}+\frac{1}{2\mu R^2}\left[1+\frac{m_e}{m_p}g_J(R)\right]\hat{J}^2$$
$$+ E_0^{(0)}(R) + E^{ad}(R) + E^{nad}(R) \tag{6.90}$$

where the rotational and vibrational g factor radial functions in terms of atomic masses are defined as

$$g_{v/J}(R) = g^{nuc} + g_{v/J}^{el}(R) \tag{6.91}$$

$$g^{nuc} = \frac{m_p}{\mu R^2}\left[Z_A(R_{A,z}-R_{CM,z})^2 + Z_B(R_{B,z}-R_{CM,z})^2\right]$$

$$= m_p\frac{Z_A M_B^2 + Z_B M_A^2}{(M_A+M_B)M_A M_B} \tag{6.92}$$

$$g_v^{el}(R) = -\frac{m_p}{m_e}\frac{2\hbar^2}{\mu}\sum_{n\neq 0}\frac{\langle\Psi_0^{(0)}(R)|\frac{\partial}{\partial R}|\Psi_n^{(0)}(R)\rangle\langle\Psi_n^{(0)}(R)|\frac{\partial}{\partial R}|\Psi_0^{(0)}(R)\rangle}{E_0^{(0)}(R)-E_n^{(0)}(R)} \tag{6.93}$$

$$= -\frac{m_p}{m_e}\frac{\hbar^2}{\mu}\langle\langle\frac{\partial}{\partial R};\frac{\partial}{\partial R}\rangle\rangle_{\omega=0} \tag{6.94}$$

$$g_J^{el}(R) = \frac{m_p}{m_e}\frac{2}{\mu R^2}\sum_{n\neq 0}\frac{\left|\langle\Psi_0^{(0)}(R)|\hat{L}_\perp(\vec{R}_{CM})|\Psi_n^{(0)}(R)\rangle\right|^2}{E_0^{(0)}(R) - E_n^{(0)}(R)} \qquad (6.95)$$

$$= \frac{m_p}{m_e}\frac{1}{\mu R^2}\langle\!\langle\,\hat{L}_\perp(\vec{R}_{CM})\,;\,\hat{L}_\perp(\vec{R}_{CM})\,\rangle\!\rangle_{\omega=0} \qquad (6.96)$$

6.8 Partitioning of the g Factors

In order to fit vibration-rotation spectra of several isotopologues of diatomic molecules to the effective Hamiltonian, Eq. (6.90), one normally partitions the corrections to the reduced masses, i.e. the g factor radial functions $g_J(R)$ and $g_v(R)$, into two isotopically independent terms that are associated with one or the other nucleus

$$g_J(R) = \frac{\mu}{M_A}g_J^A(R) + \frac{\mu}{M_B}g_J^B(R) \qquad (6.97)$$

$$g_v(R) = \frac{\mu}{M_A}g_v^A(R) + \frac{\mu}{M_B}g_v^B(R) \qquad (6.98)$$

Physically, these isotopically independent g factors correspond to the hypothetical situations, where the molecule rotates around one of the atoms, i.e. the axis of rotation goes through this atom, or where only one of the atoms moves during the vibration. Mathematically they are obtained (Watson, 1973, 1980; Sauer, 1998; Kjær and Sauer, 2009) by choosing the atoms, i.e. $\vec{R}_A$ or $\vec{R}_B$, instead of the centre of mass $\vec{R}_{CM}$ as origin in the expression for the rotational g factor and by replacing in the expression for the vibrational g factor the canonical momentum operator for the vibration $\hat{P}_R = -i\hbar\frac{\partial}{\partial R}$ with the following two isotopically invariant operators

$$\hat{P}_{zA} = \hat{P}_R + \frac{(R_{B,z} - R_{CM,z})}{R}\hat{O}_z^p \qquad (6.99)$$

$$\hat{P}_{zB} = \hat{P}_R + \frac{(R_{A,z} - R_{CM,z})}{R}\hat{O}_z^p \qquad (6.100)$$

where $\hat{O}^p$ is the total canonical momentum operator of the electrons, whose cartesian components are defined in Eq. (3.65), and the molecule AB is placed along the z-axis. The isotopically invariant contributions to the g factors are thus given as

$$g_J^A(R) = \frac{m_p}{\mu}Z_A + \frac{m_p}{m_e\mu R^2}\langle\!\langle\,\hat{L}_x(\vec{R}_B)\,;\,\hat{L}_x(\vec{R}_B)\,\rangle\!\rangle_{\omega=0} \qquad (6.101)$$

$$g_J^B(R) = \frac{m_p}{\mu}Z_B + \frac{m_p}{m_e\mu R^2}\langle\!\langle\,\hat{L}_x(\vec{R}_A)\,;\,\hat{L}_x(\vec{R}_A)\,\rangle\!\rangle_{\omega=0} \qquad (6.102)$$

$$g_v^A(R) = \frac{m_p}{\mu}Z_A + \frac{m_p}{m_e\mu}\langle\!\langle\,\hat{P}_{zA}\,;\,\hat{P}_{zA}\,\rangle\!\rangle_{\omega=0} \qquad (6.103)$$

$$g_v^B(R) = \frac{m_p}{\mu}Z_B + \frac{m_p}{m_e\mu}\langle\!\langle\,\hat{P}_{zB}\,;\,\hat{P}_{zB}\,\rangle\!\rangle_{\omega=0} \qquad (6.104)$$

where we have chosen $\hat{L}_\perp = \hat{L}_x$ here and in the rest of this section.

In the following, we are going to prove Eqs. (6.97) and (6.98) following the derivations in Section 6.3 and will derive other useful relations involving the isotopically invariant g factors. In order to do so we will reformulate $g_J^B(R)$ and $g_v^B(R)$ by adding and subtracting $\vec{R}_{CM}$ in Eq. (6.102) and inserting Eq. (6.100) in Eq. (6.104), leading to

$$g_J^B(R) = \frac{m_p}{\mu} Z_B + \frac{m_p}{m_e \mu R^2} \langle\langle \hat{L}_x(\vec{R}_{CM}) \, ; \, \hat{L}_x(\vec{R}_{CM}) \rangle\rangle_{\omega=0}$$
$$- \frac{m_p}{m_e \mu R^2}(R_{CM,z} - R_{A,z}) \left(\langle\langle \hat{O}_y^p \, ; \, \hat{L}_x(\vec{R}_{CM}) \rangle\rangle_{\omega=0} + \langle\langle \hat{L}_x(\vec{R}_{CM}) \, ; \, \hat{O}_y^p \rangle\rangle_{\omega=0} \right)$$
$$+ \frac{m_p}{m_e \mu R^2}(R_{CM,z} - R_{A,z})^2 \langle\langle \hat{O}_y^p \, ; \, \hat{O}_y^p \rangle\rangle_{\omega=0} \tag{6.105}$$

and

$$g_v^B(R) = \frac{m_p}{\mu} Z_B + \frac{m_p}{m_e \mu} \langle\langle \hat{P}_R \, ; \, \hat{P}_R \rangle\rangle_{\omega=0}$$
$$- \frac{m_p}{m_e \mu R}(R_{CM,z} - R_{A,z}) \left(\langle\langle \hat{O}_z^p \, ; \, \hat{P}_R \rangle\rangle_{\omega=0} + \langle\langle \hat{P}_R \, ; \, \hat{O}_z^p \rangle\rangle_{\omega=0} \right)$$
$$+ \frac{m_p}{m_e \mu R^2}(R_{CM,z} - R_{A,z})^2 \langle\langle \hat{O}_z^p \, ; \, \hat{O}_z^p \rangle\rangle_{\omega=0} \tag{6.106}$$

The second terms are the electronic contributions to the rotational and vibrational g factors, whereas the last three terms are static response functions involving the total electronic momentum operator $\hat{\vec{O}}^p$ and another operator $\hat{\vec{O}}$. Using Eqs. (3.64) and (3.141) we can again replace them by ground-state expectation values of commutators of the operator $\hat{\vec{O}}$ and the sum of the position operators of the electrons, $\hat{\vec{O}}^r$,

$$g_J^B(R) = g_J(R) + \frac{m_p}{\mu} Z_B - \frac{m_p}{\mu R^2} \left[Z_A (R_{A,z} - R_{CM,z})^2 + Z_B (R_{B,z} - R_{CM,z})^2 \right]$$
$$+ \frac{2m_p}{\mu R^2}(R_{CM,z} - R_{A,z}) \frac{1}{i\hbar} \langle \Psi_0^{(0)} | [\hat{O}_y^r, \hat{L}_x(\vec{R}_{CM})] | \Psi_0^{(0)} \rangle$$
$$- \frac{m_p}{\mu R^2}(R_{CM,z} - R_{A,z})^2 \frac{1}{i\hbar} \langle \Psi_0^{(0)} | [\hat{O}_y^r, \hat{O}_y^p] | \Psi_0^{(0)} \rangle \tag{6.107}$$

and

$$g_v^B(R) = g_v(R) + \frac{m_p}{\mu} Z_B - \frac{m_p}{\mu R^2} \left[Z_A (R_{A,z} - R_{CM,z})^2 + Z_B (R_{B,z} - R_{CM,z})^2 \right]$$
$$- \frac{2m_p}{\mu R}(R_{CM,z} - R_{A,z}) \left(\langle \frac{\partial}{\partial R} \Psi_0 | \sum_i \hat{r}_{i,z} | \Psi_0 \rangle + \langle \Psi_0 | \sum_i \hat{r}_{i,z} | \frac{\partial}{\partial R} \Psi_0 \rangle \right)$$
$$- \frac{m_p}{\mu R^2}(R_{CM,z} - R_{A,z})^2 \frac{1}{i\hbar} \langle \Psi_0^{(0)} | [\hat{O}_y^r, \hat{O}_y^p] | \Psi_0^{(0)} \rangle \tag{6.108}$$

Evaluating the commutators, (6.34) and (6.35), we can rewrite this as

$$g_J^B(R) = g_J(R) + \frac{m_p}{\mu}\left[Z_B - Z_A\left(\frac{R_{A,z} - R_{CM,z}}{R}\right)^2 - Z_B\left(\frac{R_{B,z} - R_{CM,z}}{R}\right)^2\right]$$

$$-\frac{2m_p}{\mu R^2}(R_{CM,z} - R_{A,z})\langle\Psi_0^{(0)}|\sum_i (r_{i,z} - R_{CM,z})|\Psi_0^{(0)}\rangle$$

$$-\frac{m_p}{\mu R^2}(R_{CM,z} - R_{A,z})^2 N \qquad (6.109)$$

and

$$g_v^B(R) = g_v(R) + \frac{m_p}{\mu}\left[Z_B - Z_A\left(\frac{R_{A,z} - R_{CM,z}}{R}\right)^2 - Z_B\left(\frac{R_{B,z} - R_{CM,z}}{R}\right)^2\right]$$

$$-\frac{2m_p}{\mu R}(R_{CM,z} - R_{A,z})\frac{\partial}{\partial R}\langle\Psi_0|\hat{O}_z^r|\Psi_0\rangle$$

$$-\frac{m_p}{\mu R^2}(R_{CM,z} - R_{A,z})^2 N \qquad (6.110)$$

Recalling that according to Eq. (4.25) the z-component of the electric dipole moment $\mu_z(\vec{R}_{CM}, R)$ for the internuclear distance R and with the origin of the coordinate system at the centre of mass, is given as

$$\mu_z(\vec{R}_{CM}, R) = e\left[Z_A(R_{A,z} - R_{CM,z}) + Z_B(R_{B,z} - R_{CM,z})\right] - e\langle 0|\sum_i (r_{i,z} - R_{CM,z})|0\rangle$$

$$(6.111)$$

and that the total charge q of the molecule was defined in Eq. (6.37) we can write

$$g_J^B(R) = g_J(R) - \frac{2m_p}{e\mu R}\mu_z(\vec{R}_{CM}, R)\frac{R_{A,z} - R_{CM,z}}{R} + \frac{m_p}{\mu}q\left(\frac{R_{A,z} - R_{CM,z}}{R}\right)^2 \quad (6.112)$$

and

$$g_v^B(R) = g_v(R) - \frac{2m_p}{e\mu}\frac{\partial}{\partial R}\mu_z(\vec{R}_{CM}, R)\frac{R_{A,z} - R_{CM,z}}{R} + \frac{m_p}{\mu}q\left(\frac{R_{A,z} - R_{CM,z}}{R}\right)^2$$

$$(6.113)$$

Choosing the coordinate system such that $R_{A,z} - R_{CM,z} = -R\,\mu/M_A$, which implies for a molecule of polarity $^+AB^-$ that $\mu_z < 0$, we can finally write

$$g_J^B(R) = g_J(R) + \frac{2m_p}{eR}\frac{\mu_z(\vec{R}_{CM}, R)}{M_A} + m_p q\frac{\mu}{M_A^2} \qquad (6.114)$$

and

$$g_v^B(R) = g_v(R) + \frac{2m_p}{eM_A}\frac{\partial}{\partial R}\mu_z(\vec{R}_{CM}, R) + m_p q\frac{\mu}{M_A^2} \qquad (6.115)$$

The analogous derivations for nucleus A would then give

$$g_J^A(R) = g_J(R) - \frac{2m_p}{eR} \frac{\mu_z(\vec{R}_{CM}, R)}{M_B} + m_p q \frac{\mu}{M_B^2} \quad (6.116)$$

and

$$g_v^A(R) = g_v(R) - \frac{2m_p}{eM_B} \frac{\partial}{\partial R} \mu_z(\vec{R}_{CM}, R) + m_p q \frac{\mu}{M_B^2} \quad (6.117)$$

For a neutral diatomic molecule, i.e. $q = 0$, we can then trivially prove the partitioning of the rotational and vibrational g factors in Eqs. (6.97) and (6.98) by inserting Eqs. (6.114)–(6.117).

On the other hand, simply averaging the two isotopically invariant g factors gives another partitioning of the g factors

$$g_J(R) = g_J^{irr}(R) - \frac{m_p}{eR} \mu_z(\vec{R}_{CM}, R) \left(\frac{1}{M_A} - \frac{1}{M_B} \right) - \frac{m_p}{2} q \left(\frac{\mu}{M_A^2} + \frac{\mu}{M_B^2} \right) \quad (6.118)$$

and

$$g_v(R) = g_v^{irr}(R) - \frac{m_p}{e} \frac{\partial}{\partial R} \mu_z(\vec{R}_{CM}, R) \left(\frac{1}{M_A} - \frac{1}{M_B} \right) - \frac{m_p}{2} q \left(\frac{\mu}{M_A^2} + \frac{\mu}{M_B^2} \right) \quad (6.119)$$

where two "irreducible" non-adiabatic contributions to the g factors were defined as

$$\begin{aligned} g_J^{irr}(R) &= \frac{1}{2} \left[g_J^B(R) + g_J^A(R) \right] \\ &= \frac{m_p}{2\mu} \Big(Z_A + Z_B + \frac{1}{m_e R^2} \langle\langle \hat{L}_x(\vec{R}_A) \,;\, \hat{L}_x(\vec{R}_A) \rangle\rangle_{\omega=0} \\ &\quad + \frac{1}{m_e R^2} \langle\langle \hat{L}_x(\vec{R}_B) \,;\, \hat{L}_x(\vec{R}_B) \rangle\rangle_{\omega=0} \Big) \end{aligned} \quad (6.120)$$

and

$$\begin{aligned} g_v^{irr}(R) &= \frac{1}{2} \left[g_v^B(R) + g_v^A(R) \right] \\ &= \frac{m_p}{2\mu} \left[Z_A + Z_B + \frac{1}{m_e} \left(\langle\langle \hat{P}_{zA} \,;\, \hat{P}_{zA} \rangle\rangle_{\omega=0} + \langle\langle \hat{P}_{zB} \,;\, \hat{P}_{zB} \rangle\rangle_{\omega=0} \right) \right] \end{aligned} \quad (6.121)$$

These relations give some physical insight in the g factors (Sauer, 1998; Kjær and Sauer, 2009). In addition to a contribution from the overall charge of the molecule there are two contributions: one due to the electric dipole moment or gradient of the dipole moment with respect to the internuclear distance and one irreducible non-adiabatic contribution consisting of the average of the isotopically invariant g factors. Interestingly, there seems to be a parallelism to the gross selection rules for rotation and vibration spectra, where a permanent electric dipole moment or a change in the electric dipole moment under vibration is required.

Finally, by subtracting the isotopically invariant g factors we can obtain expressions for the electric dipole moment

$$\mu_z(R) = \frac{eR\mu}{2m_p}\left[g_J^B(R) - g_J^A(R)\right] \quad (6.122)$$

and the gradient of the electric dipole moment with respect to the internuclear distance R

$$\frac{\partial}{\partial R}\mu_z(\vec{R}_{CM}, R) = \frac{e\mu}{2m_p}\left[g_v^B(R) - g_v^A(R)\right] \quad (6.123)$$

From a computational point of view these expressions are not interesting, but they offer an alternative route to the experimental determination of the electric dipole moment and its gradient (Ogilvie and Liao, 1994). The analysis of vibration-rotation spectra recorded without external electric fields via a fit to the effective Hamiltonian in Eq. (6.90) will produce values for the isotopically invariant g factors that can then be used in equations Eqs. (6.122) and (6.123) for this purpose.

6.9 Further Reading

Rotational g Tensor

J. F. Ogilvie, J. Oddershede and S. P. A. Sauer, *The Rotational g Factor of Diatomic Molecules in State $^1\Sigma^+$ or 0^+*, Adv. Chem. Phys. 111, 475–536 (2000).

W. H. Flygare, *Molecular Structure and Dynamics*, Prentice-Hall, Englewood Cliffs (1978): Chapter 6.8.

Spin Rotation Tensor

W. H. Flygare, *Molecular Structure and Dynamics*, Prentice-Hall, Englewood Cliffs (1978): Chapter 6.9.

Effective Vibration-Rotational Hamiltonian

J. Brown and A. Carrington, *Rotational Spectroscopy of Diatomic Molecules*, Cambridge University Press, Cambridge (2003): Chapter 7.

7
Frequency-Dependent and Spectral Properties

The discussion to this point has been limited to static electric and magnetic fields. However, molecules are often exposed to time-dependent fields, as for example in the interaction with electromagnetic radiation. Some of the properties introduced in this chapter, like the frequency-dependent polarizability are generalizations to time- or frequency-dependent fields of the properties introduced in Chapters 4 and 5. Other spectral properties like the vertical excitation energies, transition dipole moments and properties derived from them, are a completely different type of property as they cannot be defined as derivatives of the ground-state energy.

7.1 Time-Dependent Fields

Solving Maxwell's equations [see Exercise 7.1] for the vector potential of a **plane** or **linear polarized** electromagnetic wave oscillating with angular frequency ω gives

$$\vec{A}(\vec{r},t) = \vec{\mathcal{A}}^\omega e^{i(\vec{k}\cdot\vec{r}-\omega t)} + \vec{\mathcal{A}}^{\omega*} e^{-i(\vec{k}\cdot\vec{r}-\omega t)} \tag{7.1}$$

where the wave or propagation vector $\vec{k}$ points in the direction of the propagation of the wave. It has the length

$$|\vec{k}| = n_r(\omega)\frac{\omega}{c} = \frac{2\pi n_r(\omega)}{\lambda} \tag{7.2}$$

where c is the speed of light, λ the wavelength of the electromagnetic wave in vacuum and $n_r(\omega)$ the refractive index of the medium through which the wave propagates. The refractive index is the ratio of the speed of electromagnetic radiation in vacuum to the speed in a medium. The dependence of the refractive index on the frequency is called dispersion. In vacuum, the refractive index is therefore equal to 1. The amplitude $\vec{\mathcal{A}}^\omega$ is in principle a complex vector perpendicular to the propagation vector $\vec{k}$, but can be chosen to be purely imaginary by an appropriate choice of origin of the time variable. With this choice the time dependence of the vector potential reduces to

$$\vec{A}(\vec{r},t) = i2\vec{\mathcal{A}}^\omega \sin(\vec{k}\cdot\vec{r} - \omega t) \tag{7.3}$$

showing that the vector potential is then real. Using Eqs. (2.33) and (2.34) we can obtain expressions for the electric and magnetic fields of this plane monochromatic electromagnetic wave

$$\vec{\mathcal{E}}(\vec{r},t) = \vec{\mathcal{E}}^\omega \cos(\vec{k}\cdot\vec{r} - \omega t) \tag{7.4}$$

$$\vec{\mathcal{B}}(\vec{r},t) = \vec{\mathcal{B}}^\omega \cos(\vec{k}\cdot\vec{r} - \omega t) \tag{7.5}$$

where the amplitude $\vec{\mathcal{E}}^\omega = i2\omega \vec{A}^\omega$ of the electric field is also perpendicular to the propagation vector, while the amplitude $\vec{\mathcal{B}}^\omega = i2\vec{k}\times\vec{A}^\omega$ of the magnetic field is perpendicular to both the propagation vector $\vec{k}$ and the amplitude of the vector potential and thus to the electric field.

Exercise 7.1 Show that the plane-polarized electromagnetic wave in Eq. (7.1) is a solution to Maxwell's equation for the vector potential in vacuum Eq. (2.129).

A general pulse of coherent polychromatic electromagnetic radiation[1] can be described as superposition of monochromatic plane waves. The vector potential and fields of such a pulse are then given as

$$\vec{A}(\vec{r},t) = \int_0^\infty i2\vec{A}^\omega \sin(\vec{k}\cdot\vec{r} - \omega t)d\omega = \int_0^\infty \vec{A}^\omega \left[e^{i(\vec{k}\cdot\vec{r}-\omega t)} - e^{-i(\vec{k}\cdot\vec{r}-\omega t)}\right]d\omega \tag{7.6}$$

$$\vec{\mathcal{E}}(\vec{r},t) = \int_0^\infty \vec{\mathcal{E}}^\omega \cos(\vec{k}\cdot\vec{r} - \omega t)d\omega = \frac{1}{2}\int_0^\infty \vec{\mathcal{E}}^\omega \left[e^{i(\vec{k}\cdot\vec{r}-\omega t)} + e^{-i(\vec{k}\cdot\vec{r}-\omega t)}\right]d\omega \tag{7.7}$$

$$\vec{\mathcal{B}}(\vec{r},t) = \int_0^\infty \vec{\mathcal{B}}^\omega \cos(\vec{k}\cdot\vec{r} - \omega t)d\omega = \frac{1}{2}\int_0^\infty \vec{\mathcal{B}}^\omega \left[e^{i(\vec{k}\cdot\vec{r}-\omega t)} + e^{-i(\vec{k}\cdot\vec{r}-\omega t)}\right]d\omega \tag{7.8}$$

Typical molecules have diameters of 1 to 100 Å, which is thus the order of magnitude of the maximal length of the vector $\vec{r}$ in the expression for the vector potential and fields. The propagation vector $\vec{k}$, on the other hand, has length $\frac{2\pi}{\lambda}$, meaning that the product $\vec{k}\cdot\vec{r}$ is much smaller than 1 for all types of electromagnetic radiation with longer wavelengths than X-rays. For these waves we can expand $e^{i\vec{k}\cdot\vec{r}}$ in a power series

$$e^{i\vec{k}\cdot\vec{r}} = 1 + i\vec{k}\cdot\vec{r} + \frac{1}{2!}(i\vec{k}\cdot\vec{r})^2 + \cdots \approx 1 \tag{7.9}$$

and approximate it by the first term, 1. This is the same as setting $\vec{k} = \vec{0}$ and implies that we ignore the spatial variation of the vector potential across a molecule. We can therefore write for the vector potential

$$\vec{A}(t) = \int_0^\infty \frac{2}{i}\vec{A}^\omega \sin(\omega t)d\omega = \int_0^\infty \vec{A}^\omega \left(e^{-i\omega t} - e^{i\omega t}\right)d\omega \tag{7.10}$$

and for the electric field

$$\vec{\mathcal{E}}(t) = \int_0^\infty \vec{\mathcal{E}}^\omega \cos(\omega t)d\omega = \frac{1}{2}\int_0^\infty \vec{\mathcal{E}}^\omega \left(e^{-i\omega t} + e^{i\omega t}\right)d\omega \tag{7.11}$$

---

[1] For incoherent radiation one would have to introduce a phase that depends on the frequency of the radiation.

The magnetic induction $\vec{\mathcal{B}}(t)$, however, vanishes because it is the curl of the vector potential according to Eq. (2.34), which means that the amplitude $\vec{\mathcal{B}}^\omega = \imath 2\vec{k} \times \vec{\mathcal{A}}^\omega$ is zero for $\vec{k} = \vec{0}$. In this long-wavelength approximation with this purely time-dependent vector potential we can then choose the velocity or length gauge for the vector as discussed at the end of Section 2.9. Because the length gauge implies that the perturbation Hamiltonian becomes $-\hat{\vec{\mu}} \cdot \vec{\mathcal{E}}(t)$ and thus involves the electric dipole moment operator, the long-wavelength approximation is often called the **dipole approximation**.

Retaining also the second term $\imath \vec{k} \cdot \vec{r}$ in the expansion of $e^{\imath \vec{k} \cdot \vec{r}}$ leads to a spatially non-uniform vector potential and electric field, as well as to a non-vanishing magnetic induction and to an interaction with molecules via the magnetic dipole and electric quadrupole operators.

Plane or linear polarization of radiation, given in Eqs. (7.4) and (7.5), is not the only possibility. One alternative is circular polarization in the form of **left-circularly polarized** and **right-circularly polarized** waves whose electric-field vectors are given as

$$\vec{\mathcal{E}}^L(\vec{r},t) = \mathcal{E}^\omega \left[\vec{e}_i \cos\left(\frac{\omega n_r^L(\omega)}{c} \vec{e}_k \cdot \vec{r} - \omega t \right) + \vec{e}_j \sin\left(\frac{\omega n_r^L(\omega)}{c} \vec{e}_k \cdot \vec{r} - \omega t \right) \right] \quad (7.12)$$

$$\vec{\mathcal{E}}^R(\vec{r},t) = \mathcal{E}^\omega \left[\vec{e}_i \cos\left(\frac{\omega n_r^R(\omega)}{c} \vec{e}_k \cdot \vec{r} - \omega t \right) - \vec{e}_j \sin\left(\frac{\omega n_r^R(\omega)}{c} \vec{e}_k \cdot \vec{r} - \omega t \right) \right] \quad (7.13)$$

The two unit vectors $\vec{e}_i$ and $\vec{e}_j$ are perpendicular to each other and to the direction of the propagation, i.e. $\vec{e}_i \perp \vec{e}_k$, $\vec{e}_j \perp \vec{e}_k$ and $\vec{e}_i \perp \vec{e}_j$, where $\vec{e}_k$ is a unit vector in the direction of $\vec{k}$. The associated magnetic induction vectors are

$$\vec{\mathcal{B}}^L(\vec{r},t) = \mathcal{B}^\omega \left[-\vec{e}_i \sin\left(\frac{\omega n_r^L(\omega)}{c} \vec{e}_k \cdot \vec{r} - \omega t \right) - \vec{e}_j \cos\left(\frac{\omega n_r^L(\omega)}{c} \vec{e}_k \cdot \vec{r} - \omega t \right) \right] \quad (7.14)$$

$$\vec{\mathcal{B}}^R(\vec{r},t) = \mathcal{B}^\omega \left[\vec{e}_i \sin\left(\frac{\omega n_r^R(\omega)}{c} \vec{e}_k \cdot \vec{r} - \omega t \right) - \vec{e}_j \cos\left(\frac{\omega n_r^R(\omega)}{c} \vec{e}_k \cdot \vec{r} - \omega t \right) \right] \quad (7.15)$$

For right-circularly polarized radiation the electric field vector $\vec{\mathcal{E}}^R$ rotates clockwise when looking into the oncoming wave, i.e. at the source of the radiation. Circular polarization of photons corresponds to the two possible projections of the photon's spin on the direction of propagation, S_z, called **helicity**. Right-circularly polarized photons have $m_s = -1$ and thus $S_z = -\hbar$, while left-circularly polarized photons have $m_s = 1$. Plane-polarized radiation can then be expressed as a superposition of left- and right-circulary polarized waves with the same refractive index, $n_r(\omega) = n_r^R(\omega) = n_r^L(\omega)$, i.e.

$$\vec{\mathcal{E}}(\vec{r},t) = \vec{\mathcal{E}}^L(\vec{r},t) + \vec{\mathcal{E}}^R(\vec{r},t) = 2\mathcal{E}^\omega \vec{e}_i \cos\left(\frac{\omega n_r(\omega)}{c} \vec{e}_k \cdot \vec{r} - \omega t \right) \quad (7.16)$$

7.2 Frequency-Dependent Polarizability

In the presence of a pulse of coherent polychromatic electromagnetic radiation and making the dipole approximation we can generalize the expansion in Eq. (4.44) for a component of the now time-dependent electric dipole moment to (Buckingham, 1967)

$$\mu_\alpha(\vec{\mathcal{E}}(t)) = \mu_\alpha + \mu_\alpha^{ind}(\vec{\mathcal{E}}(t)) \tag{7.17}$$

$$\mu_\alpha(\vec{\mathcal{E}}(t)) = \mu_\alpha + \sum_\beta \int_0^\infty \alpha_{\alpha\beta}(-\omega_1;\omega_1)\mathcal{E}_\beta^{\omega_1}\cos(\omega_1 t)\,d\omega_1 \tag{7.18}$$

$$+ \frac{1}{2}\sum_{\beta\gamma}\int_0^\infty\int_0^\infty \beta_{\alpha\beta\gamma}(-\omega_1-\omega_2;\omega_1,\omega_2)\mathcal{E}_\beta^{\omega_1}\cos(\omega_1 t)\mathcal{E}_\gamma^{\omega_2}\cos(\omega_2 t)\,d\omega_1\,d\omega_2$$

$$+\ \ldots$$

where $\alpha_{\alpha\beta}(-\omega_1;\omega_1)$ and $\beta_{\alpha\beta\gamma}(-\omega_1-\omega_2;\omega_1,\omega_2)$ are components of the **frequency-dependent electric dipole polarizability**, also called the **dynamic electric dipole polarizability** and **first hyperpolarizability tensors**, respectively.

The isotropic frequency-dependent electric dipole polarizability $\alpha(-\omega_1;\omega_1)$ is the trace of the polarizability tensor

$$\alpha(-\omega_1;\omega_1) = \frac{1}{3}\sum_\alpha \alpha_{\alpha\alpha}(-\omega_1;\omega_1) \tag{7.19}$$

It is the molecular property underlying the refractive index $n_r(\omega_1)$ of a macroscopic sample with number density $\mathcal{N}$. For a non-polar molecule, i.e. without permanent electric dipole moment, the relation between them can be shown to be[2]

$$n_r(\omega_1) = \sqrt{\frac{1 + 2\alpha(-\omega_1;\omega_1)\frac{\mathcal{N}}{3\epsilon_0}}{1 - \alpha(-\omega_1;\omega_1)\frac{\mathcal{N}}{3\epsilon_0}}} \approx 1 + \alpha(-\omega_1;\omega_1)\frac{\mathcal{N}}{2\epsilon_0} \tag{7.20}$$

In order to derive a quantum mechanical expression for the frequency-dependent polarizability we can make use of time-dependent response theory as described in Section 3.11. We need therefore to evaluate the time-dependent expectation value of the electric dipole operator $\langle \Psi_0(\vec{\mathcal{E}}(t)) \mid \hat{\mu}_\alpha \mid \Psi_0(\vec{\mathcal{E}}(t)) \rangle$ in the presence of a time-dependent electric field, Eq. (7.11). Employing the length gauge, Eqs. (2.122) – (2.124), which implies that the time-dependent electric field enters the Hamiltonian via the scalar potential in Eq. (2.105), the perturbation Hamilton operator for the periodic and spatially uniform electric field of the electromagnetic wave is given as

$$\hat{H}^{(1)}(t) = -\sum_\beta \hat{O}_\beta^\mathcal{E}\,\mathcal{E}_\beta(t)$$

$$= -\sum_\beta \hat{\mu}_\beta \int_0^\infty \frac{\mathcal{E}_\beta^{\omega_1}}{2}\left(e^{-\imath\omega_1 t} + e^{\imath\omega_1 t}\right)d\omega_1 = -\sum_\beta \hat{\mu}_\beta \int_{-\infty}^\infty \frac{\mathcal{E}_\beta^{\omega_1}}{2}e^{-\imath\omega_1 t}d\omega_1 \tag{7.21}$$

---

[2] References to the derivation can be found in the *Further Reading* section.

Comparison with Eq. (3.78) shows that the Fourier components of the operator and the field are given as

$$\hat{O}^\omega_{\beta...} = -\hat{\mu}_\beta \qquad (7.22)$$

$$\mathcal{F}_{\beta...}(\omega) = \mathcal{E}^{\omega_1}_\beta \qquad (7.23)$$

Insertion of these operators in Eq. (3.109) yields for the expansion of the time-dependent dipole moment

$$\langle \Psi_0(\vec{\mathcal{E}}(t)) | \hat{\mu}_\alpha | \Psi_0(\vec{\mathcal{E}}(t)) \rangle$$

$$= \langle \Psi_0^{(0)} | \hat{\mu}_\alpha | \Psi_0^{(0)} \rangle + \int_{-\infty}^{\infty} \sum_\beta \langle\langle \hat{\mu}_\alpha ; -\hat{\mu}_\beta \rangle\rangle_{\omega_1} \frac{\mathcal{E}^{\omega_1}_\beta}{2} e^{-i\omega_1 t} d\omega_1 + \cdots \qquad (7.24)$$

$$= \langle \Psi_0^{(0)} | \hat{\mu}_\alpha | \Psi_0^{(0)} \rangle - \sum_\beta \int_0^\infty \langle\langle \hat{\mu}_\alpha ; \hat{\mu}_\beta \rangle\rangle_{\omega_1} \mathcal{E}^{\omega_1}_\beta \cos(\omega_1 t) d\omega_1 + \cdots \qquad (7.25)$$

Comparing this with the classical expansion of a time-dependent dipole moment in Eq. (7.18) we can identify the frequency-dependent polarizability tensor as a linear response function or polarization propagator

$$\alpha_{\alpha\beta}(-\omega_1; \omega_1) = -\langle\langle \hat{\mu}_\alpha ; \hat{\mu}_\beta \rangle\rangle_{\omega_1} \qquad (7.26)$$

Using the spectral representation of the polarization propagator, Eq. (3.110), we can alternatively write

$$\alpha_{\alpha\beta}(-\omega_1; \omega_1) = -\sum_{n\neq 0} \frac{\langle \Psi_0^{(0)} | \hat{\mu}_\alpha | \Psi_n^{(0)} \rangle \langle \Psi_n^{(0)} | \hat{\mu}_\beta | \Psi_0^{(0)} \rangle}{\hbar\omega_1 + E_0^{(0)} - E_n^{(0)}}$$

$$- \sum_{n\neq 0} \frac{\langle \Psi_0^{(0)} | \hat{\mu}_\beta | \Psi_n^{(0)} \rangle \langle \Psi_n^{(0)} | \hat{\mu}_\alpha | \Psi_0^{(0)} \rangle}{-\hbar\omega_1 + E_0^{(0)} - E_n^{(0)}} \qquad (7.27)$$

$$= 2 \sum_{n\neq 0} \frac{(E_n^{(0)} - E_0^{(0)}) \langle \Psi_0^{(0)} | \hat{\mu}_\alpha | \Psi_n^{(0)} \rangle \langle \Psi_n^{(0)} | \hat{\mu}_\beta | \Psi_0^{(0)} \rangle}{(E_n^{(0)} - E_0^{(0)})^2 - \hbar^2 \omega_1^2} \qquad (7.28)$$

which for $\omega_1 = 0$, i.e. for the static polarizability, reduces to the expression obtained by static response theory or Rayleigh–Schrödinger perturbation theory in Section 4.5.

7.3 Optical Rotation

In Section 7.1 we discussed that plane or linear polarized radiation can be expressed as the superposition of left-circularly polarized $\vec{\mathcal{E}}^L$ and right-circularly polarized waves $\vec{\mathcal{E}}^R$ with the same refractive index. If the refractive indices for left- and right-circularly polarized radiation, however, differ by

$$\Delta n_r(\omega_1) = n_r^L(\omega_1) - n_r^R(\omega_1) \qquad (7.29)$$

the superposition will still be a plane-polarized wave

$$\vec{\mathcal{E}}(\vec{r},t) = \vec{\mathcal{E}}^L(\vec{r},t) + \vec{\mathcal{E}}^R(\vec{r},t)$$
$$= 2\mathcal{E}^{\omega_1} \cos\left(\frac{\omega_1 \left[n_r^L(\omega_1) + n_r^R(\omega_1)\right]}{2c} \vec{e}_k \cdot \vec{r} - \omega_1 t\right)$$
$$\times \left[\vec{e}_i \cos\left(\frac{\omega_1 \Delta n_r(\omega_1)}{2c} \vec{e}_k \cdot \vec{r}\right) + \vec{e}_j \sin\left(\frac{\omega_1 \Delta n_r(\omega_1)}{2c} \vec{e}_k \cdot \vec{r}\right)\right] \quad (7.30)$$

but with a plane of polarization that is rotated by an angle

$$\Delta\theta = \frac{\omega_1 \Delta n_r(\omega_1)}{2c} \vec{e}_k \cdot \vec{r} \quad (7.31)$$

compared to the original wave, Eq. (7.16). This phenomenon is called **optical rotation** and a medium that has the property that left- and right-circularly polarized waves propagate with different velocity, i.e. that their refractive indices differ, is called a **circularly birefringent** medium.[3] Chiral molecules have this property, because they experience the spatial variation of the electric field vector of left- and right-circularly polarized waves as being of the same or opposite handedness as their own structure.

In the derivation of the molecular properties, which give rise to this effect, we have to take the spatial variation of the electric field vector into account and can thus not make the dipole approximation, contrary to the last section. This implies that we have to include a contribution from the interaction with the curl of the time-dependent electric-field, $\vec{\nabla} \times \vec{\mathcal{E}}(\vec{r},t)$, to the expansion of the induced dipole moment of a molecule in Eq. (7.18). However, Maxwell's third equation, Eq. (2.37) relates the curl of the electric-field vector to the time derivative $\partial \vec{B}(\vec{r},t)/\partial t$ of the magnetic induction and we can thus alternatively replace the spatial variation and expand the induced dipole moment instead in the electric field and the time derivative of the magnetic induction of a monochromatic wave (Buckingham, 1967) as

$$\mu_\alpha^{L/R}(\vec{\mathcal{E}}^{L/R}(t), \vec{B}^{L/R}(t))$$
$$= \mu_\alpha + \mu_\alpha^{L/R,ind}(\vec{\mathcal{E}}^{L/R}(t), \vec{B}^{L/R}(t)) \quad (7.32)$$
$$= \mu_\alpha + \sum_\beta \alpha_{\alpha\beta}(-\omega_1;\omega_1)\, \mathcal{E}_\beta^{L/R}(t) + \sum_\beta \frac{1}{\omega_1} G'_{\alpha\beta}(-\omega_1;\omega_1) \frac{\partial \vec{B}_\beta^{L/R}(t)}{\partial t} + \ldots \quad (7.33)$$

where $\mathbf{G'}(-\omega_1;\omega_1)$ is the **mixed frequency-dependent electric dipole magnetic dipole polarizability tensor**, whose isotropic value $G'(-\omega_1;\omega_1)$ is again the trace of the tensor

$$G'(-\omega_1;\omega_1) = \frac{1}{3} \sum_\alpha G'_{\alpha\alpha}(-\omega_1;\omega_1) \quad (7.34)$$

---

[3] **Birefringence** is in general the property that the refractive indices of radiation along two directions in a medium differ.

Extending the relation between the refractive index and the polarizability, Eq. (7.20), the refractive indices of a macroscopic sample with number density $\mathcal{N}$ for left- and right-circularly polarized waves become

$$n_r^{L/R}(\omega_1) \approx 1 + \alpha(-\omega_1;\omega_1)\frac{\mathcal{N}}{2\epsilon_0} \mp G'(-\omega_1;\omega_1)\frac{\mathcal{N}}{2c\epsilon_0} \qquad (7.35)$$

and we can finally express the difference in refractive indices

$$\Delta n_r(\omega_1) = -G'(-\omega_1;\omega_1)\frac{\mathcal{N}}{c\epsilon_0} \qquad (7.36)$$

and the angle of rotation

$$\Delta\theta = -\frac{\omega_1\, G'(-\omega_1;\omega_1)\mathcal{N}}{2\epsilon_0 c^2}\vec{e}_k \cdot \vec{r} \qquad (7.37)$$

in terms of the molecular mixed frequency-dependent electric dipole magnetic dipole polarizability $G'(-\omega_1;\omega_1)$.

In order to derive a quantum mechanical expression for the mixed dynamic electric dipole magnetic dipole polarizability tensor we have to evaluate the time-dependent expectation value of the electric dipole operator $\langle\Psi_0(t)|\hat{\mu}_\alpha|\Psi_0(t)\rangle$ in the presence of the time-dependent magnetic induction of left- or right-circularly polarized radiation

$$\vec{\mathcal{B}}^{L/R}(t) = \mathcal{B}^{\omega_1}\left[\mp\vec{e}_i\sin(\omega_1 t) - \vec{e}_j\cos(\omega_1 t)\right] \qquad (7.38)$$

$$= \pm\vec{e}_i\frac{\mathcal{B}^{\omega_1}}{2\imath}\left(e^{-\imath\omega_1 t} - e^{\imath\omega_1 t}\right) - \vec{e}_j\frac{\mathcal{B}^{\omega_1}}{2}\left(e^{-\imath\omega_1 t} + e^{\imath\omega_1 t}\right) \qquad (7.39)$$

For a closed-shell molecule the perturbation Hamiltonian operator then becomes to first order

$$\hat{H}^{(1)}_{L/R}(t) = -\hat{\vec{O}}^{lB} \cdot \vec{\mathcal{B}}^{L/R}(t)$$

$$= \mp\hat{\vec{m}}^l \cdot \vec{e}_i\, \frac{\mathcal{B}^{\omega_1}}{2\imath}\left(e^{-\imath\omega_1 t} - e^{\imath\omega_1 t}\right) + \hat{\vec{m}}^l \cdot \vec{e}_j\, \frac{\mathcal{B}^{\omega_1}}{2}\left(e^{-\imath\omega_1 t} + e^{\imath\omega_1 t}\right) \qquad (7.40)$$

Comparison with Eq. (3.78) shows that the Fourier components of the operator and the field are given as

$$\hat{O}^\omega_1 = \mp\hat{\vec{m}}^l \cdot \vec{e}_i \qquad (7.41)$$

$$\mathcal{F}_1(\omega_1) = \frac{\mathcal{B}^{\omega_1}}{2\imath}\left[\delta(\omega-\omega_1) - \delta(\omega+\omega_1)\right] \qquad (7.42)$$

$$\hat{O}^\omega_2 = \hat{\vec{m}}^l \cdot \vec{e}_j \qquad (7.43)$$

$$\mathcal{F}_2(\omega_1) = \frac{\mathcal{B}^{\omega_1}}{2}\left[\delta(\omega-\omega_1) + \delta(\omega+\omega_1)\right] \qquad (7.44)$$

160 Frequency-Dependent and Spectral Properties

Insertion of these operators in Eq. (3.109) yields for the expansion of the time-dependent dipole moment

$$\langle\Psi_0(t)|\hat{\mu}_\alpha|\Psi_0(t)\rangle^{L/R}$$
$$=\langle\Psi_0^{(0)}|\hat{\mu}_\alpha|\Psi_0^{(0)}\rangle + \left(\langle\langle\hat{\mu}_\alpha\,;\,\mp\hat{\vec{m}}^l\cdot\vec{e}_i\,\rangle\rangle_{\omega_1}\,e^{-i\omega_1 t} - \langle\langle\hat{\mu}_\alpha\,;\,\mp\hat{\vec{m}}^l\cdot\vec{e}_i\,\rangle\rangle_{-\omega_1}\,e^{i\omega_1 t}\right)\frac{\mathcal{B}^{\omega_1}}{2i}$$
$$+\left(\langle\langle\hat{\mu}_\alpha\,;\,\hat{\vec{m}}^l\cdot\vec{e}_j\,\rangle\rangle_{\omega_1}\,e^{-i\omega_1 t} + \langle\langle\hat{\mu}_\alpha\,;\,\hat{\vec{m}}^l\cdot\vec{e}_j\,\rangle\rangle_{-\omega_1}\,e^{i\omega_1 t}\right)\frac{\mathcal{B}^{\omega_1}}{2}$$
$$+\cdots \tag{7.45}$$

The electric dipole moment operator $\hat{\vec{\mu}}$ is a hermitian and real operator, whereas $\hat{\vec{m}}^l$ is hermitian and purely imaginary. The linear response function of such operators is thus purely imaginary and according to Eq. (3.113) antisymmetric with respect to a change in the sign of the frequency ω_1. We can therefore rewrite the expansion as

$$\langle\Psi_0(t)|\hat{\mu}_\alpha|\Psi_0(t)\rangle^{L/R}$$
$$=\langle\Psi_0^{(0)}|\hat{\mu}_\alpha|\Psi_0^{(0)}\rangle \mp \langle\langle\hat{\mu}_\alpha\,;\,\hat{\vec{m}}^l\cdot\vec{e}_i\,\rangle\rangle_{\omega_1}\left(e^{-i\omega_1 t}+e^{i\omega_1 t}\right)\frac{\mathcal{B}^{\omega_1}}{2i}$$
$$+ \langle\langle\hat{\mu}_\alpha\,;\,\hat{\vec{m}}^l\cdot\vec{e}_j\,\rangle\rangle_{\omega_1}\left(e^{-i\omega_1 t}-e^{i\omega_1 t}\right)\frac{\mathcal{B}^{\omega_1}}{2}+\cdots \tag{7.46}$$
$$=\langle\Psi_0^{(0)}|\hat{\mu}_\alpha|\Psi_0^{(0)}\rangle \pm \langle\langle\hat{\mu}_\alpha\,;\,\hat{\vec{m}}^l\cdot\vec{e}_i\,\rangle\rangle_{\omega_1}\cos(\omega_1 t)\,i\,\mathcal{B}^{\omega_1}$$
$$- \langle\langle\hat{\mu}_\alpha\,;\,\hat{\vec{m}}^l\cdot\vec{e}_j\,\rangle\rangle_{\omega_1}\sin(\omega_1 t)\,i\,\mathcal{B}^{\omega_1}+\cdots \tag{7.47}$$

Using the fact that the time derivative of the magnetic induction in Eq. (7.39) becomes

$$\frac{\partial\vec{B}^{L/R}(t)}{\partial t} = \omega_1\mathcal{B}^{\omega_1}\left[\mp\vec{e}_i\cos(\omega_1 t)+\vec{e}_j\sin(\omega_1 t)\right] \tag{7.48}$$

we finally obtain for the expansion of the time-dependent dipole moment

$$\langle\Psi_0(t)|\hat{\mu}_\alpha|\Psi_0(t)\rangle^{L/R} = \langle\Psi_0^{(0)}|\hat{\mu}_\alpha|\Psi_0^{(0)}\rangle - \frac{i}{\omega_1}\langle\langle\hat{\mu}_\alpha\,;\,\hat{\vec{m}}^l\,\rangle\rangle_{\omega_1}\cdot\frac{\partial\vec{B}^{L/R}(t)}{\partial t}+\cdots \tag{7.49}$$

Comparing this with the classical expansion of a time-dependent dipole moment in Eq. (7.33) we can identify the frequency-dependent mixed electric dipole magnetic dipole polarizability tensor as a linear response function or polarization propagator

$$G'_{\alpha\beta}(-\omega_1;\omega_1) = -i\langle\langle\hat{\mu}_\alpha\,;\,\hat{m}^l_\beta\,\rangle\rangle_{\omega_1} \tag{7.50}$$

As mentioned before the $\langle\langle\hat{\mu}_\alpha\,;\,\hat{m}^l_\beta\,\rangle\rangle_{\omega_1}$ response function is purely imaginary, but the **G′** tensor is real.

Using the spectral representation of the polarization propagator, Eq. (3.110), we can alternatively write

$$G'_{\alpha\beta}(-\omega_1;\omega_1) = -i\sum_{n\neq 0}\frac{\langle\Psi_0^{(0)}|\hat{\mu}_\alpha|\Psi_n^{(0)}\rangle\langle\Psi_n^{(0)}|\hat{m}_\beta^l|\Psi_0^{(0)}\rangle}{\hbar\omega_1 + E_0^{(0)} - E_n^{(0)}}$$

$$- i\sum_{n\neq 0}\frac{\langle\Psi_0^{(0)}|\hat{m}_\beta^l|\Psi_n^{(0)}\rangle\langle\Psi_n^{(0)}|\hat{\mu}_\alpha|\Psi_0^{(0)}\rangle}{-\hbar\omega_1 + E_0^{(0)} - E_n^{(0)}} \quad (7.51)$$

$$= 2i\hbar\sum_{n\neq 0}\frac{\omega_1\langle\Psi_0^{(0)}|\hat{\mu}_\alpha|\Psi_n^{(0)}\rangle\langle\Psi_n^{(0)}|\hat{m}_\beta^l|\Psi_0^{(0)}\rangle}{(E_n^{(0)} - E_0^{(0)})^2 - \hbar^2\omega_1^2} \quad (7.52)$$

and can see that for $\omega_1 = 0$ the frequency-dependent mixed electric dipole magnetic dipole polarizability tensor vanishes.

7.4 Electronic Excitation Energies and Transition Moments

In the previous sections it was shown that frequency-dependent linear response properties, such as frequency-dependent polarizabilities, can be obtained as the value of the polarization propagator for the appropriate operators. Furthermore, all static second-order properties discussed in Chapters 4 and 5 can be calculated as the value of a polarization propagator for zero frequency.

In addition to these properties, which are all related to the value of a particular polarization propagator, we can get further information about a molecule by studying the poles and residues of the linear response function or polarization propagator. We can see from Eq. (3.110), that the polarization propagator has a singularity or pole, if the frequency ω of the perturbation takes one of the following values

$$\omega = \pm\frac{E_n^{(0)} - E_0^{(0)}}{\hbar} \quad (7.53)$$

However, $E_n^{(0)} - E_0^{(0)} = \Delta E_{n0}$ is the difference in energy between the unperturbed reference state $|\Psi_0^{(0)}\rangle$ and one of the other unperturbed states $|\Psi_n^{(0)}\rangle$ and thus equal to the **vertical electronic excitation energy from state** $\Psi_0^{(0)}$ **to state** $\Psi_n^{(0)}$. Finding the poles of the polarization propagator is thus a way of directly calculating the vertical electronic excitation energies of a system.

Furthermore, the residuum corresponding to a pole, $\hbar\omega_{n0} = E_n^{(0)} - E_0^{(0)}$, defined as

$$\lim_{\omega\to\omega_{n0}} \hbar(\omega - \omega_{n0})\langle\langle\hat{O}^\omega_{\alpha\ldots};\hat{O}^\omega_{\alpha\ldots}\rangle\rangle_\omega = \langle\Psi_0^{(0)}|\hat{O}^\omega_{\alpha\ldots}|\Psi_n^{(0)}\rangle\langle\Psi_n^{(0)}|\hat{O}^\omega_{\alpha\ldots}|\Psi_0^{(0)}\rangle$$

$$= |\langle\Psi_n^{(0)}|\hat{O}^\omega_{\alpha\ldots}|\Psi_0^{(0)}\rangle|^2 \quad (7.54)$$

is the square of the norm of the **electronic transition moment** $M_{n0,\alpha\ldots}$ of operator $\hat{O}^\omega_{\alpha\ldots}$ from state $|\Psi_0^{(0)}\rangle$ to state $|\Psi_n^{(0)}\rangle$ as defined in Section 3.10.

In the following, we want to derive expressions for the operators $\hat{O}^{\omega}_{\alpha\ldots}$. For the general vector potential given in Eq. (7.6) the first-order perturbation Hamiltonian takes the form

$$\hat{H}^{(1)}(t) = \frac{e}{m_e} \sum_i^N \vec{A}(\vec{r}_i, t) \cdot \hat{\vec{p}}_i \tag{7.55}$$

and the transition rate for absorption of a photon becomes

$$W_{n0}^{(1)} = \frac{\pi}{2}\left(\frac{e}{\hbar m_e}\right)^2 \sum_\alpha |M_{n0,\alpha\ldots}|^2 \, (A_\alpha^{\omega_{n0}})^2$$

$$= \frac{\pi}{2}\left(\frac{e}{\hbar m_e}\right)^2 \sum_\alpha \left|\langle \Psi_n^{(0)} | \sum_i^N e^{i\vec{k}\cdot\vec{r}_i}\hat{p}_{i,\alpha} | \Psi_0^{(0)}\rangle\right|^2 \, (A_\alpha^{\omega_{n0}})^2 \tag{7.56}$$

However, instead of evaluating directly the transition moments of the $\sum_i^N e^{i\vec{k}\cdot\vec{r}_i}\hat{p}_i$ interaction operator one expands the exponential according to Eq. (7.9). The first term gives again the dipole approximation, i.e. one ignores the spatial variation of the vector potential. This reduces the expression for the transition rate to

$$W_{n0}^{(1)} = \frac{\pi}{2}\left(\frac{e}{\hbar m_e}\right)^2 \sum_\alpha \left|\langle \Psi_n^{(0)} | \hat{O}^p_\alpha | \Psi_0^{(0)}\rangle\right|^2 \, (A_\alpha^{\omega_{n0}})^2 \tag{7.57}$$

where $\hat{O}^p$ is the total canonical momentum operator of the electrons defined in Eq. (3.65). But according to the off-diagonal hypervirial relation, Eq. (3.66), we can replace the transition moment of $\hat{O}^p$ by a transition moment of $\hat{O}^r$ and obtain for the transition rate

$$W_{n0}^{(1)} = \frac{\pi}{2}\left(\frac{e}{\hbar}\right)^2 \left(E_n^{(0)} - E_0^{(0)}\right)^2 \sum_\alpha \left|\langle \Psi_n^{(0)} | \hat{O}^r_\alpha | \Psi_0^{(0)}\rangle\right|^2 \, (A_\alpha^{\omega_{n0}})^2 \tag{7.58}$$

or in terms of the electric dipole moment operator $\hat{\vec{\mu}}$, Eq. (4.30),

$$W_{n0}^{(1)} = \frac{\pi}{2\hbar^2}\left(E_n^{(0)} - E_0^{(0)}\right)^2 \sum_\alpha \left|\langle \Psi_n^{(0)} | \hat{\mu}_\alpha | \Psi_0^{(0)}\rangle\right|^2 \, (A_\alpha^{\omega_{n0}})^2 \tag{7.59}$$

In the dipole approximation one thus arrives at a transition moment of the dipole operator, $\langle \Psi_n^{(0)} | \hat{\vec{\mu}} | \Psi_0^{(0)}\rangle$, which is called the **electric dipole transition moment** $\vec{M}_{n0}^{E1}$ and that explains why it is called the dipole approximation.

Going beyond the dipole approximation we consider now the next term in the expansion, Eq. (7.9), $i\vec{k}\cdot\vec{r}$. The next contribution to the transition moment then becomes

$$\langle \Psi_n^{(0)} | i \sum_i^N \vec{k}\cdot\vec{r}_i \, \hat{p}_{i,\alpha} | \Psi_0^{(0)}\rangle \tag{7.60}$$

This contribution depends not only on the direction of the polarization of the radiation, here α, but also on the direction of the propagation of the wave, given by the direction in which the propagation vector $\vec{k}$ points. For the derivation of the detailed form of

Electronic Excitation Energies and Transition Moments   163

the operators in this contribution, we will consider radiation traveling along the z-axis whose vector potential is polarized in the x-direction, which implies that $\alpha = x$ and that the propagation vector is $\vec{k} = (0, 0, \frac{\omega_{n0}}{c})$ according to Eq. (7.2). The contribution to the transition moment then becomes

$$M_{n0,zx} = \frac{i\omega_{n0}}{c} \langle \Psi_n^{(0)} | \sum_i^N z_i\, \hat{p}_{i,x} | \Psi_0^{(0)} \rangle \qquad (7.61)$$

where we can recognise the operator $z_i\, \hat{p}_{i,x}$ as one half of the y-component of the electronic angular momentum operator, Eq. (5.23). Using the same trick as in Eq. (5.7), we can write it as the sum of its symmetric and antisymmetric part

$$\sum_i^N z_i\, \hat{p}_{i,x} = \frac{1}{2}\sum_i^N (z_i\, \hat{p}_{i,x} + x_i\, \hat{p}_{i,z}) + \frac{1}{2}\sum_i^N (z_i\, \hat{p}_{i,x} - x_i\, \hat{p}_{i,z}) \qquad (7.62)$$

where the second, antisymmetric part is the y-component of the total orbital angular momentum operator of the electrons $\vec{\hat{L}}$, Eq. (5.23). Recalling that x_i and $\hat{p}_{i,z}$ commute and using the commutator relation Eq. (3.64) we can rewrite the first, symmetric part as well, giving

$$\sum_i^N z_i\, \hat{p}_{i,x} = \frac{1}{2}\frac{m_e}{i\hbar}\sum_i^N \left(z_i\, [x_i, \hat{H}^{(0)}] + [z_i, \hat{H}^{(0)}]\, x_i \right) + \frac{1}{2}\hat{L}_y \qquad (7.63)$$

The contribution to the transition moment then becomes

$$M_{n0,zx} = \frac{\omega_{n0}}{2c}\frac{m_e}{\hbar} \langle \Psi_n^{(0)} | \sum_i^N \left(z_i x_i \hat{H}^{(0)} - \hat{H}^{(0)} z_i x_i \right) | \Psi_0^{(0)} \rangle + \frac{i\omega_{n0}}{2c} \langle \Psi_n^{(0)} | \hat{L}_y | \Psi_0^{(0)} \rangle \qquad (7.64)$$

However, the states $\Psi_0^{(0)}$ and $\Psi_n^{(0)}$ are eigenstates of the Hamiltonian $\hat{H}^{(0)}$ and we get therefore

$$M_{n0,zx} = -\frac{\omega_{n0}}{2c}\frac{m_e}{\hbar}\left(E_n^{(0)} - E_0^{(0)} \right) \langle \Psi_n^{(0)} | \sum_i^N z_i x_i | \Psi_0^{(0)} \rangle + \frac{i\omega_{n0}}{2c} \langle \Psi_n^{(0)} | \hat{L}_y | \Psi_0^{(0)} \rangle$$

$$= -\frac{\omega_{n0}^2}{2c}\frac{m_e}{\hbar} \langle \Psi_n^{(0)} | \sum_i^N z_i x_i | \Psi_0^{(0)} \rangle + \frac{i\omega_{n0}}{2c} \langle \Psi_n^{(0)} | \hat{L}_y | \Psi_0^{(0)} \rangle \qquad (7.65)$$

Defining a second electric moment operator $\hat{O}_{\alpha\beta}^{rr}$ as

$$\hat{O}_{\alpha\beta}^{rr} = \sum_i^N \hat{r}_{i,\alpha}\hat{r}_{i,\beta} \qquad (7.66)$$

and using the definition of the magnetic dipole moment operator, Eq. (5.21), we can write a general $\alpha\beta$ element of this contribution to the transition moment as

$$M_{n0,\alpha\beta} = M^{E2}_{n0,\alpha\beta} + \sum_\gamma \epsilon_{\alpha\beta\gamma} M^{M1}_{n0,\gamma} \qquad (7.67)$$

$$= -\frac{\omega_{n0}^2}{2c}\frac{m_e}{\hbar}\langle\Psi_n^{(0)}|\hat{O}^{rr}_{\alpha\beta}|\Psi_0^{(0)}\rangle - \sum_\gamma \epsilon_{\alpha\beta\gamma}\frac{\imath\omega_{n0}}{2c}\frac{m_e}{e}\langle\Psi_n^{(0)}|\hat{m}^l_\gamma|\Psi_0^{(0)}\rangle \qquad (7.68)$$

where we have defined implicitly a component of the **electric quadrupole transition moment** $\mathbf{M}^{E2}_{n0}$ and of the **magnetic dipole transition moment** $\vec{M}^{M1}_{n0}$.

All these transition moments, $\vec{M}^{E1}_{n0}$, $\mathbf{M}^{E2}_{n0}$, and $\vec{M}^{M1}_{n0}$ can be obtained as residua of the appropriate polarization propagators according to Eq. (7.54), i.e. the frequency-dependent dipole $\alpha(-\omega;\omega)$ and quadrupole polarizability $C(-\omega;\omega)$ and a frequency-dependent paramagnetic contribution to the magnetizability $\xi^p(-\omega;\omega)$.

The intensity of a measured absorption band is usually reported in terms of the dimensionless **dipole oscillator strength**, which is defined in terms of the electric dipole transition moments as

$$f^l_{n0} = \frac{2}{3}\frac{m_e}{\hbar^2 e^2}\left(E_n^{(0)} - E_0^{(0)}\right)|\langle\Psi_n^{(0)}|\hat{\vec{\mu}}|\Psi_0^{(0)}\rangle|^2 = \frac{2}{3}\frac{m_e}{\hbar^2}\left(E_n^{(0)} - E_0^{(0)}\right)|\langle\Psi_n^{(0)}|\hat{\vec{O}}^r|\Psi_0^{(0)}\rangle|^2 \qquad (7.69)$$

Due to the appearance of the position operator, this is called the **dipole oscillator strength in the length representation**. One can consider the oscillator strength as the trace of a tensor of cartesian components

$$f^l_{n0,\alpha\beta} = 2\frac{m_e}{\hbar^2 e^2}\left(E_n^{(0)} - E_0^{(0)}\right)\langle\Psi_0^{(0)}|\hat{\mu}_\alpha|\Psi_n^{(0)}\rangle\langle\Psi_n^{(0)}|\hat{\mu}_\beta|\Psi_0^{(0)}\rangle \qquad (7.70)$$

Using the off-diagonal hypervirial relation, Eq. (3.66), one can define two alternative formulations of the oscillator strength (Hansen, 1967), a **mixed representation**

$$f^m_{n0} = \frac{2}{3}\frac{1}{\imath\hbar e}\langle\Psi_0^{(0)}|\hat{\vec{O}}^p|\Psi_n^{(0)}\rangle\langle\Psi_n^{(0)}|\hat{\vec{\mu}}|\Psi_0^{(0)}\rangle = \frac{2}{3}\frac{1}{\imath\hbar}\langle\Psi_0^{(0)}|\hat{\vec{O}}^p|\Psi_n^{(0)}\rangle\langle\Psi_n^{(0)}|\hat{\vec{O}}^r|\Psi_0^{(0)}\rangle \qquad (7.71)$$

and a **velocity representation**

$$f^v_{n0} = \frac{2}{3}\frac{1}{m_e}\frac{|\langle\Psi_n^{(0)}|\hat{\vec{O}}^p|\Psi_0^{(0)}\rangle|^2}{E_n^{(0)} - E_0^{(0)}} \qquad (7.72)$$

The mixed representation is particular interesting because it does not involve the excitation energies explicitly. It can alternatively also be written in the following two ways

$$f^m_{n0} = -\frac{2}{3}\frac{1}{\imath\hbar e}\langle\Psi_0^{(0)}|\hat{\vec{\mu}}|\Psi_n^{(0)}\rangle\langle\Psi_n^{(0)}|\hat{\vec{O}}^p|\Psi_0^{(0)}\rangle$$

$$= \frac{1}{3}\frac{1}{\imath\hbar e}\left[\langle\Psi_0^{(0)}|\hat{\vec{O}}^p|\Psi_n^{(0)}\rangle\langle\Psi_n^{(0)}|\hat{\vec{\mu}}|\Psi_0^{(0)}\rangle - \langle\Psi_0^{(0)}|\hat{\vec{\mu}}|\Psi_n^{(0)}\rangle\langle\Psi_n^{(0)}|\hat{\vec{O}}^p|\Psi_0^{(0)}\rangle\right] \qquad (7.73)$$

For optically active, i.e. chiral, molecules the intensity of the bands of circular dichroism (CD) spectra is expressed in terms of a **rotational strength** that is defined as

$$R_{n0} = \imath \langle \Psi_0^{(0)} | \hat{\vec{\mu}} | \Psi_n^{(0)} \rangle \langle \Psi_n^{(0)} | \hat{\vec{m}}^l | \Psi_0^{(0)} \rangle \tag{7.74}$$

and that can be calculated as residuum of the mixed frequency-dependent electric dipole magnetic dipole polarizability tensor $\mathbf{G}'(-\omega;\omega)$, Eq. (7.50) and thus of the $\langle\langle \hat{\mu}_\alpha ; \hat{m}^l_\beta \rangle\rangle_\omega$ polarization propagator.

The calculation of electronic vertical excitation energies $\Delta E_{n0} = \hbar\omega_{n0}$ and corresponding transitions moments or oscillator strengths f_{n0} from the linear response functions or polarization propagators is a very interesting alternative to the usual approach because it is done in a direct way. Neither the wavefunctions $|\Psi_{0/n}^{(0)}\rangle$ nor the energies $E_{0/n}^{(0)}$ of the initial state 0 or final state n have to be calculated explicitly in order to obtain these spectral properties, because ΔE_{n0} and f_{n0} are obtained directly as poles and residues of the polarization propagator that is calculated by approximations to Eq. (3.159). The response theory approach is therefore predestinate to the calculation of electronic spectra.

Finding the poles of the polarization propagator as given e.g. in Eq. (3.159) implies finding the values of the frequency ω for which the matrix $\left(\mathbf{E}^{[2]} - \hbar\omega \mathbf{S}^{[2]}\right)$ becomes singular. This could in principle be done by a pole search where one tries to determine the frequency of the pole by repeatedly evaluating the response function. However, this is cumbersome and unnecessary because singularity of this matrix is also the necessary condition for that the set of linear equations

$$\left(\mathbf{E}^{[2]} - \hbar\omega \mathbf{S}^{[2]}\right) \mathbf{X} = 0 \tag{7.75}$$

has a non-trivial solution for $\mathbf{X}$, i.e. $\mathbf{X} \neq 0$. This, on the other hand, is simply the generalized eigenvalue equation for the electronic hessian matrix $\mathbf{E}^{[2]}$

$$\mathbf{E}^{[2]} \mathbf{X} = \hbar\omega \mathbf{S}^{[2]} \mathbf{X} \tag{7.76}$$

Finding the poles of the propagator corresponds therefore to solving the **generalized eigenvalue problem** for the electronic Hessian matrix or the principal propagator, which is written out here in more detail

$$\left[\begin{pmatrix} \langle \Psi_0^{(0)} | [\hat{h}_i^\dagger, [\hat{H}^{(0)}, \hat{h}_j]] | \Psi_0^{(0)} \rangle & \cdots \\ \vdots & \ddots \end{pmatrix} \right.$$

$$\left. -\hbar\omega_{n0} \begin{pmatrix} \langle \Psi_0^{(0)} | [\hat{h}_i^\dagger, \hat{h}_j] | \Psi_0^{(0)} \rangle & \cdots \\ \vdots & \ddots \end{pmatrix} \right] \begin{pmatrix} X_j^{n0} \\ \vdots \end{pmatrix} = 0 \tag{7.77}$$

The vertical excitation energies ΔE_{n0} are thus obtained as eigenvalues $\hbar\omega_{n0}$ and $\{X_j^{n0}\}$ are the elements of the corresponding eigenvectors. The transition moments, finally, can be calculated from the eigenvectors $\{X_j^{n0}\}$ and the property gradient vectors

$$T_j(\hat{O}_\alpha) = \langle \Psi_0^{(0)} | [\hat{h}_j^\dagger, \hat{O}_\alpha] | \Psi_0^{(0)} \rangle \text{ as}$$

$$\langle \Psi_n^{(0)} | \hat{O}_\alpha | \Psi_0^{(0)} \rangle = \sum_j X_j^{n0} \langle \Psi_0^{(0)} | [\hat{h}_j^\dagger, \hat{O}_\alpha] | \Psi_0^{(0)} \rangle \tag{7.78}$$

7.5 Dipole Oscillator Strength Sums

The set of dipole oscillator strengths $\{f_{n0}\}$, defined in Eq. (7.69), is often called the **dipole oscillator strength distribution (DOSD)**. Summed over all excited states, bound as well as continuum states, they are related to several other molecular properties, as will be shown in the following. One defines two types of energy-weighted moments of the dipole oscillator strength distribution[4]

$$S(k) = \sum_{n \neq 0} \left(E_n^{(0)} - E_0^{(0)} \right)^k f_{n0} \tag{7.79}$$

$$L(k) = \sum_{n \neq 0} \left(E_n^{(0)} - E_0^{(0)} \right)^k \ln(E_n^{(0)} - E_0^{(0)}) f_{n0} \tag{7.80}$$

also called **dipole oscillator strength sums**. Depending on whether one sums the oscillator strengths in their length, mixed or velocity representation one thus obtains the sums in the three representations. In the following we will only distinguish between the three representations when necessary by adding the superscripts l, m or v. As for the oscillator strengths, Eq. (7.70), one can also define sums for the cartesian components of the dipole oscillator strengths as

$$S_{\alpha\beta}(k) = \sum_{n \neq 0} \left(E_n^{(0)} - E_0^{(0)} \right)^k f_{n0,\alpha\beta} \tag{7.81}$$

$$L_{\alpha\beta}(k) = \sum_{n \neq 0} \left(E_n^{(0)} - E_0^{(0)} \right)^k \ln(E_n^{(0)} - E_0^{(0)}) f_{n0,\alpha\beta} \tag{7.82}$$

Several dipole oscillator strength sums are related to other molecular properties by so-called **dipole oscillator strength sum rules**. The best known is the **Thomas–Reiche–Kuhn sum rule** that relates the $S(0)$ sum to the number of electrons N of the system, i.e.

$$S(0) = \sum_{n \neq 0} f_{n0} = N \tag{7.83}$$

as can be shown easily [see Exercise 7.2].

Exercise 7.2 Derive the Thomas-Reiche-Kuhn sum rule Eq. (7.83).

Hint: Start with the mixed representation of the oscillator strengths in Eq. (7.73). Use the fact that the set of excited states $\Psi_n^{(0)}$ is complete, i.e.

---

[4] For the unbound continuum states the summation should be replaced by an integration.

Dipole Oscillator Strength Sums

$$\sum_n |\Psi_n^{(0)}\rangle\langle\Psi_n^{(0)}| = 1$$

and use the commutator relation Eq. (6.34).

Comparing the definition of the components of the oscillator strength in the length representation, Eq. (7.70), with the expression for a component of the frequency-dependent polarizability in Eq. (7.28) we can see that the polarizability can be written in terms of the oscillator strengths as

$$\alpha_{\alpha\beta}(-\omega;\omega) = \frac{\hbar^2 e^2}{m_e} \sum_{n\neq 0} \frac{f^l_{n0,\alpha\beta}}{(E_n^{(0)} - E_0^{(0)})^2 - \hbar^2\omega^2}$$

$$= \frac{\hbar^2 e^2}{m_e} \sum_{n\neq 0} \frac{f^l_{n0,\alpha\beta}}{(E_n^{(0)} - E_0^{(0)})^2} \frac{1}{1 - \frac{\hbar^2\omega^2}{(E_n^{(0)} - E_0^{(0)})^2}} \quad (7.84)$$

For frequencies smaller than the lowest excitation energy, i.e. $\left|\hbar\omega/(E_n^{(0)} - E_0^{(0)})\right| < 1$, we can expand the last term in a Taylor series and obtain

$$\alpha_{\alpha\beta}(-\omega;\omega) = \frac{\hbar^2 e^2}{m_e} \sum_{k=0}^{\infty} (\hbar\omega)^{2k} \sum_{n\neq 0} \frac{f^l_{n0,\alpha\beta}}{(E_n^{(0)} - E_0^{(0)})^{2k+2}} \quad (7.85)$$

or in terms of the dipole oscillator strength sums

$$\alpha_{\alpha\beta}(-\omega;\omega) = \frac{e^2\hbar^2}{m_e} \sum_{k=0}^{\infty} (\hbar\omega)^{2k} S^l_{\alpha\beta}(-2k-2) \quad (7.86)$$

This is often called the **Cauchy moment expansion** of the frequency-dependent polarizability and the sums $S(k)$ for even but negative values of k are called Cauchy moments. $S(-2)$ in particular turns out to be proportional to the static polarizability

$$\alpha_{\alpha\beta}(0;0) = \frac{e^2\hbar^2}{m_e} S^l_{\alpha\beta}(-2) \quad (7.87)$$

which is another well-known example of a dipole oscillator strength sum rule. The other Cauchy moments, i.e. even and negative sums, $S(-4)$, $S(-6)$, $\cdots$, describe the frequency dependence or dispersion of the frequency-dependent polarizability and can therefore be defined either as even derivatives of the frequency-dependent polarizability

$$S^l_{\alpha\beta}(-m-2) = \frac{m_e}{e^2\hbar^2} \frac{1}{\hbar^m m!} \left(\frac{d^m}{d\omega^m} \alpha_{\alpha\beta}(-\omega;\omega)\right)_{\omega=0} \quad (7.88)$$

for $m = 2, 4, 6, \cdots$ being even positive numbers or alternatively as both odd and even derivatives of the frequency-dependent polarizability

$$S^l_{\alpha\beta}(-2k-2) = \frac{m_e}{e^2\hbar^2} \frac{1}{2^k \hbar^{2k} k!} \lim_{\omega \to 0} \left(\frac{1}{\omega}\frac{d}{d\omega}\right)^k \alpha_{\alpha\beta}(-\omega;\omega) \qquad (7.89)$$

for $k = 1, 2, 3, \cdots$ being positive numbers. However, the positive even dipole oscillator strength sums can also be obtained as derivatives of the frequency-dependent polarizability [see Exercise 7.3]

$$S^l_{\alpha\beta}(2k) = (-1)^{k-1} \frac{m_e}{e^2\hbar^2} \frac{1}{2^k \hbar^{2k} k!} \lim_{\omega \to \infty} \left(\omega^3 \frac{d}{d\omega}\right)^k \omega^2 \alpha_{\alpha\beta}(-\omega;\omega) \qquad (7.90)$$

for $k = 0, 1, 2, 3, \cdots$

Exercise 7.3 Derive the Thomas–Reiche–Kuhn and $S(2)$ sum rules from Eq. (7.90).

Recalling that the frequency-dependent polarizability is related to the $\langle\langle \hat{\mu}_\alpha ; \hat{\mu}_\beta \rangle\rangle_\omega$ propagator, Eq. (7.26), we can express the even dipole oscillator strength sums also as derivatives of this polarization propagator, i.e.

$$S^l_{\alpha\beta}(2k) = (-1)^k \frac{m_e}{e^2\hbar^2} \frac{1}{2^k \hbar^{2k} k!} \lim_{\omega \to \infty} \left(\omega^3 \frac{d}{d\omega}\right)^k \omega^2 \langle\langle \hat{\mu}_\alpha ; \hat{\mu}_\beta \rangle\rangle_\omega$$

$$\text{for } k = 0, 1, 2, 3, \cdots \qquad (7.91)$$

$$S^l_{\alpha\beta}(-2k-2) = -\frac{m_e}{e^2\hbar^2} \frac{1}{2^k \hbar^{2k} k!} \lim_{\omega \to 0} \left(\frac{1}{\omega}\frac{d}{d\omega}\right)^k \langle\langle \hat{\mu}_\alpha ; \hat{\mu}_\beta \rangle\rangle_\omega$$

$$\text{for } k = 1, 2, 3, \cdots \qquad (7.92)$$

Similar relations between the even sums in mixed, S^m, and velocity representation, S^v, and the $\langle\langle \hat{\mu}_\alpha ; \hat{O}^p_\beta \rangle\rangle_\omega$ and $\langle\langle \hat{O}^p_\alpha ; \hat{O}^p_\beta \rangle\rangle_\omega$ polarization propagators can also be derived (Jørgensen et al., 1978).

The dipole oscillator strengths and their sums play not only an important role in the description of the interaction of molecules with electromagnetic radiation but also in the description of the interaction of molecules with beams of charged particles, i.e. ions. A beam of ions with charge Z passing with velocity v through matter is scattered by the medium molecules and loses part of its kinetic energy E_{kin}. This is normally expressed in terms of the linear **stopping power** or energy loss per unit path length x defined as

$$S(v) = -\frac{1}{\mathcal{N}} \frac{dE_{kin}}{dx} \qquad (7.93)$$

where $\mathcal{N}$ is the density of molecules in the target.

In the case of fast ions moving through a medium the main contribution to the energy loss comes from the inelastic collision with the electrons of the molecules in the medium that will be exited or even ionised. The simplest expression describing this process is the **Bethe formula** derived via first-order perturbation theory (Bethe, 1930)

$$S(v) = \frac{Ze^2 Ne^2}{m_e v^2 4\pi\epsilon_0^2} \ln\left(\frac{2m_e v^2}{I(0)}\right) \tag{7.94}$$

where N is the number of electrons in the target molecules and $I(0)$ is called the **mean excitation energy** of the target molecules. The mean excitation energy is not a simple mean value of all electronic excitation energies $(E_n^{(0)} - E_0^{(0)})$, but is defined in terms of the energy-weighted moments or sums of the dipole oscillator strength distribution as

$$\ln I(0) = \frac{L(0)}{S(0)} = \frac{\sum_{n\neq 0} \ln(E_n^{(0)} - E_0^{(0)}) f_{n0}}{\sum_{n\neq 0} f_{n0}} \tag{7.95}$$

7.6 van der Waals Coefficients

In Sections 4.1 and 4.2 we discussed the fact that the electric moments of molecules play an important role in the description of the intermolecular forces between two molecules separated by a large distance. Their contribution to the interaction energy is of purely classical, i.e. electrostatic nature. Here, we want to show now that also the contribution from quantum mechanical **dispersion or London forces**, i.e. the **dispersion energy** E_{AB}^{disp}, can be related to molecular properties of the two interacting molecules. In particular, we will see that it is related to the frequency-dependent polarizabilities, which is in line with the physical interpretation of the dispersion forces as arising from the interaction of induced dipole moments, which implies that both charge distributions are perturbed by their interaction.

The dispersion energy can thus be derived by perturbation theory where the perturbation Hamiltonian consists of the interaction potential of the two charge distributions. One expands both charge distributions in multipole series and keeps the first term, which for two uncharged molecules A and B separated by a distance $|\vec{R}_{AB}|$ is the dipole–dipole interaction term

$$\hat{H}_{AB}^{(1)} = \frac{1}{4\pi\epsilon_0 |R_{AB}|^3} \left[\hat{\vec{\mu}}_A \cdot \hat{\vec{\mu}}_B - \frac{3(\hat{\vec{\mu}}_A \cdot \vec{R}_{AB})(\vec{R}_{AB} \cdot \hat{\vec{\mu}}_B)}{R_{AB}^2} \right] \tag{7.96}$$

where $\hat{\vec{\mu}}_A$ and $\hat{\vec{\mu}}_B$ are the dipole moment operators of the two molecules. The unperturbed Hamiltonian of the complex is then the Hamiltonian of two non-interacting molecules and thus just the sum of the Hamiltonian operators of the two separate molecules with the unperturbed complex energies $E_{n_A,m_B}^{(0)}$ being the sum of the energies of the separate molecules

$$E_{n_A,m_B}^{(0)} = E_{n_A}^{(0)} + E_{m_B}^{(0)} \tag{7.97}$$

and the unperturbed complex wavefunctions $\Psi_{n_A,m_B}^{(0)}$ being the product of the corresponding molecular wavefunctions

$$\Psi_{n_A,m_B}^{(0)} = \Psi_{n_A}^{(0)} \Psi_{m_B}^{(0)} \tag{7.98}$$

The first-order correction to the energy of the ground state of the complex, Eq. (3.29), then becomes

$$E^{(1)}_{0_A,0_B} = \langle \Psi^{(0)}_{0_A,0_B} | \hat{H}^{(1)}_{AB} | \Psi^{(0)}_{0_A,0_B} \rangle = \langle \Psi^{(0)}_{n_A} \Psi^{(0)}_{m_B} | \hat{H}^{(1)}_{AB} | \Psi^{(0)}_{n_A} \Psi^{(0)}_{m_B} \rangle \quad (7.99)$$

or after inserting the perturbation Hamiltonian, Eq. (7.96),

$$E^{(1)}_{0_A,0_B} = \frac{1}{4\pi\epsilon_0 |R_{AB}|^3} \left[\vec{\mu}_A \cdot \vec{\mu}_B - \frac{3(\vec{\mu}_A \cdot \vec{R}_{AB})(\vec{R}_{AB} \cdot \vec{\mu}_B)}{R^2_{AB}} \right] \quad (7.100)$$

This is the electrostatic interaction between the permanent dipole moments $\vec{\mu}_A$ and $\vec{\mu}_B$ of the two molecules.

The second-order energy correction, Eq. (3.33), becomes

$$E^{(2)}_{0_A,0_B} = \sum_{n_A,m_B \neq 0_A,0_B} \frac{\langle \Psi^{(0)}_{0_A,0_B} | \hat{H}^{(1)}_{AB} | \Psi^{(0)}_{n_A,m_B} \rangle \langle \Psi^{(0)}_{n_A,m_B} | \hat{H}^{(1)}_{AB} | \Psi^{(0)}_{0_A,0_B} \rangle}{E^{(0)}_{0_A,0_B} - E^{(0)}_{n_A,m_B}} \quad (7.101)$$

where the double sum runs over all complex states in which at least one of the molecules is excited. We can therefore distinguish between terms where one molecule is excited, e.g. molecule A

$$E^{ind}_{0_A,0_B} = \sum_{n_A \neq 0_A} \frac{\langle \Psi^{(0)}_{0_A,0_B} | \hat{H}^{(1)}_{AB} | \Psi^{(0)}_{n_A,0_B} \rangle \langle \Psi^{(0)}_{n_A,0_B} | \hat{H}^{(1)}_{AB} | \Psi^{(0)}_{0_A,0_B} \rangle}{E^{(0)}_{0_A,0_B} - E^{(0)}_{n_A,0_B}} \quad (7.102)$$

and terms where both are excited

$$E^{disp}_{0_A,0_B} = \sum_{n_A \neq 0_A} \sum_{m_B \neq 0_B} \frac{\langle \Psi^{(0)}_{0_A,0_B} | \hat{H}^{(1)}_{AB} | \Psi^{(0)}_{n_A,m_B} \rangle \langle \Psi^{(0)}_{n_A,m_B} | \hat{H}^{(1)}_{AB} | \Psi^{(0)}_{0_A,0_B} \rangle}{E^{(0)}_{0_A,0_B} - E^{(0)}_{n_A,m_B}} \quad (7.103)$$

The former is the **induction energy** contribution to the intermolecular forces and can be shown to consist of the static polarizability of molecule A and the permanent electric dipole moment of molecule B. We will not consider this term any further here.

Inserting the unperturbed complex energies and wavefunctions, Eqs. (7.97) and (7.98), in the latter term gives

$$E^{disp}_{0_A,0_B} = \sum_{n_A \neq 0_A} \sum_{m_B \neq 0_B} \frac{\langle \Psi^{(0)}_{0_A} \Psi^{(0)}_{0_B} | \hat{H}^{(1)}_{AB} | \Psi^{(0)}_{n_A} \Psi^{(0)}_{m_B} \rangle \langle \Psi^{(0)}_{n_A} \Psi^{(0)}_{m_B} | \hat{H}^{(1)}_{AB} | \Psi^{(0)}_{0_A} \Psi^{(0)}_{0_B} \rangle}{E^{(0)}_{0_A} - E^{(0)}_{n_A} + E^{(0)}_{0_B} - E^{(0)}_{m_B}} \quad (7.104)$$

where the product of transition moments of $\hat{H}^{(1)}_{AB}$ can be written more explicitly as

$$\left| \langle \Psi^{(0)}_{0_A} \Psi^{(0)}_{0_B} | \hat{H}^{(1)}_{AB} | \Psi^{(0)}_{n_A} \Psi^{(0)}_{m_B} \rangle \right|^2$$

$$= \frac{1}{(4\pi\epsilon_0)^2 |R_{AB}|^{10}} \left| \vec{R}^T_{AB} \left(\langle \Psi^{(0)}_{0_A} | \hat{\vec{\mu}}_A | \Psi^{(0)}_{n_A} \rangle \cdot \langle \Psi^{(0)}_{0_B} | \hat{\vec{\mu}}_B | \Psi^{(0)}_{m_B} \rangle \, I_3 \right. \right.$$

$$\left. \left. - 3 \langle \Psi^{(0)}_{0_A} | \hat{\vec{\mu}}_A | \Psi^{(0)}_{n_A} \rangle \otimes \langle \Psi^{(0)}_{0_B} | \hat{\vec{\mu}}_B | \Psi^{(0)}_{m_B} \rangle \right) \vec{R}_{AB} \right|^2 \quad (7.105)$$

However, in the gas or liquid phase the molecules can have all possible orientations with respect to each other. Therefore, for the isotropic interaction between the two molecules one has to average over all molecular orientations, which reduces the absolute square of the transition matrix element to

$$\left|\langle \Psi^{(0)}_{0_A}\Psi^{(0)}_{0_B}|\hat{H}^{(1)}_{AB}|\Psi^{(0)}_{n_A}\Psi^{(0)}_{m_B}\rangle\right|^2 = \frac{\left|\langle \Psi^{(0)}_{0_A}|\hat{\vec{\mu}}_A|\Psi^{(0)}_{n_A}\rangle\right|^2 \left|\langle \Psi^{(0)}_{0_B}|\hat{\vec{\mu}}_B|\Psi^{(0)}_{m_B}\rangle\right|^2}{24\pi^2\epsilon_0^2|R_{AB}|^6} \qquad (7.106)$$

The dipole–dipole contribution to the isotropic **dispersion energy** between two neutral molecules is thus given as

$$E^{disp}_{0_A,0_B} = -\frac{1}{24\pi^2\epsilon_0^2|R_{AB}|^6}\sum_{n_A\neq 0_A}\sum_{m_B\neq 0_B}\frac{\left|\langle \Psi^{(0)}_{0_A}|\hat{\vec{\mu}}_A|\Psi^{(0)}_{n_A}\rangle\right|^2\left|\langle \Psi^{(0)}_{0_B}|\hat{\vec{\mu}}_B|\Psi^{(0)}_{m_B}\rangle\right|^2}{(E^{(0)}_{n_A} - E^{(0)}_{0_A}) + (E^{(0)}_{m_B} - E^{(0)}_{0_B})} \qquad (7.107)$$

In order to evaluate this contribution one needs only all excitation energies and corresponding transition dipole moments for molecule A and also for molecule B. Both can be obtained from the poles and residues of a polarization propagator for molecule A and separately for molecule B as described in Section 7.4. However, it is preferable to avoid the simultaneous summation over all states and express the dispersion energy in terms of molecular properties. This can be achieved by using the following integral transform

$$\frac{1}{x+y} = \frac{2}{\pi}\int_0^\infty \frac{x}{x^2+z^2}\frac{y}{y^2+z^2}\,dz \qquad (7.108)$$

for the denominator of the dispersion energy, i.e.

$$E^{disp}_{0_A,0_B} = -\frac{1}{12\pi^3\epsilon_0^2|R_{AB}|^6}\int_0^\infty dz \sum_{n_A\neq 0_A}\frac{(E^{(0)}_{n_A} - E^{(0)}_{0_A})\left|\langle \Psi^{(0)}_{0_A}|\hat{\vec{\mu}}_A|\Psi^{(0)}_{n_A}\rangle\right|^2}{(E^{(0)}_{n_A} - E^{(0)}_{0_A})^2 + z^2}$$

$$\times \sum_{m_B\neq 0_B}\frac{(E^{(0)}_{m_B} - E^{(0)}_{0_B})\left|\langle \Psi^{(0)}_{0_B}|\hat{\vec{\mu}}_B|\Psi^{(0)}_{m_B}\rangle\right|^2}{(E^{(0)}_{m_B} - E^{(0)}_{0_B})^2 + z^2} \qquad (7.109)$$

Choosing $z = \sqrt{-1}\hbar\omega = \imath\hbar\omega$ and then comparing the two summations with Eq. (7.28) we can see that they correspond to frequency-dependent polarizabilities for imaginary frequencies, giving

$$E^{disp}_{0_A,0_B} = -\frac{3\hbar}{16\pi^3\epsilon_0^2|R_{AB}|^6}\int_0^\infty d\omega\, \alpha^A(-\imath\omega;\imath\omega)\,\alpha^B(-\imath\omega;\imath\omega) \qquad (7.110)$$

Commonly, the dipole–dipole contribution to the dispersion energy is written as

$$E^{disp}_{0_A,0_B} = -\frac{C^{AB}_6}{|R_{AB}|^6} \qquad (7.111)$$

where the C_6 or **van der Waals dispersion coefficient** is then defined as

$$C_6^{AB} = \frac{3\hbar}{16\pi^3 \epsilon_0^2} \int_0^\infty d\omega \, \alpha^A(-\imath\omega;\imath\omega) \, \alpha^B(-\imath\omega;\imath\omega) \qquad (7.112)$$

which is often called the Casimir–Polder formula (Casimir and Polder, 1948). As discussed before this expression is for the isotropic interaction. Similar, but more complicated, expressions have been derived for the situation, where the orientation of the two molecules to each other is important (Visser et al., 1983, 1984; Visser and Wormer, 1984; Rijks and Wormer, 1988; Hettema et al., 1994).

As for real frequencies, Eq. (7.26), one can also obtain components of the polarizability for imaginary frequencies from correspondingly complex linear response functions (Norman et al., 2003)

$$\alpha_{\alpha\beta}(-\imath\omega;\imath\omega) = -\langle\langle\, \hat{\mu}_\alpha \,;\, \hat{\mu}_\beta \,\rangle\rangle_{\imath\omega} \qquad (7.113)$$

Alternatively, one can make use of the fact that the frequency dependence of the polarizability can be expressed in terms of dipole oscillator strength sums Eq. (7.86). This expansion, however, converges only for frequencies below the first excitation energy, i.e. $\hbar\omega < \min\{E_n^{(0)} - E_0^{(0)}\}$. Nevertheless, the expansion can be extended beyond this convergence radius and in particular into the complex plane by using well-known analytical continuation techniques based on Padé approximants $[n,m]_\alpha$ to the frequency-dependent polarizability α (Langhoff and Karplus, 1970). In particular, the $[n, n-1]_\alpha$ Padé approximant to $\alpha(\imath\omega)$ expressed by the dipole oscillator strength sums, Eq. (7.86), can be used as a lower bound

$$\alpha(\imath\omega) \geq [n, n-1]_\alpha \qquad (7.114)$$

An upper bound can be obtained either from the $[n,n]_\alpha$ Padé approximant to $\alpha(\imath\omega)$

$$\alpha(\imath\omega) \leq [n,n]_\alpha \qquad (7.115)$$

or via the same type of Padé approximant $[n, n-1]$ as for the lower bound, but now to $S^l(0) - \omega^2\alpha(\imath\omega)$ instead of to $\alpha(\imath\omega)$. This approximant is usually denoted as $[n, n-1]_\beta$ and an upper bound to $\alpha(\imath\omega)$ is then given as

$$\alpha(\imath\omega) \leq \frac{S^l(0) - [n, n-1]_\beta}{\omega^2} \qquad (7.116)$$

7.7 Further Reading

Refractive Index and Optical Rotation

P. Atkins and R. Friedman, *Molecular Quantum Mechanics*, 4th edn. Oxford University Press, Oxford (2005): Chapters 12.9–12.12.

A. Hinchliffe and R. W. Munn, *Molecular Electromagnetism*, John Wiley and Sons Ltd, Chichester (1985): Chapter 15.6.

Transition Moments

B. H. Bransden and C. J. Joachain, *Physics of Atoms and Molecules*, 2nd ed. Pearson Education, Harlow (2003): Chapters 2.8, 4.1–4.3, 9.1–9.3.

Stopping Power

B. H. Bransden and C. J. Joachain, *Physics of Atoms and Molecules*, 2nd edn. Pearson Education, Harlow (2003): Chapter 13.3.

van der Waals Coefficients

P. Atkins and R. Friedman, *Molecular Quantum Mechanics*, 4th edn. Oxford University Press, Oxford (2005): Chapter 12.5.

A. Hinchliffe and R. W. Munn, *Molecular Electromagnetism*, John Wiley and Sons Ltd, Chichester (1985): Chapter 15.7.

8
Vibrational Contributions to Molecular Properties

The expressions for the molecular properties given in Chapters 4–7 were derived for a set of fixed nuclear coordinates. However, this is not a realistic description of a molecule, since even at 0 K a molecule vibrates. In order to obtain agreement with experimental data it is necessary to take the effects of nuclear motion into account. Their contribution is in general not negligible especially for NMR spin-spin coupling constants and higher-order polarizabilities. For example, calculated first and second hyperpolarizabilities, which do not include any vibrational corrections (Bishop, 1990) are of questionable relevance to experiment, even though they may be of value for benchmarking purposes. Furthermore, experimentally observable effects like temperature dependence and isotope shifts of e.g. NMR parameters are solely due to differences in these nuclear motion corrections.

In this chapter we will therefore discuss the contributions from the nuclear wavefunction to the molecular properties derived in the previous chapters. However, in doing so we will still make use of the Born–Oppenheimer approximation. In the following, we will use the static polarizability as example and illustrate how these vibrational corrections can be incorporated (Bishop and Cheung, 1980; Bishop et al., 1980). The expression, which we are going to derive, can then easily be transferred to all linear response properties. A detailed description of vibrational corrections to static and frequency-dependent hyperpolarizabilities can be found in the reviews by Bishop (1990; 1998).

In order to incorporate the effects of nuclear motion we have to go back to the Hamiltonian, Eq. (2.1), which includes the kinetic energy operators for the nuclei. The corresponding eigenfunctions are the so-called vibronic wavefunctions[1] $\Phi_{kv}^{(0)}(\{\vec{r}_i\}, \{\vec{R}_K\})$ with energy $E_{kv}^{(0)}$ and are characterized by the electronic, k, and vibrational, v, quantum numbers, where v stands throughout the chapter collectively for the vibrational quantum numbers of all vibrational modes of the molecule. The proper approach for the treatment of the nuclear motion effects would be to use these unperturbed vibronic wavefunctions $\Phi_{kv}^{(0)}(\{\vec{r}_i\}, \{\vec{R}_K\})$ instead of the unperturbed electronic wavefunctions $\Psi_k^{(0)}(\{\vec{r}_i\}; \{\vec{R}_K\})$ in the derivation of expression for the molecular properties in Chapters 4–7. However, we still want to make use

---

[1] We neglect in this section the rotational motion of the whole molecule. The corresponding subscript J is therefore also missing here compared to the expressions in Section 2.2.

of the Born–Oppenheimer approximation and have therefore the choice of applying it before or after the effect of the external perturbation is introduced in the Hamiltonian.

8.1 Sum-over-States Treatment

In the first approach, the so-called **sum-over-states treatment** of the corrections due to molecular vibration, the effects of the perturbation on the electronic and vibrational part of the wavefunction are treated simultaneously. This means that perturbation theory as developed in Section 3.2 is applied to the vibronic wavefunctions. In the presence of an external electric field with component $\mathcal{E}_\beta$ the perturbed vibronic wavefunction for, e.g., the electronic ground state $k = 0$ and an arbitrary vibrational state v, $\Phi_{0v}(\vec{\mathcal{E}})$, is thus obtained through first order, Eq. (3.27), as[2]

$$\Phi_{0v}(\vec{\mathcal{E}}) = \Phi_{0v}^{(0)} + \Phi_{0v}^{(1)}(\vec{\mathcal{E}})$$

$$= \Phi_{0v} + \sum_{nv' \neq 0v} |\Phi_{nv'}^{(0)}\rangle \frac{\langle \Phi_{nv'}^{(0)} | -\sum_\beta \left(\hat{\mu}_\beta + \hat{\Omega}_\beta^\mathcal{E}\right) \mathcal{E}_\beta | \Phi_{0v}^{(0)} \rangle}{E_{0v}^{(0)} - E_{nv'}^{(0)}} \tag{8.1}$$

where the first-order perturbation Hamiltonian, Eq. (4.29), for the perturbation by a homogenous electric field was already inserted. The summation now includes vibronic wavefunctions for all vibrational levels v' of all electronic states n with the exception of the particular vibrational level v of the electronic ground state $n = 0$. Using this wavefunction in the expression for the second-order energy correction, Eq. (3.30), we can evaluate the components of the static polarizability tensor, $\alpha_{\alpha\beta}$, from their definition as second derivatives of the perturbed energy, Eq. (4.65), and obtain

$$\alpha_{\alpha\beta} = -2 \sum_{nv' \neq 0v} \frac{\langle \Phi_{0v}^{(0)} | \hat{\mu}_\alpha + \hat{\Omega}_\alpha^\mathcal{E} | \Phi_{nv'}^{(0)} \rangle \langle \Phi_{nv'}^{(0)} | \hat{\mu}_\beta + \hat{\Omega}_\beta^\mathcal{E} | \Phi_{0v}^{(0)} \rangle}{E_{0v}^{(0)} - E_{nv'}^{(0)}} \tag{8.2}$$

The summation is typically split into two parts, defining the components $\alpha_{\alpha\beta}^v$ of the **vibrational polarizability**, unfortunately sometimes also called **atomic polarizability**, where the summation goes over all vibrational states $v' \neq v$ of the same electronic state $n = 0$

$$\alpha_{\alpha\beta}^v = -2 \sum_{v' \neq v} \frac{\langle \Phi_{0v}^{(0)} | \hat{\mu}_\alpha + \hat{\Omega}_\alpha^\mathcal{E} | \Phi_{0v'}^{(0)} \rangle \langle \Phi_{0v'}^{(0)} | \hat{\mu}_\beta + \hat{\Omega}_\beta^\mathcal{E} | \Phi_{0v}^{(0)} \rangle}{E_{0v}^{(0)} - E_{0v'}^{(0)}} \tag{8.3}$$

---

[2] Here and in most equations of the following sections we will not show explicitly that the vibronic wavefunctions Φ_{kv} depend on the electronic $\{\vec{r}_i\}$ and nuclear position vectors $\{\vec{R}_K\}$ nor that the vibrational wavefunctions Θ_v depend on the nuclear position vectors $\{\vec{R}_K\}$ and the electronic wavefunctions Ψ_k on the position vectors of the electrons $\{\vec{r}_i\}$ and parametrically on the nuclear position vectors $\{\vec{R}_K\}$.

and an electronic-vibrational polarizability

$$\alpha_{\alpha\beta}^{e,v} = -2 \sum_{n\neq 0, v'} \frac{\langle \Phi_{0v}^{(0)} | \hat{\mu}_\alpha + \hat{\Omega}_\alpha^{\mathcal{E}} | \Phi_{nv'}^{(0)} \rangle \langle \Phi_{nv'}^{(0)} | \hat{\mu}_\beta + \hat{\Omega}_\beta^{\mathcal{E}} | \Phi_{0v}^{(0)} \rangle}{E_{0v}^{(0)} - E_{nv'}^{(0)}} \tag{8.4}$$

where the summation goes over all other electronic states $n \neq 0$ and all their vibrational levels v'.

Now, one can make use of the the Born-Oppenheimer approximation, as discussed in Section 2.2, and approximate the unperturbed vibronic wavefunctions as a simple product of vibrational $\Theta_v^{(0)}(\{\vec{R}_K\})$ and electronic wavefunction $\Psi_n^{(0)}(\{\vec{r}_i\}; \{\vec{R}_K\})$, i.e.

$$\Phi_{nv}^{(0)}(\{\vec{r}_i\}, \{\vec{R}_K\}) = \Psi_n^{(0)}(\{\vec{r}_i\}; \{\vec{R}_K\}) \, \Theta_v^{(0)}(\{\vec{R}_K\}) \tag{8.5}$$

Using this ansatz in the expressions for the polarizabilities one obtains

$$\alpha_{\alpha\beta}^{v} = -2 \sum_{v'\neq v} \frac{\langle \Theta_v^{(0)} | \langle \Psi_0^{(0)} | \hat{\mu}_\alpha + \hat{\Omega}_\alpha^{\mathcal{E}} | \Psi_0^{(0)} \rangle | \Theta_{v'}^{(0)} \rangle \langle \Theta_{v'}^{(0)} | \langle \Psi_0^{(0)} | \hat{\mu}_\beta + \hat{\Omega}_\beta^{\mathcal{E}} | \Psi_0^{(0)} \rangle | \Theta_v^{(0)} \rangle}{E_{0v}^{(0)} - E_{0v'}^{(0)}}$$

$$\tag{8.6}$$

$$\alpha_{\alpha\beta}^{e,v} = -2 \sum_{v'} \sum_{n\neq 0} \frac{\langle \Theta_v^{(0)} | \langle \Psi_0^{(0)} | \hat{\mu}_\alpha | \Psi_n^{(0)} \rangle | \Theta_{v'}^{(0)} \rangle \langle \Theta_{v'}^{(0)} | \langle \Psi_n^{(0)} | \hat{\mu}_\beta | \Psi_0^{(0)} \rangle | \Theta_v^{(0)} \rangle}{E_{0v}^{(0)} - E_{nv'}^{(0)}} \tag{8.7}$$

The electronic ground-state expectation values in the numerator of the vibrational polarizability $\boldsymbol{\alpha}^v$ are components of the permanent electric dipole moment, Eq. (4.40), and we can therefore write the vibrational polarizability more compactly as

$$\alpha_{\alpha\beta}^{v} = -2 \sum_{v'\neq v} \frac{\langle \Theta_v^{(0)} | \mu_\alpha(\{\vec{R}_K\}) | \Theta_{v'}^{(0)} \rangle \langle \Theta_{v'}^{(0)} | \mu_\beta(\{\vec{R}_K\}) | \Theta_v^{(0)} \rangle}{E_{0v}^{(0)} - E_{0v'}^{(0)}} \tag{8.8}$$

In the electronic-vibrational polarizability $\alpha_{\alpha\beta}^{e,v}$, on the other hand, the nuclear part of the electric dipole moment operator $\hat{\Omega}_\alpha^{\mathcal{E}}$, Eq. (4.32), cannot contribute because the electronic states $|\Psi_0^{(0)}\rangle$ and $|\Psi_n^{(0)}\rangle$ are orthogonal.

For all the magnetic linear response properties derived in Chapters 5 and 6 one would obtain expressions similar to the electronic-vibrational polarizability, Eq. (8.7). On the other hand, the diamagnetic contributions to the magnetic properties as well as all first-order properties, i.e. properties defined as first derivatives of the energy, will take the following simple expectation value form

$$\langle \Theta_v^{(0)} | \langle \Psi_0^{(0)} | \hat{O} | \Psi_0^{(0)} \rangle | \Theta_v^{(0)} \rangle \tag{8.9}$$

where $\hat{O}$ would be of the $\hat{O}_{\alpha\beta\ldots}^{\mathcal{FF}}$ type for the diamagnetic contributions to the magnetic properties and of the $\hat{O}_{\alpha\ldots}^{\mathcal{F}}$ type for the first-order properties. Finally, for closed-shell molecules there is no pure vibrational contribution to the magnetic properties similar to Eq. (8.8), because the permanent magnetic moment or molecular magnetic induction vanishes due to the quenching of the angular momentum operator, as discussed in Section 5.3.

In order to evaluate the vibrational polarizability, Eq. (8.8), one needs the energies of all vibrational states of the electronic ground state and the corresponding vibrational dipole transition moments, which requires knowledge of the potential energy and electric dipole moment surface of this single electronic state. For the electronic-vibrational polarizability, Eq. (8.7), however, one would need to know not only all excited electronic states, $\Psi_n^{(0)}$ and the electronic dipole transition moments to them but also all the vibrational states, $\Theta_{v'}^{(0)}$, of these excited states, which makes this approach rather difficult to apply in actual calculations.

However, we can make the approximation in Eq. (8.7) that the differences between the vibrational energies are much smaller than the differences between the electronic energies, i.e.

$$E_{0v}^{(0)} - E_{nv'}^{(0)} \approx E_{00}^{(0)} - E_{n0}^{(0)} \tag{8.10}$$

which removes the dependence on the vibrational states from the denominator. Consequently we can use in the numerator the fact that the vibrational wavefunctions form a complete set, i.e. that

$$1 = \sum_{v'} |\Theta_{v'}^{(0)}\rangle\langle\Theta_{v'}^{(0)}| \tag{8.11}$$

and obtain from Eq. (8.7) a simplified expression for the electronic-vibrational polarizability

$$\alpha_{\alpha\beta}^{e,v} \approx -2 \sum_{n \neq 0} \frac{\langle\Theta_v^{(0)}|\langle\Psi_0^{(0)}|\hat{\mu}_\alpha|\Psi_n^{(0)}\rangle\langle\Psi_n^{(0)}|\hat{\mu}_\beta|\Psi_0^{(0)}\rangle|\Theta_v^{(0)}\rangle}{E_{00}^{(0)} - E_{n0}^{(0)}} \tag{8.12}$$

8.2 Clamped-Nucleus Treatment

In the second approach, the so-called **clamped-nucleus treatment**, the effect of the perturbation on the electronic and nuclear motion is treated sequentially. First, the Born–Oppenheimer approximation is applied to the vibronic wavefunction of the ground state, $\Phi_{0v}(\{\vec{r}_i\}, \{\vec{R}_K\})$, which is therefore expressed as a product of an electronic wavefunction $\Psi_0(\{\vec{r}_i\}; \{\vec{R}_K\})$ and a vibrational wavefunction $\Theta_v(\{\vec{R}_K\})$

$$\Phi_{0v}(\{\vec{r}_i\}, \{\vec{R}_K\}) = \Psi_0(\{\vec{r}_i\}; \{\vec{R}_K\}) \, \Theta_v(\{\vec{R}_K\}) \tag{8.13}$$

Secondly, in the presence of an external electric field with component $\mathcal{E}_\beta$, the field gives rise to a first-order perturbation Hamiltonian, Eq. (4.29), and the electronic wavefunction can be expanded in a perturbation series Eq. (3.16). To first order the electronic wavefunction, Eq. (3.27), and the electronic energy including the nuclear repulsion, Eq. (3.29), are then given as

$$\Psi_0(\vec{\mathcal{E}}) = \Psi_0^{(0)} + \Psi_0^{(1)}(\vec{\mathcal{E}}) = \Psi_0^{(0)} + \sum_{n \neq 0} |\Psi_n^{(0)}\rangle \frac{\langle\Psi_n^{(0)}|-\sum_\beta \left(\hat{\mu}_\beta + \hat{\Omega}_\beta^{\mathcal{E}}\right)\mathcal{E}_\beta|\Psi_0^{(0)}\rangle}{E_0^{(0)}(\{\vec{R}_K\}) - E_n^{(0)}(\{\vec{R}_K\})} \tag{8.14}$$

and

$$E_0(\{\vec{R}_K\}, \vec{\mathcal{E}}) = E_0^{(0)}(\{\vec{R}_K\}) + E_0^{(1)}(\{\vec{R}_K\}, \vec{\mathcal{E}})$$
$$= E_0^{(0)}(\{\vec{R}_K\}) + \langle\Psi_0^{(0)}| -\sum_\beta \hat{\mu}_\beta \mathcal{E}_\beta |\Psi_0^{(0)}\rangle - \sum_\beta \hat{\Omega}_\beta^{\mathcal{E}} \mathcal{E}_\beta \quad (8.15)$$

where $E_0^{(0)}(\{\vec{R}_K\})$ indicates that the electronic energies are for a given set of nuclear coordinates $\{\vec{R}_K\}$.[3] Recognising the last two terms as the permanent electric dipole moment $\vec{\mu}(\{\vec{R}_K\})$, from Eq. (4.40), again we can write the energy as

$$E_0(\{\vec{R}_K\}, \vec{\mathcal{E}}) = E_0^{(0)}(\{\vec{R}_K\}) - \vec{\mu}(\{\vec{R}_K\}) \cdot \vec{\mathcal{E}} \quad (8.16)$$

With this energy as potential energy for the nuclear motion the nuclear Schrödinger equation in the Born–Oppenheimer approximation, Eq. (2.12), becomes

$$\left[\sum_K \frac{1}{2m_K}\hat{p}_K^2 + E_0(\{\vec{R}_K\}, \vec{\mathcal{E}})\right]|\Theta_v(\vec{\mathcal{E}})\rangle = E_{0v}(\vec{\mathcal{E}})|\Theta_v(\vec{\mathcal{E}})\rangle \quad (8.17)$$

which means that the external electric field, $\mathcal{E}_\beta$, enters the nuclear Hamiltonian together with an operator that is an expectation value over the electronic wavefunction but depends on the nuclear position vectors $\{\vec{R}_K\}$

$$\left[\sum_K \frac{1}{2m_K}\hat{p}_K^2 + E_0^{(0)}(\{\vec{R}_K\}) - \vec{\mu}(\{\vec{R}_K\}) \cdot \vec{\mathcal{E}}\right]|\Theta_v(\vec{\mathcal{E}})\rangle = E_{0v}(\vec{\mathcal{E}})|\Theta_v(\vec{\mathcal{E}})\rangle \quad (8.18)$$

This equation is now also solved with perturbation theory assuming that the unperturbed equation, i.e. without the $-\vec{\mu}(\{\vec{R}_K\}) \cdot \vec{\mathcal{E}}$ term, has been solved and a complete set of eigenfunctions $\{\Theta_v^{(0)}\}$ has been obtained. The vibrational wavefunction for an arbitrary vibrational state v of the given electronic state is then also expanded in a perturbation series and is to first order given as

$$\Theta_v(\vec{\mathcal{E}}) = \Theta_v^{(0)} + \Theta_v^{(1)}(\vec{\mathcal{E}}) = \Theta_v^{(0)} + \sum_{v' \neq v} |\Theta_{v'}^{(0)}\rangle \frac{\langle\Theta_{v'}^{(0)}| -\vec{\mu}(\{\vec{R}_K\}) \cdot \vec{\mathcal{E}} |\Theta_v^{(0)}\rangle}{E_{0v}^{(0)} - E_{0v'}^{(0)}} \quad (8.19)$$

An expression for the polarizability can finally be obtained by time-independent response theory as described in Sections 3.3 and 3.4, i.e. Eq. (4.79),

---

[3] See the footnote of the previous section for the implied dependence of the wavefunctions on the electronic and/or nuclear position vectors.

$$\alpha_{\alpha\beta} = \frac{\partial}{\partial \mathcal{E}_\beta} \langle \Phi_{0v}(\vec{\mathcal{E}}) | \hat{\mu}_\alpha + \hat{\Omega}_\alpha^\mathcal{E} | \Phi_{0v}(\vec{\mathcal{E}}) \rangle^{(1)} \tag{8.20}$$

$$= \frac{\partial}{\partial \mathcal{E}_\beta} \Big[\langle \Theta_v^{(0)} \Psi_0^{(1)}(\vec{\mathcal{E}}) | \hat{\mu}_\alpha | \Psi_0^{(0)} \Theta_v^{(0)} \rangle + \langle \Theta_v^{(0)} \Psi_0^{(0)} | \hat{\mu}_\alpha | \Psi_0^{(1)}(\vec{\mathcal{E}}) \Theta_v^{(0)} \rangle$$

$$+ \langle \Theta_v^{(1)}(\vec{\mathcal{E}}) \Psi_0^{(0)} | \hat{\mu}_\alpha + \hat{\Omega}_\alpha^\mathcal{E} | \Psi_0^{(0)} \Theta_v^{(0)} \rangle$$

$$+ \langle \Theta_v^{(0)} \Psi_0^{(0)} | \hat{\mu}_\alpha + \hat{\Omega}_\alpha^\mathcal{E} | \Psi_0^{(0)} \Theta_v^{(1)}(\vec{\mathcal{E}}) \rangle \Big] \tag{8.21}$$

where we have used that the first-order correction to a vibronic wavefunction like Eq. (8.13), i.e. a simple product function, is given as

$$\Phi_{0v}^{(1)}(\{\vec{r}_i\}, \{\vec{R}_K\}, \vec{\mathcal{E}}) = \Psi_0^{(1)}(\{\vec{r}_i\}; \{\vec{R}_K\}, \vec{\mathcal{E}}) \, \Theta_v^{(0)}(\{\vec{R}_K\})$$

$$+ \Psi_0^{(0)}(\{\vec{r}_i\}; \{\vec{R}_K\}) \, \Theta_v^{(1)}(\{\vec{R}_K\}, \vec{\mathcal{E}}) \tag{8.22}$$

This then gives again two contributions: a vibrationally averaged electronic polarizability

$$\alpha_{\alpha\beta}^{e,av} = -2 \, \langle \Theta_v^{(0)} | \sum_{n \neq 0} \frac{\langle \Psi_0^{(0)} | \hat{\mu}_\alpha | \Psi_n^{(0)} \rangle \langle \Psi_n^{(0)} | \hat{\mu}_\beta | \Psi_0^{(0)} \rangle}{E_0^{(0)}(\{\vec{R}_K\}) - E_n^{(0)}(\{\vec{R}_K\})} | \Theta_v^{(0)} \rangle$$

$$= \langle \Theta_v^{(0)}(\{\vec{R}_K\}) | \alpha_{\alpha\beta}(\{\vec{R}_K\}) | \Theta_v^{(0)}(\{\vec{R}_K\}) \rangle \tag{8.23}$$

and the vibrational polarizability, given in Eq. (8.6) or Eq. (8.8).

Although the expression for the vibrational polarizability is thus the same as the one obtained with the *sum-over-states* treatment, the expression for the electronic contribution differs significantly from Eq. (8.7) or Eq. (8.12). The averaged electronic contribution in Eq. (8.23) is simply the pure electronic polarizability as given in Eq. (4.74) but averaged with the unperturbed vibrational wavefunction $\Theta_v^{(0)}(\{\vec{R}_K\})$ of the electronic ground state. In the approximate form of the electronic-vibrational polarizability in Eq. (8.12), on the other hand, the transition moments to each excited electronic state are averaged individually with the unperturbed vibrational wavefunction $\Theta_v^{(0)}(\{\vec{R}_K\})$ of the electronic ground state and then divided by the difference between the energies of the vibrational ground state of the respective excited electronic state $E_{n0}^{(0)}$ and of the vibrational ground state of the electronic ground state $E_{00}^{(0)}$.

8.3 Vibrational and Thermal Averaging

In this section, we will describe in more detail how the vibrational averaging of the pure electronic polarizability and the calculation of the vibrational polarizability is carried out. We will hereby distinguish between diatomic and polyatomic molecules.

180 Vibrational Contributions to Molecular Properties

For diatomic molecules the vibration-rotation wavefunctions can be obtained numerically as solutions of the one-dimensional Schrödinger equation

$$\left\{-\frac{\hbar^2}{2\mu}\left[\frac{d^2}{dR^2}+\frac{J(J+1)}{R^2}\right]+E_0(R)\right\}|\Theta_{v,J}^{(0)}\rangle = E_{v,J}^{(0)}|\Theta_{v,J}^{(0)}\rangle \qquad (8.24)$$

where R is the internuclear distance, J is the rotational quantum number and $E_0^{(0)}(R)$ is the Born–Oppenheimer electronic energy including the nuclear repulsion.[4] The vibrational or more precisely vibration-rotational averaging in the clamped-nucleus treatment in Eq. (8.23) can then be carried out numerically, if one calculates the polarizability as given in Eq. (4.74) pointwise as a function of the internuclear distance R. The calculation of the vibrational polarizability in Eq. (8.8) requires correspondingly the pointwise calculation of the electric dipole moment as a function of the internuclear distance and the numerical calculation of all vibrational wavefunctions and corresponding vibrational energies supported by the potential energy surface of the electronic ground state.

For polyatomic molecules the electronic polarizability in the *clamped-nucleus* treatment is frequently expressed as the polarizability evaluated at an equilibrium geometry $\{\vec{R}_{K,e}\}$ plus a vibrational correction $\Delta\alpha^v$

$$\alpha_{\alpha\beta}^{e,av} = \langle\Theta_v^{(0)}(\{\vec{R}_K\})|\alpha_{\alpha\beta}(\{\vec{R}_K\})|\Theta_v^{(0)}(\{\vec{R}_K\})\rangle = \alpha_{\alpha\beta}(\{\vec{R}_{K,e}\}) + \Delta\alpha_{\alpha\beta}^v \qquad (8.25)$$

When one is interested in the correction for the vibrational ground state "$v = 0$", i.e. the state where the vibrational quantum numbers of all the vibrational modes of the molecule are equal to zero, one calls this the **zero-point vibrational correction (ZPVC)** $\Delta\alpha^{\text{ZPVC}} = \Delta\alpha^{v=0}$. In order to calculate a vibrational correction one expands the polarizability normally in a Taylor series in the set of normal coordinates[5] $\{Q_a\}$

$$\alpha_{\alpha\beta}(\{\vec{R}_K\}) = \alpha_{\alpha\beta}(\{\vec{R}_{K,e}\}) + \sum_a \left(\frac{\partial\alpha_{\alpha\beta}}{\partial Q_a}\right)Q_a + \frac{1}{2}\sum_{ab}\left(\frac{\partial^2\alpha_{\alpha\beta}}{\partial Q_a \partial Q_b}\right)Q_a Q_b + \cdots \qquad (8.26)$$

The vibrational correction is then obtained as a series of expectation values of increasing powers of the normal coordinates $\{Q_a\}$

$$\Delta\alpha_{\alpha\beta}^v = \sum_a \left(\frac{\partial\alpha_{\alpha\beta}}{\partial Q_a}\right)\langle\Theta_v^{(0)}|Q_a|\Theta_v^{(0)}\rangle + \frac{1}{2}\sum_{ab}\left(\frac{\partial^2\alpha_{\alpha\beta}}{\partial Q_a \partial Q_b}\right)\langle\Theta_v^{(0)}|Q_a Q_b|\Theta_v^{(0)}\rangle + \cdots$$
$$(8.27)$$

The unperturbed, i.e. field-free, vibrational wavefunctions $\Theta_v^{(0)}$ are in principle found by solving the unperturbed vibrational Schrödinger equation Eq. (2.12), in

---

[4] For highly accurate calculations of the vibration-rotation wavefunctions and energies one should of course employ the effective Hamiltonian in Eq. (6.67).
[5] For M nuclei there are $3M - 6$ normal coordinates or $3M - 5$ in the case of linear molecules.

which the nuclear potential energy $E_0^{(0)}(\{\vec{R}_K\})$ is also expanded in a Taylor series in the normal coordinates $\{Q_a\}$

$$E_0^{(0)}(\{\vec{R}_K\}) = E_0^{(0)}(\{\vec{R}_{K,e}\}) + \frac{1}{2}\sum_a \omega_a^2 Q_a^2 + \frac{1}{6}\sum_{abc} K_{abc} Q_a Q_b Q_c + \cdots \qquad (8.28)$$

where $\omega_a = \sqrt{K_{aa}}$ and K_{abc} are the harmonic vibrational frequency and force constant and the cubic force constant, respectively, defined as

$$K_{ab} = \frac{\partial^2 E_0^{(0)}(\{\vec{R}_K\})}{\partial Q_a \partial Q_b} \qquad (8.29)$$

$$K_{abc} = \frac{\partial^3 E_0^{(0)}(\{\vec{R}_K\})}{\partial Q_a \partial Q_b \partial Q_c} \qquad (8.30)$$

Terminating the expansion after the quadratic term and choosing $E_0^{(0)}(\{\vec{R}_{K,e}\})$ as the zero point of the potential energy one obtains an harmonic potential for which the vibrational Schrödinger equation

$$\frac{1}{2}\sum_a \left[-\hbar^2 \frac{\partial^2}{\partial Q_a^2} + \omega_a^2 Q_a^2\right] |\Theta_v^{(0,0)}(\{Q_a\})\rangle = E_v^{(0,0)} |\Theta_v^{(0,0)}(\{Q_a\})\rangle \qquad (8.31)$$

can be separated into equations for each normal mode Q_a

$$\frac{1}{2}\left[-\hbar^2 \frac{\partial^2}{\partial Q_a^2} + \omega_a^2 Q_a^2\right] |\vartheta_{v_a}(Q_a)\rangle = E_{v_a} |\vartheta_{v_a}(Q_a)\rangle \qquad (8.32)$$

with the energy

$$E_{v_a} = \left(v_a + \frac{1}{2}\right)\hbar\omega_a \qquad (8.33)$$

and $\vartheta_{v_a}(Q_a)$ being a one-mode harmonic oscillator wavefunction. The total many-mode vibrational wavefunction and energy in this harmonic approximation are then given as

$$\Theta_v^{(0,0)}(\{Q_a\}) = \prod_a \vartheta_{v_a}(Q_a) \qquad (8.34)$$

$$E_v^{(0,0)} = \sum_a E_{v_a} \qquad (8.35)$$

The anharmonicity of the potential energy function, to first order expressed by the cubic force constants $\{K_{abc}\}$, leads to a perturbation operator $\frac{1}{6}\sum_{abc} K_{abc} Q_a Q_b Q_c$. The field-free vibrational wavefunction $\Theta_v^{(0)}(\{Q_a\})$ is thus expanded in the usual perturbation series

$$\Theta_v^{(0)}(\{Q_a\}) = \Theta_v^{(0,0)}(\{Q_a\}) + \Theta_v^{(0,1)}(\{Q_a\}) + \cdots \qquad (8.36)$$

182 *Vibrational Contributions to Molecular Properties*

where the zeroth-order wavefunction is given in Eq. (8.34). For the expansion of the first-order correction $\Theta_v^{(0,1)}(\{Q_a\})$ we need according to Eq. (3.27) a complete set of functions, which in general consists of all many-mode vibrational wavefunctions obtained by exciting one, two and so forth up to all one-mode harmonic oscillator functions $\vartheta_{v_a}(Q_a)$ to all possible higher (or lower) vibrational levels. However, in the following we will derive only an expression for zero point vibrational corrections and restrict ourselves therefore to corrections to the ground state vibrational wavefunction $\Theta_{v=0}^{(0,0)}(\{Q_a\})$, which implies that we can restrict ourselves to many-mode vibrational wavefunctions $\Theta_{v_b=1}^{(0,0)}(\{Q_a\})$, where only one of the one-mode harmonic oscillator function $\vartheta_{v_b}(Q_b)$ was excited to the $v_b = 1$ level, while the other modes $a \neq b$ remain in the lowest level $v_a = 0$ (Kern and Matcha, 1968), i.e.

$$\Theta_{v_b=1}^{(0,0)}(\{Q_a\}) = \vartheta_{v_b=1}(Q_b) \prod_{a \neq b} \vartheta_{v_a=0}(Q_a) \qquad (8.37)$$

The first-order correction to the ground-state vibrational wavefunction then reads

$$\Theta_{v=0}^{(0,1)}(\{Q_a\}) = -\sum_b |\Theta_{v_b=1}^{(0,0)}(\{Q_a\})\rangle \frac{\langle \Theta_{v_b=1}^{(0,0)}(\{Q_a\}) | \frac{1}{6}\sum_{cde} K_{cde} Q_c Q_d Q_e | \Theta_v^{(0,0)}(\{Q_a\})\rangle}{\hbar \omega_b} \qquad (8.38)$$

Using the following expectation values of the normal coordinates over one-mode harmonic oscillator wavefunctions

$$\langle \vartheta_{v_a} | Q_a | \vartheta_{v_a'}\rangle = 0 \text{ if } v_a' \neq v_a \pm 1 \qquad (8.39)$$

$$\langle \vartheta_{v_a} | Q_a | \vartheta_{v_a+1}\rangle = \sqrt{\frac{\hbar}{2\omega_a}(v_a+1)} \qquad (8.40)$$

$$\langle \vartheta_{v_a} | Q_a Q_a | \vartheta_{v_a}\rangle = \frac{\hbar}{\omega_a}(v_a + \frac{1}{2}) \qquad (8.41)$$

we can evaluate the transition matrix element in Eq. (8.38) [see Exercise 8.1] and then obtain for the first-order correction to the perturbed wavefunction

$$\Theta_{v=0}^{(0,1)}(\{Q_a\}) = -\frac{1}{4}\sum_b |\Theta_{v_b=1}^{(0,0)}(\{Q_a\})\rangle \sqrt{\frac{\hbar}{2\omega_b^3}} \sum_c \frac{K_{bcc}}{\omega_c} \qquad (8.42)$$

Exercise 8.1 Prove that

$$\frac{1}{6}\sum_{cde} K_{cde} \langle \Theta_{v_b=1}^{(0,0)}(\{Q_a\}) | Q_c Q_d Q_e | \Theta_v^{(0,0)}(\{Q_a\})\rangle = \frac{\hbar}{4}\sqrt{\frac{\hbar}{2\omega_b}}\sum_c \frac{K_{bcc}}{\omega_c}$$

using the expectation values in Eqs. (8.39) to (8.41).

Now, we are finally ready to evaluate the matrix elements of the normal coordinates in the expression for the zero-point vibrational correction. To the lowest non-vanishing order in perturbation theory this gives

$$\Delta\alpha_{\alpha\beta}^{\text{ZPVC}} = \sum_a \left(\frac{\partial\alpha_{\alpha\beta}}{\partial Q_a}\right)\langle\Theta_{v=0}^{(0)}|Q_a|\Theta_{v=0}^{(0)}\rangle^{(1)} + \frac{1}{2}\sum_{ab}\left(\frac{\partial^2\alpha_{\alpha\beta}}{\partial Q_a \partial Q_b}\right)\langle\Theta_{v=0}^{(0)}|Q_a Q_b|\Theta_{v=0}^{(0)}\rangle^{(0)} \tag{8.43}$$

as a consequence of Eqs. (8.39) and (8.41). Inserting Eqs. (8.34) and (8.42) into Eq. (8.43) and using the properties (8.39) to (8.41) of the one-mode harmonic oscillator functions again, one obtains finally for the zero-point-vibrational correction to the static polarizability tensor [see Exercise 8.2]

$$\Delta\alpha_{\alpha\beta}^{\text{ZPVC}} = -\frac{\hbar}{4}\sum_a \frac{1}{\omega_a^2}\left(\frac{\partial\alpha_{\alpha\beta}}{\partial Q_a}\right)\left(\sum_b \frac{K_{abb}}{\omega_b}\right) + \frac{\hbar}{4}\sum_a \frac{1}{\omega_a}\left(\frac{\partial^2\alpha_{\alpha\beta}}{\partial Q_a^2}\right) \tag{8.44}$$

where the first term thus arises from the anharmonic term in the potential, Eq. (8.28), and the second term comes from the non-linear term in the expansion of the polarizability, Eq. (8.26). The are therefore sometimes called the mechanical and electrical anharmonic contributions. Equivalent expressions for higher vibrational levels have also been derived (Toyama et al., 1964).

Exercise 8.2 Derive the expression for the zero-point-vibrational correction (ZPVC) in Eq. (8.44).

The effect of temperature, T, can finally be included by Boltzmann averaging the averaged electronic polarizability over several vibrational states of energy $E_v^{(0)}$

$$\alpha_{\alpha\beta}^e(T) = \frac{\sum_v \alpha_{\alpha\beta}^{e,av} e^{-\frac{E_v^{(0)}}{kT}}}{\sum_v e^{-\frac{E_v^{(0)}}{kT}}} \tag{8.45}$$

or after inserting the expansion of the vibrationally averaged polarizability, Eq. (8.25),

$$\alpha_{\alpha\beta}^e(T) = \alpha_{\alpha\beta}(\{\vec{R}_{K,e}\}) + \frac{\sum_v \Delta\alpha_{\alpha\beta}^v e^{-\frac{E_v^{(0)}}{kT}}}{\sum_v e^{-\frac{E_v^{(0)}}{kT}}} \tag{8.46}$$

For the evaluation of the vibrational polarizability, Eq. (8.8), one needs to calculate the vibrational energies, $E_{0v}^{(0)}$, as well as vibrational transition moments, $\langle\Theta_v^{(0)}|\mu_\alpha|\Theta_{v'}^{(0)}\rangle$, of the electric dipole moment. Both can be obtained by a perturbation theory treatment similar to the one used here for the evaluation of the ZPVC to electronic polarizability (Bishop and Kirtman, 1991).

8.4 Further Reading

Vibrational Corrections to NMR Parameters

T. A. Ruden and K. Ruud, in M. Kaupp, M. Bühl and V. G. Malkin, ed. *Calculation of NMR and EPR Parameters Theory and Applications*, Wiley-VCH, Weinheim (2004): Chapter 10, pages 153–173.

Vibrational Corrections to Optical Parameters

D. M. Bishop, *Molecular Vibrational and Rotational Motion in Static and Dynamic Electric Fields*, Rev. Mod. Phys. 62, 343–374 (1990).

D. M. Bishop, *Molecular Vibration and Nonlinear Optics*, Adv. Chem. Phys. 104, 1–40 (1998).

Part III

Computational Methods for the Calculation of Molecular Properties

In this final part we want to discuss quantum chemical methods that can be used to calculate values of all the electric, magnetic and spectral properties that were defined in the Chapters 4, 5, 6 and 7. The emphasis will hereby be more on the conceptual aspects, on what the methods have in common and on how they differ than on the technical and computational details of the methods. Explicit formulas for these methods ready to be implemented in computer programs will not be presented here in general.

In principle, one could start from the perturbation theory expressions that were derived in these chapters and insert the appropriate ground and excited-state wavefunctions, $|\Psi_0^{(0)}\rangle$ and $|\Psi_n^{(0)}\rangle$. However, we have to remember that perturbation theory as outlined in Section 3.2 and thus all the expressions for molecular properties derived in the previous chapters are based on the assumption that we have solved the unperturbed Schrödinger equation, Eq. (3.14). Already in Section 3.2 it was mentioned that this is not possible, but we had postponed it to the present part to draw the necessary consequences of this problem. There are at least three possible ways out of this dilemma and thus, starting points for approximate methods:

1. We can make approximations to the "exact" perturbation theory expressions derived in Chapters 4–7. This means that we will insert approximate unperturbed energies and wavefunctions in the expressions derived in Chapters 4–7 or approximate unperturbed wavefunctions and an incomplete set of operators in the propagators derived in Section 3.12. Methods based on this approach are discussed in Chapter 10.
2. We can keep the perturbation theory approach, but use approximate solutions to the unperturbed Schrödinger equation as zeroth-order wavefunction and energy, i.e. we can rederive expression for the corrections to the energy and wavefunctions using perturbation theory but now with approximate wavefunctions. Methods based on this approach are discussed in Chapter 11.

3. We can abandon the perturbation theory approach all together and go back to the definitions of the properties as derivatives of the energy. These definitions are then applied to approximate expressions for the energy of a molecule in the presence of the perturbations. Methods based on this approach are briefly discussed in Chapter 12.

Consequently, the present part is organised in such a way that all methods that fall in one of these three categories are discussed together in a chapter. There are cases where identical working equations can be derived from two or all three of the starting points. The most prominent cases are the methods obtained with self-consistent field (SCF) or multiconfigurational self-consistent field (MCSCF) wavefunctions. The first-order polarization propagator approximation (FOPPA) better known as the random-phase approximation (RPA), the coupled Hartree–Fock method (CHF), the SCF linear response theory and the analytical second derivative of the SCF energy will all give identical equations and results for e.g. a static polarizability and correspondingly their multiconfigurational generalisations.

In addition to the question of approximate perturbation theory there are at least two other criteria that can be used to classify practical quantum chemical methods for the calculation of electromagnetic properties:

Fulfillment of the Hellmann–Feynman theorem: In Chapters 4–7 we have seen that molecular properties can be defined either as derivatives of the electronic energy or as derivatives of molecular electromagnetic moments and fields. For exact states, which obviously fulfill the Hellmann–Feynman theorem, both definitions lead to the same expressions for the properties, but for an approximate method the same expression will only be obtained if the approximate method fulfills the Hellmann–Feynman theorem.

Possibility for the calculation of time- or frequency-dependent properties: Response theory methods obviously fulfill this condition and static properties are actually obtained as the zero-frequency limit of response function. Methods based on derivatives of the energy, on the other hand, can *a priori* not be used to calculate time- or frequency-dependent properties, as the energy is not part of the time-dependent Schrödinger equation. However, derivative techniques have become so popular that several types of quasi- or pseudo-energies have been defined also for the time-dependent case.

The methods that will be discussed in the following are all of the *ab initio* type. Given the molecular field free Hamiltonian in Eq. (2.9), with the nuclear coordinates and charges and the electronic mass given as parameters, in these methods all integrals over this Hamiltonian or parts of it are evaluated *ab initio*, i.e. by strict application of the appropriate mathematical rules and without using further data from experiment or otherwise. The emphasis will be in particular on the SCF and on so-called correlated methods. Semi-empirical or density functional theory (DFT) methods on the other hand will not be mentioned explicitly. However, most of what will be said about SCF-based methods for the calculation of properties will also apply to semi-empirical or DFT methods, because one can consider to a certain extent the semi-empirical and DFT methods as variants of SCF, just with a slightly different Hamiltonian.

Throughout this part only closed-shell molecules and thus closed-shell wavefunctions will be discussed. This implies that permanent magnetic moments and fields as defined in Sections 5.3 and 5.6 will be zero.

We will start in the following chapter with a short review of *ab initio* methods for the calculation of ground-state energies and wavefunctions, before we discuss how these methods are employed in the calculation of electromagnetic properties.

9
Short Review of Electronic Structure Methods

In the present chapter a very brief review of closed-shell *ab initio* methods for the calculation of the energy and wavefunction in the absence of perturbations is given. We will cover only those methods whose application to the calculation of electromagnetic properties will be discussed in the following sections. The main purpose of this section is thus to introduce the concepts and notation of the different methods.

In the previous chapters we have in the notation carefully distinguished between the ground state $|\Psi_0^{(0)}\rangle$ and exited states $|\Psi_n^{(0)}\rangle$ and between perturbed and unperturbed wavefunctions. In this and the following chapter, however, we will work with unperturbed wavefunctions and energies. For the sake of a simpler notation we will drop the superscript "(0)" for the unperturbed, i.e. field-free, problem. It will be only used for the field-free Hamiltonian $\hat{H}^{(0)}$ in these chapters.

For all the methods covered in this chapter it holds that the approximations[1] $|\Phi_0\rangle$ to the ground-state N-electron wavefunction $|\Psi_0^{(0)}\rangle$ can be expressed as a linear combination of Slater determinants $\{|\Phi_n\rangle\}$

$$|\Psi_0^{(0)}\rangle \approx |\Phi_0\rangle = \sum_n |\Phi_n\rangle \, C_{n0} \qquad (9.1)$$

Throughout this chapter we will assume that the approximate wavefunctions $|\Phi_0\rangle$ are always properly normalized. A **Slater determinant** is an antisymmetrized product

$$|\Phi_n\rangle = \frac{1}{\sqrt{N!}} \det \begin{vmatrix} \psi_1(\vec{x}_1) & \psi_2(\vec{x}_1) & \cdots & \psi_N(\vec{x}_1) \\ \psi_1(\vec{x}_2) & \psi_2(\vec{x}_2) & \cdots & \psi_N(\vec{x}_2) \\ \vdots & \vdots & \ddots & \vdots \\ \psi_1(\vec{x}_N) & \psi_2(\vec{x}_N) & \cdots & \psi_N(\vec{x}_N) \end{vmatrix} \qquad (9.2)$$

of one-electron molecular **spin-orbitals** $\{\psi_p(\vec{x})\}$, which are the product of a **spatial molecular orbital** $\phi_p(\vec{r})$ and the appropriate abstract one-electron **spin functions** $\alpha(s)$ or $\beta(s)$.

---

[1] To indicate via the notation that we are dealing with approximate wavefunctions, contrary to the previous chapters, Φ is used as the symbol for all approximate N-electron wavefunctions instead of Ψ. Furthermore, a superscript such as SCF, MCSCF, MPn or CC is added for particular approximate wavefunctions.

$$\psi_p(\vec{x}) = \phi_p(\vec{r}) \begin{cases} \alpha(s) \\ \beta(s) \end{cases} \qquad (9.3)$$

where s denotes the abstract spin variable and $\vec{x}$ stands for spatial and spin variables of an electron together. In restricted methods the same set of spatial molecular orbitals $\{\phi_p(\vec{r})\}$ is used for α and β spin orbitals.

Although it is not necessary in general, we will assume here that the molecular orbitals (MO) are expanded in a basis of one-electron functions, $\{\chi_\mu\}$, denoted by Greek indices and called atomic orbitals (AO) (although there is no restriction on their position within the molecule)

$$\phi_p = \sum_\mu \chi_\mu \, c_{\mu p} \qquad (9.4)$$

where $\{c_{\mu p}\}$ are the molecular orbital coefficients. The spin orbitals in Eq. (9.3) and therefore the Slater determinants in Eq. (9.2) depend consequently on the molecular orbital coefficients $\{c_{\mu p}\}$.

The approximate methods differ then in how the energy is calculated and how the molecular orbital coefficients, $\{c_{\mu p}\}$, and the coefficients $\{C_{n0}\}$ in the expansion in Slater determinants are determined. In this context, we can distinguish between

Variational methods such as the self-consistent field (SCF), multiconfigurational self-consistent field (MCSCF) or full configuration interaction (full CI) methods, where the energy is calculated as an expectation value

$$E_0(\{C_{n0}\},\{c_{\mu p}\}) = \langle \Phi_0(\{C_{n0}\},\{c_{\mu p}\}) | \hat{H}^{(0)} | \Phi_0(\{C_{n0}\},\{c_{\mu p}\}) \rangle \qquad (9.5)$$

and the wavefunction parameters $\{C_{n0}\}$ and $\{c_{\mu p}\}$ are obtained variationally, i.e. from the condition that the energy should be a minimum with respect to the wavefunction parameters:

$$\frac{\partial E_0(\{C_{n0}\},\{c_{\mu p}\})}{\partial C_{n0}} = \frac{\partial E_0(\{C_{n0}\},\{c_{\mu p}\})}{\partial c_{\mu p}} = 0 \qquad (9.6)$$

Non-variational methods such as Møller–Plesset perturbation theory (MP) or the coupled cluster (CC) method, where the energy can be expressed as a transition or asymmetric expectation value

$$E_0(\{C_{n0}\},\{c_{\mu p}\}) = \langle \Phi_0'(\{C_{n0}\},\{c_{\mu p}\}) | \hat{H}^{(0)} | \Phi_0(\{C_{n0}\},\{c_{\mu p}\}) \rangle \qquad (9.7)$$

and the determinant expansion coefficients $\{C_{n0}\}$ are not obtained variationally, but by projecting the corresponding Schrödinger equation against appropriate determinants, while the molecular orbital coefficients $\{c_{\mu p}\}$ are still obtained variationally from the SCF energy

$$\frac{\partial E_0^{SCF}(\{c_{\mu p}\})}{\partial c_{\mu p}} = 0 \qquad (9.8)$$

In a few cases explicit expressions of determinant expansion coefficients or other matrix elements are given in this chapter in terms of two electron repulsion integrals over spatial orbitals in the so-called Mulliken or chemical notation

$$(\phi_i(\vec{r}_1)\phi_k(\vec{r}_1)|\phi_j(\vec{r}_2)\phi_l(\vec{r}_2)) = \frac{e^2}{4\pi\epsilon_0} \int_{\vec{r}_1}\int_{\vec{r}_2} \phi_i^*(\vec{r}_1)\,\phi_j^*(\vec{r}_2)\frac{1}{|\vec{r}_1-\vec{r}_2|}\,\phi_k(\vec{r}_1)\,\phi_l(\vec{r}_2)\,d\vec{r}_1\,d\vec{r}_2 \tag{9.9}$$

9.1 Hartree–Fock Theory

All the *ab initio* methods discussed here are based on the Hartree–Fock (HF) or self-consistent field method. In closed-shell HF theory the unperturbed many-electron wavefunction $|\Psi_0^{(0)}\rangle$ is approximated by a single Slater determinant

$$|\Psi_0^{(0)}\rangle \approx |\Phi_0^{\mathrm{SCF}}\rangle \tag{9.10}$$

The spatial orbitals are solutions to the Hartree–Fock equations

$$\hat{f}(i)\,\phi_p(\vec{r}_i) = \epsilon_p\,\phi_p(\vec{r}_i) \tag{9.11}$$

which are derived from the condition that the Hartree–Fock energy, E_0^{SCF},

$$E_0^{\mathrm{SCF}} = \langle \Phi_0^{\mathrm{SCF}}|\hat{H}^{(0)}|\Phi_0^{\mathrm{SCF}}\rangle \tag{9.12}$$

has to be stationary with respect to a variation of the spin orbitals $\delta\psi_p$

$$\delta E_0^{\mathrm{SCF}} = 0 \tag{9.13}$$

under the constraint that the orbitals have to remain orthonormal

$$\langle \psi_p | \psi_q \rangle = \delta_{pq} \tag{9.14}$$

The Hartree–Fock Hamiltonian $\hat{F}$ is the sum of the one-electron Fock operators $\hat{f}(i)$ and is defined as

$$\hat{F} = \sum_i^N \hat{f}(i) = \sum_i^N \left[\hat{h}^{(0)}(i) + \hat{v}^{\mathrm{HF}}(i)\right] \tag{9.15}$$

where $\hat{h}^{(0)}(i)$ is defined in Eq. (2.101) and $\hat{v}^{\mathrm{HF}}(i)$ is an effective one-electron potential, called the Hartree–Fock potential, defined as

$$\hat{v}^{\mathrm{HF}}(i) = \frac{e^2}{4\pi\epsilon_0} \sum_j^{\mathrm{occ}} \int_{\vec{r}_2} \phi_j^*(\vec{r}_2)\frac{2-\hat{P}_{12}}{|\vec{r}_i-\vec{r}_2|}\,\phi_j(\vec{r}_2)\,d\vec{r}_2 \tag{9.16}$$

where $\hat{P}_{12}$ is a permutation operator, which permutes electron 1 with 2. $\hat{v}^{\mathrm{HF}}(i)$ is the potential that an electron experiences when it moves in the averaged field of all the other electrons. In order to calculate this averaged field of the other electrons one needs molecular orbitals $\phi_j(\vec{r}_2)$ that describe the other electrons. Obviously they are also solutions of the Hartree–Fock equations Eq. (9.11). The Fock operator thus depends on its own eigenfunctions and the Hartree–Fock equations have to be solved iteratively until self-consistency of the Hartree–Fock potential $\hat{v}^{\text{HF}}(i)$ is obtained. This is why the Hartree–Fock method is also called the **self-consistent field method**.

The eigenvalues of the Fock operator are the orbital energies ϵ_p

$$\epsilon_p = \langle \psi_p | \hat{f} | \psi_p \rangle = \langle \psi_p | \hat{h}^{(0)} | \psi_p \rangle + \sum_j^{\text{occ}} \Big[\big(\psi_p(\vec{r}_1)\,\psi_p(\vec{r}_1) | \psi_j(\vec{r}_2)\,\psi_j(\vec{r}_2) \big)$$

$$- \big(\psi_p(\vec{r}_1)\,\psi_j(\vec{r}_1) | \psi_j(\vec{r}_2)\,\psi_p(\vec{r}_2) \big) \Big] \quad (9.17)$$

or in terms of the spatial orbitals

$$\epsilon_p = \langle \phi_p | \hat{f} | \phi_p \rangle = \langle \phi_p | \hat{h}^{(0)} | \phi_p \rangle + \sum_j^{\text{occ}} \Big[2\big(\phi_p(\vec{r}_1)\,\phi_p(\vec{r}_1) | \phi_j(\vec{r}_2)\,\phi_j(\vec{r}_2)\big)$$

$$- \big(\phi_p(\vec{r}_1)\,\phi_j(\vec{r}_1) | \phi_j(\vec{r}_2)\,\phi_p(\vec{r}_2) \big) \Big] \quad (9.18)$$

The N spin orbitals with the lowest energy, or $N/2$ spatial orbitals, are then used to construct the N-electron Slater determinant $|\Phi_0^{\text{SCF}}\rangle$, i.e. the Hartree–Fock wavefunction or SCF determinant. These spin or spatial orbitals are therefore called the **occupied orbitals** and are denoted with the Latin indices $i, j, k, \ldots$. Solutions to the Hartree–Fock equation with higher orbital energies are called **unoccupied or virtual orbitals** and are denoted by the indices $a, b, c, \ldots$, while general spatial orbitals have indices $p, q, r, \ldots$.

The Hartree–Fock wavefunction $|\Phi_0^{\text{SCF}}\rangle$ is an eigenfunction of the Hartree–Fock Hamiltonian $\hat{F}$

$$\hat{F}|\Phi_0^{\text{SCF}}\rangle = \sum_i^N \epsilon_i |\Phi_0^{\text{SCF}}\rangle \quad (9.19)$$

However, the eigenvalue is the sum of the orbital energies of the occupied orbitals and not the Hartree–Fock energy, which is the expectation value of the full Hamilton operator $\hat{H}^{(0)}$, Eq. (9.12). When evaluated, this expectation value for the Hartree–Fock energy becomes

$$E_0^{\text{SCF}} = \sum_i^{\text{occ}} \langle \psi_i | \hat{h}^{(0)} | \psi_i \rangle + \frac{1}{2} \sum_i^{\text{occ}} \sum_j^{\text{occ}} \Big[\big(\psi_i(\vec{r}_1)\,\psi_i(\vec{r}_1) | \psi_j(\vec{r}_2)\,\psi_j(\vec{r}_2) \big)$$

$$- \big(\psi_i(\vec{r}_1)\,\psi_j(\vec{r}_1) | \psi_j(\vec{r}_2)\,\psi_i(\vec{r}_2) \big) \Big] \quad (9.20)$$

or using the definition of the orbital energies Eq. (9.17)

$$E_0^{SCF} = \sum_i^{occ} \epsilon_i - \frac{1}{2} \sum_i^{occ} \sum_j^{occ} \Big[\big(\psi_i(\vec{r}_1)\,\psi_i(\vec{r}_1) \big| \psi_j(\vec{r}_2)\,\psi_j(\vec{r}_2)\big)$$

$$- \big(\psi_i(\vec{r}_1)\,\psi_j(\vec{r}_1) \big| \psi_j(\vec{r}_2)\,\psi_i(\vec{r}_2)\big) \Big] \quad (9.21)$$

In terms of the spatial orbitals the Hartree–Fock energy is given as

$$E_0^{SCF} = 2 \sum_i^{occ} \langle \phi_i | \hat{h}^{(0)} | \phi_i \rangle + \sum_i^{occ} \sum_j^{occ} \Big[2\big(\phi_i(\vec{r}_1)\,\phi_i(\vec{r}_1) \big| \phi_j(\vec{r}_2)\,\phi_j(\vec{r}_2)\big)$$

$$- \big(\phi_i(\vec{r}_1)\,\phi_j(\vec{r}_1) \big| \phi_j(\vec{r}_2)\,\phi_i(\vec{r}_2)\big) \Big] \quad (9.22)$$

or

$$E_0^{SCF} = 2 \sum_i^{occ} \epsilon_i - \sum_i^{occ} \sum_j^{occ} \Big[2\big(\phi_i(\vec{r}_1)\,\phi_i(\vec{r}_1) \big| \phi_j(\vec{r}_2)\,\phi_j(\vec{r}_2)\big)$$

$$- \big(\phi_i(\vec{r}_1)\,\phi_j(\vec{r}_1) \big| \phi_j(\vec{r}_2)\,\phi_i(\vec{r}_2)\big) \Big] \quad (9.23)$$

In the Roothaan–Hartree–Fock approach (Roothaan, 1951) the molecular orbitals are expanded in atomic orbitals, $\{\chi_\mu\}$, Eq. (9.4). Solving the Hartree–Fock equations for this ansatz then corresponds to finding the molecular orbital coefficients.

The variational condition for the Hartree–Fock energy, Eq. (9.13), is then similarly

$$\frac{\partial E_0^{SCF}(\{c_{\mu p}\})}{\partial c_{\mu p}} = 0 \quad (9.24)$$

which applied to Eq. (9.12) implies also

$$\frac{\partial}{\partial c_{\mu p}} |\Phi_0^{SCF}(\{c_{\mu p}\})\rangle = 0 \quad (9.25)$$

9.2 Excited Determinants and Excitation Operators

Additional eigenfunctions of the Hartree–Fock Hamiltonian can be generated by including virtual orbitals in the determinants instead of some or all of the occupied orbitals. They are normally classified according to their relation to the Hartree–Fock wavefunction, also called the SCF determinant, $|\Phi_0^{SCF}\rangle$. A determinant in which one of the orbitals of the SCF determinant $|\Phi_0^{SCF}\rangle$, i.e. an occupied orbital i, is replaced by another unoccupied orbital a is called a **singly excited determinant** $|\Phi_i^a\rangle$. In **doubly excited determinants** $|\Phi_{ij}^{ab}\rangle$ two occupied orbitals i and j are replaced by two unoccupied orbitals a and b and so forth up to N-tuply excited determinants, where N is the number of electrons in the system.[2]

It is often convenient to express this in terms of general excitation operators ${}^e\hat{h}_{i_\mu}$, which act on the Hartree–Fock determinant. The excitation level is indicated by the subscript i and μ refers to a particular operator of this general excitation level. The whole set of excitation operators of level i is often collected in a column vector denoted by $\hat{\bm{h}}_i$. Alternatively, one often expresses excitation operators of a particular level in terms of the **single excitation or orbital rotation operators** $\hat{q}^\dagger_{ai}$, where the subscript ai then refers to the involved virtual and occupied orbitals.

The effect of single, double, etc. excitation operators acting on the Hartree–Fock determinant can then be expressed in both notations as

$$ {}^e\hat{h}_{1_\mu} |\Phi_0^{\text{SCF}}\rangle = \hat{q}^\dagger_{ai} |\Phi_0^{\text{SCF}}\rangle = |\Phi_i^a\rangle \tag{9.26} $$

$$ {}^e\hat{h}_{2_\mu} |\Phi_0^{\text{SCF}}\rangle = \hat{q}^\dagger_{ai}\hat{q}^\dagger_{bj} |\Phi_0^{\text{SCF}}\rangle = |\Phi_{ij}^{ab}\rangle \tag{9.27} $$

$$ \vdots $$

With these operators we can reformulate the expansion of the general approximate N-electron wavefunction $|\Phi_0\rangle$ in a linear combination of Slater determinants, Eq. (9.1), alternatively as

$$ |\Phi_0\rangle = C\left(1 + \hat{T}_1 + \hat{T}_2 + \cdots + \hat{T}_i + \cdots + \hat{T}_N\right) |\Phi_0^{\text{SCF}}\rangle \tag{9.28} $$

where the excitation $\hat{T}_i$ operators are defined as

$$ \hat{T}_1 = \sum_{ai} t_i^a \, \hat{q}^\dagger_{ai} \tag{9.29} $$

$$ \hat{T}_2 = \sum_{\substack{a>b \\ i>j}} t_{ij}^{ab} \, \hat{q}^\dagger_{ai} \, \hat{q}^\dagger_{bj} \tag{9.30} $$

$$ \vdots $$

$$ \hat{T}_i = \sum_\mu t_{i_\mu} \, {}^e\hat{h}_{i_\mu} \tag{9.31} $$

$$ \vdots $$

$$ \hat{T}_N = \sum_\mu t_{N_\mu} \, {}^e\hat{h}_{N_\mu} \tag{9.32} $$

---

[2] These commonly used names are quite unfortunate, because *a priori* these determinants are not related to the excited states of a molecule. However, in the simplest possible treatment of excited states they can be used as a crude approximation for the excited states of a molecule, as discussed in Section 10.2.

The effect of e.g. $\hat{T}_1$ acting on $|\Phi_0^{\text{SCF}}\rangle$ in Eq. (9.28) is thus

$$\hat{T}_1|\Phi_0^{\text{SCF}}\rangle = \sum_{ai} t_i^a \, \hat{q}_{ai}^\dagger |\Phi_0^{\text{SCF}}\rangle = \sum_{ai} t_i^a |\Phi_i^a\rangle \tag{9.33}$$

and similar for $\hat{T}_2$ etc. The expansion coefficients t_i^a, t_{ij}^{ab} or in general t_{i_μ} will in the following bear different names like correlation coefficients or amplitudes, depending on how they are determined.

Analogously, one also expands a general approximate N-electron bra state $\langle\Phi_0|$ in a complete set of operators acting on the bra Hartree–Fock determinant

$$\langle\Phi_0| = \langle\Phi_0^{\text{SCF}}| \left(1 + \hat{\Lambda}_1 + \hat{\Lambda}_2 + \cdots + \hat{\Lambda}_i + \cdots + \hat{\Lambda}_N\right) C \tag{9.34}$$

where the de-excitation operators $\hat{\Lambda}_i$ are defined as

$$\hat{\Lambda}_1 = \sum_{ai} \lambda_i^a \, \hat{q}_{ai} \tag{9.35}$$

$$\hat{\Lambda}_2 = \sum_{\substack{a>b \\ i>j}} \lambda_{ij}^{ab} \, \hat{q}_{ai} \, \hat{q}_{bj} \tag{9.36}$$

$$\vdots$$

$$\hat{\Lambda}_{i_\mu} = \sum_{i_\mu} \lambda_{i_\mu} \, {}^d\hat{h}_{i_\mu} \tag{9.37}$$

$$\vdots$$

$$\hat{\Lambda}_{N_\mu} = \sum_{N_\mu} \lambda_{N_\mu} \, {}^d\hat{h}_{N_\mu} \tag{9.38}$$

The $\hat{q}_{ai}$, $\hat{q}_{ai}\hat{q}_{bj}$ or general ${}^d\hat{h}_{i_\mu}$ operators are the hermitian conjugate or adjoint of the excitation operators

$$\hat{q}_{ai} = (\hat{q}_{ai}^\dagger)^\dagger \tag{9.39}$$

$$\hat{q}_{ai}\hat{q}_{bj} = (\hat{q}_{bj}^\dagger \hat{q}_{ai}^\dagger)^\dagger \tag{9.40}$$

$$\vdots$$

$${}^d\hat{h}_{i_\mu} = {}^e\hat{h}_{i_\mu}^\dagger \tag{9.41}$$

They are therefore the **single de-excitation, double de-excitation** and so forth operators and their effect is best described by letting them act on the Hartree–Fock determinant as a bra state

$$\langle\Phi_0^{\text{SCF}}|\, {}^d\hat{h}_{1_\mu} = \langle\Phi_0^{\text{SCF}}|\hat{q}_{ai} = \langle\Phi_i^a| \tag{9.42}$$

$$\langle\Phi_0^{\text{SCF}}|\, {}^d\hat{h}_{2_\mu} = \langle\Phi_0^{\text{SCF}}|\hat{q}_{bj}\hat{q}_{ai} = \langle\Phi_{ij}^{ab}| \tag{9.43}$$

$$\vdots$$

The expansion coefficients λ_{i_μ} are normally just called "λ" amplitudes and will not necessarily be related to the t_{i_μ} amplitudes.

The whole set of excitation operators $\{{}^e\hat{h}_{i_\mu}\}$ forms a complete set of operators, meaning that acting on the Hartree–Fock wavefunction they generate all excited determinants and form therefore a resolution of the identity

$$1 = |\Phi_0^{\text{SCF}}\rangle\langle\Phi_0^{\text{SCF}}| + \sum_{i_\mu} {}^e\hat{h}_{i_\mu}|\Phi_0^{\text{SCF}}\rangle\langle\Phi_0^{\text{SCF}}|\,{}^d\hat{h}_{i_\mu} \quad (9.44)$$

9.3 Multiconfigurational Self-Consistent Field Method

The multiconfigurational self-consistent field method (MCSCF) is the generalization of the Hartree–Fock method in the sense that the wavefunction

$$|\Psi_0^{(0)}\rangle \approx |\Phi_0^{\text{MCSCF}}\rangle = \sum_n |\Phi_n\rangle\, C_{n0} \quad (9.45)$$

is now a linear combination of several Slater determinants or configuration state functions, $|\Phi_n\rangle$. The latter are spin- or symmetry-adapted linear combinations of a few determinants.

The molecular orbital coefficients, $\{c_{\mu p}\}$, as well as the configuration expansion coefficients, $\{C_{n0}\}$, are simultaneously determined variationally

$$\frac{\partial E_0^{\text{MCSCF}}(\{C_{n0}\},\{c_{\mu p}\})}{\partial c_{\mu p}} = \frac{\partial E_0^{\text{MCSCF}}(\{C_{n0}\},\{c_{\mu p}\})}{\partial C_{n0}} = 0 \quad (9.46)$$

An alternative formulation of the MCSCF wavefunction makes use of an exponential unitary transformation of the orbitals and also of the configuration state functions in a given initial wavefunction $|\Phi_0^{\text{MCSCF}}\rangle$

$$|\Phi_0^{\text{MCSCF}}(\{\kappa_{pq}\},\{S_{n0}\})\rangle = e^{-\hat{\kappa}}e^{-\hat{S}}|\Phi_0^{\text{MCSCF}}\rangle \quad (9.47)$$

The operators $\hat{S}$ and $\hat{\kappa}$ are defined as

$$\hat{S} = \sum_{n\neq 0}\left(S_{n0}\,\hat{R}_{n0}^\dagger - S_{n0}^*\,\hat{R}_{0n}\right) \quad (9.48)$$

$$\hat{\kappa} = \sum_{p>q}\kappa_{pq}\left(\hat{q}_{pq}^\dagger - \hat{q}_{qp}^\dagger\right) \quad (9.49)$$

where $\{\hat{R}_{n0}^\dagger\}$ and $\{\hat{R}_{0n}\}$ are **state transfer operators**

$$\hat{R}_{n0}^\dagger = |\Phi_n^{\text{MCSCF}}\rangle\langle\Phi_0^{\text{MCSCF}}| \quad (9.50)$$

$$\hat{R}_{0n} = |\Phi_0^{\text{MCSCF}}\rangle\langle\Phi_n^{\text{MCSCF}}| \quad (9.51)$$

the wavefunctions $|\Phi_n^{\text{MCSCF}}\rangle$ are the orthogonal complement states to $|\Phi_0^{\text{MCSCF}}\rangle$ and $\hat{q}_{pq}^\dagger$ are the single excitation operators of Eq. (9.26) but here for general orbitals p and q.

The same formulation can also be used for the single determinant of the Hartree–Fock wavefunction. All coefficients S_{n0} then vanish and only the orbitals are unitarily transformed, i.e.

$$|\Phi_0^{\text{SCF}}(\{\kappa_{ai}\})\rangle = e^{-\hat{\kappa}}|\Phi_0^{\text{SCF}}\rangle \tag{9.52}$$

The anti-hermitian operator $\hat{\kappa}$ is here defined as

$$\hat{\kappa} = \sum_{ai} \kappa_{ai} \left(\hat{q}_{ai}^\dagger - \hat{q}_{ia}^\dagger\right) \tag{9.53}$$

where $\hat{q}_{ai}^\dagger$ are the proper single excitation operators, Eq. (9.26), which explains why they are also called the **orbital rotation operators**. Rotations between the virtual orbitals vanish obviously, while rotations between the occupied orbitals correspond to linear combinations of the columns in the Slater determinant and leave the value of the determinant, i.e. the wavefunction, unchanged. One of the advantages of this formulation of the Hartree–Fock wavefunction is that the orthonormality of the orbitals, Eq. (9.14), is always preserved in a unitary transformation.

The formulation of the MCSCF wavefunction in Eq. (9.47) will later be the starting point for the derivation of MCSCF linear response functions in Section 11.2.

9.4 Configuration Interaction

In the multiconfigurational self-consistent field method both the configuration coefficients $\{C_{n0}\}$ as well as the molecular orbital coefficients $\{c_{\mu p}\}$ are varied until the energy becomes minimal. If one keeps the latter fixed and optimizes the energy only with respect to the configuration coefficients $\{C_{n0}\}$, i.e.

$$\frac{\partial E_0^{\text{CI}}(\{C_{n0}\})}{\partial C_{n0}} = \frac{\partial \langle \Phi_0^{\text{CI}}(\{C_{n0}\}) | \hat{H}^{(0)} | \Phi_0^{\text{CI}}(\{C_{n0}\}) \rangle}{\partial C_{n0}} = 0 \tag{9.54}$$

one obtains the **configuration interaction (CI)** method. The wavefunction in the CI method then takes the same form as in MCSCF with the difference that the molecular orbital coefficients $\{c_{\mu p}\}$ are kept fixed

$$|\Psi_0^{(0)}\rangle \approx |\Phi_0^{\text{CI}}\rangle = \sum_n |\Phi_n\rangle\, C_{n0} \tag{9.55}$$

However, normally one expresses the CI wavefunction like in Eq. (9.28)[3] in terms of the Hartree–Fock wavefunction and the excited determinants $\{|\Phi_{i\cdots}^{a\cdots}\rangle\}$ as

$$|\Phi_0^{\text{CI}}\rangle = |\Phi_0^{\text{SCF}}\rangle C_0 + \sum_{\substack{a \\ i}} |\Phi_i^a\rangle\, C_i^a + \sum_{\substack{a>b \\ i>j}} |\Phi_{ij}^{ab}\rangle\, C_{ij}^{ab} + \sum_{\substack{a>b>c \\ i>j>k}} |\Phi_{ijk}^{abc}\rangle\, C_{ijk}^{abc} + \cdots \tag{9.56}$$

Application of the variational condition, Eq. (9.54), then leads to a set of linear equations for the configuration coefficients, which are conveniently written as the following matrix eigenvalue equation

---

[3] The C and t coefficients are then related as $C_{i\cdots}^{a\cdots} = C t_{i\cdots}^{a\cdots}$ and $C_0 = C$.

$$\hat{H}C = E^{CI}C \tag{9.57}$$

where $\hat{H}$ is the **configuration interaction** or **CI matrix**, i.e. the matrix of the unperturbed molecular Hamiltonian $\hat{H}^{(0)}$ in the basis of the Hartree–Fock and excited determinants $\{|\Phi_0^{SCF}\rangle, |\Phi_{i\ldots}^{a\ldots}\rangle\}$, and C is the configuration coefficients collected in a column vector.

Solving the eigenvalue equation for the energy E^{CI} one has to evaluate matrix elements of the Hamiltonian $\hat{H}^{(0)}$ between the Hartree–Fock Slater determinant Φ_0^{SCF} and excited Slater determinants or between two excited Slater determinants. In this context the so-called **Slater–Condon rules** become very useful, which state that

- The matrix element between two Slater determinants, which differ by only one spin orbital, i.e. a Slater determinant Φ and another Φ_p^r, where the spin orbital ψ_p is replaced by a spin orbital ψ_r, is equal to

$$\langle \Phi | \hat{H}^{(0)} | \Phi_p^r \rangle = \langle \psi_p | \hat{h}^{(0)} | \psi_r \rangle \tag{9.58}$$
$$+ \sum_s \left[\left(\psi_p(\vec{r}_1)\,\psi_r(\vec{r}_1) \big| \psi_s(\vec{r}_2)\,\psi_s(\vec{r}_2) \right) - \left(\psi_p(\vec{r}_1)\,\psi_s(\vec{r}_1) \big| \psi_s(\vec{r}_2)\,\psi_r(\vec{r}_2) \right) \right]$$

where the summation over s runs over all spin orbitals that are included in both Slater determinants.

- The matrix element between two Slater determinants, which differ by two spin orbitals, i.e. a Slater determinant Φ and another Φ_{pq}^{rs}, where the spin orbital ψ_p is replaced by a spin orbital ψ_r and the spin orbital ψ_q is replaced by a spin orbital ψ_s, is equal to

$$\langle \Phi | \hat{H}^{(0)} | \Phi_{pq}^{rs} \rangle = \left(\psi_p(\vec{r}_1)\,\psi_r(\vec{r}_1) \big| \psi_q(\vec{r}_2)\,\psi_s(\vec{r}_2) \right) - \left(\psi_p(\vec{r}_1)\,\psi_s(\vec{r}_1) \big| \psi_q(\vec{r}_2)\,\psi_r(\vec{r}_2) \right) \tag{9.59}$$

- All matrix elements between two Slater determinants, which differ by more than two spin orbitals, i.e. a Slater determinant Φ and another $\Phi_{pq\ldots}^{rs\ldots}$ vanish

$$\langle \Phi | \hat{H}^{(0)} | \Phi_{pq\ldots}^{rs\ldots} \rangle = 0 \tag{9.60}$$

However, the matrix element $\langle \Phi_0^{SCF} | \hat{H}^{(0)} | \Phi_i^a \rangle$ between the Hartree–Fock determinant and any singly excited determinant is special and vanishes

$$\langle \Phi_0^{SCF} | \hat{H}^{(0)} | \Phi_i^a \rangle = \langle \psi_i | \hat{f} | \psi_a \rangle = 0 \tag{9.61}$$

because it is equal to an occupied-virtual off-diagonal element of the Fock matrix, which is zero, if the orbitals are solutions of the Hartree–Fock equations Eq. (9.11). This is called the **Brillouin theorem**.

9.5 Møller–Plesset Perturbation Theory

One of the most widely used methods for treating the electron correlation missing in the Hartree–Fock wavefunction is Møller–Plesset (MP) perturbation theory (Møller and Plesset, 1934; Pople *et al.*, 1976), an application of Rayleigh–Schrödinger perturbation theory as outlined in Section 3.2 to the electron correlation problem. The field-free Hamiltonian $\hat{H}^{(0)}$ is thereby partitioned

$$\hat{H}^{(0)} = \hat{F} + \hat{V} \tag{9.62}$$

in the Hartree–Fock Hamiltonian $\hat{F}$ and the so-called fluctuation potential $\hat{V}$. The latter is the difference between the correct electron–electron repulsion term of the Hamiltonian, Eq. (2.101), and the effective electron repulsion as expressed by the sum over the Hartree–Fock potentials

$$\hat{V} = \sum_{i<j} \hat{g}(i,j) - \sum_i \hat{v}^{\mathrm{HF}}(i) \tag{9.63}$$

The fluctuation potential takes over the role of the perturbation and the wavefunction and energy are thus expanded in a perturbation series in $\hat{V}$

$$|\Psi_0^{(0)}\rangle \approx |\Phi_0^{\mathrm{MP}}\rangle = C\left(|\Phi_0^{\mathrm{SCF}}\rangle + |\Phi^{\mathrm{MP1}}\rangle + |\Phi^{\mathrm{MP2}}\rangle + \cdots\right) \tag{9.64}$$

$$E_0^{(0)} \approx E_0^{\mathrm{MP}} = E^{\mathrm{MP0}} + E^{\mathrm{MP1}} + E^{\mathrm{MP2}} + \cdots$$

$$= E_0^{\mathrm{SCF}} + E^{\mathrm{MP2}} + \cdots \tag{9.65}$$

where C is a normalization constant and the zeroth-order wavefunction is the single determinant SCF wavefunction, $|\Phi_0^{\mathrm{SCF}}\rangle$. The notation without and with the subscript "0" tries to distinguish between the MP second-order correction to the energy E^{MP2} and the MP second-order (MP2) energy $E_0^{\mathrm{MP2}} = E_0^{\mathrm{SCF}} + E^{\mathrm{MP2}}$. Equation (9.65) states that the Hartree–Fock energy is the sum of the zeroth and first-order energy and that one has to go at least to second order for the first correction to the Hartree–Fock energy.[4] This can easily be seen [see Exercise 9.1], if one compares the definition of the Hartree–Fock energy, Eq. (9.12), as expectation value of the total Hamiltonian with the partitioning of the Hamiltonian in MP perturbation theory, Eq. (9.63), and the perturbation theory expression for the zeroth and first-order energy, Eqs. (3.14) and (3.29). The zeroth-order energy is the eigenvalue of the Hartree–Fock Hamiltonian $\hat{F}$ and thus only the sum of the orbital energies of the occupied orbitals.

Exercise 9.1 Show that the Hartree-Fock energy is equal to $E^{\mathrm{MP0}} + E^{\mathrm{MP1}}$.

The first- and higher-order corrections to the wavefunction, Eqs. (3.23) and (3.28), are as usual expanded in the eigenfunctions of the unperturbed Hamiltonian $\hat{F}$, which

---

[4] This was actually the question investigated in the original paper by Møller and Plesset (Møller and Plesset, 1934). Møller–Plesset perturbation theory as a systematic approach for calculating electronic energies and wavefunctions was first extensively explored by Pople and coworkers (Pople *et al.*, 1976).

are the **singly excited** $|\Phi_i^a\rangle$, **doubly excited** $|\Phi_{ij}^{ab}\rangle$, ... N-tuply excited determinants, Eqs. (9.26), (9.27), *etc.* With this definition of the many-electron basis functions the first-order correction to the wavefunction, Eq. (3.27), becomes[5]

$$|\Phi^{MP1}\rangle = \sum_{n\neq 0} |\Phi_n\rangle \frac{\langle \Phi_n|\hat{V}|\Phi_0^{SCF}\rangle}{\langle \Phi_0^{SCF}|\hat{F}|\Phi_0^{SCF}\rangle - \langle \Phi_n|\hat{F}|\Phi_n\rangle} = \hat{T}_2[1]|\Phi_0^{SCF}\rangle = \sum_{\substack{a>b \\ i>j}} t_{ij}^{ab}[1]\,|\Phi_{ij}^{ab}\rangle$$

(9.66)

because the matrix element $\langle \Phi_n|\hat{V}|\Phi_0^{SCF}\rangle$ vanishes for all but doubly excited determinants $|\Phi_{ij}^{ab}\rangle$ [see Exercise 9.2]. The **first-order doubles correlation coefficients** are thus given as

$$t_{ij}^{ab}[1] = \frac{\bigl(\phi_a(\vec{r}_1)\,\phi_i(\vec{r}_1)\bigl|\phi_b(\vec{r}_2)\,\phi_j(\vec{r}_2)\bigr)}{\epsilon_i + \epsilon_j - \epsilon_a - \epsilon_b}$$

(9.67)

where the two-electron repulsion integral is defined in Eq. (9.9). Correlation coefficients is the name commonly given in MP perturbation theory to the determinant expansion coefficients C_{n0} of the approximate wavefunction Eq. (9.1).

Exercise 9.2 Show that only doubly excited determinants can contribute to the first-order correction to the wavefunction.

Hint: Use the Slater–Condon rules in Eqs. (9.58) to (9.60) and Brillouin's theorem Eq. (9.61).

The second-order MP correction to the energy is then given as an asymmetric expectation value

$$E^{MP2} = \langle \Phi_0^{SCF}|\hat{V}|\Phi^{MP1}\rangle = \langle \Phi_0^{SCF}|\hat{V}\hat{T}_2[1]|\Phi_0^{SCF}\rangle$$
$$= \frac{1}{2}\sum_{\substack{ab \\ ij}} \bigl(\phi_i(\vec{r}_1)\,\phi_a(\vec{r}_1)\bigl|\phi_j(\vec{r}_2)\,\phi_b(\vec{r}_2)\bigr)\bigl\{4\,t_{ij}^{ab}[1] - 2\,t_{ji}^{ab}[1]\bigr\}$$

(9.68)

The second-order MP correction to the wavefunction

$$|\Phi^{MP2}\rangle = \sum_{n\neq 0} |\Phi_n\rangle \Biggl(\frac{\langle \Phi_n|\hat{V}|\Phi^{MP1}\rangle}{\langle \Phi_0^{SCF}|\hat{F}|\Phi_0^{SCF}\rangle - \langle \Phi_n|\hat{F}|\Phi_n\rangle}$$
$$- \langle \Phi_0^{SCF}|\hat{V}|\Phi_0^{SCF}\rangle \frac{\langle \Phi_n\,|\,\Phi^{MP1}\rangle}{\langle \Phi_0^{SCF}|\hat{F}|\Phi_0^{SCF}\rangle - \langle \Phi_n|\hat{F}|\Phi_n\rangle} \Biggr)$$

(9.69)

---

[5] The "[1]" in $\hat{T}_2[1]$ or $t_{ij}^{ab}[1]$ indicates here and in the following that one is dealing with MP1 correlation coefficients. Correspondingly, "[2]" will denote MP2 correlation coefficients.

consists of singly, doubly, triply and quadruply excited determinants

$$|\Phi^{\text{MP2}}\rangle = \sum_{ai} t_i^a[2]\,|\Phi_i^a\rangle + \sum_{\substack{a>b\\i>j}} t_{ij}^{ab}[2]\,|\Phi_{ij}^{ab}\rangle + \sum_{\substack{a>b>c\\i>j>k}} t_{ijk}^{abc}[2]\,|\Phi_{ijk}^{abc}\rangle + \sum_{\substack{a>b>c>d\\i>j>k>l}} t_{ijkl}^{abcd}[2]\,|\Phi_{ijkl}^{abcd}\rangle \tag{9.70}$$

where, e.g., the second-order singles correlation coefficients are given as

$$t_i^a[2] = \frac{1}{\sqrt{2}}\frac{1}{\epsilon_i - \epsilon_a}\left(\sum_{jbc}\left(\phi_a(\vec{r}_1)\,\phi_b(\vec{r}_1)|\phi_j(\vec{r}_2)\,\phi_c(\vec{r}_2)\right)\left\{4\,t_{ij}^{bc}[1] - 2\,t_{ji}^{bc}[1]\right\}\right.$$

$$\left.- \sum_{jkb}\left(\phi_k(\vec{r}_1)\,\phi_i(\vec{r}_1)|\phi_j(\vec{r}_2)\,\phi_b(\vec{r}_2)\right)\left\{4\,t_{jk}^{ba}[1] - 2\,t_{kj}^{ba}[1]\right\}\right) \tag{9.71}$$

9.6 Coupled Cluster Theory

In coupled cluster theory (CC) the N-electron wavefunction is given as

$$|\Psi_0^{(0)}\rangle \approx |\Phi_0^{\text{CC}}\rangle = e^{\hat{T}}|\Phi_0^{\text{SCF}}\rangle \tag{9.72}$$

where the **cluster operator** $\hat{T}$ consists of single, double and so forth excitation operators

$$\hat{T} = \hat{T}_1 + \hat{T}_2 + \cdots + \hat{T}_i + \cdots + \hat{T}_N \tag{9.73}$$

defined in Eqs. (9.29) to (9.32). The determinant expansion coefficients t_i^a, t_{ij}^{ab}, t_{i_μ} and so forth of the singly, doubly, i-tuply, etc. excited determinants are in coupled cluster theory called **singles**, **doubles**, **i-tuple**, etc. **amplitudes**. Different coupled cluster methods are obtained by truncating the expansion of $\hat{T}$ in Eq. (9.73). In the popular **coupled cluster singles and doubles model (CCSD)**, the cluster operator consists of $\hat{T}_1$ and $\hat{T}_2$.

Inserting the coupled cluster wavefunction, Eq. (9.72), in the Schrödinger equation gives the coupled cluster Schrödinger equation

$$\hat{H}^{(0)}|\Phi_0^{\text{CC}}\rangle = E_0^{\text{CC}}|\Phi_0^{\text{CC}}\rangle \tag{9.74}$$

Similarly to the MP2 energy, Eq. (9.68), the coupled cluster energy is obtained as a transition expectation value by projecting against the Hartree–Fock wavefunction

$$E_0^{\text{CC}} = \langle\Phi_0^{\text{SCF}}|\hat{H}^{(0)}|\Phi_0^{\text{CC}}\rangle \tag{9.75}$$

where it was used that $|\Phi_0^{\text{SCF}}\rangle$ is orthogonal to all excited determinants. Alternatively, one could also have projected against $\langle\Phi_0^{\text{SCF}}|e^{-T}$

$$E_0^{\text{CC}} = \langle\Phi_0^{\text{SCF}}|e^{-\hat{T}}\hat{H}^{(0)}|\Phi_0^{\text{CC}}\rangle \tag{9.76}$$

Again, similarly to Møller–Plesset perturbation theory, the amplitudes can be obtained by solving the set of coupled non-linear equations that are generated by projecting the coupled cluster Schrödinger equation against $\langle \Phi_i^a | e^{-\hat{T}}$, $\langle \Phi_{ij}^{ab} | e^{-\hat{T}}$, etc.

$$\langle \Phi_i^a | e^{-\hat{T}} \hat{H}^{(0)} | \Phi_0^{CC} \rangle = 0 \tag{9.77}$$

$$\langle \Phi_{ij}^{ab} | e^{-\hat{T}} \hat{H}^{(0)} | \Phi_0^{CC} \rangle = 0 \tag{9.78}$$

$$\vdots$$

$$\langle \Phi^{SCF} | {}^d\hat{h}_{N_\mu} e^{-\hat{T}} \hat{H}^{(0)} | \Phi_0^{CC} \rangle = 0 \tag{9.79}$$

where the right-hand sides vanish because of

$$\langle \Phi^{SCF} | {}^d\hat{h}_{i_\mu} e^{-\hat{T}} | \Phi_0^{CC} \rangle = \langle \Phi^{SCF} | {}^d\hat{h}_{i_\mu} e^{-\hat{T}} e^{\hat{T}} | \Phi_0^{SCF} \rangle = \langle \Phi^{SCF} | {}^d\hat{h}_{i_\mu} | \Phi_0^{SCF} \rangle = 0 \tag{9.80}$$

The coupled cluster amplitude equations are often collectively called the **coupled cluster vector function** e_i with elements,

$$e_{i_\mu} = \langle \Phi^{SCF} | {}^d\hat{h}_{i_\mu} e^{-\hat{T}} \hat{H}^{(0)} | \Phi_0^{CC} \rangle = 0 \tag{9.81}$$

For the popular **CCSD** model recalling the definition of an exponential operator Eq. (3.71) and using the Møller–Plesset partitioning of the Hamiltonian, Eq. (9.62), the expression for the energy and the amplitude equations then become

$$E_0^{CCSD} = \langle \Phi_0^{SCF} | e^{-\hat{T}} \hat{H}^{(0)} | \Phi_{CCSD} \rangle = \langle \Phi_0^{SCF} | e^{-(\hat{T}_1+\hat{T}_2)} \hat{H}^{(0)} e^{-(\hat{T}_1+\hat{T}_2)} | \Phi_0^{SCF} \rangle$$

$$= E_0^{SCF} + \langle \Phi_0^{SCF} | \hat{V}(\frac{1}{2}\hat{T}_1^2 + \hat{T}_2) | \Phi_0^{SCF} \rangle \tag{9.82}$$

and

$$\langle \Phi_i^a | e^{-(\hat{T}_1+\hat{T}_2)} \hat{H}^{(0)} | \Phi_0^{CCSD} \rangle = \langle \Phi_i^a | [\hat{F}, \hat{T}_1] + \hat{V}^{T_1} + [\hat{V}^{T_1}, \hat{T}_2] | \Phi_0^{SCF} \rangle = 0 \tag{9.83}$$

$$\langle \Phi_{ij}^{ab} | e^{-(\hat{T}_1+\hat{T}_2)} \hat{H}^{(0)} | \Phi_0^{CCSD} \rangle$$
$$= \langle \Phi_{ij}^{ab} | [\hat{F}, \hat{T}_2] + \hat{V}^{T_1} + [\hat{V}^{T_1}, \hat{T}_2] + \frac{1}{2}[[\hat{V}^{T_1}, \hat{T}_2], \hat{T}_2] | \Phi_0^{SCF} \rangle = 0 \tag{9.84}$$

where we have introduced the $\hat{T}_1$ transformed operators (Koch et al., 1994)

$$\hat{O}^{T_1} = e^{-\hat{T}_1} \hat{O} \, e^{\hat{T}_1} \tag{9.85}$$

in analogy to the interaction picture in Eq. (3.70).

In the **CC2** model (Christiansen et al., 1995b) these amplitude equations are approximated based on Møller-Plesset perturbation theory arguments, however, with the slight difference that the single excitations and thus the $\hat{T}_1$ operator are treated as being of zeroth order, while in normal Møller–Plesset perturbation theory they enter first in the second-order wavefunction, Eq. (9.69), due to the Brillouin theorem, Eq. (9.61), and thus are of second order. The reason for this choice is simply that contrary to MP2 the CC2 method was constructed for the calculation of not only electronic energies and wavefunctions but primarily for the calculation of molecular properties via the response theory approach. In the presence of an external perturbation, however, the single excitations will already enter the wavefunction in zeroth-order, as will be shown later in Section 11.1. On the other hand, the double excitation and thus the $\hat{T}_2$ operator are treated as first order like in Møller–Plesset perturbation theory, Eq. (9.66). The amplitude equations to second order, defined in this way, i.e. the CC2 amplitude equations then become

$$\langle \Phi_i^a | e^{-(\hat{T}_1+\hat{T}_2)} \hat{H}^{(0)} | \Phi_0^{CC2} \rangle = \langle \Phi_i^a | [\hat{F}, \hat{T}_1] + \hat{V}^{T_1} + [\hat{V}^{T_1}, \hat{T}_2] | \Phi_0^{SCF} \rangle = 0 \quad (9.86)$$

$$\langle \Phi_{ij}^{ab} | e^{-(\hat{T}_1+\hat{T}_2)} \hat{H}^{(0)} | \Phi_0^{CC2} \rangle = \langle \Phi_{ij}^{ab} | [\hat{F}, \hat{T}_2] + \hat{V}^{T_1} | \Phi_0^{SCF} \rangle = 0 \quad (9.87)$$

The next-higher method would be **CCSDT** (Noga and Bartlett, 1987, 1988; Scuseria and Schaefer, 1988), where triple excitations $\hat{T}_3$ are also included in the wavefunction. This is therefore a rather expensive method and not yet employed on a regular basis. However, one can make the same type of approximation to the equations for the triples amplitudes, i.e. the triples coupled cluster vector function e_3, as were made in CC2 to the doubles amplitude equations. This leads then to the **CC3** model (Christiansen et al., 1995a). Nevertheless, one still has to solve the equations for the triples amplitudes iteratively. An non-iterative alternative is the **CCSD(T)** model (Raghavachari et al., 1989), where the triples correction to the CCSD energy is obtained from the triples contribution to the fourth-order Møller–Plesset perturbation theory energy and from one fifth-order term describing the coupling between singles and triples. Both contributions are, however, evaluated with the CCSD amplitudes.

9.7 The Hellmann–Feynman Theorem for Approximate Wavefunctions

The approximate wavefunctions $|\Phi_0\rangle$ described in the previous sections of this chapter depend on molecular orbital coefficients $\{c_{\mu p}\}$ and possibly also on some kind of configuration or determinant coefficients $\{C_{n0}\}$ that together are here denoted as $\{C_i\}$. The energy of a molecule in all these approximate methods can be expressed as the following asymmetric expectation value

$$E_0 = \langle \Phi_0' | \hat{H}^{(0)} | \Phi_0 \rangle \quad (9.88)$$

In the case of the variational methods, SCF, MCSCF and CI, $|\Phi_0'\rangle = |\Phi_0\rangle$ and we have the normal expectation value. For the non-variational methods such as Møller–Plesset perturbation or coupled cluster theory, the energy is calculated as a transition expectation value, where $|\Phi_0'\rangle = |\Phi_0^{SCF}\rangle$.

Let us now consider again the case of a Hamiltonian $\hat{H}(\lambda)$, which depends on a perturbation symbolized by the real parameter λ. Both sets of wavefunction coefficients will depend on λ and the wavefunction thus indirectly also

$$|\Phi_0(\lambda)\rangle = |\Phi_0(\{C_i(\lambda)\})\rangle \quad (9.89)$$

In addition to the wavefunction parameters, $\{c_{\mu p}\}$ and $\{C_{0k}\}$, also the basis functions χ_μ can depend on the perturbation. This will be the case when the perturbation corresponds to a change in the geometry or when perturbation-dependent basis functions such as the gauge including atomic orbitals (GIAO) (London, 1937) are used. In the following, however, we will ignore the dependence of the basis functions on the perturbation.

The derivative of the electronic energy $E_0(\lambda, \{C_i(\lambda)\})$ with respect to the real parameter λ is then

$$\frac{dE_0(\lambda, \{C_i(\lambda)\})}{d\lambda} = \frac{\partial E_0(\lambda, \{C_i(\lambda)\})}{\partial \lambda} + \sum_i \left(\frac{\partial (E_0(\lambda, \{C_i(\lambda)\}))}{\partial C_i(\lambda)}\right)\left(\frac{\partial C_i(\lambda)}{\partial \lambda}\right) \quad (9.90)$$

or

$$\frac{dE_0(\lambda, \{C_i(\lambda), C_i'(\lambda)\})}{d\lambda} = \langle \Phi_0'(\{C_i'(\lambda)\})| \frac{\partial \hat{H}(\lambda)}{\partial \lambda} |\Phi_0(\{C_i(\lambda)\})\rangle$$

$$+ \sum_i \langle \frac{\partial \Phi_0'(\{C_i'(\lambda)\})}{\partial C_i'(\lambda)} | \hat{H}(\lambda) | \Phi_0(\{C_i(\lambda)\})\rangle \left(\frac{\partial C_i'(\lambda)}{\partial \lambda}\right) \quad (9.91)$$

$$+ \sum_i \langle \Phi_0'(\{C_i'(\lambda)\})| \hat{H}(\lambda) | \frac{\partial \Phi_0(\{C_i(\lambda)\})}{\partial C_i(\lambda)}\rangle \left(\frac{\partial C_i(\lambda)}{\partial \lambda}\right)$$

in the case that the energy can be written as the asymmetric expectation value, Eq. (9.88). The molecular orbital coefficients will actually be the same in $|\Phi_0'\rangle$ and $|\Phi_0\rangle$, i.e. $c'_{\mu p} = c_{\mu p}$ for MPn and CC wavefunctions.

If the wavefunction is variationally optimised with respect to all parameters, i.e.

$$\frac{\partial E_0(\lambda, \{C_i(\lambda), C_i'(\lambda)\})}{\partial C_i(\lambda)} = \frac{\partial E_0(\lambda, \{C_i(\lambda), C_i'(\lambda)\})}{\partial C_i'(\lambda)} = 0 \quad (9.92)$$

and thus

$$\frac{\partial}{\partial C_i(\lambda)} |\Phi_0(\{C_i(\lambda)\})\rangle = \frac{\partial}{\partial C_i'(\lambda)} \langle \Phi_0(\{C_i'(\lambda)\})| = 0 \quad (9.93)$$

the Hellmann–Feynman theorem is fulfilled again

$$\frac{dE_0(\lambda, \{C_i(\lambda), C_i'(\lambda)\})}{d\lambda} = \frac{\partial E_0(\lambda, \{C_i(\lambda), C_i'(\lambda)\})}{\partial \lambda}$$

$$= \langle \Phi_0'(\{C_i'(\lambda)\})| \frac{\partial \hat{H}(\lambda)}{\partial \lambda} |\Phi_0(\{C_i(\lambda)\})\rangle \quad (9.94)$$

This is always the case for a SCF and MCSCF wavefunction, because they are optimized with respect to all wavefunction parameters. Truncated CI wavefunctions, by contrast, are not variationally optimized with respect to the molecular orbital coefficients. The Hellmann–Feynman theorem is therefore satisfied only in the limit of a full CI wavefunction, when the molecular orbital coefficients are redundant.

In non-variational approaches such as Møller–Plesset perturbation theory or coupled cluster methods the wavefunction is not at all variationally optimized. However, it is possible to choose $\langle \Phi_0'|$ in such a way that the Hellmann–Feynman theorem is fulfilled to a certain extent, while the transition expectation value in Eq. (9.88) still gives the energy.

Arponen (1983) defined such a transition expectation value for the coupled cluster energy

$$E_0^{CC,\Lambda} = \langle \Phi_0^\Lambda | \hat{H}^{(0)} | \Phi_0^{CC} \rangle \qquad (9.95)$$

between the coupled cluster state and a dual bra or "Λ" state $\langle \Phi_0^\Lambda |$ defined as

$$\langle \Phi_0^\Lambda | = \langle \Phi_0^{SCF} | (1 + \hat{\Lambda}) \, e^{-\hat{T}} \qquad (9.96)$$

The $\hat{\Lambda}$ operator is defined in complete analogy to the $\hat{T}$ operator, Eq. (9.73) as the sum of the de-excitation operators

$$\hat{\Lambda} = \hat{\Lambda}_1 + \hat{\Lambda}_2 + \cdots + \hat{\Lambda}_i + \cdots + \hat{\Lambda}_N \qquad (9.97)$$

given in Eqs. (9.35) to (9.38).

The transition expectation value in Eq. (9.95) is an example of the much more general concept of Lagrangians (Jørgensen and Helgaker, 1988; Helgaker and Jørgensen, 1988; Helgaker et al., 1989; Christiansen et al., 1998b) for non-variational wavefunctions. A Lagrangian is the normal expression for the energy of a non-variational wavefunction augmented with constraints that are simply the equations from which the wavefunction parameters are determined, multiplied with some Lagrangian multipliers. The goal with this formulation of the Lagrangian is that it can be made stationary with respect to the wavefunction parameters contrary to the normal energy expression of non-variational wavefunctions.

In the case of the coupled cluster wavefunction the equations for the wavefunction parameters, i.e. for the coupled cluster amplitudes t_{i_μ}, are simply the equations for the coupled cluster vector function e_{i_μ} in Eq. (9.81). The constraints are then $e_{i_\mu} = 0$ and the coupled cluster Langrangian L_0^{CC} (Christiansen et al., 1995a, 1998b) is given as

$$L_0^{CC} = E_0^{CC} + \sum_{i_\mu} \lambda_{i_\mu} e_{i_\mu} \qquad (9.98)$$

where λ_{i_μ} are Lagrangian multipliers. Comparison with Eq. (9.95) [see Exercise 9.3] shows that Arponen's transition expectation value is just an alternative way of writing the coupled cluster Langrangian

$$E_0^{CC,\Lambda} = L_0^{CC} \qquad (9.99)$$

and that the Lagrangian multipliers in the coupled cluster Lagrangian are the λ_{i_μ} amplitudes defined in Eqs. (9.35) to (9.38). The definition of the dual bra or "Λ" state $\langle \Phi_0^\Lambda |$ is thus just a convenient way to build these constraints into an expectation value expression. The variational condition with respect to the λ_{i_μ} amplitudes is trivially fulfilled for the coupled cluster Lagrangian $E_0^{CC,\Lambda} = L_0^{CC}$

$$\frac{\partial L_0^{CC}}{\partial \lambda_{i_\mu}} = \frac{\partial E_0^{CC,\Lambda}}{\partial \lambda_{i_\mu}} = 0 \qquad (9.100)$$

because the derivative with respect to the λ_{i_μ} amplitudes is

$$\frac{\partial \langle \Phi_0^\Lambda | \hat{H}^{(0)} | \Phi_{CC} \rangle}{\partial \lambda_{i_\mu}} = \langle \Phi^{SCF} | {}^d h_{i_\mu} e^{-\hat{T}} \hat{H}^{(0)} | \Phi_0^{CC} \rangle = e_{i_\mu} = 0 \qquad (9.101)$$

i.e. the coupled cluster vector function e_{i_μ}, Eq. (9.81), and thus the equations that the coupled cluster amplitudes anyway fulfill. On the other hand, equations for the CC λ_{i_μ} amplitudes are obtained from the condition that the coupled cluster Lagrangian $E_0^{CC,\Lambda} = L_0^{CC}$ is stationary with respect to the t_{i_μ} amplitudes, i.e.

$$\frac{\partial L_0^{CC}}{\partial t_{i_\mu}} = \frac{\partial E_0^{CC,\Lambda}}{\partial t_{i_\mu}} = \frac{\partial \langle \Phi_0^\Lambda | \hat{H}^{(0)} | \Phi_{CC} \rangle}{\partial t_{i_\mu}} = \langle \Phi_0^\Lambda | [\hat{H}^{(0)}, {}^e h_{i_\mu}] | \Phi_{CC} \rangle = 0 \quad (9.102)$$

Exercise 9.3 Insert the expressions for the "Λ" state $\langle \Phi_0^\Lambda |$, Eq. (9.96), and the $\hat{\Lambda}$ operator, Eq. (9.97), in the expression for Arponen's transition expectation value, Eq. (9.95), and show that is equal to the coupled cluster Langrangian L_0^{CC} in Eq. (9.98).

The transition expectation value $E_0^{CC,\Lambda}$, or coupled cluster Lagrangian L_0^{CC}, is thus stationary with respect to the configuration or determinant coefficients and therefore satisfies partially the Hellmann–Feynman theorem

$$\frac{\partial E_0^{CC,\Lambda}}{\partial \lambda} = \frac{\partial L_0^{CC}}{\partial \lambda} = \frac{\partial \langle \Phi_0^\Lambda(\lambda) | \hat{H}(\lambda) | \Phi_0^{CC}(\lambda) \rangle}{\partial \lambda} = \langle \Phi_0^\Lambda | \frac{\partial \hat{H}(\lambda)}{\partial \lambda} | \Phi_0^{CC} \rangle \quad (9.103)$$

which implies that identical expressions for first-order properties can be obtained as first derivatives of the energy or as asymmetric expectation value with the Λ state.

Similar transition expectation values can also be defined for other non-variational methods like Møller–Plesset perturbation theory, where one defines a Lagrangian by adding the equations for the correlation coefficients as extra conditions multiplied with Lagrangian multipliers to the respective MP energy expression (Hättig and Heß, 1995; Aiga and Itoh, 1996).

In Section 12.2 it will be discussed that this approach for the calculation of expectation values is called the **unrelaxed** method, because the conditions for the molecular orbital coefficients were not included as additional constraints in the coupled cluster Lagrangian given in Eq. (9.95) or Eq. (9.98). A coupled cluster Lagrangian including orbital relaxation takes the following form

$$L_0^{CC,\text{relax}} = \langle \Phi_0^{SCF} | e^{-\hat{\kappa}} \hat{H}^{(0)} e^{\hat{\kappa}} e^{\hat{T}} | \Phi_0^{SCF} \rangle$$
$$+ \sum_{i_\mu} \lambda_{i_\mu} \langle \Phi_0^{SCF} | {}^d\hat{h}_{i_\mu} e^{-\hat{T}} e^{-\hat{\kappa}} \hat{H}^{(0)} e^{\hat{\kappa}} e^{\hat{T}} | \Phi_0^{SCF} \rangle$$
$$+ \sum_{pq} \tau_{pq} \langle \Phi_0^{SCF} | \left[q_{pq}^\dagger, e^{-\hat{\kappa}} \hat{H}^{(0)} e^{\hat{\kappa}} \right] | \Phi_0^{SCF} \rangle \quad (9.104)$$

where $\hat{\kappa}$ is the orbital rotation operator from Eq. (9.49) and the τ_{pq} coefficients are the Lagrangian multipliers for the conditions on the molecular orbital coefficients, i.e. the Brillouin theorem Eq. (9.61).

Although it is not directly related to the Hellmann–Feynman theorem, we list also the second derivatives of the energy for an approximate wavefunction here

$$\frac{d^2}{d\lambda^2} E_0(\lambda, \{C_i(\lambda)\}) = \frac{\partial^2 E_0(\lambda, \{C_i(\lambda)\})}{\partial \lambda^2} + 2 \sum_i \frac{\partial^2 E_0(\lambda, \{C_i(\lambda)\})}{\partial \lambda\, \partial C_i} \frac{\partial C_i}{\partial \lambda} \qquad (9.105)$$

$$+ \sum_i \frac{\partial^2 E_0(\lambda, \{C_i(\lambda)\})}{\partial C_i^2} \left(\frac{\partial C_i}{\partial \lambda}\right)^2 + \sum_i \frac{\partial E_0(\lambda, \{C_i(\lambda)\})}{\partial C_i} \frac{\partial^2 C_i}{\partial \lambda^2}$$

For a variational energy the last term vanishes again and the second derivative of the energy for variational methods is given as

$$\frac{d^2}{d\lambda^2} E_0(\lambda, \{C_i(\lambda)\}) = \frac{\partial^2 E_0(\lambda, \{C_i(\lambda)\})}{\partial \lambda^2} + 2 \sum_i \frac{\partial^2 E_0(\lambda, \{C_i(\lambda)\})}{\partial \lambda\, \partial C_i} \frac{\partial C_i}{\partial \lambda}$$

$$+ \sum_i \frac{\partial^2 E_0(\lambda, \{C_i(\lambda)\})}{\partial C_i^2} \left(\frac{\partial C_i}{\partial \lambda}\right)^2 \qquad (9.106)$$

9.8 Approximate Density Matrices

In Section 2.3 the electron density $P(\vec{r})$ and a reduced one-electron density matrix $P(\vec{r}, \vec{r}\,')$ were defined in Eqs. (2.17), (2.20) and (2.22). In Section 3.5 it was then shown how the electron density can be used in the calculation of expectation values.

In the present section we want to derive now approximations to the electron density and reduced one-electron density matrix using two of the approximate wavefunctions presented in the previous sections: the SCF wavefunction $|\Phi_0^{SCF}\rangle$ and the Møller–Plesset perturbation theory wavefunction through second order, $|\Phi_0^{MP1}\rangle + |\Phi_0^{MP2}\rangle$.

But before doing so we want to introduce two more entities that are very convenient in the context of approximate methods, the **density matrices** D_{pq} and $D_{\mu\nu}$ in the molecular orbital and atomic orbital basis. They are the coefficients in the expansion of the electron density $P(\vec{r})$ in the set of molecular orbitals $\{\phi_p\}$ or atomic orbitals $\{\chi_\mu\}$

$$P(\vec{r}) = \sum_{pq} \phi_p(\vec{r})\, D_{pq}\, \phi_q^*(\vec{r}) \qquad (9.107)$$

$$= \sum_{\mu\nu} \chi_\mu(\vec{r})\, D_{\mu\nu}\, \chi_\nu^*(\vec{r}) \qquad (9.108)$$

and in the expansion of the reduced one-electron density matrix $P(\vec{r}, \vec{r}\,')$

$$P(\vec{r}, \vec{r}\,') = \sum_{pq} \phi_p(\vec{r})\, D_{pq}\, \phi_q^*(\vec{r}\,') \qquad (9.109)$$

$$= \sum_{\mu\nu} \chi_\mu(\vec{r})\, D_{\mu\nu}\, \chi_\nu^*(\vec{r}\,') \qquad (9.110)$$

The electron density $P^{\text{SCF}}(\vec{r})$ for an SCF wavefunction can be obtained by simple application of Eq. (2.20) and the Slater–Condon rules Eq. (9.58)

$$P^{\text{SCF}}(\vec{r}) = \langle \Phi_0^{\text{SCF}} | \hat{P}(\vec{r}) | \Phi_0^{\text{SCF}} \rangle = \sum_i^{N/2} 2 \langle \phi_i(\vec{r}_1) | \delta(\vec{r}_1 - \vec{r}) | \phi_i(\vec{r}_1) \rangle = \sum_i^{N/2} 2\, \phi_i^*(\vec{r})\, \phi_i(\vec{r})$$

(9.111)

Comparison with Eq. (9.107) shows that the SCF density matrix in the molecular orbital basis is given as

$$D_{ij}^{\text{SCF}} = 2\, \delta_{ij}, \qquad D_{ia}^{\text{SCF}} = 0, \qquad D_{ab}^{\text{SCF}} = 0 \qquad (9.112)$$

Transforming the molecular orbitals in Eq. (9.111) to the atomic orbital basis

$$P^{\text{SCF}}(\vec{r}) = \sum_i^{N/2} 2\, \phi_i^*(\vec{r})\, \phi_i(\vec{r}) = \sum_i^{N/2} \sum_{\mu\nu} 2\, \chi_\nu^*(\vec{r})\, c_{\nu i}^*\, \chi_\mu(\vec{r})\, c_{\mu i} \qquad (9.113)$$

and comparison with Eq. (9.108) gives the SCF density matrix in the atomic orbital basis

$$D_{\mu\nu}^{\text{SCF}} = 2 \sum_i^{N/2} c_{\nu i}^*\, c_{\mu i} \qquad (9.114)$$

Similarly, the MP second-order correction to the electron density can be defined as (Jensen et al., 1988a,b)

$$P^{MP2}(\vec{r}) = \langle \Phi^{\text{MP1}} | \hat{P}(\vec{r}) | \Phi^{\text{MP1}} \rangle + \langle \Phi_0^{\text{SCF}} | \hat{P}(\vec{r}) | \Phi^{\text{MP2}} \rangle + \langle \Phi^{\text{MP2}} | \hat{P}(\vec{r}) | \Phi_0^{\text{SCF}} \rangle$$

$$= \sum_{pq} \phi_p(\vec{r})\, \phi_q^*(\vec{r})\, D_{pq}^{MP2} \qquad (9.115)$$

which gives for the second-order correction to the density matrix in the molecular orbital basis

$$D_{ij}^{MP2} = -\sum_{abk} t_{ik}^{ab}[1]\, \{4\, t_{jk}^{ab}[1] - 2\, t_{kj}^{ab}[1]\} \qquad (9.116)$$

$$D_{ab}^{MP2} = \sum_{cij} t_{ij}^{ac}[1]\, \{4\, t_{ij}^{bc}[1] - 2\, t_{ji}^{bc}[1]\} \qquad (9.117)$$

$$D_{ia}^{MP2} = D_{ai}^{MP2} = \sqrt{2}\, t_i^a[2] \qquad (9.118)$$

This is often called the **unrelaxed second-order correction to the density matrix** in order to distinguish it from the relaxed density matrix, which will be defined in Section 12.2.

In Section 3.5 also the perturbed and in particular the first-order electron density $P_\alpha^{(1)}(\vec{r})$ and first-order reduced one-electron density matrix $P_\alpha^{(1)}(\vec{r}, \vec{r}')$ were introduced. For these we can define corresponding **first-order density matrices** $D_{\alpha,pq}^{(1)}$ and $D_{\alpha,\mu\nu}^{(1)}$

in the molecular orbital and atomic orbital basis as coefficients in the expansion of the first-order reduced one-electron density matrix $P_\alpha^{(1)}(\vec{r},\vec{r}\,')$

$$P_\alpha^{(1)}(\vec{r},\vec{r}\,') = \sum_{pq} \phi_p(\vec{r}) \, D_{\alpha,pq}^{(1)} \, \phi_q^*(\vec{r}\,') \qquad (9.119)$$

$$= \sum_{\mu\nu} \chi_\mu(\vec{r}) \, D_{\alpha,\mu\nu}^{(1)} \, \chi_\nu^*(\vec{r}\,') \qquad (9.120)$$

Approximate expressions for these first-order density matrices will be derived later in Chapters 10 to 12.

9.9 Further Reading

Electronic Structure Theory

A. Szabo and N. S. Ostlund, *Modern Quantum Chemistry: Introduction to Advanced Electronic Structure Theory*, McGraw-Hill, New York (1989).

T. Helgaker, P. Jørgensen and J. Olsen, *Molecular Electronic-Structure Theory*, John Wiley & Sons, Chichester (2000).

F. Jensen, *Introduction to Computational Chemistry*, 2nd edn John Wiley & Sons, Chichester (2007): Chapters 3 and 4.

10
Approximations to Exact Perturbation and Response Theory Expressions

In the beginning of this part it was discussed that one way to obtain practical expressions for the electromagnetic properties, which can be implemented in computer programs, is to make approximations to the "exact" expressions derived with time-independent and time-dependent perturbation theory in Chapter 3.

In Section 10.1 we will illustrate this for ground-state expectation values such as Eq. (4.25) and many others and in Section 10.2 for sum-over-states expressions such as Eq. (4.74) and many others. In the rest of the chapter we will discuss methods in which approximations are made to the exact matrix representation of the linear response function or polarization propagator given in Eq. (3.159). This equation is exact as long as a complete set of excitation and de-excitation operators $\{h_n\}$ is used and the reference state $|\Psi_0^{(0)}\rangle$ is an eigenfunction of the unperturbed Hamiltonian. Approximate polarization propagator methods are thus obtained by truncating the set of operators and by using an approximate reference state $|\Psi_0^{(0)}\rangle$. Møller–Plesset (MP) perturbation theory, linearized coupled cluster and multiconfigurational self-consistent field (MCSCF) wavefunctions are commonly employed as approximate reference states in polarization propagator approximations and will be discussed in Sections 10.3 and 10.4.

10.1 Ground-State Expectation Values

All approximate *ab initio* methods presented in Chapter 9 are based on Slater determinants built with molecular orbitals. In this section we will therefore derive expressions for the expectation value $\langle \Phi_0 | \hat{O} | \Phi_0 \rangle$ of a general one-electron but spin-free operator $\hat{O} = \sum_i \hat{o}(i)$ with an approximate closed-shell wavefunction $|\Phi_0\rangle$ in terms of the molecular spatial orbitals $\{\phi_p\}$.

Starting from the expression for the ground-state expectation value, Eq. (3.46), as integral over the reduced one-electron density matrix $P(\vec{r}, \vec{r}')$ and using the expansion of it in molecular orbitals, Eq. (9.109), we obtain an expression for the expectation value as a contraction of the density matrix in the molecular orbital basis and molecular property integrals

$$\langle \Phi_0 | \hat{O} | \Phi_0 \rangle = \sum_{pq} D_{pq} \int \phi_q^*(\vec{r}_1) \hat{o}(1) \, \phi_p(\vec{r}_1) \, d\vec{r}_1 = \sum_{pq} D_{pq} \, \langle \phi_q | \hat{o} | \phi_p \rangle \qquad (10.1)$$

or alternatively with the AO density matrix

$$\langle \Phi_0 | \hat{O} | \Phi_0 \rangle = \sum_{\mu\nu} D_{\mu\nu} \langle \chi_\nu | \hat{o} | \chi_\mu \rangle \tag{10.2}$$

This expression is general and we can therefore use it together with, e.g. the SCF or MP2 density matrices derived in Section 9.8.

Turning now to the first-order correction to the field-dependent expectation value $\langle \Psi_0(\vec{\mathcal{F}}) | \hat{O} | \Psi_0(\vec{\mathcal{F}}) \rangle^{(1)}$ of this operator, we can analogously obtain an expression in terms of molecular orbitals and the first-order density matrix by using Eq. (9.119), i.e.

$$\langle \Psi_0(\vec{\mathcal{F}}) | \hat{O} | \Psi_0(\vec{\mathcal{F}}) \rangle^{(1)} = \sum_\alpha \mathcal{F}_\alpha \int_{\vec{r}\,'_1 = \vec{r}_1} \hat{o}(1) \, P_\alpha^{(1)}(\vec{r}_1, \vec{r}\,'_1) \, d\vec{r}_1$$

$$= \sum_\alpha \mathcal{F}_\alpha \sum_{pq} D^{(1)}_{\alpha,pq} \langle \phi_q | \hat{o} | \phi_p \rangle \tag{10.3}$$

$$= \sum_\alpha \mathcal{F}_\alpha \sum_{\mu\nu} D^{(1)}_{\alpha,\mu\nu} \langle \chi_\nu | \hat{o} | \chi_\mu \rangle \tag{10.4}$$

10.2 Sum-over-States Methods

The sum-over-states method for the calculation of second-or higher-order properties is based on equations like (3.33), (3.110), (3.114) or (3.125), to name a few. The main task is thus to obtain a set of excitation energies $E_n^{(0)} - E_0^{(0)}$ and transition moments $\langle \Psi_0^{(0)} | \hat{O} | \Psi_n^{(0)} \rangle$ with the appropriate operator $\hat{O} = \sum_i^N \hat{o}_i$ or alternatively a ground-state wavefunction $\Psi_0^{(0)}$ and a set of excited-state wavefunctions $\{\Psi_n^{(0)}\}$ from which the excitation energies and transition moments can be calculated.

In Section 9.2 it was mentioned that the simplest approximation for an excited state $|\Psi_n^{(0)}\rangle$ is to represent it by one singly excited determinant Φ_i^a. Approximating at the same time the ground-state wavefunction $\Psi_0^{(0)}$ with the Hartree–Fock determinant Φ_0^{SCF} and the Hamiltonian by the Hartree–Fock Hamiltonian $\hat{F}$, Eq. (9.15), the excitation energies $E_n^{(0)} - E_0^{(0)}$ become equal to orbital energy differences $\epsilon_a - \epsilon_i$ and the transition moments $\langle \Psi_0^{(0)} | \hat{O} | \Psi_n^{(0)} \rangle$ become simple matrix elements of the corresponding one-electron operator $\hat{o}_i$ in the molecular orbital basis $\langle \phi_i | \hat{o} | \phi_a \rangle$ [see Exercise 10.1]. The spectral representation of the polarization propagator, Eq. (3.110), thus becomes approximated as

$$\langle\langle \hat{O}_{\alpha\ldots} ; \hat{O}_{\beta\ldots} \rangle\rangle_\omega \approx \sum_{ia} \left(\frac{\langle \phi_i | \hat{o}_{\alpha\ldots} | \phi_a \rangle \langle \phi_a | \hat{o}_{\beta\ldots} | \phi_i \rangle}{\hbar\omega - \epsilon_a + \epsilon_i} + \frac{\langle \phi_i | \hat{o}_{\beta\ldots} | \phi_a \rangle \langle \phi_a | \hat{o}_{\alpha\ldots} | \phi_i \rangle}{-\hbar\omega - \epsilon_a + \epsilon_i} \right) \tag{10.5}$$

In the static limit, $\omega = 0$, this is called the **uncoupled Hartree–Fock approximation (UCHF)** (Dalgarno, 1959), which played an important role in the early days of calculations of molecular properties.

Exercise 10.1 Show that
$$\langle \Phi_0^{SCF} | \hat{O} | \Phi_i^a \rangle = \langle \phi_i | \hat{o} | \phi_a \rangle$$
and
$$\langle \Phi_i^a | \hat{F} | \Phi_i^a \rangle - \langle \Phi_0^{SCF} | \hat{F} | \Phi_0^{SCF} \rangle = \epsilon_a - \epsilon_i$$
using the Slater–Condon rules Eq. (9.58) and the expression for the orbital energy Eq. (9.17).

Excitation energies and transition moments can in principle be obtained as poles and residua of polarization propagators as discussed in Section 7.4. However, only in the case that the set of operators $\{\hat{h}_n\}$ in Eq. (7.77) is restricted to single excitation and de-excitation operators $\{q_{ai}^\dagger, q_{ai}\}$ is it computationally feasible to determine all excitation energies. This restricts this approach to single-excitation-based methods like the random phase approximation (RPA) discussed in Sections 10.3 and 11.1 or time-dependent density functional theory (TD-DFT).

Nowadays, the sum-over-states method is thus mostly used in three cases.

- The first is benchmark studies of two-electron systems using explicitly correlated wavefunctions (see, e.g. Bishop (1994)).
- The second is the study of hyperpolarizabilities of larger systems using semi-empirical methods.
- Finally, it is used in the analysis of contributions to a molecular property like a polarizability or NMR spin-spin coupling constant from excitations between individual, typically localized, molecular orbitals (see, e.g. Hansen and Bouman (1985), Packer and Pickup (1995), Sauer and Provasi (2008) or Provasi and Sauer (2009)). This is normally done at the level of the random phase approximation or time-dependent density functional theory.

10.3 Møller–Plesset Perturbation Theory Polarization Propagator

In the polarization propagator approximations based on Møller–Plesset perturbation theory the reference state $|\Psi_0^{(0)}\rangle$ in Eqs. (3.160) to (3.163) is approximated by the Møller–Plesset perturbation theory wavefunction in Eq. (9.64). The complete set of operators $\hat{h}$ consists of all possible single excitation and de-excitation operators $\hat{h}_1$, all possible double excitation and de-excitation operators $\hat{h}_2$ up to all possible N-tuple excitation and de-excitation operators $\hat{h}_N$ with respect to the SCF wavefunction $|\Phi_0^{SCF}\rangle$ (Dalgaard, 1979; Olsen et al., 2005) as defined in Eqs. (9.26) to (9.43), i.e.

$$\hat{h}_1 = \begin{pmatrix} {}^e\hat{h}_1 \\ {}^d\hat{h}_1 \end{pmatrix} = \begin{pmatrix} \hat{q}^\dagger \\ \hat{q} \end{pmatrix} \tag{10.6}$$

$$\hat{h}_2 = \begin{pmatrix} {}^e\hat{h}_2 \\ {}^d\hat{h}_2 \end{pmatrix} = \begin{pmatrix} \hat{q}^\dagger \hat{q}^\dagger \\ \hat{q}\hat{q} \end{pmatrix} \tag{10.7}$$

$$\vdots$$

The matrix form of the polarization propagator, (3.159), can thus be written as

$$\langle\langle \hat{P}_\alpha ; \hat{O}^\omega_{\beta...} \rangle\rangle_\omega = \begin{pmatrix} \boldsymbol{T}_1^T(\hat{P}_\alpha) & \boldsymbol{T}_2^T(\hat{P}_\alpha) & \cdots \end{pmatrix} \begin{pmatrix} \boldsymbol{M}_{11} & \boldsymbol{M}_{12} & \cdots \\ \boldsymbol{M}_{21} & \boldsymbol{M}_{22} & \cdots \\ \vdots & \vdots & \ddots \end{pmatrix}^{-1} \begin{pmatrix} \boldsymbol{T}_1(\hat{O}^\omega_{\beta...}) \\ \boldsymbol{T}_2(\hat{O}^\omega_{\beta...}) \\ \vdots \end{pmatrix}$$
(10.8)

where the $\boldsymbol{M}_{ij}$ matrices in the principal propagator are defined as

$$\boldsymbol{M}_{ij} = \hbar\omega \boldsymbol{S}_{ij} - \boldsymbol{E}_{ij} \tag{10.9}$$

The property gradient vectors $\boldsymbol{T}_i^T(\hat{P}_\alpha)$ and $\boldsymbol{T}_j(\hat{O}^\omega_{\beta...})$ are given as

$$\boldsymbol{T}_i^T(\hat{P}_\alpha) = \begin{pmatrix} {}^e\boldsymbol{T}_i^T(\hat{P}_\alpha) & {}^d\boldsymbol{T}_i^T(\hat{P}_\alpha) \end{pmatrix} = \begin{pmatrix} \langle\Psi_0^{(0)}|[\hat{P}_\alpha, {}^e\boldsymbol{h}_i^T]|\Psi_0^{(0)}\rangle & \langle\Psi_0^{(0)}|[\hat{P}_\alpha, {}^d\boldsymbol{h}_i^T]|\Psi_0^{(0)}\rangle \end{pmatrix}$$
(10.10)

$$\boldsymbol{T}_i(\hat{O}^\omega_{\beta...}) = \begin{pmatrix} {}^e\boldsymbol{T}_i(\hat{O}^\omega_{\beta...}) \\ {}^d\boldsymbol{T}_i(\hat{O}^\omega_{\beta...}) \end{pmatrix} = \begin{pmatrix} \langle\Psi_0^{(0)}|[{}^e\boldsymbol{h}_i^\dagger, \hat{O}^\omega_{\beta...}]|\Psi_0^{(0)}\rangle \\ \langle\Psi_0^{(0)}|[{}^d\boldsymbol{h}_i^\dagger, \hat{O}^\omega_{\beta...}]|\Psi_0^{(0)}\rangle \end{pmatrix} \tag{10.11}$$

and the overlap $\boldsymbol{S}_{ij}$ and electronic Hessian matrices $\boldsymbol{E}_{ij}$ are defined as

$$\boldsymbol{S}_{ij} = \begin{pmatrix} {}^{ee}\boldsymbol{S}_{ij} & {}^{ed}\boldsymbol{S}_{ij} \\ {}^{de}\boldsymbol{S}_{ij} & {}^{dd}\boldsymbol{S}_{ij} \end{pmatrix} = \begin{pmatrix} \langle\Psi_0^{(0)}|[{}^e\boldsymbol{h}_i^\dagger, {}^e\boldsymbol{h}_j^T]|\Psi_0^{(0)}\rangle & \langle\Psi_0^{(0)}|[{}^e\boldsymbol{h}_i^\dagger, {}^d\boldsymbol{h}_j^T]|\Psi_0^{(0)}\rangle \\ \langle\Psi_0^{(0)}|[{}^d\boldsymbol{h}_i^\dagger, {}^e\boldsymbol{h}_j^T]|\Psi_0^{(0)}\rangle & \langle\Psi_0^{(0)}|[{}^d\boldsymbol{h}_i^\dagger, {}^d\boldsymbol{h}_j^T]|\Psi_0^{(0)}\rangle \end{pmatrix} \tag{10.12}$$

$$\boldsymbol{E}_{ij} = \begin{pmatrix} {}^{ee}\boldsymbol{E}_{ij} & {}^{ed}\boldsymbol{E}_{ij} \\ {}^{de}\boldsymbol{E}_{ij} & {}^{dd}\boldsymbol{E}_{ij} \end{pmatrix}$$

$$= \begin{pmatrix} \langle\Psi_0^{(0)}|[{}^e\boldsymbol{h}_i^\dagger, [\hat{F}+\hat{V}, {}^e\boldsymbol{h}_j^T]]|\Psi_0^{(0)}\rangle & \langle\Psi_0^{(0)}|[{}^e\boldsymbol{h}_i^\dagger, [\hat{F}+\hat{V}, {}^d\boldsymbol{h}_j^T]]|\Psi_0^{(0)}\rangle \\ \langle\Psi_0^{(0)}|[{}^d\boldsymbol{h}_i^\dagger, [\hat{F}+\hat{V}, {}^e\boldsymbol{h}_j^T]]|\Psi_0^{(0)}\rangle & \langle\Psi_0^{(0)}|[{}^d\boldsymbol{h}_i^\dagger, [\hat{F}+\hat{V}, {}^d\boldsymbol{h}_j^T]]|\Psi_0^{(0)}\rangle \end{pmatrix} \tag{10.13}$$

A series of approximations of increasing order n is then obtained by requiring that the matrix elements in $\boldsymbol{S}_{ij}$, $\boldsymbol{E}_{ij}$ as well as $\boldsymbol{T}_i^T(\hat{P}_\alpha)$ and $\boldsymbol{T}_i(\hat{O}^\omega_{\beta...})$ are evaluated through order n in the fluctuation potential. However, in the definition of a polarization propagator approximation to a particular order n one concentrates on the single excitations and considers the higher-excited contributions only as corrections to the former. This is most easily done in a partitioned form of the principal propagator matrix $\boldsymbol{M}$. Using the following relation for the inverse of a blocked matrix [see Exercise 10.2]

$$\begin{pmatrix} \boldsymbol{U} & \boldsymbol{V} \\ \boldsymbol{W} & \boldsymbol{Z} \end{pmatrix}^{-1} = \begin{pmatrix} (\boldsymbol{U}-\boldsymbol{V}\boldsymbol{Z}^{-1}\boldsymbol{W})^{-1} & (\boldsymbol{W}-\boldsymbol{Z}\boldsymbol{V}^{-1}\boldsymbol{U})^{-1} \\ (\boldsymbol{V}-\boldsymbol{U}\boldsymbol{W}^{-1}\boldsymbol{Z})^{-1} & (\boldsymbol{Z}-\boldsymbol{W}\boldsymbol{U}^{-1}\boldsymbol{V})^{-1} \end{pmatrix} \tag{10.14}$$

$$= \begin{pmatrix} (\boldsymbol{U}-\boldsymbol{V}\boldsymbol{Z}^{-1}\boldsymbol{W})^{-1} & -\boldsymbol{U}^{-1}\boldsymbol{V}\left(\boldsymbol{Z}-\boldsymbol{W}\boldsymbol{U}^{-1}\boldsymbol{V}\right)^{-1} \\ -\boldsymbol{Z}^{-1}\boldsymbol{W}\left(\boldsymbol{U}-\boldsymbol{V}\boldsymbol{Z}^{-1}\boldsymbol{W}\right)^{-1} & (\boldsymbol{Z}-\boldsymbol{W}\boldsymbol{U}^{-1}\boldsymbol{V})^{-1} \end{pmatrix}$$

one can rewrite the matrix representation of polarization propagator, Eq. (10.8), in a partitioned form as [see Exercise 10.3]

$$\langle\langle \hat{P}_\alpha ; \hat{O}^\omega_{\beta\ldots} \rangle\rangle_\omega = \left[T_1^T(\hat{P}_\alpha) - T_{2\ldots}^T(\hat{P}_\alpha)(M_{2\ldots 2\ldots})^{-1} M_{2\ldots 1} \right] \tag{10.15}$$

$$\times \left[M_{11} - M_{12\ldots}(M_{2\ldots 2\ldots})^{-1} M_{2\ldots 1} \right]^{-1} \left[T_1(\hat{O}^\omega_{\beta\ldots}) - M_{12\ldots}(M_{2\ldots 2\ldots})^{-1} T_{2\ldots}(\hat{O}^\omega_{\beta\ldots}) \right]$$

where the "$2\cdots$" notation is meant to indicate that all contributions from the h_2 and higher operators are included in these matrices or vectors. In an nth-order polarization propagator approximation one includes then all matrix element of order up to and including n in the M_{11} matrix and $T_1^T(\hat{P}_\alpha)$ and $T_1(\hat{O}^\omega_{\beta\ldots})$ vectors. In addition, one includes in the matrices $M_{12\ldots}$, $M_{2\ldots 1}$ and $M_{2\ldots 2\ldots}$ as well as in the vectors $T_{2\ldots}^T(\hat{P}_\alpha)$ and $T_{2\ldots}(\hat{O}^\omega_{\beta\ldots})$ only terms up to such an order as necessary for making the products $T_{2\ldots}^T(\hat{P}_\alpha)(M_{2\ldots 2\ldots})^{-1} M_{2\ldots 1}$ or $M_{12\ldots}(M_{2\ldots 2\ldots})^{-1} T_{2\ldots}(\hat{O}^\omega_{\beta\ldots})$ and $M_{12\ldots}(M_{2\ldots 2\ldots})^{-1} M_{2\ldots 1}$ to be correct through nth order.

Exercise 10.2 Prove Eq. (10.14).

Exercise 10.3 Derive the partitioned form of the matrix representation of the polarization propagator, Eq. (10.15), using the relation for the inverse of a blocked matrix, Eq. (10.14).

10.3.1 First Order and Zeroth Order

In the **first-order polarization propagator approximation (FOPPA)** the reference state is then the Hartree–Fock wavefunction $|\Phi_0^{SCF}\rangle$ and one needs to include only the h_1 operators in the set of operators. In principle, one could also include the h_2 and higher operators, because there are contributions to the M_{22} matrices already in zeroth order and to the M_{12}, M_{21} matrices in first order. However, due to the partitioning in Eq. (10.15), it is the order of the whole $M_{12}(M_{22})^{-1}M_{21}$ contribution that counts and this is of second order at least, because the lowest non-vanishing order of the M_{12} and M_{21} matrices is first order.

This approximation is better known as the **time-dependent Hartree–Fock approximation (TDHF)** (McLachlan and Ball, 1964) (see Section 11.1) or **random phase approximation (RPA)** (Rowe, 1968) and can also be derived as the linear response of an SCF wavefunction, as described in Section 11.2. Furthermore, the structure of the equations is the same as in time-dependent density functional theory (TD-DFT), although they differ in the expressions for the elements of the Hessian matrix E_{22}. The polarization propagator in the RPA is then given as

$$\langle\langle \hat{P}_\alpha ; \hat{O}^\omega_{\beta\ldots} \rangle\rangle_\omega = \begin{pmatrix} {}^e P^{T(0)}_\alpha & {}^d P^{T(0)}_\alpha \end{pmatrix} \begin{pmatrix} {}^e X_{\beta\ldots}(\omega) \\ {}^d X_{\beta\ldots}(\omega) \end{pmatrix} \tag{10.16}$$

with the so-called solution vector given as

$$\begin{pmatrix} {}^e X_{\beta\ldots}(\omega) \\ {}^d X_{\beta\ldots}(\omega) \end{pmatrix} = \left[\hbar\omega \begin{pmatrix} 1 & 0 \\ 0 & -1 \end{pmatrix} - \begin{pmatrix} A^{(0,1)} & B^{(1)*} \\ B^{(1)} & A^{(0,1)*} \end{pmatrix} \right]^{-1} \begin{pmatrix} {}^e O^{\omega(0)}_{\beta\ldots} \\ {}^d O^{\omega(0)}_{\beta\ldots} \end{pmatrix} \tag{10.17}$$

where the elements of the RPA $\boldsymbol{A}^{(0,1)}$ and $\boldsymbol{B}^{(1)}$ matrices and of the ${}^e\boldsymbol{P}_\alpha^{(0)}$, ${}^d\boldsymbol{P}_\alpha^{(0)}$, ${}^e\boldsymbol{O}_{\beta\ldots}^{\omega(0)}$ and ${}^d\boldsymbol{O}_{\beta\ldots}^{\omega(0)}$ vectors are in terms of real spin-orbitals $\{\psi_p\}$ given as

$$A_{ai,bj}^{(0,1)} = \left({}^{ee}\boldsymbol{E}_{11}\right)_{ai,bj} = \langle \Phi_0^{\mathrm{SCF}} | [q_{ai}, [\hat{F}+\hat{V}, q_{bj}^\dagger]] | \Phi_0^{\mathrm{SCF}} \rangle \tag{10.18}$$

$$= \langle \Phi_i^a | \hat{F}+\hat{V} | \Phi_j^b \rangle - \delta_{ij}\delta_{ab}\langle \Phi_0^{\mathrm{SCF}} | \hat{F}+\hat{V} | \Phi_0^{\mathrm{SCF}} \rangle \tag{10.19}$$

$$= (\epsilon_a - \epsilon_i)\delta_{ij}\delta_{ab}$$
$$+ \left(\psi_a(\vec{r}_1)\,\psi_i(\vec{r}_1) \middle| \psi_j(\vec{r}_2)\,\psi_b(\vec{r}_2)\right) - \left(\psi_a(\vec{r}_1)\,\psi_b(\vec{r}_1) \middle| \psi_j(\vec{r}_2)\,\psi_i(\vec{r}_2)\right)$$

$$B_{ai,bj}^{(1)} = \left({}^{de}\boldsymbol{E}_{11}\right)_{ai,bj} = \langle \Phi_0^{\mathrm{SCF}} | [q_{ai}, [\hat{F}+\hat{V}, q_{bj}]] | \Phi_0^{\mathrm{SCF}} \rangle \tag{10.20}$$

$$= -\langle \Phi_0^{\mathrm{SCF}} | \hat{H}^{(0)} | \Phi_{ij}^{ab} \rangle$$
$$= \left(\psi_a(\vec{r}_1)\,\psi_j(\vec{r}_1) \middle| \psi_b(\vec{r}_2)\,\psi_i(\vec{r}_2)\right) - \left(\psi_a(\vec{r}_1)\,\psi_i(\vec{r}_1) \middle| \psi_b(\vec{r}_2)\,\psi_j(\vec{r}_2)\right)$$

$${}^e P_{\alpha,ai}^{(0)} = \left({}^e \boldsymbol{T}_1^T(\hat{P}_\alpha)\right)_{ai} = \langle \Phi_0^{\mathrm{SCF}} | [\hat{P}_\alpha, q_{ai}^\dagger] | \Phi_0^{\mathrm{SCF}} \rangle = \langle \psi_i | \hat{p}_\alpha | \psi_a \rangle \tag{10.21}$$

$${}^d P_{\alpha,ai}^{(0)} = \left({}^d \boldsymbol{T}_1^T(\hat{P}_\alpha)\right)_{ai} = \langle \Phi_0^{\mathrm{SCF}} | [\hat{P}_\alpha, q_{ai}] | \Phi_0^{\mathrm{SCF}} \rangle = -\langle \psi_a | \hat{p}_\alpha | \psi_i \rangle \tag{10.22}$$

$${}^e O_{\beta\ldots,ai}^{\omega(0)} = \left({}^e \boldsymbol{T}_1(\hat{O}_{\beta\ldots}^\omega)\right)_{ai} = \langle \Phi_0^{\mathrm{SCF}} | [q_{ai}, \hat{O}_{\beta\ldots}^\omega] | \Phi_0^{\mathrm{SCF}} \rangle = \langle \psi_a | \hat{o}_{\beta\ldots}^\omega | \psi_i \rangle \tag{10.23}$$

$${}^d O_{\beta\ldots,ai}^{\omega(0)} = \left({}^d \boldsymbol{T}_1(\hat{O}_{\beta\ldots}^\omega)\right)_{ai} = \langle \Phi_0^{\mathrm{SCF}} | [q_{ai}^\dagger, \hat{O}_{\beta\ldots}^\omega] | \Phi_0^{\mathrm{SCF}} \rangle = -\langle \psi_i | \hat{o}_{\beta\ldots}^\omega | \psi_a \rangle \tag{10.24}$$

and the first-order contributions to the property gradients ${}^e\boldsymbol{P}_\alpha^{T(1)}$, ${}^d\boldsymbol{P}_\alpha^{T(1)}$, ${}^e\boldsymbol{O}_{\beta\ldots}^{\omega(1)}$ and ${}^d\boldsymbol{O}_{\beta\ldots}^{\omega(1)}$ vanish [see Exercise 10.4].

Exercise 10.4 Explain why there is no first-order contribution to the property gradient ${}^e\boldsymbol{P}_\alpha^{T(1)}$.

Using the explicit RPA expressions for the property gradients, Eqs. (10.21) and (10.22), the RPA polarization propagator can be written as

$$\langle\langle \hat{P}_\alpha ; \hat{O}_{\beta\ldots}^\omega \rangle\rangle_\omega = \sum_{ai} \left({}^e X_{\beta\ldots,ai}(\omega)\langle \psi_i | \hat{p}_\alpha | \psi_a \rangle - {}^d X_{\beta\ldots,ai}(\omega)\langle \psi_a | \hat{p}_\alpha | \psi_i \rangle\right) \tag{10.25}$$

Recalling furthermore Eq. (3.118) and Eq. (9.119) we can identify then the occupied-virtual and virtual-occupied blocks of the RPA first-order density matrix $D_{\beta,pq}^{(1),\mathrm{RPA}}$ as

$$D_{\beta,ai}^{(1),\mathrm{RPA}} = {}^e X_{\beta\ldots,ai}(\omega) \tag{10.26}$$

$$D_{\beta,ia}^{(1),\mathrm{RPA}} = -{}^d X_{\beta\ldots,ai}(\omega) \tag{10.27}$$

In Section 11.1 we will see that the elements of the solution vector are actually the coefficients of the first-order correction to the molecular orbitals.

A frequent approximation to the RPA, which is also employed nowadays in the context of TD-DFT, is obtained by setting the $\boldsymbol{B}^{(1)}$ matrix zero. This is often called the **Tamm–Dancoff** approximation or **mono-excited CI** as the remaining $\boldsymbol{A}^{(0,1)}$ matrix is simply the CI matrix of the Hamiltonian minus the Hartree–Fock energy in the basis of the singly excited determinants, Eq. (10.19). This corresponds to describing the excited states as a linear combination of singly excited determinants $\{\Phi_i^a\}$ while the ground state is the simple Hartree–Fock determinant. The $\boldsymbol{B}^{(1)}$ matrix, on the other hand, is the "Hartree–Fock–doubly excited determinant" part of the CI matrix and introduces correlation in the ground state, which in the RPA is described with the Hartree–Fock determinant plus a linear combination of all doubly excited determinants (Hansen and Bouman, 1979). However, the expansion coefficients in this linear combination are not variationally optimized but can be derived from the condition that the off-diagonal hypervirial relation, Eq. (3.62), is fulfilled for the RPA transition moments (Hansen and Bouman, 1979).

Going one step further and retaining only the zeroth-order contribution to the hessian matrix, i.e. $\boldsymbol{A}^{(0)}$ brings us back to the frequency-dependent version of uncoupled Hartree–Fock, Eq. (10.5), sometimes also called the zeroth-order polarization propagator approximation (ZOPPA).

10.3.2 Second Order

In the **second-order polarization propagator approximation (SOPPA)** (Nielsen et al., 1980) all terms in the partitioned form of the polarization propagator, Eq. (10.15), are evaluated through second order.[1] This implies that we have to include now also contributions from the $\hat{h}_2$ operators, because the first non-vanishing term in the $\boldsymbol{T}_{2\ldots}^T(\hat{P}_\alpha)(\boldsymbol{M}_{2\ldots2\ldots})^{-1}\boldsymbol{M}_{2\ldots1}$ or $\boldsymbol{M}_{12\ldots}(\boldsymbol{M}_{2\ldots2\ldots})^{-1}\boldsymbol{T}_{2\ldots}(\hat{O}_{\beta\ldots}^\omega)$ and $\boldsymbol{M}_{12}(\boldsymbol{M}_{22})^{-1}\boldsymbol{M}_{21}$ contributions is of second order, as discussed earlier and has to be included now.

The polarization propagator in the SOPPA is then given as

$$\langle\langle \hat{P}_\alpha ; \hat{O}_{\beta\ldots}^\omega \rangle\rangle_\omega = \begin{pmatrix} {}^e\boldsymbol{P}_\alpha^{T(0,2)} & {}^d\boldsymbol{P}_\alpha^{T(0,2)} & {}^e\boldsymbol{\Pi}_\alpha^{T(1)} & {}^d\boldsymbol{\Pi}_\alpha^{T(1)} \end{pmatrix} \begin{pmatrix} {}^e\boldsymbol{X}_{\beta\ldots}(\omega) \\ {}^d\boldsymbol{X}_{\beta\ldots}(\omega) \\ {}^e\boldsymbol{\Xi}_{\beta\ldots}(\omega) \\ {}^d\boldsymbol{\Xi}_{\beta\ldots}(\omega) \end{pmatrix} \quad (10.28)$$

with the solution vector given as

$$\begin{pmatrix} {}^e\boldsymbol{X}_{\beta\ldots}(\omega) \\ {}^d\boldsymbol{X}_{\beta\ldots}(\omega) \\ {}^e\boldsymbol{\Xi}_{\beta\ldots}(\omega) \\ {}^d\boldsymbol{\Xi}_{\beta\ldots}(\omega) \end{pmatrix} = \left[\omega \begin{pmatrix} \boldsymbol{\Sigma}^{(0,2)} & 0 & 0 & 0 \\ 0 & -\boldsymbol{\Sigma}^{(0,2)*} & 0 & 0 \\ 0 & 0 & 1 & 0 \\ 0 & 0 & 0 & -1 \end{pmatrix} - \begin{pmatrix} \boldsymbol{A}^{(0,1,2)} & \boldsymbol{B}^{(1,2)*} & \tilde{\boldsymbol{C}}^{(1)} & 0 \\ \boldsymbol{B}^{(1,2)} & \boldsymbol{A}^{(0,1,2)*} & 0 & \tilde{\boldsymbol{C}}^{(1)*} \\ \boldsymbol{C}^{(1)} & 0 & \boldsymbol{D}^{(0)} & 0 \\ 0 & \boldsymbol{C}^{(1)*} & 0 & \boldsymbol{D}^{(0)*} \end{pmatrix} \right]^{-1}$$

$$\times \begin{pmatrix} {}^e\boldsymbol{O}_{\beta\ldots}^{\omega(0,2)} \\ {}^d\boldsymbol{O}_{\beta\ldots}^{\omega(0,2)} \\ {}^e\boldsymbol{\Omega}_{\beta\ldots}^{\omega(1)} \\ {}^d\boldsymbol{\Omega}_{\beta\ldots}^{\omega(1)} \end{pmatrix} \quad (10.29)$$

---

[1] "Through second order" means that the zeroth-, first- and second-order terms are included.

An analysis of the matrix elements using the Slater Condon rules shows that besides the first-order correction to the wavefunction $|\Phi^{\text{MP1}}\rangle$ only the single excitation part, $\sum_{ai} t_i^a[2] |\Phi_i^a\rangle$, is required from the second-order correction $|\Phi^{\text{MP2}}\rangle$, Eq. (9.70). In the following, explicit expressions for the elements of the SOPPA matrices and vectors in terms of spatial orbitals $\{\phi_p\}$ for two spin-free operators $\hat{P}_\alpha$ and $\hat{O}_{\beta...}^{\omega}$ are given.

The $\boldsymbol{A}$-, $\boldsymbol{B}$- and $\boldsymbol{\Sigma}$-matrix, which are already present in RPA then become in SOPPA

$$A_{ai,bj}^{(0,1,2)} = \left({}^{ee}\boldsymbol{E}_{11}\right)_{ai,bj}^{(0,1,2)} = \langle \Phi_0^{\text{MP}} | [q_{ai}, [\hat{F} + \hat{V}, q_{bj}^\dagger]] | \Phi_0^{\text{MP}} \rangle^{(0,1,2)} \qquad (10.30)$$

$$= (\epsilon_a - \epsilon_i)\delta_{ij}\delta_{ab}$$
$$+ 2\big(\phi_a(\vec{r}_1)\,\phi_i(\vec{r}_1)\big|\phi_j(\vec{r}_2)\,\phi_b(\vec{r}_2)\big) - \big(\phi_a(\vec{r}_1)\,\phi_b(\vec{r}_1)\big|\phi_j(\vec{r}_2)\,\phi_i(\vec{r}_2)\big)$$
$$+ \frac{1}{2}(\epsilon_b - \epsilon_j)\left(\delta_{ab}\,D_{ij}^{\text{MP2}} - \delta_{ij}\,D_{ba}^{\text{MP2}}\right)$$
$$- \frac{1}{2}\delta_{ab}\sum_{cdk}\big(\phi_j(\vec{r}_1)\,\phi_c(\vec{r}_1)\big|\phi_k(\vec{r}_2)\,\phi_d(\vec{r}_2)\big)\left\{4\,t_{ik}^{cd}[1] - 2\,t_{ki}^{cd}[1]\right\}$$
$$- \frac{1}{2}\delta_{ij}\sum_{ckl}\big(\phi_k(\vec{r}_1)\,\phi_b(\vec{r}_1)\big|\phi_l(\vec{r}_2)\,\phi_c(\vec{r}_2)\big)\left\{4\,t_{kl}^{ac}[1] - 2\,t_{lk}^{ac}[1]\right\}$$

$$B_{ai,bj}^{(1,2)} = \left({}^{de}\boldsymbol{E}_{11}\right)_{ai,bj}^{(1,2)} = \langle \Phi_0^{\text{MP}} | [q_{ai}, [\hat{F} + \hat{V}, q_{bj}]] | \Phi_0^{\text{MP}} \rangle^{(1,2)} \qquad (10.31)$$

$$= -2\big(\phi_a(\vec{r}_1)\,\phi_i(\vec{r}_1)\big|\phi_b(\vec{r}_2)\,\phi_j(\vec{r}_2)\big) + \big(\phi_a(\vec{r}_1)\,\phi_j(\vec{r}_1)\big|\phi_b(\vec{r}_2)\,\phi_i(\vec{r}_2)\big)$$
$$+ \frac{1}{2}\sum_{ck}\Big[\big(\phi_a(\vec{r}_1)\,\phi_j(\vec{r}_1)\big|\phi_k(\vec{r}_2)\,\phi_c(\vec{r}_2)\big)\left\{4\,t_{ik}^{bc}[1] - 2\,t_{ki}^{bc}[1]\right\}$$
$$+ \big(\phi_b(\vec{r}_1)\,\phi_i(\vec{r}_1)\big|\phi_k(\vec{r}_2)\,\phi_c(\vec{r}_2)\big)\left\{4\,t_{jk}^{ac}[1] - 2\,t_{kj}^{ac}[1]\right\}$$
$$+ \big(\phi_a(\vec{r}_1)\,\phi_c(\vec{r}_1)\big|\phi_k(\vec{r}_2)\,\phi_j(\vec{r}_2)\big)\left\{4\,t_{ki}^{bc}[1] - 2\,t_{ik}^{bc}[1]\right\}$$
$$+ \big(\phi_b(\vec{r}_1)\,\phi_c(\vec{r}_1)\big|\phi_k(\vec{r}_2)\,\phi_i(\vec{r}_2)\big)\left\{4\,t_{kj}^{ac}[1] - 2\,t_{jk}^{ac}[1]\right\}\Big]$$
$$- \frac{1}{2}\sum_{kl}\big(\phi_k(\vec{r}_1)\,\phi_i(\vec{r}_1)\big|\phi_l(\vec{r}_2)\,\phi_j(\vec{r}_2)\big)\left\{4\,t_{kl}^{ab}[1] - 2\,t_{lk}^{ab}[1]\right\}$$
$$- \frac{1}{2}\sum_{cd}\big(\phi_a(\vec{r}_1)\,\phi_c(\vec{r}_1)\big|\phi_b(\vec{r}_2)\,\phi_d(\vec{r}_2)\big)\left\{4\,t_{ij}^{cd}[1] - 2\,t_{ji}^{cd}[1]\right\}$$

$$\Sigma_{ai,bj}^{(0,2)} = \left({}^{ee}\boldsymbol{S}_{11}\right)_{ai,bj}^{(0,2)} = \langle \Phi_0^{\text{MP}} | [q_{ai}, q_{bj}^\dagger] | \Phi_0^{\text{MP}} \rangle^{(0,2)} \qquad (10.32)$$

$$= \delta_{ab}\delta_{ij} + \frac{1}{2}\delta_{ab}\,D_{ij}^{\text{MP2}} - \frac{1}{2}\delta_{ij}\,D_{ba}^{\text{MP2}}$$

The second-order correction to the $\boldsymbol{A}$ matrix, as given here, is not Hermitian. Therefore, one normally uses a symmetrized second-order correction $A'^{(2)}_{ai,bj}$ defined as

$$A'^{(2)}_{ai,bj} = \frac{1}{2}\left(A^{(2)}_{ai,bj} + A^{(2)}_{bj,ai}\right) = A^{(2)}_{ai,bj} + \frac{1}{2}\left(A^{(2)}_{bj,ai} - A^{(2)}_{ai,bj}\right) \qquad (10.33)$$

Compared to the **A**-matrix in RPA, Eq. (10.18), one obtains in second order two additional contributions, which consist of contractions of two-electron repulsion integrals with the first-order doubles correlation coefficients defined in Eq. (9.67), and one term that contains the second-order correction to the density matrix, Eqs. (9.116) and (9.117). The latter contribution

$$\langle \Phi_0^{MP} | [q_{ai}, [\hat{F} + \hat{V}, q_{bj}^\dagger]] | \Phi_0^{MP} \rangle^{(0)} \langle \Phi_0^{MP} | \Phi_0^{MP} \rangle^{(2)} \tag{10.34}$$

is a renormalisation term, which arises because the Møller–Plesset perturbation theory wavefunction, Eq. (9.64), has to be normalized at each order.

The second-order correction to the **B**-matrix consists also of contractions of two-electron repulsion integrals with the first-order doubles correlation coefficients. But contrary to the **A**-matrix, where only two-electron repulsion integrals with two occupied and two virtual molecular orbitals contribute, the **B**-matrix also includes integrals with four occupied or four virtual molecular orbitals. On the other hand, there is no renormalization term in the **B**-matrix, as the zeroth-order **B**-matrix vanishes.

As in the case of the RPA we can analyze which kind of matrix elements of the Hamiltonian the second-order contributions to the **A**- and **B**-matrices correspond to. The new contributions are all matrix elements between the Hartree–Fock determinant and the first-order Møller–Plesset perturbation theory wavefunction. As the latter consist of doubly excited determinants, the new contributions to **A** will include matrix elements between triply and singly excited determinants, $\langle \Phi_{ikl}^{acd} | \hat{F} + \hat{V} | \Phi_j^b \rangle$, as well as between doubly excited determinants and the Hartree–Fock determinant, $\langle \Phi_{jk}^{bc} | \hat{F} + \hat{V} | \Phi_0^{SCF} \rangle$. Similar in the **B**-matrix one obtains additional matrix elements between two singly excited determinants, $\langle \Phi_i^a | \hat{F} + \hat{V} | \Phi_k^c \rangle$, and between two doubly excited determinants, $\langle \Phi_{ij}^{ab} | \hat{F} + \hat{V} | \Phi_{kl}^{cd} \rangle$ [see Exercise 10.5]. This implies that in SOPPA more electron correlation is included in the ground state in terms of triply excited determinants, but also that the excited states are correlated by the admixture of doubly excited determinants.

Exercise 10.5 Explain why the second-order contributions to the molecular Hessian matrix consist of the following type of matrix elements: $\langle \Phi_{ikl}^{acd} | \hat{F} + \hat{V} | \Phi_j^b \rangle$, $\langle \Phi_{jk}^{bc} | \hat{F} + \hat{V} | \Phi_0^{SCF} \rangle$, $\langle \Phi_i^a | \hat{F} + \hat{V} | \Phi_k^c \rangle$ and $\langle \Phi_{ij}^{ab} | \hat{F} + \hat{V} | \Phi_{kl}^{cd} \rangle$.

Hint : Expand the commutator in the matrix element and insert the first-order Møller–Plesset perturbation theory wavefunction.

The second-order correction to the overlap matrix Σ consists of a similar term with the second-order correction to the density matrix, Eqs. (9.116) and (9.117), as the renormalization contribution the **A**-matrix. This is to be expected as the overlap matrix is related to the norm of the second-order Møller–Plesset perturbation theory wavefunction, which can be expressed in terms of the second-order correction to the density matrix as

$$\langle \Phi_0^{MP} | \Phi_0^{MP} \rangle^{(2)} = \sum_a D_{aa}^{MP2} - \sum_i D_{ii}^{MP2} \tag{10.35}$$

In addition to these second-order corrections to the RPA matrices there are three new matrices due to the $\hat{h}_2$ operators. In the following, we present explicit expressions for them in terms of spatial orbitals $\{\phi_p\}$ and for two spin-free operators $\hat{P}_\alpha$ and $\hat{O}^\omega_{\beta\ldots}$ using a biorthogonal set of double excitation operators $\{\hat{q}^\dagger\hat{q}^\dagger, \hat{q}\hat{q}\}$ (Bak et al., 2000).

$$D^{(0)}_{aibj,ckdl} = \left(^{ee}\boldsymbol{E}_{22}\right)^{(0)}_{aibj,ckdl} = \langle \Phi^{\text{SCF}}_0 | [q_{ai}q_{bj}, [\hat{F} + \hat{V}, q^\dagger_{ck}q^\dagger_{dl}]] | \Phi^{\text{SCF}}_0 \rangle^{(0)} \quad (10.36)$$

$$= \frac{1}{(1+\delta_{ab}\delta_{ij})} (\epsilon_a + \epsilon_b - \epsilon_i - \epsilon_j)(\delta_{ac}\delta_{ik}\delta_{bd}\delta_{jl} + \delta_{ad}\delta_{il}\delta_{bc}\delta_{jk})$$

$$C^{(1)}_{aibj,ck} = \left(^{ee}\boldsymbol{E}_{21}\right)^{(1)}_{aibj,ck} = \langle \Phi^{\text{SCF}}_0 | [q_{ai}q_{bj}, [\hat{F} + \hat{V}, q^\dagger_{ck}]] | \Phi^{\text{SCF}}_0 \rangle^{(1)} \quad (10.37)$$

$$= \frac{\sqrt{2}}{(1+\delta_{ab}\delta_{ij})} \Big\{ \delta_{ik}\big(\phi_a(\vec{r}_1)\,\phi_c(\vec{r}_1)\big|\phi_b(\vec{r}_2)\,\phi_j(\vec{r}_2)\big)$$

$$+ \delta_{jk}\big(\phi_a(\vec{r}_1)\,\phi_i(\vec{r}_1)\big|\phi_b(\vec{r}_2)\,\phi_c(\vec{r}_2)\big)$$

$$- \delta_{ac}\big(\phi_k(\vec{r}_1)\,\phi_i(\vec{r}_1)\big|\phi_b(\vec{r}_2)\,\phi_j(\vec{r}_2)\big)$$

$$- \delta_{bc}\big(\phi_a(\vec{r}_1)\,\phi_i(\vec{r}_1)\big|\phi_k(\vec{r}_2)\,\phi_j(\vec{r}_2)\big) \Big\}$$

$$\tilde{C}^{(1)}_{ck,aibj} = \left(^{ee}\boldsymbol{E}_{12}\right)^{(1)}_{ck,aibj} = \langle \Phi^{\text{SCF}}_0 | [q_{ck}, [\hat{F} + \hat{V}, q^\dagger_{ai}q^\dagger_{bj}]] | \Phi^{\text{SCF}}_0 \rangle^{(1)} \quad (10.38)$$

$$= \frac{1}{\sqrt{2}} \Big[\delta_{ik}\Big\{ 2\big(\phi_j(\vec{r}_1)\,\phi_b(\vec{r}_1)\big|\phi_c(\vec{r}_2)\,\phi_a(\vec{r}_2)\big) - \big(\phi_j(\vec{r}_1)\,\phi_a(\vec{r}_1)\big|\phi_c(\vec{r}_2)\,\phi_b(\vec{r}_2)\big) \Big\}$$

$$+ \delta_{jk}\Big\{ 2\big(\phi_i(\vec{r}_1)\,\phi_a(\vec{r}_1)\big|\phi_c(\vec{r}_2)\,\phi_b(\vec{r}_2)\big) - \big(\phi_i(\vec{r}_1)\,\phi_b(\vec{r}_1)\big|\phi_c(\vec{r}_2)\,\phi_a(\vec{r}_2)\big) \Big\}$$

$$- \delta_{ac}\Big\{ 2\big(\phi_i(\vec{r}_1)\,\phi_k(\vec{r}_1)\big|\phi_j(\vec{r}_2)\,\phi_b(\vec{r}_2)\big) - \big(\phi_j(\vec{r}_1)\,\phi_k(\vec{r}_1)\big|\phi_i(\vec{r}_2)\,\phi_b(\vec{r}_2)\big) \Big\}$$

$$- \delta_{bc}\Big\{ 2\big(\phi_j(\vec{r}_1)\,\phi_k(\vec{r}_1)\big|\phi_i(\vec{r}_2)\,\phi_a(\vec{r}_2)\big) - \big(\phi_i(\vec{r}_1)\,\phi_k(\vec{r}_1)\big|\phi_j(\vec{r}_2)\,\phi_a(\vec{r}_2)\big) \Big\} \Big]$$

These matrices are very similar to the contributions of same order to the $\boldsymbol{A}$ matrix. The $\boldsymbol{D}^{(0)}$ matrix, being zeroth order, consists only of molecular orbital energy differences. This means that pure double excitations are treated in SOPPA only in zeroth order and thus not more accurate than in a uncoupled Hartree–Fock calculation. The $\boldsymbol{C}^{(1)}$ matrices, which couple double excitations with the single excitations are first order and consist therefore of two-electron repulsion integrals like the $\boldsymbol{A}^{(1)}$ and $\boldsymbol{B}^{(1)}$ matrices. Although the pure double excitations are only treated through zeroth order, the effect of the doubles excitations on the singly excited states is still correct through second order as one can see from the partitioned form of the polarization propagator Eq. (10.15).

Turning to the property gradients, there are again additional second-order corrections to the $\boldsymbol{P}^{T(0)}_\alpha$ and $\boldsymbol{O}^{\omega(1)}_{\beta\ldots}$ contributions of the RPA and additional new contributions $\boldsymbol{\Pi}^{T(1)}_\alpha$ and $\boldsymbol{\Omega}^{\omega(1)}_{\beta\ldots}$ due to the $\hat{h}_2$ operators.

220 *Approximations to Exact Perturbation and Response Theory Expressions*

$$^eP_{\alpha,ai}^{(0,2)} = \left(^e\mathbf{T}_1^T(\hat{P}_\alpha)\right)_{ai}^{(0,2)} = \langle\Phi_0^{\text{MP}}|[\hat{P}_\alpha, q_{ai}^\dagger]|\Phi_0^{\text{MP}}\rangle^{(0,2)} \tag{10.39}$$

$$= \sqrt{2}\langle\phi_i|\hat{p}_\alpha|\phi_a\rangle + \frac{1}{\sqrt{2}}\sum_j \left(\langle\phi_j|\hat{p}_\alpha|\phi_a\rangle\, D_{ji}^{\text{MP2}} - \langle\phi_i|\hat{p}_\alpha|\phi_j\rangle\, D_{aj}^{\text{MP2}}\right)$$

$$+ \frac{1}{\sqrt{2}}\sum_b \left(\langle\phi_b|\hat{p}_\alpha|\phi_a\rangle\, D_{bi}^{\text{MP2}} - \langle\phi_i|\hat{p}_\alpha|\phi_b\rangle\, D_{ab}^{\text{MP2}}\right)$$

$$^dP_{\alpha,ai}^{(0,2)} = \left(^d\mathbf{T}_1^T(\hat{P}_\alpha)\right)_{ai}^{(0,2)} = \langle\Phi_0^{\text{MP}}|[\hat{P}_\alpha, q_{ai}]|\Phi_0^{\text{MP}}\rangle^{(0,2)} \tag{10.40}$$

$$= -\sqrt{2}\langle\phi_a|\hat{p}_\alpha|\phi_i\rangle - \frac{1}{\sqrt{2}}\sum_j \left(\langle\phi_a|\hat{p}_\alpha|\phi_j\rangle\, D_{ij}^{\text{MP2}} - \langle\phi_j|\hat{p}_\alpha|\phi_i\rangle\, D_{ja}^{\text{MP2}}\right)$$

$$- \frac{1}{\sqrt{2}}\sum_b \left(\langle\phi_a|\hat{p}_\alpha|\phi_b\rangle\, D_{ib}^{\text{MP2}} - \langle\phi_b|\hat{p}_\alpha|\phi_i\rangle\, D_{ba}^{\text{MP2}}\right)$$

$$^eO_{\beta\ldots,ai}^{\omega(0,2)} = \left(^e\mathbf{T}_1(\hat{O}_{\beta\ldots}^\omega)\right)_{ai}^{(0,2)} = \langle\Phi_0^{\text{MP}}|[q_{ai},\hat{O}_{\beta\ldots}^\omega]|\Phi_0^{\text{MP}}\rangle^{(0,2)}$$

$$= \sqrt{2}\langle\phi_a|\hat{o}_{\beta\ldots}^\omega|\phi_i\rangle + \frac{1}{\sqrt{2}}\sum_j \left(\langle\phi_a|\hat{o}_{\beta\ldots}^\omega|\phi_j\rangle\, D_{ij}^{\text{MP2}} - \langle\phi_j|\hat{o}_{\beta\ldots}^\omega|\phi_i\rangle\, D_{ja}^{\text{MP2}}\right)$$

$$+ \frac{1}{\sqrt{2}}\sum_b \left(\langle\phi_a|\hat{o}_{\beta\ldots}^\omega|\phi_b\rangle\, D_{ib}^{\text{MP2}} - \langle\phi_b|\hat{o}_{\beta\ldots}^\omega|\phi_i\rangle\, D_{ba}^{\text{MP2}}\right)$$

$$^dO_{\beta\ldots,ai}^{\omega(0,2)} = \left(^d\mathbf{T}_1(\hat{O}_{\beta\ldots}^\omega)\right)_{ai}^{(0,2)} = \langle\Phi_0^{\text{MP}}|[q_{ai}^\dagger,\hat{O}_{\beta\ldots}^\omega]|\Phi_0^{\text{MP}}\rangle^{(0,2)} \tag{10.41}$$

$$= -\sqrt{2}\langle\phi_i|\hat{o}_{\beta\ldots}^\omega|\phi_a\rangle - \frac{1}{\sqrt{2}}\sum_j \left(\langle\phi_j|\hat{o}_{\beta\ldots}^\omega|\phi_a\rangle\, D_{ji}^{\text{MP2}} - \langle\phi_i|\hat{o}_{\beta\ldots}^\omega|\phi_j\rangle\, D_{aj}^{\text{MP2}}\right)$$

$$- \frac{1}{\sqrt{2}}\sum_b \left(\langle\phi_b|\hat{o}_{\beta\ldots}^\omega|\phi_a\rangle\, D_{bi}^{\text{MP2}} - \langle\phi_i|\hat{o}_{\beta\ldots}^\omega|\phi_b\rangle\, D_{ab}^{\text{MP2}}\right)$$

$$^e\Pi_{\alpha,aibj}^{(1)} = \left(^e\mathbf{T}_2^T(\hat{P}_\alpha)\right)_{aibj}^{(1)} = \langle\Phi_0^{\text{MP}}|[\hat{P}_\alpha, q_{ai}^\dagger q_{bj}^\dagger]|\Phi_0^{\text{MP}}\rangle^{(1)} \tag{10.42}$$

$$= -\frac{1}{2}\sum_k \left(\langle\phi_i|\hat{p}_\alpha|\phi_k\rangle\,\{4\,t_{kj}^{ab}[1] - 2\,t_{jk}^{ab}[1]\} + \langle\phi_j|\hat{p}_\alpha|\phi_k\rangle\,\{4\,t_{ik}^{ab}[1] - 2\,t_{ki}^{ab}[1]\}\right)$$

$$+ \frac{1}{2}\sum_c \left(\langle\phi_c|\hat{p}_\alpha|\phi_a\rangle\,\{4\,t_{ij}^{cb}[1] - 2\,t_{ji}^{cb}[1]\} + \langle\phi_c|\hat{p}_\alpha|\phi_b\rangle\,\{4\,t_{ij}^{ac}[1] - 2\,t_{ji}^{ac}[1]\}\right)$$

$$^d\Pi_{\alpha,aibj}^{(1)} = \left(^d\mathbf{T}_2^T(\hat{P}_\alpha)\right)_{aibj}^{(1)} = \langle\Phi_0^{\text{MP}}|[\hat{P}_\alpha, q_{ai}q_{bj}]|\Phi_0^{\text{MP}}\rangle^{(1)} \tag{10.43}$$

$$= \frac{1}{2}\sum_k \left(\langle\phi_k|\hat{p}_\alpha|\phi_i\rangle\,\{4\,t_{jk}^{ba}[1] - 2\,t_{kj}^{ba}[1]\} + \langle\phi_k|\hat{p}_\alpha|\phi_j\rangle\,\{4\,t_{ki}^{ba}[1] - 2\,t_{ik}^{ba}[1]\}\right)$$

$$- \frac{1}{2}\sum_c \left(\langle\phi_a|\hat{p}_\alpha|\phi_c\rangle\,\{4\,t_{ji}^{bc}[1] - 2\,t_{ij}^{bc}[1]\} + \langle\phi_b|\hat{p}_\alpha|\phi_c\rangle\,\{4\,t_{ji}^{ca}[1] - 2\,t_{ij}^{ca}[1]\}\right)$$

$$^e\Omega^{\omega(1)}_{\beta\cdots,aibj} = \left(^eT_2(\hat{O}^\omega_{\beta\cdots})\right)^{(1)}_{aibj} = \langle\Phi^{\text{MP}}_0|[q_{ai}q_{bj},\hat{O}^\omega_{\beta\cdots}]|\Phi^{\text{MP}}_0\rangle^{(1)} \tag{10.44}$$

$$= -\frac{2}{(1+\delta_{ab}\delta_{ij})}\sum_k\left\{\langle\phi_k|\hat{o}^\omega_{\beta\cdots}|\phi_i\rangle\, t^{ba}_{jk}[1] + \langle\phi_k|\hat{o}^\omega_{\beta\cdots}|\phi_j\rangle\, t^{ba}_{ki}[1]\right\}$$

$$+ \frac{2}{(1+\delta_{ab}\delta_{ij})}\sum_c\left\{\langle\phi_a|\hat{o}^\omega_{\beta\cdots}|\phi_c\rangle\, t^{bc}_{ji}[1] + \langle\phi_b|\hat{o}^\omega_{\beta\cdots}|\phi_c\rangle\, t^{ca}_{ji}[1]\right\}$$

$$^d\Omega^{\omega(1)}_{\beta\cdots,aibj} = \left(^dT_2(\hat{O}^\omega_{\beta\cdots})\right)^{(1)}_{aibj} = \langle\Phi^{\text{MP}}_0|[q^\dagger_{ai}q^\dagger_{bj},\hat{O}^\omega_{\beta\cdots}]|\Phi^{\text{MP}}_0\rangle^{(1)} \tag{10.45}$$

$$= \frac{2}{(1+\delta_{ab}\delta_{ij})}\sum_k\left\{\langle\phi_i|\hat{o}^\omega_{\beta\cdots}|\phi_k\rangle\, t^{ab}_{kj}[1] + \langle\phi_j|\hat{o}^\omega_{\beta\cdots}|\phi_k\rangle\, t^{ab}_{ik}[1]\right\}$$

$$- \frac{2}{(1+\delta_{ab}\delta_{ij})}\sum_c\left\{\langle\phi_c|\hat{o}^\omega_{\beta\cdots}|\phi_a\rangle\, t^{cb}_{ij}[1] + \langle\phi_c|\hat{o}^\omega_{\beta\cdots}|\phi_b\rangle\, t^{ac}_{ij}[1]\right\}$$

We see that the first-order $\mathbf{\Pi}^{T(1)}_\alpha$ and $\mathbf{\Omega}^{\omega(1)}_{\beta\cdots}$ vectors consist of contractions of the property integrals in molecular orbital basis with the first-order doubles correlation coefficients, while the second-order corrections to $\mathbf{P}^{T(0)}_\alpha$ and $\mathbf{O}^{\omega(0)}_{\beta\cdots}$ consist of contractions of the property integrals with the second-order correction to the density matrix. Furthermore, one should note that contrary to the renormalization term in the $\mathbf{A}^{(2)}$ matrix and the second-order correction to the overlap matrix the corrections to the $\mathbf{P}^{T(0)}_\alpha$ and $\mathbf{O}^{\omega(0)}_{\beta\cdots}$ property gradients also have contributions from the occupied-virtual and virtual-occupied off-diagonal blocks of the density matrix Eq. (9.118). This is therefore the only place where the second-order correction to the wavefunction contributes in SOPPA.

Finally, one might wonder whether it is possible to write the SOPPA polarization propagator as the contraction of property integrals of the operator $\hat{P}$ in the molecular orbital basis with a SOPPA first-order density matrix in analogy to Eq. (10.25). However, this is not immediately possible for two reasons:

1. The second-order corrections to the $\hat{\mathbf{h}}_1$ part of the property gradient imply that a SOPPA first-order density matrix will not be contracted with property integrals in the molecular orbital basis. However, defining a kind of MP2 correction to the molecular orbitals as

$$\phi^{\text{MP2}}_p = \frac{1}{2}\left(\sum_i \phi_i\, D^{\text{MP2}}_{ip} - \sum_a \phi_a\, D^{\text{MP2}}_{ap}\right) \tag{10.46}$$

and requiring that the MP2 correction to the property integrals is linear in this correction, i.e.

$$\langle\phi_i|\hat{p}_\alpha|\phi_a\rangle^{\text{MP2}} = \langle\phi_i|\hat{p}_\alpha|\phi^{\text{MP2}}_a\rangle + \langle\phi^{\text{MP2}}_i|\hat{p}_\alpha|\phi_a\rangle \tag{10.47}$$

one could express the $\hat{\mathbf{h}}_1$ part of the propagator as a contraction of these correlated property integrals with a SOPPA first-order density matrix.

2. The $\hat{\mathbf{h}}_2$ contributions to the property gradients $\mathbf{\Pi}^{T(1)}_\alpha$ as well as to the solution vector $\mathbf{\Xi}_{\beta\cdots}(\omega)$ are both 4 index quantities, which prevents us from formulating

their contribution as a contraction of property integrals with a first-order one-particle density matrix.

10.3.3 Higher-Order and Mixed Methods

Evaluating all the terms in the partitioned form of the polarization propagator, Eq. (10.15), through third order one obtains the **third-order polarization propagator approximation (TOPPA)**. The expressions for all matrix elements have been derived but only parts have been implemented (Geertsen et al., 1991a).

However, other attempts have been made to improve on the treatment of electron correlation in SOPPA. Three SOPPA-like methods have thus been presented. All are based on the fact that a coupled cluster wavefunction gives a better description than the Møller–Plesset first- and second-order wavefunctions, Eqs. (9.66) and (9.70). In the second-order polarization propagator with coupled cluster singles and doubles amplitudes–**SOPPA(CCSD)**–method (Sauer, 1997), the reference state $|\Psi_0^{(0)}\rangle$ in Eqs. (3.160) to (3.163) is approximated by a linearized CCSD wavefunction

$$|\Psi_0^{(0)}\rangle \approx \left(1 + \hat{T}_1 + \hat{T}_2\right)|\Phi_0^{SCF}\rangle \tag{10.48}$$

This keeps essentially the structure of the SOPPA equations but replaces in all matrix elements the first-order MP doubles correlation coefficients, Eq. (9.67), and the second-order MP singles correlation coefficient, Eq. (9.71), by coupled cluster singles and doubles amplitudes. In the earlier coupled cluster singles and doubles polarization propagator approximation (**CCSDPPA**) (Geertsen et al., 1991a), a precursor to SOPPA(CCSD), this was done only partially and in particular not in the second-order correction to the density matrix $\boldsymbol{D}^{MP2}$. Very recently, a third method (Kjær et al., 2010), **SOPPA(CC2)**, was proposed in which the Møller–Plesset correlation coefficients are replaced by the corresponding CC2 amplitudes instead of the CCSD amplitudes. This reduces the computational cost in the calculation of the amplitudes to the same as in the SOPPA calculation.

10.3.4 Iterative and Non-Iterative Doubles Correction

In a previous section, we have discussed that in the SOPPA method the description of excited states, which in the RPA approximation are described as a linear combination of singly excited determinants, is improved by including electron correlation in the reference state and by including also double excitation operators $\hat{h}_2$. As a consequence of counting orders in the partitioned form of the polarization propagator the main purpose of the double excitations is to improve the description of the same single excitation-dominated states as in RPA. However, this correction is rather costly, because in SOPPA the corrections from the double excitation operators $\hat{h}_2$ are included in the principal propagator whose dimension is therefore increased from two times the number of single excitations squared as in RPA to two times the number of single and double excitations squared.

It is thus worthwhile to consider alternatives for the calculation of single-excitation-dominated excited states with double excitation corrections in SOPPA without having to diagonalize a matrix of the full dimension of single and double excitations squared. The original implementations of the SOPPA method were therefore based on the partitioned form of the polarization propagator, Eq. (10.15). However, the eigenvalue equation for the partitioned principal propagator in SOPPA becomes

$$\begin{pmatrix} \boldsymbol{A}^{(0,1,2)} + \tilde{\boldsymbol{C}}^{(1)}(\hbar\omega_n - \boldsymbol{D}^{(0)})^{-1}\boldsymbol{C}^{(1)} & \boldsymbol{B}^{(1,2)} \\ \boldsymbol{B}^{(1,2)} & \boldsymbol{A}^{(0,1,2)} + \tilde{\boldsymbol{C}}^{(1)}(-\hbar\omega_n - \boldsymbol{D}^{(0)})^{-1}\boldsymbol{C}^{(1)} \end{pmatrix} \begin{pmatrix} {}^e\boldsymbol{X}_n \\ {}^d\boldsymbol{X}_n \end{pmatrix}$$

$$= \hbar\omega_n \begin{pmatrix} \boldsymbol{\Sigma}^{(0,2)} & \boldsymbol{0} \\ \boldsymbol{0} & -\boldsymbol{\Sigma}^{(0,2)} \end{pmatrix} \begin{pmatrix} {}^e\boldsymbol{X}_n \\ {}^d\boldsymbol{X}_n \end{pmatrix} \qquad (10.49)$$

which means that it is frequency dependent and depends on its own eigenvalues $\hbar\omega_n$, i.e. the excitation energies. One has to solve this equation iteratively for one excitation energy at a time by inserting the eigenvalue obtained in one iteration as frequency in the principal propagator for the next iteration until this eigenvalue remains the same and one has obtained self-consistency like for the Hartree–Fock equations, Eq. (9.11). This procedure has to be repeated for each excitation energy, i.e. for each eigenvalue of the principal propagator, and is therefore no longer applied in the context of SOPPA. However, it is nowadays used in the calculation of excitation energies from coupled cluster response functions at the CC3 level, see Section 11.4.

As an approximate alternative one does not directly include the doubles corrections in the principal propagator but corrects the RPA excitation energies, obtained by solving the RPA eigenvalue problem, with a non-iterative doubles correction. This approach is called **doubles corrected random-phase approximation–RPA(D)** (Christiansen et al., 1998a) and is based on **pseudo-perturbation theory** that was described in Section 3.13.

Applied to SOPPA this means that the electronic Hessian and overlap matrices are each split in three contributions

$$\boldsymbol{E} = \boldsymbol{E}^{(0)} + \boldsymbol{E}^{(1)} + \boldsymbol{E}^{(2)} \qquad (10.50)$$

with

$$\boldsymbol{E}^{(0)} = \begin{pmatrix} \boldsymbol{A}^{(0,1)} & \boldsymbol{B}^{(1)} & 0 & 0 \\ \boldsymbol{B}^{(1)} & \boldsymbol{A}^{(0,1)} & 0 & 0 \\ 0 & 0 & \boldsymbol{D}^{(0)} & 0 \\ 0 & 0 & 0 & \boldsymbol{D}^{(0)} \end{pmatrix} \qquad (10.51)$$

$$\boldsymbol{E}^{(1)} = \begin{pmatrix} 0 & 0 & \tilde{\boldsymbol{C}}^{(1)} & 0 \\ 0 & 0 & 0 & \tilde{\boldsymbol{C}}^{(1)} \\ \boldsymbol{C}^{(1)} & 0 & 0 & 0 \\ 0 & \boldsymbol{C}^{(1)} & 0 & 0 \end{pmatrix} \qquad (10.52)$$

$$\boldsymbol{E}^{(2)} = \begin{pmatrix} \boldsymbol{A}^{(2)} & \boldsymbol{B}^{(2)} & 0 & 0 \\ \boldsymbol{B}^{(2)} & \boldsymbol{A}^{(2)} & 0 & 0 \\ 0 & 0 & 0 & 0 \\ 0 & 0 & 0 & 0 \end{pmatrix} \qquad (10.53)$$

and

$$S = S^{(0)} + S^{(1)} + S^{(2)}$$
$$= \begin{pmatrix} 1 & 0 & 0 & 0 \\ 0 & -1 & 0 & 0 \\ 0 & 0 & 1 & 0 \\ 0 & 0 & 0 & -1 \end{pmatrix} + \begin{pmatrix} 0 & 0 & 0 & 0 \\ 0 & 0 & 0 & 0 \\ 0 & 0 & 0 & 0 \\ 0 & 0 & 0 & 0 \end{pmatrix} + \begin{pmatrix} \Sigma^{(2)} & 0 & 0 & 0 \\ 0 & -\Sigma^{(2)} & 0 & 0 \\ 0 & 0 & 0 & 0 \\ 0 & 0 & 0 & 0 \end{pmatrix} \quad (10.54)$$

Choosing the zeroth-order eigenvectors to be the RPA eigenvectors

$$L_n^{(0)\dagger} = R_n^{(0)} = \begin{pmatrix} {}^e X_n^{\text{RPA}} \\ {}^d X_n^{\text{RPA}} \\ 0 \\ 0 \end{pmatrix} \quad (10.55)$$

implies that the zeroth-order eigenvalues are the RPA excitation energies

$$\omega_n^{(0)} = \omega_n^{\text{RPA}} \quad (10.56)$$

and that the $E^{(1)}$ and $S^{(1)}$ matrices do not contribute in first order, because

$$L_n^{(0)} E^{(1)} R_n^{(0)} = 0 \quad (10.57)$$

as assumed in Eqs. (3.187) and (3.188).

The second-order correction to the eigenvalues, Eq. (3.194), then becomes

$$\hbar\omega_n^{(2)} = \begin{pmatrix} {}^e X_n^{\text{RPA}\dagger} & {}^d X_n^{\text{RPA}\dagger} \end{pmatrix} \begin{pmatrix} A^{(2)} - \hbar\omega_n^{\text{RPA}} \Sigma^{(2)} & B^{(2)} \\ B^{(2)} & A^{(2)} + \hbar\omega_n^{\text{RPA}} \Sigma^{(2)} \end{pmatrix} \begin{pmatrix} {}^e X_n^{\text{RPA}} \\ {}^d X_n^{\text{RPA}} \end{pmatrix}$$
$$- \begin{pmatrix} {}^e X_n^{\text{RPA}\dagger} & {}^d X_n^{\text{RPA}\dagger} \end{pmatrix} \begin{pmatrix} \tilde{C}^{(1)} & 0 \\ 0 & \tilde{C}^{(1)} \end{pmatrix} \begin{pmatrix} D^{(0)} - \hbar\omega_n^{\text{RPA}} & 0 \\ 0 & D^{(0)} + \hbar\omega_n^{\text{RPA}} \end{pmatrix}^{-1}$$
$$\times \begin{pmatrix} C^{(1)} & 0 \\ 0 & C^{(1)} \end{pmatrix} \begin{pmatrix} {}^e X_n^{\text{RPA}} \\ {}^d X_n^{\text{RPA}} \end{pmatrix} \quad (10.58)$$

and the eigenvalues and thus excitation energies in the RPA(D) method are finally defined as the sum of the zeroth- and second-order contributions

$$\hbar\omega_n^{\text{RPA(D)}} = \hbar\omega_n^{\text{RPA}} + \hbar\omega_n^{(2)} \quad (10.59)$$

Compared with the partitioned SOPPA eigenvalue problem in Eq. (10.49) we can see that only the RPA eigenvectors are necessary and that thus only the RPA eigenvalue problem has to be solved. The second-order and doubles correction, on the other hand, are evaluated only once, meaning that we have a non-iterative doubles correction in RPA(D).

10.4 Multiconfigurational Polarization Propagator

In the multiconfigurational polarization propagator approximation, normally called **multiconfigurational random phase approximation (MCRPA)**, the set of operators contains state transfer operators $\{\boldsymbol{R}^\dagger, \boldsymbol{R}\}$, Eq. (9.51), in addition to the non-redundant single excitation $\boldsymbol{q}^\dagger$ and de-excitation $\boldsymbol{q}$ operators (Yeager and Jørgensen, 1979). The expression for the polarization propagator in MCRPA can be obtained from Eq. (10.8) if one identifies $\boldsymbol{h}_2$ with $\{\boldsymbol{R}^\dagger, \boldsymbol{R}\}$ and $|\Psi_0^{(0)}\rangle$ with $|\Phi_0^{\mathrm{MCSCF}}\rangle$.

$$\langle\langle\, \hat{P}_\alpha\, ;\, \hat{O}_{\beta\ldots}^\omega\,\rangle\rangle_\omega = \boldsymbol{T}^T(\hat{P}_\alpha)\,(\hbar\omega\boldsymbol{S} - \boldsymbol{E})^{-1}\,\boldsymbol{T}(\hat{O}_{\beta\ldots}^\omega) \tag{10.60}$$

$$= \begin{pmatrix} {}^e\boldsymbol{T}^T(\hat{P}_\alpha) & {}^d\boldsymbol{T}^T(\hat{P}_\alpha) \end{pmatrix} \left[\hbar\omega \begin{pmatrix} \boldsymbol{\Sigma} & \boldsymbol{\Delta} \\ -\boldsymbol{\Delta} & -\boldsymbol{\Sigma} \end{pmatrix} - \begin{pmatrix} \boldsymbol{A} & \boldsymbol{B} \\ \boldsymbol{B} & \boldsymbol{A} \end{pmatrix}\right]^{-1} \begin{pmatrix} {}^e\boldsymbol{T}(\hat{O}_{\beta\ldots}^\omega) \\ {}^d\boldsymbol{T}(\hat{O}_{\beta\ldots}^\omega) \end{pmatrix}$$

where the matrices and vectors are commonly defined as

$$\boldsymbol{A} = \begin{pmatrix} \langle\Phi_0^{\mathrm{MCSCF}}|[\boldsymbol{q},[\hat{H}^{(0)},\boldsymbol{q}^{\dagger T}]]|\Phi_0^{\mathrm{MCSCF}}\rangle & \langle\Phi_0^{\mathrm{MCSCF}}|[\boldsymbol{R}^\dagger,[\hat{H}^{(0)},\boldsymbol{q}^T]]|\Phi_0^{\mathrm{MCSCF}}\rangle \\ \langle\Phi_0^{\mathrm{MCSCF}}|[\boldsymbol{R},[\hat{H}^{(0)},\boldsymbol{q}^{\dagger T}]]|\Phi_0^{\mathrm{MCSCF}}\rangle & \langle\Phi_0^{\mathrm{MCSCF}}|[\boldsymbol{R},[\hat{H}^{(0)},\boldsymbol{R}^{\dagger T}]]|\Phi_0^{\mathrm{MCSCF}}\rangle \end{pmatrix} \tag{10.61}$$

$$\boldsymbol{B} = \begin{pmatrix} \langle\Phi_0^{\mathrm{MCSCF}}|[\boldsymbol{q},[\hat{H}^{(0)},\boldsymbol{q}^T]]|\Phi_0^{\mathrm{MCSCF}}\rangle & \langle\Phi_0^{\mathrm{MCSCF}}|[\boldsymbol{R},[\hat{H}^{(0)},\boldsymbol{q}^T]]|\Phi_0^{\mathrm{MCSCF}}\rangle \\ \langle\Phi_0^{\mathrm{MCSCF}}|[\boldsymbol{R},[\hat{H}^{(0)},\boldsymbol{q}^{\dagger T}]]|\Phi_0^{\mathrm{MCSCF}}\rangle & \langle\Phi_0^{\mathrm{MCSCF}}|[\boldsymbol{R},[\hat{H}^{(0)},\boldsymbol{R}^T]]|\Phi_0^{\mathrm{MCSCF}}\rangle \end{pmatrix} \tag{10.62}$$

$$\boldsymbol{\Sigma} = \begin{pmatrix} \langle\Phi_0^{\mathrm{MCSCF}}|[\boldsymbol{q},\boldsymbol{q}^{\dagger T}]|\Phi_0^{\mathrm{MCSCF}}\rangle & \langle\Phi_0^{\mathrm{MCSCF}}|[\boldsymbol{q},\boldsymbol{R}^{\dagger T}]|\Phi_0^{\mathrm{MCSCF}}\rangle \\ \langle\Phi_0^{\mathrm{MCSCF}}|[\boldsymbol{R},\boldsymbol{q}^{\dagger T}]|\Phi_0^{\mathrm{MCSCF}}\rangle & \langle\Phi_0^{\mathrm{MCSCF}}|[\boldsymbol{R},\boldsymbol{R}^{\dagger T}]|\Phi_0^{\mathrm{MCSCF}}\rangle \end{pmatrix} \tag{10.63}$$

$$\boldsymbol{\Delta} = \begin{pmatrix} \langle\Phi_0^{\mathrm{MCSCF}}|[\boldsymbol{q},\boldsymbol{q}^T]|\Phi_0^{\mathrm{MCSCF}}\rangle & \langle\Phi_0^{\mathrm{MCSCF}}|[\boldsymbol{q},\boldsymbol{R}^T]|\Phi_0^{\mathrm{MCSCF}}\rangle \\ \langle\Phi_0^{\mathrm{MCSCF}}|[\boldsymbol{R},\boldsymbol{q}^T]|\Phi_0^{\mathrm{MCSCF}}\rangle & \langle\Phi_0^{\mathrm{MCSCF}}|[\boldsymbol{R},\boldsymbol{R}^T]|\Phi_0^{\mathrm{MCSCF}}\rangle \end{pmatrix} \tag{10.64}$$

$${}^e\boldsymbol{T}^T(\hat{P}_\alpha) = \begin{pmatrix} \langle\Phi_0^{\mathrm{MCSCF}}|[\hat{P}_\alpha,\boldsymbol{q}^{\dagger T}]|\Phi_0^{\mathrm{MCSCF}}\rangle & \langle\Phi_0^{\mathrm{MCSCF}}|[\hat{P}_\alpha,\boldsymbol{R}^{\dagger T}]|\Phi_0^{\mathrm{MCSCF}}\rangle \end{pmatrix} \tag{10.65}$$

$${}^d\boldsymbol{T}^T(\hat{P}_\alpha) = -{}^e\boldsymbol{T}^{T*}(\hat{P}_\alpha) \tag{10.66}$$

$${}^d\boldsymbol{T}^T(\hat{P}_\alpha) = -{}^e\boldsymbol{T}^{T*}(\hat{P}_\alpha)\,{}^e\boldsymbol{T}(\hat{O}_{\beta\ldots}^\omega) = \begin{pmatrix} \langle\Phi_0^{\mathrm{MCSCF}}|[\boldsymbol{q},\hat{O}_{\beta\ldots}^\omega]|\Phi_0^{\mathrm{MCSCF}}\rangle \\ \langle\Phi_0^{\mathrm{MCSCF}}|[\boldsymbol{R},\hat{O}_{\beta\ldots}^\omega]|\Phi_0^{\mathrm{MCSCF}}\rangle \end{pmatrix} \tag{10.67}$$

$${}^d\boldsymbol{T}(\hat{O}_{\beta\ldots}^\omega) = -{}^e\boldsymbol{T}^*(\hat{O}_{\beta\ldots}^\omega) \tag{10.68}$$

Since in MCRPA the reference wavefunction $|\Psi_0^{(0)}\rangle$ is a variational MCSCF wavefunction, one can derive the MCRPA also by application of linear response theory, Section 11.2, or of the quasi-energy derivative method, Section 12.3, to this MCSCF state.

10.5 Further Reading

P. Jørgensen and J. Simons, *Second Quantization-Based Methods in Quantum Chemistry*, Academic Press, New York (1981): Chapter 6.

R. McWeeny, *Methods of Molecular Quantum Mechanics*, Academic Press, London, 2nd edn (1992): Chapters 12 and 13.

J. Oddershede, P. Jørgensen and D. L. Yeager, *Polarization Propagator Methods in Atomic and Molecular Calculations*, Comput. Phys. Rep. 2, 33–92 (1991).

11
Perturbation and Response Theory with Approximate Wavefunctions

In this chapter we will follow now the second approach, which means that we will apply time-independent and time-dependent perturbation theory from Chapter 3 to approximate solutions of the unperturbed molecular Hamiltonian. In particular, we will illustrate this in the following for Hartree–Fock, MCSCF and coupled cluster wavefunctions.

11.1 Coupled and Time-Dependent Hartree–Fock

In the **coupled Hartree–Fock method (CHF)**, which was probably derived the first time by Peng (1941) and rederived many times (Stevens et al., 1963; Gerratt and Mills, 1968), second- and higher-order static properties are obtained by solving the Hartree–Fock equations

$$\hat{f}(1,\vec{\mathcal{F}})\,\psi_p(\vec{r}_1,\vec{\mathcal{F}}) = \epsilon_p(\vec{\mathcal{F}})\,\psi_p(\vec{r}_1,\vec{\mathcal{F}}) \tag{11.1}$$

self-consistently in the presence of a perturbing field $\vec{\mathcal{F}}$ under the condition that the perturbed occupied[1] spin orbitals $\{\psi_i(\vec{\mathcal{F}})\}$ remain orthonormal

$$\langle \psi_i(\vec{r}_1,\vec{\mathcal{F}}) \mid \psi_j(\vec{r}_1,\vec{\mathcal{F}}) \rangle = \delta_{ij} \tag{11.2}$$

Contrary to the unperturbed Hartree–Fock theory, where the molecular orbitals are expanded in atomic one-electron basis functions Eq. (9.4), one normally expands the perturbed occupied spin orbitals in the set of orthonormalized unperturbed molecular spin orbitals $\{\psi_q\}$

$$\psi_i(\vec{r}_1,\vec{\mathcal{F}}) = \sum_q^{all} \psi_q(\vec{r}_1)\,U_{qi}(\vec{\mathcal{F}}) \tag{11.3}$$

One of the consequences of this choice is that the unperturbed Fock matrix becomes diagonal in this basis.

---

[1] We are only interested in the occupied orbitals here.

Inserting this ansatz in the perturbed Hartree–Fock equations, (11.1) and (11.2), gives

$$\hat{f}(1,\vec{\mathcal{F}}) \sum_q^{all} \psi_q(\vec{r}_1)\, U_{qi}(\vec{\mathcal{F}}) = \epsilon_i(\vec{\mathcal{F}}) \sum_q^{all} \psi_q(\vec{r}_1)\, U_{qi}(\vec{\mathcal{F}}) \qquad (11.4)$$

and for the orthonormality condition

$$\sum_q^{all} U^*_{jq}(\vec{\mathcal{F}})\, U_{qi}(\vec{\mathcal{F}}) = \delta_{ji} \qquad (11.5)$$

Multiplying the perturbed Hartree–Fock equations, Eq. (11.4), from the left with another basis function, an unperturbed molecular orbital ψ_r, followed by integration one obtains a matrix form of the perturbed Hartree–Fock equations

$$\sum_q^{all} F_{rq}(\vec{\mathcal{F}})\, U_{qi}(\vec{\mathcal{F}}) = \epsilon_i(\vec{\mathcal{F}}) \sum_q^{all} \delta_{rq}\, U_{qi}(\vec{\mathcal{F}}) \qquad (11.6)$$

where an element of the perturbed Fock matrix $F_{pq}(\vec{\mathcal{F}})$ is given as

$$\begin{aligned}F_{pq}(\vec{\mathcal{F}}) &= \langle \psi_p | \hat{f}(\vec{\mathcal{F}}) | \psi_q \rangle \\&= \langle \psi_p | \hat{h}^{(0)} + \hat{h}^{(1)} + \hat{h}^{(2)} | \psi_q \rangle \\&\quad + \sum_j^{occ} \sum_{st}^{all} U^*_{sj}(\vec{\mathcal{F}}) \left\{ (\psi_p \psi_q | \psi_s \psi_t) - (\psi_p \psi_t | \psi_s \psi_q) \right\} U_{tj}(\vec{\mathcal{F}})\end{aligned} \qquad (11.7)$$

and the first- and second-order one-electron perturbation Hamiltonians $\hat{h}^{(1)} + \hat{h}^{(2)}$ are according to Eqs. (2.108), (2.109) and (2.110) scalar or tensor products of the general perturbation $\mathcal{F}_{\alpha\cdots}$ and two one-electron perturbation operators $\hat{o}^{\mathcal{F}}_{\alpha\cdots}$ and $\hat{o}^{\mathcal{F}\mathcal{F}}_{\alpha\beta\cdots}$

$$\hat{h}^{(1)}(i) + \hat{h}^{(2)}(i) = \sum_{\alpha\cdots} \mathcal{F}_{\alpha\cdots}\, \hat{o}^{\mathcal{F}}_{i,\alpha\cdots} + \sum_{\alpha,\beta,\cdots} \mathcal{F}_{\alpha\cdots}\, \hat{o}^{\mathcal{F}\mathcal{F}}_{i,\alpha\beta\cdots}\, \mathcal{F}_{\beta\cdots} \qquad (11.8)$$

Exercise 11.1 Derive the orthonormality condition, Eq. (11.5) for the expansion coefficients $U_{qi}(\vec{\mathcal{F}})$ of the perturbed molecular spin orbitals $\{\psi_i(\vec{\mathcal{F}})\}$.

Exercise 11.2 Derive the expression for the perturbed Fock matrix $F_{rq}(\vec{\mathcal{F}})$, Eq. (11.7).

The perturbed Fock matrix in the basis of the unperturbed molecular spin orbitals, $F_{pq}(\vec{\mathcal{F}})$, the perturbed orbital energies $\epsilon_i(\vec{\mathcal{F}})$ and the coefficients $U_{qi}(\vec{\mathcal{F}})$ of the perturbed orbitals are then expanded in orders of this perturbing field

$$F_{pq}(\vec{\mathcal{F}}) = F^{(0)}_{pq} + \sum_{\alpha\cdots} F^{(1)}_{\alpha\cdots,pq}\, \mathcal{F}_{\alpha\cdots} + \cdots \qquad (11.9)$$

$$\epsilon_i(\vec{\mathcal{F}}) = \epsilon_i^{(0)} + \sum_{\alpha\cdots} \epsilon_{\alpha\cdots,i}^{(1)} \mathcal{F}_{\alpha\cdots} + \cdots \tag{11.10}$$

$$U_{qi}(\vec{\mathcal{F}}) = U_{qi}^{(0)} + \sum_{\alpha\cdots} U_{\alpha\cdots,qi}^{(1)} \mathcal{F}_{\alpha\cdots} + \cdots \tag{11.11}$$

The zeroth-order Fock matrix and the matrix of the zeroth-order coefficients are both diagonal

$$U_{qp}^{(0)} = \delta_{qp} \tag{11.12}$$

$$F_{pq}^{(0)} = \epsilon_p^{(0)} \delta_{pq} \tag{11.13}$$

which is a consequence of the fact that we used the unperturbed molecular spin orbitals $\{\psi_q\}$ as basis. The perturbed spin orbitals, in Eq. (11.3), are thus expanded in orders of the perturbation

$$\psi_i(\vec{r}_1, \vec{\mathcal{F}}) = \psi_i(\vec{r}_1) + \sum_q^{\text{all}} \psi_q(\vec{r}_1) \sum_{\alpha\cdots} U_{\alpha\cdots,qi}^{(1)} \mathcal{F}_{\alpha\cdots} + \cdots \tag{11.14}$$

Also, the SCF density matrix is perturbed, $D_{\mu\nu}^{\text{CHF}}(\vec{\mathcal{F}})$, and is expanded in orders of the perturbation

$$D_{\mu\nu}^{\text{SCF}}(\vec{\mathcal{F}}) = D_{\mu\nu}^{\text{SCF}} + \sum_{\alpha\cdots} D_{\alpha\cdots,\mu\nu}^{(1),\text{CHF}} \mathcal{F}_{\alpha\cdots} + \cdots \tag{11.15}$$

From the definition of the unperturbed density matrix $D_{\mu\nu}^{\text{SCF}}$ in Eq. (9.114) it follows that for a closed-shell molecule the first-order correction to the SCF density matrix in the atomic orbital basis is then

$$D_{\alpha\cdots,\mu\nu}^{(1),\text{CHF}} = 2 \sum_i \left(c_{\alpha\cdots,\mu i}^{(1)*} c_{\nu i} + c_{\mu i}^* c_{\alpha\cdots,\nu i}^{(1)} \right) = 2 \left(U_{\alpha\cdots,\mu\nu}^{(1)*} + U_{\alpha\cdots,\nu\mu}^{(1)} \right) \tag{11.16}$$

Inserting these expansions in the perturbed Hartree–Fock equations, Eqs. (11.6) and (11.5), and separating orders, we obtain the first-order Hartree–Fock equations

$$\sum_q^{\text{all}} (F_{rq}^{(0)} U_{\alpha\cdots,qi}^{(1)} + F_{\alpha\cdots,rq}^{(1)} U_{qi}^{(0)}) = \epsilon_i^{(0)} \sum_q^{\text{all}} \delta_{rq} U_{\alpha\cdots,qi}^{(1)} + \epsilon_{\alpha\cdots,i}^{(1)} \sum_q^{\text{all}} \delta_{rq} U_{qi}^{(0)} \tag{11.17}$$

and

$$\sum_q^{\text{all}} \left(U_{jq}^{(0)*} U_{\alpha\cdots,qi}^{(1)} + U_{\alpha\cdots,jq}^{(1)*} U_{qi}^{(0)} \right) = 0 \tag{11.18}$$

which after evaluation of the zeroth-order terms, Eqs. (11.12) and (11.13), become

$$(\epsilon_r^{(0)} - \epsilon_i^{(0)}) U_{\alpha\cdots,ri}^{(1)} = \epsilon_{\alpha\cdots,i}^{(1)} \delta_{ri} - F_{\alpha\cdots,ri}^{(1)} \tag{11.19}$$

$$U_{\alpha\cdots,ji}^{(1)} + U_{\alpha\cdots,ji}^{(1)*} = 0 \tag{11.20}$$

For $r \neq i$ one finally obtains for the expansion coefficients of the first-order correction to the orbitals

$$U^{(1)}_{\alpha\cdots,ri} = -\frac{F^{(1)}_{\alpha\cdots,ri}}{\epsilon^{(0)}_r - \epsilon^{(0)}_i} \quad (11.21)$$

an equation that must be solved iteratively since the first-order Fock matrix, $F^{(1)}_{\alpha\cdots,ri}$, depends on $\mathbf{U}^{(1)}_{\alpha\cdots}$.

$$F^{(1)}_{\alpha\cdots,ri} = \langle \psi_r | \hat{o}^{\mathcal{F}}_{\alpha\cdots} | \psi_i \rangle \quad (11.22)$$
$$+ \sum_j^{occ} \sum_{st} \{ (\psi_r \psi_i | \psi_s \psi_t) - (\psi_r \psi_t | \psi_s \psi_i) \} \left(U^{(1)*}_{\alpha\cdots,sj} \delta_{tj} + \delta_{sj} U^{(1)}_{\alpha\cdots,tj} \right)$$

Exercise 11.3 Derive the expression for the first-order correction to the Fock matrix $F^{(1)}_{\alpha\cdots,ri}$ in Eq. (11.22).

In principle, the summation over q in Eq. (11.14), runs over all spin-orbitals. However, mixing the occupied orbitals among themselves does not change the total wavefunction (Stevens et al., 1963). Therefore, we only need the virtual-occupied blocks of the matrices of the first- and higher-order coefficients $U^{(n)}_{\alpha\cdots,ai}$. The first-order Hartree–Fock equations then read

$$\sum_j^{occ} \sum_b^{vir} \left[(\epsilon^{(0)}_a - \epsilon^{(0)}_i) \delta_{ab} \delta_{ij} + (\psi_a \psi_i | \psi_j \psi_b) - (\psi_a \psi_b | \psi_j \psi_i) \right] U^{(1)}_{\alpha\cdots,bj}$$
$$+ \sum_j^{occ} \sum_b^{vir} \left[(\psi_a \psi_i | \psi_b \psi_j) - (\psi_a \psi_j | \psi_b \psi_i) \right] U^{(1)*}_{\alpha\cdots,bj}$$
$$= -\langle \psi_a | \hat{o}^{\mathcal{F}}_{\alpha\cdots} | \psi_i \rangle \quad (11.23)$$

The two expressions in the parentheses are the RPA $\mathbf{A}^{(0,1)}$ and $\mathbf{B}^{(1)}$ matrices given in Eqs. (10.18) and (10.20) and we can rewrite Eq. (11.23) therefore as

$$\sum_j^{occ} \sum_b^{vir} \left[A^{(0,1)}_{ai,bj} U^{(1)}_{\alpha\cdots,bj} - B^{(1)}_{ai,bj} U^{(1)*}_{\alpha\cdots,bj} \right] = -\langle \psi_a | \hat{o}^{\mathcal{F}}_{\alpha\cdots} | \psi_i \rangle \quad (11.24)$$

For real (−) or imaginary (+) perturbations we have that $U^{(1)*}_{\alpha\cdots,bj} = U^{(1)}_{\alpha\cdots,bj}$ or $U^{(1)*}_{\alpha\cdots,bj} = -U^{(1)}_{\alpha\cdots,bj}$, respectively, and Eq. (11.24) simplifies to two sets of linear equations

$$\sum_j^{occ} \sum_b^{vir} \left(A^{(0,1)}_{ai,bj} \mp B^{(1)}_{ai,bj} \right) U^{(1)}_{\alpha\cdots,bj} = -\langle \psi_a | \hat{o}^{\mathcal{F}}_{\alpha\cdots} | \psi_i \rangle \quad (11.25)$$

Once the expansion coefficients of the first-order correction to the orbitals, $U^{(1)}_{\alpha\cdots,bj}$, are obtained as solutions of these equations, we can calculate the corrections to the Hartree–Fock energy and thus molecular properties.

From Eq. (9.21) we can deduce that in the presence of the perturbing field $\vec{\mathcal{F}}$ the perturbed Hartree–Fock energy is given as

$$E_0^{\mathrm{SCF}}(\vec{\mathcal{F}}) = \sum_i^{occ} \epsilon_i(\vec{\mathcal{F}}) \tag{11.26}$$

$$-\frac{1}{2}\sum_{ij}^{occ}\sum_{abcd}^{vir} U^*_{ai}(\vec{\mathcal{F}})U^*_{cj}(\vec{\mathcal{F}})\left[(\psi_a\psi_b|\psi_c\psi_d) - (\psi_a\psi_d|\psi_c\psi_b)\right] U_{bi}(\vec{\mathcal{F}})U_{dj}(\vec{\mathcal{F}})$$

Inserting the perturbation series, Eqs. (11.10) and (11.11), and separating orders the second-order correction to the energy can be shown to be [see Exercise 11.4]

$$E_0^{(2),\mathrm{SCF}} = \sum_{\alpha,\beta,\cdots}\sum_i^{occ} \mathcal{F}_{\alpha\cdots}\langle\psi_i|\hat{o}^{\mathcal{F}\mathcal{F}}_{\alpha\beta\cdots}|\psi_i\rangle \mathcal{F}_{\beta\cdots} \tag{11.27}$$

$$+ \sum_{\alpha,\beta,\cdots}\sum_i^{occ}\sum_a^{vir} \mathcal{F}_{\alpha\cdots}\left(U^{(1)*}_{\alpha\cdots,ai}\langle\psi_a|\hat{o}^{\mathcal{F}}_{\beta\cdots}|\psi_i\rangle + \langle\psi_i|\hat{o}^{\mathcal{F}}_{\beta\cdots}|\psi_a\rangle U^{(1)}_{\alpha\cdots,ai}\right)\mathcal{F}_{\beta\cdots}$$

which should be compared with the expression for the second-order energy correction of exact wavefunctions in Eq. (3.34). Second-order molecular properties that are defined as second derivatives of the energy are thus obtained at the coupled Hartree–Fock level as

$$\frac{\partial^2 E_0^{(2),\mathrm{SCF}}}{\partial \mathcal{F}_{\alpha\cdots}\partial \mathcal{F}_{\beta\cdots}} = \sum_i^{occ}\langle\psi_i|\hat{o}^{\mathcal{F}\mathcal{F}}_{\alpha\beta\cdots}|\psi_i\rangle$$

$$+ \sum_i^{occ}\sum_a^{vir}\left(U^{(1)*}_{\alpha\cdots,ai}\langle\psi_a|\hat{o}^{\mathcal{F}}_{\beta\cdots}|\psi_i\rangle + \langle\psi_i|\hat{o}^{\mathcal{F}}_{\beta\cdots}|\psi_a\rangle U^{(1)}_{\alpha\cdots,ai}\right) \tag{11.28}$$

Exercise 11.4 Derive the second-order correction to the Hartree–Fock energy, Eq. (11.27) starting from Eq. (11.26) as shown by Peng (1941) or Stevens et al. (1963).

The relation to the random phase approximation (RPA) in Section 10.3 becomes clear, when one considers also the complex conjugate equation of Eq. (11.24)

$$\sum_j^{occ}\sum_b^{vir}\left[-B^{(1)}_{ai,bj} U^{(1)}_{\alpha\cdots,bj} + A^{(0,1)}_{ai,bj} U^{(1)*}_{\alpha\cdots,bj}\right] = -\langle\psi_i|\hat{o}^{\mathcal{F}}_{\alpha\cdots}|\psi_a\rangle \tag{11.29}$$

realises that the molecular orbital integrals $\langle\psi_a|\hat{o}^{\mathcal{F}}_{\alpha\cdots}|\psi_i\rangle$ over the first-order perturbation operator $\hat{o}^{\mathcal{F}}_{\alpha\cdots}$ in these equations are the elements of the property gradients

$^e\boldsymbol{O}^{\omega(0)}_{\alpha\cdots}$ and $^d\boldsymbol{O}^{\omega(0)}_{\alpha\cdots}$ defined in Eqs. (10.23) and (10.24) and combines Eqs. (11.24) and Eq. (11.29) in one matrix equation

$$\begin{pmatrix} \boldsymbol{A}^{(0,1)} & \boldsymbol{B}^{(1)} \\ \boldsymbol{B}^{(1)} & \boldsymbol{A}^{(0,1)} \end{pmatrix} \begin{pmatrix} \boldsymbol{U}^{(1)}_{\alpha\cdots} \\ -\boldsymbol{U}^{(1)*}_{\alpha\cdots} \end{pmatrix} = - \begin{pmatrix} ^e\boldsymbol{O}^{\omega(0)}_{\alpha\cdots} \\ ^d\boldsymbol{O}^{\omega(0)}_{\alpha\cdots} \end{pmatrix} \tag{11.30}$$

or

$$\begin{pmatrix} \boldsymbol{U}^{(1)}_{\alpha\cdots} \\ -\boldsymbol{U}^{(1)*}_{\alpha\cdots} \end{pmatrix} = - \begin{pmatrix} \boldsymbol{A}^{(0,1)} & \boldsymbol{B}^{(1)} \\ \boldsymbol{B}^{(1)} & \boldsymbol{A}^{(0,1)} \end{pmatrix}^{-1} \begin{pmatrix} ^e\boldsymbol{O}^{\omega(0)}_{\alpha\cdots} \\ ^d\boldsymbol{O}^{\omega(0)}_{\alpha\cdots} \end{pmatrix} \tag{11.31}$$

Comparing this with Eq. (10.17) we can see that for zero frequency of the perturbation the RPA solution vectors $^e\boldsymbol{X}_{\alpha\cdots}(\omega=0)$ and $^d\boldsymbol{X}_{\alpha\cdots}(\omega=0)$ are identical to the expansion coefficients $\boldsymbol{U}^{(1)}_{\alpha}$ and $-\boldsymbol{U}^{(1)*}_{\alpha}$ of the first-order correction to the orbitals and have therefore a simple, physical interpretation.

In **Time-dependent Hartree–Fock theory (TDHF)** (Langhoff et al., 1972) this derivation is generalized to the time-dependent field of a monochromatic linear polarized radiation in the dipole approximation Eq. (3.76). The molecular orbitals are then also time dependent and are again expanded in the unperturbed orbitals

$$\psi_i(\vec{r}_1,t,\vec{\mathcal{F}}) = \psi_i(\vec{r}_1) + \frac{1}{2}\sum_p \psi_p(\vec{r}_1) \sum_{\alpha\cdots} \mathcal{F}_{\alpha\cdots}(\omega) \left[U^{(1)}_{\alpha\cdots,pi}(\omega)e^{\imath\omega t} + U^{(1)}_{\alpha\cdots,pi}(-\omega)e^{-\imath\omega t} \right]$$
$$+ \cdots \tag{11.32}$$

The frequency-dependent expansion coefficients $U^{(1)}_{\alpha\cdots,pi}(\omega)$ and $U^{(1)}_{\alpha\cdots,pi}(-\omega)$ are then determined by inserting the time-dependent orbitals in the time-dependent version of the Hartree–Fock equation,

$$\left[\hat{f}(1,\vec{\mathcal{F}}) - \imath\frac{\partial}{\partial t}\right]\psi_p(\vec{r}_1,t,\vec{\mathcal{F}}) = \epsilon_p(\vec{\mathcal{F}})\,\psi_p(\vec{r}_1,t,\vec{\mathcal{F}}) \tag{11.33}$$

which can be derived from Frenkel's variational principle (Frenkel, 1934).

In analogy to the derivation of the coupled Hartree–Fock equations one can then derive the time-dependent Hartree–Fock equations for the first-order expansion coefficients

$$\left[\hbar\omega\begin{pmatrix}1 & 0\\0 & -1\end{pmatrix} - \begin{pmatrix}\boldsymbol{A}^{(0,1)} & \boldsymbol{B}^{(1)*}\\ \boldsymbol{B}^{(1)} & \boldsymbol{A}^{(0,1)*}\end{pmatrix}\right]\begin{pmatrix}\boldsymbol{U}^{(1)}_{\alpha\cdots}(-\omega)\\ -\boldsymbol{U}^{(1)}_{\alpha\cdots}(\omega)\end{pmatrix} = \begin{pmatrix}^e\boldsymbol{O}^{\omega(0)}_{\alpha\cdots}\\ ^d\boldsymbol{O}^{\omega(0)}_{\alpha\cdots}\end{pmatrix} \tag{11.34}$$

which are just the RPA equations Eq. (10.17) with

$$\begin{pmatrix} \boldsymbol{U}^{(1)}_{\alpha\cdots}(-\omega) \\ -\boldsymbol{U}^{(1)}_{\alpha\cdots}(\omega) \end{pmatrix} = \begin{pmatrix} ^e\boldsymbol{X}_{\alpha\cdots}(\omega) \\ ^d\boldsymbol{X}_{\alpha\cdots}(\omega) \end{pmatrix} \tag{11.35}$$

As has already been mentioned, the variational nature of the Hartree–Fock wavefunction means that the CHF/TDHF equations are equivalent to the RPA equations. Unlike RPA and its correlated extensions, however, an atomic-orbital-based solution of the iterative CHF equations cannot give excitation energies and transition moments.

Historically, CHF, was favoured over RPA since it could be solved in the atomic orbital basis (Diercksen and McWeeny, 1966), rather than requiring a transformation to the molecular orbital basis. The need for an inverse Hessian in RPA/SOPPA also restricted the size of system that could be studied. However, the use of direct atomic-orbital-driven methods for RPA response properties (Feyereisen et al., 1992) and for SOPPA (Bak et al., 2000; Christiansen et al., 1998a), coupled with iterative methods for solving the inverse Hessian, mean that they can now be applied as widely as CHF/TDHF and provide far more information about excited states and properties.

11.2 Multiconfigurational Linear Response Functions

In the application of response theory to an SCF wavefunction $|\Phi_0^{SCF}\rangle$, or to an MCSCF wavefunction $|\Phi_0^{MCSCF}\rangle$, presented in the following way first by Olsen and Jørgensen (1985), the time-dependent MCSCF state $|\Phi_0^{MCSCF}(t)\rangle$ is usually expressed as (Olsen and Jørgensen, 1985; Fuchs et al., 1993)

$$|\Phi_0^{MCSCF}(t)\rangle = e^{i\hat{\kappa}(t)} e^{i\hat{S}(t)} |\Phi_0^{MCSCF}\rangle \tag{11.36}$$

where

$$\hat{\kappa}(t) = \sum_{p>q} \left[\kappa_{pq}(t) \, \hat{q}^\dagger_{pq} + \kappa^*_{pq}(t) \, \hat{q}_{pq} \right] \tag{11.37}$$

$$\hat{S}(t) = \sum_{n\neq 0} \left[S_{n0}(t) \, \hat{R}^\dagger_{n0} + S^*_{n0}(t) \, \hat{R}_{0n} \right] \tag{11.38}$$

are the time-dependent versions of the operators in Eqs. (9.48) and (9.49). For the SCF case we have $S_{n0}(t) = S^*_{n0}(t) = 0$ and the orbital rotation parameters $\kappa_{pq}(t)$ and $\kappa^*_{pq}(t)$ become equal to the Fourier transforms of the TDHF expansion coefficients $U^{(1)}_{\alpha,pi}(\omega)$ and $U^{(1)}_{\alpha,pi}(-\omega)$ in Eq. (11.32).

In order to keep the equations more compact it is advantageous to collect the operators in one row vector $\hat{\bm{h}}^T = (\{\hat{q}^\dagger_{pq}\}, \{\hat{q}_{pq}\}, \{\hat{R}^\dagger_{n0}\}, \{\hat{R}_{0n}\})$ and correspondingly the time-dependent orbital rotation and state-transfer parameters in one column vector $\bm{\gamma}(t) = (\{\kappa_{pq}(t)\}, \{\kappa^*_{pq}(t)\}, \{S_{n0}(t)\}, \{S^*_{n0}(t)\})^T$. The latter are then expanded in orders of the time-dependent perturbation Hamiltonian $\hat{H}^{(1)}(t)$, Eq. (3.78),

$$\bm{\gamma}(t) = \bm{\gamma}^{(1)}(t) + \bm{\gamma}^{(2)}(t) + \cdots \tag{11.39}$$

Contrary to response theory for exact states, in Section 3.11, or for coupled cluster wavefunctions, in Section 11.4, in MCSCF response theory the time dependence of the wavefunction is not determined directly from the time-dependent Schrödinger equation in the presence of the perturbation $\hat{H}^{(1)}(t)$, Eq. (3.74). Instead, one applies the Ehrenfest theorem, Eq. (3.58), to the operators, which determine the time dependence of the MCSCF wavefunction, i.e. the operators $\{\hat{h}_j\}$

$$\frac{d}{dt}\langle\Phi_0^{\text{MCSCF}}(t)|\hat{h}_j|\Phi_0^{\text{MCSCF}}(t)\rangle + \frac{i}{\hbar}\langle\Phi_0^{\text{MCSCF}}(t)|[\hat{h}_j, \hat{H}^{(0)} + \hat{H}^{(1)}(t)]|\Phi_0^{\text{MCSCF}}(t)\rangle = 0$$

(11.40)

Inserting the expression for the time-dependent MCSCF wavefunction, Eq. (11.36), and the perturbation expansion of the wavefunction parameters, Eq. (11.39), and separating the orders one finds for the first-order equation [see Exercise 11.5]

$$i\hbar\langle\Phi_0^{\text{MCSCF}}|[\hat{h}_j, \frac{d}{dt}\hat{\kappa}(t)^{(1)} + \frac{d}{dt}\hat{S}(t)^{(1)}]|\Phi_0^{\text{MCSCF}}\rangle$$
$$- \langle\Phi_0^{\text{MCSCF}}|[[\hat{h}_j, \hat{H}^{(0)}], \hat{\kappa}(t)^{(1)} + \hat{S}(t)^{(1)}]|\Phi_0^{\text{MCSCF}}\rangle$$
$$= -i\langle\Psi_0^{\text{MCSCF}}|[\hat{h}_j, \hat{H}^{(1)}(t)]|\Psi_0^{\text{MCSCF}}\rangle \quad (11.41)$$

where the first-order operators are defined as

$$\hat{\kappa}^{(1)}(t) = \sum_{p>q}\left[\kappa_{pq}^{(1)}(t)\hat{q}_{pq}^\dagger + \kappa_{pq}^{(1)*}(t)\hat{q}_{pq}\right] \quad (11.42)$$

$$\hat{S}^{(1)}(t) = \sum_{n\neq 0}\left[S_{n0}^{(1)}(t)\hat{R}_{n0}^\dagger + S_{n0}^{(1)*}(t)\hat{R}_{0n}\right] \quad (11.43)$$

Exercise 11.5 Derive the first-order equation Eq. (11.41) from the Ehrenfest theorem Eq. (11.40).

Using the implicit definitions, Eq. (10.60), of the MCRPA Hessian $\boldsymbol{E}$ and overlap $\boldsymbol{S}$ matrices one can write the first-order equation more compactly as

$$i\hbar\boldsymbol{S}\frac{d}{dt}\boldsymbol{\gamma}^{(1)}(t) - \boldsymbol{E}\boldsymbol{\gamma}^{(1)}(t) = \boldsymbol{T}(\hat{H}^{(1)}(t)) \quad (11.44)$$

This linear system of ordinary differential equations is normally solved by Fourier transforming it to the frequency domain (Fuchs et al., 1993). We insert therefore the Fourier transform of $\hat{H}^{(1)}(t)$ from Eq. (3.78) and an analogous Fourier transform of the first-order time-dependent wavefunction parameters $\boldsymbol{\gamma}^{(1)}(t)$

$$\boldsymbol{\gamma}^{(1)}(t) = \int_{-\infty}^{\infty} d\omega\, \boldsymbol{\gamma}^{(1)}(\omega) = \frac{1}{2}\sum_{\beta...}\int_{-\infty}^{\infty} d\omega\, \boldsymbol{X}(\hat{O}_{\beta...}^\omega)\, \mathcal{F}_{\beta...}(\omega)\, e^{-i\omega t} \quad (11.45)$$

and obtain [see Exercise 11.6]

$$(\hbar\omega\boldsymbol{S} - \boldsymbol{E})\,\boldsymbol{X}(\hat{O}_{\beta...}^\omega) = \boldsymbol{T}(\hat{O}_{\beta...}^\omega) \quad (11.46)$$

which are the MCRPA response equations (10.60) again. However, contrary to Section 10.4 here we have derived them from response theory applied to an approximated MCSCF wavefunction and not by approximating the expressions derived for exact wavefunctions using the superoperator formalism.

Exercise 11.6 Fill in the missing steps in the derivation of Eq. (11.46).

11.3 Second-Order Polarization Propagator Approximation

The formulation of approximate response theory based on an exponential parametrization of the time-dependent wavefunction, Eq. (11.36), and the Ehrenfest theorem, Eq. (11.40), can also be used to derive SOPPA and higher-order Møller–Plesset perturbation theory polarization propagator approximations (Olsen et al., 2005). Contrary to the approach employed in Chapter 10, which is based on the superoperator formalism from Section 3.12 and that could not yet be extended to higher than linear response functions, the Ehrenfest-theorem-based approach can be used to derive expressions also for quadratic and higher-order response functions. In the following, it will briefly be shown how the SOPPA linear response equations, Eq. (10.29), can be derived with this approach.

The key step is to bring the Møller–Plesset perturbation theory wavefunction, Eq. (9.64), into a form, that resembles a MCSCF wavefunction, i.e.

$$|\Phi_0^{MP}\rangle = C \left(|\Phi_0^{SCF}\rangle + \sum_{ai} t_i^a |\Phi_i^a\rangle + \sum_{\substack{a>b \\ i>j}} t_{ij}^{ab} |\Phi_{ij}^{ab}\rangle + \cdots \right) \quad (11.47)$$

$$= C \left(1 + \sum_{ai} t_i^a\, \hat{R}_i^{a\dagger} + \sum_{\substack{a>b \\ i>j}} t_{ij}^{ab}\, \hat{R}_{ij}^{ab\dagger} + \cdots \right) |\Phi_0^{SCF}\rangle \quad (11.48)$$

where the correlation coefficients are then expanded in Møller–Plesset perturbation theory orders of the fluctuation potential Eq. (9.63) as

$$t_{i\cdots}^{a\cdots} = t_{i\cdots}^{a\cdots}[1] + t_{i\cdots}^{a\cdots}[2] + \cdots \quad (11.49)$$

and the MP state-transfer operators are defined with respect to the Hartree–Fock wavefunction as

$$\hat{R}_{i\cdots}^{a\cdots\dagger} = |\Phi_{i\cdots}^{a\cdots}\rangle\langle\Phi_0^{SCF}| \quad (11.50)$$

$$\hat{R}_{i\cdots}^{a\cdots} = |\Phi_0^{SCF}\rangle\langle\Phi_{i\cdots}^{a\cdots}| \quad (11.51)$$

The time-dependent Møller–Plesset wavefunction $|\Phi_0^{MP}(t)\rangle$ is then written as

$$|\Phi_0^{MP}(t)\rangle = e^{\imath\hat{\kappa}(t)} e^{\imath\hat{S}(t)} |\Phi_0^{MP}\rangle \quad (11.52)$$

where the operators governing the time dependence of the wavefunction are defined as

$$\hat{\kappa}(t) = \sum_{p>q} \left[\kappa_{pq}(t)\, \hat{q}_{pq}^\dagger + \kappa_{pq}^*(t)\, \hat{q}_{pq} \right] \quad (11.53)$$

$$\hat{S}(t) = \sum_{\substack{ab\cdots \\ ij\cdots}} \left[S_{ij\cdots}^{ab\cdots}(t)\, \hat{R}_{ij\cdots}^{ab\cdots\dagger} + S_{ij\cdots}^{ab\cdots *}(t)\, \hat{R}_{ij\cdots}^{ab\cdots} \right] \quad (11.54)$$

One should note that single excitations and de-excitations are not included in the time-dependent MP state-transfer operator $\hat{S}(t)$ because they are already included as orbital rotations $\hat{\kappa}(t)$. With this ansatz for the wavefunction and the Ehrenfest theorem applied to it

$$\frac{d}{dt}\langle\Phi_0^{\rm MP}(t)|\hat{h}_j|\Phi_0^{\rm MP}(t)\rangle + \frac{\imath}{\hbar}\langle\Phi_0^{\rm MP}(t)|[\hat{h}_j,\hat{H}^{(0)}+\hat{H}^{(1)}(t)]|\Phi_0^{\rm MP}(t)\rangle = 0 \qquad (11.55)$$

one can follow the derivation in the previous chapter and derive the equation for the SOPPA solution vector Eq. (10.29) [see Exercise 11.7].

Exercise 11.7 Derive the SOPPA Hessian and overlap matrices using the Ehrenfest theorem Eq. (11.55).

11.4 Coupled Cluster Linear Response Functions

Coupled cluster response functions were derived by Koch and Jørgensen (1990) starting from a time-dependent version of the transition expectation value Eq. (9.95) of Arponen (1983)

$$\langle\Phi_0^\Lambda(t)|\hat{P}|\Phi_0^{\rm CC}(t)\rangle \qquad (11.56)$$

where the time-dependent coupled cluster state $|\Phi_0^{\rm CC}(t)\rangle$ and dual or "Λ" state $\langle\Phi_0^\Lambda(t)|$ are defined as

$$|\Phi_0^{\rm CC}(t)\rangle = e^{\hat{T}(t)}|\Phi_0^{\rm SCF}\rangle \qquad (11.57)$$

$$\langle\Phi_0^\Lambda(t)| = \langle\Phi_0^{\rm SCF}|\left[1+\hat{\Lambda}(t)\right]e^{-\hat{T}(t)} \qquad (11.58)$$

The time-dependent cluster and Λ operator consist, like their time-independent versions, Eqs. (9.73) and (9.97), of single, double and so forth excitation operators

$$\hat{T}(t) = \hat{T}_1(t) + \hat{T}_2(t) + \hat{T}_3(t) + \cdots + \hat{T}_N(t) \qquad (11.59)$$

$$\hat{\Lambda}(t) = \hat{\Lambda}_1(t) + \hat{\Lambda}_2(t) + \hat{\Lambda}_3(t) + \cdots + \hat{\Lambda}_N(t) \qquad (11.60)$$

with

$$\hat{T}_1(t) = \sum_{ai} t_i^a(t)\, \hat{q}_{ai}^\dagger \qquad (11.61)$$

$$\hat{T}_2(t) = \sum_{\substack{a>b \\ i>j}} t_{ij}^{ab}(t)\, \hat{q}_{ai}^\dagger\, \hat{q}_{bj}^\dagger \qquad (11.62)$$

$$\vdots$$

and

$$\hat{\Lambda}_1(t) = \sum_{ai} \lambda_i^a(t)\, \hat{q}_{ai} \qquad (11.63)$$

$$\hat{\Lambda}_2(t) = \sum_{\substack{a>b \\ i>j}} \lambda_{ij}^{ab}(t)\,\hat{q}_{ai}\,\hat{q}_{bj} \tag{11.64}$$

$$\vdots$$

Using again the shorthand notation $\{{}^e\hat{h}_{i_\mu}\}$ and $\{{}^d\hat{h}_{i_\mu}\}$, from Eq. (9.41), for all excitation and de-excitation operators one can write the time-dependent cluster and Λ operator more compactly as

$$\hat{T}(t) = \sum_{i_\mu} t_{i_\mu}(t)\,{}^e\hat{h}_{i_\mu} \tag{11.65}$$

$$\hat{\Lambda}(t) = \sum_{i_\mu} \lambda_{i_\mu}(t)\,{}^d\hat{h}_{i_\mu} \tag{11.66}$$

where $t_{i_\mu}(t)$ and $\lambda_{i_\mu}(t)$ are the corresponding time-dependent amplitudes.

The time-dependent amplitudes, $t_{i_\mu}(t)$ and $\lambda_{i_\mu}(t)$, are then determined from the coupled cluster time-dependent Schrödinger equations

$$e^{-\hat{T}(t)}\,\imath\hbar\frac{d}{dt}|\Phi_0^{\rm CC}(t)\rangle = e^{-\hat{T}(t)}\,\hat{H}(t)\,|\Phi_0^{\rm CC}(t)\rangle \tag{11.67}$$

$$\left(-\imath\hbar\frac{d}{dt}\langle\Phi_0^\Lambda(t)|\right)e^{\hat{T}(t)} = \langle\Phi_0^\Lambda(t)|\,\hat{H}(t)\,e^{\hat{T}(t)} \tag{11.68}$$

by projecting them on $\langle\Phi_0^{\rm SCF}|{}^d\hat{h}_{i_\mu}$ and ${}^e\hat{h}_{i_\mu}|\Phi_0^{\rm SCF}\rangle$, respectively, yielding two systems of ordinary linear differential Eqs. [see Exercise 11.8]

$$\imath\hbar\frac{dt_{i_\mu}(t)}{dt} = \langle\Phi_0^{\rm SCF}|{}^d\hat{h}_{i_\mu}\,e^{-\hat{T}(t)}\hat{H}(t)|\Phi_0^{\rm CC}(t)\rangle \tag{11.69}$$

$$-\imath\hbar\frac{d\lambda_{i_\mu}(t)}{dt} = \langle\Phi_0^\Lambda(t)|[\hat{H}(t),{}^e\hat{h}_{i_\mu}]|\Phi_0^{\rm CC}(t)\rangle \tag{11.70}$$

Exercise 11.8 Derive the equations for the time-dependent amplitudes, Eqs. (11.69) and (11.70).

Hint: In the derivation of the $\lambda_{i_\mu}(t)$ amplitudes one should make use of the resolution of identity as given in Eq. (9.44).

In the presence of a time-dependent perturbation $\hat{H}^{(1)}(t)$, Eq. (3.75), the amplitudes $t_{i_\mu}(t)$ and $\lambda_{i_\mu}(t)$ in Eqs. (11.69) and (11.70) are expanded in a perturbation series

$$t_{i_\mu}(t) = t_{i_\mu} + t_{i_\mu}^{(1)}(t) + \cdots \tag{11.71}$$

$$\lambda_{i_\mu}(t) = \lambda_{i_\mu} + \lambda_{i_\mu}^{(1)}(t) + \cdots \tag{11.72}$$

where t_{i_μ} and λ_{i_μ} are the unperturbed and time-independent amplitudes from Sections 9.6 and 9.7. Inserting these expansions in the differential equations and separating orders one obtains in first order

$$i\hbar \frac{dt_{i_\mu}^{(1)}(t)}{dt} = \langle \Phi_0^{\text{SCF}} | {}^d\hat{h}_{i_\mu} e^{-\hat{T}} \hat{H}^{(1)}(t) | \Phi_0^{\text{CC}} \rangle + \sum_{j_\nu} t_{j_\nu}^{(1)}(t) A_{i_\mu j_\nu} \quad (11.73)$$

$$-i\hbar \frac{d\lambda_{i_\mu}^{(1)}(t)}{dt} = \langle \Phi_0^{\Lambda} | [\hat{H}^{(1)}(t), {}^e\hat{h}_{i_\mu}] | \Phi_0^{\text{CC}} \rangle + \sum_{j_\nu} t_{j_\nu}^{(1)}(t) F_{i_\mu j_\nu} + \sum_{j_\nu} \lambda_{j_\nu}^{(1)}(t) A_{j_\nu i_\mu} \quad (11.74)$$

where $|\Phi_0^{\text{CC}}\rangle$ and $\langle\Phi_0^\Lambda|$ are the time-independent, unperturbed coupled cluster, Eq. (9.72), and "Λ" state, Eq. (9.96) wavefunctions, respectively, and $\hat{T}$ is the time-independent, unperturbed cluster operator. The elements of the coupled cluster Jacobian $\boldsymbol{A}$ matrix and of the $\boldsymbol{F}$ matrix are defined as

$$A_{i_\mu j_\nu} = \langle \Phi_0^{\text{SCF}} | {}^d\hat{h}_{i_\mu} e^{-\hat{T}} [\hat{H}^{(0)}, {}^e\hat{h}_{j_\nu}] | \Phi_0^{\text{CC}} \rangle \quad (11.75)$$

$$F_{i_\mu j_\nu} = \langle \Phi_0^\Lambda | [[\hat{H}^{(0)}, {}^e\hat{h}_{i_\mu}], {}^e\hat{h}_{j_\nu}] | \Phi_0^{\text{CC}} \rangle \quad (11.76)$$

The Jacobian matrix $\boldsymbol{A}$ can be shown [see Exercise 11.9] to be the first derivative of the time-independent coupled cluster amplitude equations, i.e. the coupled cluster vector function $\boldsymbol{e}$, Eq. (9.81), with respect to the time-independent amplitudes t_{j_ν}, i.e.

$$A_{i_\mu j_\nu} = \frac{\partial}{\partial t_{j_\nu}} e_{i_\mu} = \frac{\partial}{\partial t_{j_\nu}} \langle \Phi_0^{\text{SCF}} | {}^d\hat{h}_{i_\mu} e^{-\hat{T}} \hat{H}^{(0)} | \Phi_0^{\text{CC}} \rangle \quad (11.77)$$

Exercise 11.9 Derive the coupled cluster Jacobian in Eq. (11.75) as a derivative of the coupled cluster amplitude equations, i.e. prove Eq. (11.77).

Like in the previous two chapters the two differential equations, (11.73) and (11.74), for the amplitudes are solved by transforming them to the frequency domain. We define therefore Fourier components $X_{i_\mu}(\hat{O}_\beta^\omega...)$ and $Y_{i_\mu}(\hat{O}_\beta^\omega...)$ of the first-order amplitudes as

$$t_{i_\mu}^{(1)}(t) = \int_{-\infty}^{\infty} d\omega \, t_{i_\mu}^{(1)}(\omega) = \frac{1}{2} \sum_{\beta...} \int_{-\infty}^{\infty} d\omega \, X_{i_\mu}(\hat{O}_\beta^\omega...) \mathcal{F}_{\beta...}(\omega) \, e^{-\imath\omega t} \quad (11.78)$$

$$\lambda_{i_\mu}^{(1)}(t) = \int_{-\infty}^{\infty} d\omega \, \lambda_{i_\mu}^{(1)}(\omega) = \frac{1}{2} \sum_{\beta...} \int_{-\infty}^{\infty} d\omega \, Y_{i_\mu}(\hat{O}_\beta^\omega...) \mathcal{F}_{\beta...}(\omega) \, e^{-\imath\omega t} \quad (11.79)$$

Inserting them together with the Fourier transform of $\hat{H}^{(1)}(t)$, Eq. (3.78), in the two differential equations and evaluating the derivatives gives

$$\sum_{\beta\ldots}\int_{-\infty}^{\infty}d\omega\ \hbar\omega\ X_{i_\mu}(\hat{O}^\omega_{\beta\ldots})\ \mathcal{F}_{\beta\ldots}(\omega)\ e^{-\imath\omega t}$$

$$=\sum_{\beta\ldots}\int_{-\infty}^{\infty}d\omega\ \langle\Phi_0^{\mathrm{SCF}}|{}^d\hat{h}_{i_\mu}\ e^{-\hat{T}}\hat{O}^\omega_{\beta\ldots}|\Phi_0^{\mathrm{CC}}\rangle\mathcal{F}_{\beta\ldots}(\omega)\ e^{-\imath\omega t}$$

$$+\sum_{\beta\ldots}\int_{-\infty}^{\infty}d\omega\ \sum_{j_\nu}X_{j_\nu}(\hat{O}^\omega_{\beta\ldots})\ A_{i_\mu j_\nu}\ \mathcal{F}_{\beta\ldots}(\omega)\ e^{-\imath\omega t} \qquad (11.80)$$

$$-\sum_{\beta\ldots}\int_{-\infty}^{\infty}d\omega\ \hbar\omega\ Y_{i_\mu}(\hat{O}^\omega_{\beta\ldots})\ \mathcal{F}_{\beta\ldots}(\omega)\ e^{-\imath\omega t}$$

$$=\sum_{\beta\ldots}\int_{-\infty}^{\infty}d\omega\ \langle\Phi_0^\Lambda|[\hat{O}^\omega_{\beta\ldots},{}^e\hat{h}_{i_\mu}]|\Phi_0^{\mathrm{CC}}\rangle\ \mathcal{F}_{\beta\ldots}(\omega)\ e^{-\imath\omega t}$$

$$+\sum_{\beta\ldots}\int_{-\infty}^{\infty}d\omega\ \sum_{j_\nu}F_{i_\mu j_\nu}\ X_{j_\nu}(\hat{O}^\omega_{\beta\ldots})\ \mathcal{F}_{\beta\ldots}(\omega)\ e^{-\imath\omega t}$$

$$+\sum_{\beta\ldots}\int_{-\infty}^{\infty}d\omega\ \sum_{j_\nu}Y_{j_\nu}(\hat{O}^\omega_{\beta\ldots})\ A_{j_\nu i_\mu}\ \mathcal{F}_{\beta\ldots}(\omega)\ e^{-\imath\omega t} \qquad (11.81)$$

Both equations have to be fulfilled for any frequency ω and strength of the field component $\mathcal{F}_{\beta\ldots}(\omega)$, which implies that

$$\hbar\omega\ X_{i_\mu}(\hat{O}^\omega_{\beta\ldots}) = \langle\Phi_0^{\mathrm{SCF}}|{}^d\hat{h}_{i_\mu}\ e^{-\hat{T}}\hat{O}^\omega_{\beta\ldots}|\Phi_0^{\mathrm{CC}}\rangle + \sum_{j_\nu}X_{j_\nu}(\hat{O}^\omega_{\beta\ldots})\ A_{i_\mu j_\nu} \qquad (11.82)$$

$$\hbar\omega\ Y_{i_\mu}(\hat{O}^\omega_{\beta\ldots}) = -\langle\Phi_0^\Lambda|[\hat{O}^\omega_{\beta\ldots},{}^e\hat{h}_{i_\mu}]|\Phi_0^{\mathrm{CC}}\rangle - \sum_{j_\nu}F_{i_\mu j_\nu}\ X_{j_\nu}(\hat{O}^\omega_{\beta\ldots}) - \sum_{j_\nu}Y_{j_\nu}(\hat{O}^\omega_{\beta\ldots})\ A_{j_\nu i_\mu}$$

$$(11.83)$$

Isolating the unknown Fourier components $X_{i_\mu}(\hat{O}^\omega_{\beta\ldots})$ and $Y_{i_\mu}(\hat{O}^\omega_{\beta\ldots})$ and collecting them in two column vectors $\boldsymbol{X}(\hat{O}^\omega_{\beta\ldots})$ and $\boldsymbol{Y}(\hat{O}^\omega_{\beta\ldots})$ we obtain two sets of coupled linear equations

$$\boldsymbol{X}(\hat{O}^\omega_{\beta\ldots}) = (\hbar\omega\ \mathbf{1} - \boldsymbol{A})^{-1}\ \langle\Phi_0^{\mathrm{SCF}}|{}^d\boldsymbol{h}e^{-\hat{T}}\hat{O}^\omega_{\beta\ldots}|\Phi_0^{\mathrm{CC}}\rangle \qquad (11.84)$$

$$\boldsymbol{Y}^T(\hat{O}^\omega_{\beta\ldots}) = -\left[\langle\Phi_0^\Lambda|[\hat{O}^\omega_{\beta\ldots},{}^e\boldsymbol{h}^T]|\Phi_0^{\mathrm{CC}}\rangle + \boldsymbol{F}\boldsymbol{X}(\hat{O}^\omega_{\beta\ldots})\right](\hbar\omega\ \mathbf{1} + \boldsymbol{A})^{-1} \qquad (11.85)$$

where $\mathbf{1}$ is a unit matrix of the same dimension as $\boldsymbol{A}$. One should note that the frequency-dependent cluster amplitudes $\boldsymbol{X}(\hat{O}^\omega_{\beta\ldots})$ have to be determined before one can solve for the amplitudes $\boldsymbol{Y}(\hat{O}^\omega_{\beta\ldots})$ of the "Λ" state.

These are the analogous equations to the response equations for Møller–Plesset perturbation theory polarization propagators or MCSCF linear response functions in Eqs. (10.29) and (11.46). However, there are a few important differences. First, in coupled cluster linear response theory one has to solve two sets of equations, one for each type of amplitudes. On the other hand, the coupled cluster Jacobian matrix $\boldsymbol{A}$ is only half the size of corresponding Hessian matrices in Møller–Plesset polarization propagator methods described in Sections 10.3 and 11.3, because the set of operators $\{\hat{h}_{i_\mu}\}$ consists of only excitation or de-excitation operators in coupled cluster response theory and not both, as in Møller–Plesset polarization propagator theory. Finally, the coupled cluster Jacobian matrix $\boldsymbol{A}$ is inherently asymmetric, which implies that the left and right eigenvectors of it will not be the same.

Having determined the time dependence of the coupled cluster and "Λ" wavefunctions to first order, i.e. having derived expressions for the Fourier components $X_{i_\mu}(\hat{O}^\omega_{\beta...})$ and $Y_{i_\mu}(\hat{O}^\omega_{\beta...})$ of the first-order time-dependent amplitudes in Eqs. (11.78) and (11.79), we can insert them now in an expansion of the time-dependent transition expectation value, Eq. (11.56), in orders of the perturbation

$$\langle \Phi^\Lambda_0(t) | \hat{P} | \Phi^{CC}_0(t) \rangle = \langle \Phi^\Lambda_0 | \hat{P} | \Phi^{CC}_0 \rangle$$
$$+ \sum_{i_\mu} t^{(1)}_{i_\mu}(t) \langle \Phi^\Lambda_0 | [\hat{P}, {}^e h_{i_\mu}] | \Phi^{CC}_0 \rangle$$
$$+ \sum_{i_\mu} \lambda^{(1)}_{i_\mu}(t) \langle \Phi^{SCF}_0 | {}^d h_{i_\mu} e^{-\hat{T}} \hat{P} | \Phi^{CC}_0 \rangle + \cdots \quad (11.86)$$

and obtain

$$\langle \Phi^\Lambda_0(t) | \hat{P} | \Phi^{CC}_0(t) \rangle = \langle \Phi^\Lambda_0 | \hat{P} | \Phi^{CC}_0 \rangle$$
$$+ \frac{1}{2} \sum_{\beta...} \int_{-\infty}^\infty d\omega \sum_{i_\mu} X_{i_\mu}(\hat{O}^\omega_{\beta...}) \langle \Phi^\Lambda_0 | [\hat{P}, {}^e h_{i_\mu}] | \Phi^{CC}_0 \rangle \mathcal{F}_{\beta...}(\omega) e^{-i\omega t} + \cdots$$
$$+ \frac{1}{2} \sum_{\beta...} \int_{-\infty}^\infty d\omega \sum_{i_\mu} Y_{i_\mu}(\hat{O}^\omega_{\beta...}) \langle \Phi^{SCF}_0 | {}^d h_{i_\mu} e^{-\hat{T}} \hat{P} | \Phi^{CC}_0 \rangle \mathcal{F}_{\beta...}(\omega) e^{-i\omega t} \quad (11.87)$$

Comparison with the analogous expansion of an expectation value for exact states, Eq. (3.109), shows that the coupled cluster linear response function is given as

$$\langle\langle \hat{P} ; \hat{O}^\omega_{\beta...} \rangle\rangle_\omega = \boldsymbol{T}^{\Lambda,T}(\hat{P}) \boldsymbol{X}(\hat{O}^\omega_{\beta...}) + \boldsymbol{Y}^T(\hat{O}^\omega_{\beta...}) \boldsymbol{T}^{CC}(\hat{P}) \quad (11.88)$$

where the elements of the "property gradient" vectors $\boldsymbol{T}^\Lambda(\hat{O})$ and $\boldsymbol{T}^{CC}(\hat{O})$ for an operator $\hat{O}$ are defined as

$$T^\Lambda_{i_\mu}(\hat{O}) = \langle \Phi^\Lambda_0 | [\hat{O}, {}^e h_{i_\mu}] | \Phi^{CC}_0 \rangle \quad (11.89)$$

$$T^{CC}_{i_\mu}(\hat{O}) = \langle \Phi^{SCF}_0 | {}^d h_{i_\mu} e^{-\hat{T}} \hat{O} | \Phi^{CC}_0 \rangle \quad (11.90)$$

Inserting the two solutions vectors $\boldsymbol{X}(\hat{O}^\omega_{\beta...})$ and $\boldsymbol{Y}(\hat{O}^\omega_{\beta...})$ we finally obtain

$$\langle\langle \hat{P} ; \hat{O}^{\omega}_{\beta\ldots}\rangle\rangle_{\omega} = \boldsymbol{T}^{\Lambda,T}(\hat{P})\,(\hbar\omega\,\mathbf{1} - \boldsymbol{A})^{-1}\,\boldsymbol{T}^{\mathrm{CC}}(\hat{O}^{\omega}_{\beta\ldots})$$
$$+ \boldsymbol{T}^{\Lambda,T}(\hat{O}^{\omega}_{\beta\ldots})\,(-\hbar\omega\,\mathbf{1} - \boldsymbol{A})^{-1}\,\boldsymbol{T}^{\mathrm{CC}}(\hat{P})$$
$$+ \left[\boldsymbol{F}\,(\hbar\omega\,\mathbf{1} - \boldsymbol{A})^{-1}\,\boldsymbol{T}^{\mathrm{CC}}(\hat{O}^{\omega}_{\beta\ldots})\right]^{T}(-\hbar\omega\,\mathbf{1} - \boldsymbol{A})^{-1}\,\boldsymbol{T}^{\mathrm{CC}}(\hat{P}) \quad (11.91)$$

In coupled cluster response theory the poles of the linear response function and thus the vertical excitation energies are then found as eigenvalues of the coupled cluster Jacobian

$$(\boldsymbol{A} - \hbar\omega\mathbf{1})\,\boldsymbol{R} = 0 \tag{11.92}$$

For the most common coupled cluster response function methods, CCSD (Koch and Jørgensen, 1990) and CC2 (Christiansen et al., 1995b), the Jacobian can be derived as derivatives with respect to the t_{i_ν} amplitudes of the CCSD and CC2 vector functions in Eqs. (9.83), (9.84) and (9.86), (9.87)

$$\boldsymbol{A}^{\mathrm{CCSD}} = \tag{11.93}$$

$$\begin{pmatrix} \langle \Phi_0^{\mathrm{SCF}}|\,^d\boldsymbol{h}_1[\hat{H}^{T_1} + [\hat{H}^{T_1},\hat{T}_2],\,^e\boldsymbol{h}_1^T]\,|\Phi_0^{\mathrm{SCF}}\rangle & \langle \Phi_0^{\mathrm{SCF}}|\,^d\boldsymbol{h}_1[\hat{H}^{T_1},\,^e\boldsymbol{h}_2^T]\,|\Phi_0^{\mathrm{SCF}}\rangle \\ \langle \Phi_0^{\mathrm{SCF}}|\,^d\boldsymbol{h}_2[\hat{H}^{T_1} + [\hat{H}^{T_1},\hat{T}_2],\,^e\boldsymbol{h}_1^T]\,|\Phi_0^{\mathrm{SCF}}\rangle & \langle \Phi_0^{\mathrm{SCF}}|\,^d\boldsymbol{h}_2[\hat{H}^{T_1},\,^e\boldsymbol{h}_2^T]\,|\Phi_0^{\mathrm{SCF}}\rangle \end{pmatrix}$$

$$\boldsymbol{A}^{\mathrm{CC2}} = \tag{11.94}$$

$$\begin{pmatrix} \langle \Phi_0^{\mathrm{SCF}}|\,^d\boldsymbol{h}_1[\hat{H}^{T_1} + [\hat{H}^{T_1},\hat{T}_2],\,^e\boldsymbol{h}_1^T]\,|\Phi_0^{\mathrm{SCF}}\rangle & \langle \Phi_0^{\mathrm{SCF}}|\,^d\boldsymbol{h}_1[\hat{H}^{T_1},\,^e\boldsymbol{h}_2^T]\,|\Phi_0^{\mathrm{SCF}}\rangle \\ \langle \Phi_0^{\mathrm{SCF}}|\,^d\boldsymbol{h}_2[\hat{H}^{T_1},\,^e\boldsymbol{h}_1^T]\,|\Phi_0^{\mathrm{SCF}}\rangle & \langle \Phi_0^{\mathrm{SCF}}|\,^d\boldsymbol{h}_2[\hat{F},\,^e\boldsymbol{h}_2^T]\,|\Phi_0^{\mathrm{SCF}}\rangle \end{pmatrix}$$

where $\hat{H}^{T_1}$ is the T_1 transformed unperturbed Hamiltonian

$$\hat{H}^{T_1} = e^{-\hat{T}_1}\hat{H}^{(0)}\,e^{\hat{T}_1} \tag{11.95}$$

CC2 linear response theory as well as SOPPA, in Section 10.3, are in principle both second-order response function methods, although there are significant differences. Nevertheless, one can compare the blocks of elements in the CC2 Jacobian with the corresponding matrices in the SOPPA Hessian. The "$^d\boldsymbol{h}_2,\,^e\boldsymbol{h}_2^T$" block in CC2 and the $\boldsymbol{D}^{(0)}$ matrix in SOPPA have essentially the same elements. Also the "$^d\boldsymbol{h}_1,\,^e\boldsymbol{h}_2^T$" and "$^d\boldsymbol{h}_2,\,^e\boldsymbol{h}_1^T$" blocks in CC2 contain elements similar to the $\tilde{\boldsymbol{C}}^{(1)}$ and $\boldsymbol{C}^{(1)}$ matrices. However, there are also additional contributions in CC2 due to the T_1 transformed Hamiltonian. The same holds also for the "$^d\boldsymbol{h}_1,\,^e\boldsymbol{h}_1^T$" block in CC2, which corresponds to the $\boldsymbol{A}^{0,1,2}$ matrix in SOPPA.

Response functions have also been derived for the CC3 model (Christiansen et al., 1995a), which is an approximation to the CCSDT model (Hald et al., 2001) in the same way as CC2 is to CCSD. The immense number of triple excitations included in the CC3 Jacobian makes it necessary to formulate it in a partitioned form using Eq. (10.14). However, the partitioned CC3 Jacobian then depends also on its own eigenvalues similar to the partitioned SOPPA Hessian in Eq. (10.49), which implies that one has to iterate on each eigenvalue separately. As an alternative, an approximation to CC3 has been developed by Christiansen et al. (1996) using the pseudo-perturbation theory from Section 3.13, where the triples excitations are included as non-iterative corrections to CCSD excitation energies. This **CCSDR(3)** method closely reproduces the results of CC3 calculations for vertical excitation energies (Sauer et al., 2009; Falden et al., 2009; Silva-Junior et al., 2010).

A method closely related to the CCSD linear response function approach but derived differently is the **equation-of-motion coupled cluster approach (EOM-CCSD)** (Sekino and Bartlett, 1984; Geertsen et al., 1989; Stanton and Bartlett, 1993). The EOM-CCSD excitation energies are identical to the excitation energies obtained from the CCSD linear response function, but the transition moments and second-order properties, like frequency-dependent polarizabilities of spin-spin coupling constants, differ somewhat.

11.5 Further Reading

MCSCF Response Theory

J. Olsen and P. Jørgensen, *Time-dependent response theory with applications to self-consistent field and multiconfigurational self-consistent field wave functions*, in *Modern Electronic Structure Theory* ed. D. R. Yarkony, World Scientific, Singapore (1995): Chapter 13.

12
Derivative Methods

In this chapter we will finally follow the third approach, which means that we abandon the perturbation-theory approach all together and go back to the definitions of the properties as derivatives of the energy in the presence of the perturbation. We will illustrate with a few examples how this approach can be applied to approximate expressions for the energy in the presence of both static as well as time-dependent perturbations. However, the presentation will be very brief and restricted to Møller–Plesset perturbation theory and coupled cluster energies as nothing new is obtained for variational methods compared to the response theory approaches in Chapters 10 and 11.

12.1   The Finite-Field Method

The finite-field method of Cohen and Roothaan (1965) and Pople, *et al.* (1968) involves numerical evaluation of derivatives of the electronic energy, of first- or higher-order properties, or in general of a property, P, in the presence of a perturbation operator, $\hat{O}_\alpha^\mathcal{F} \mathcal{F}_\alpha$. Calculations of P are performed for various values of field strength $\mathcal{F}_\alpha$. The desired derivative, at zero field strength, can then be obtained either by finite differences or by fitting the calculated values of P to a Taylor expansion in the field strength $\mathcal{F}_\alpha$.

In a finite-field calculation of the $\alpha_{\alpha\alpha}$ component of the static dipole polarizability tensor, for example, the perturbation operator, $-\hat{O}_\alpha^\mathcal{E} \mathcal{E}_\alpha$, is added to the Hamiltonian, $H^{(0)}$, and the electronic energy $E(\mathcal{E}_\alpha)$ or the electronic contribution to the α component of the dipole moment $\mu_\alpha(\mathcal{E}_\alpha)$ calculated for various finite values of the electric field strength $\mathcal{E}_\alpha$. $\alpha_{\alpha\alpha}$ is then obtained as the numerical first derivative with respect to $\mathcal{E}_\alpha$ of $\mu_\alpha(\mathcal{E}_\alpha)$ or as the numerical second derivative of the electric-field-dependent electronic energy $E(\mathcal{E}_\alpha)$. Off-diagonal elements $\alpha_{\beta\alpha}$ of the polarizability can be similarly obtained from the $\mu_\beta(\mathcal{E}_\alpha)$, whereas the calculation as second derivative of the energy requires a 2-dimensional fit.

The property P of which derivatives can be taken need not be a static property, but can also be a frequency-dependent polarizability $\alpha(-\omega;\omega)$, as e.g. done by Jaszuński (1987). Finite-field calculations on $\alpha(-\omega;\omega)$ facilitate calculation of $\beta(-\omega;\omega,0)$, $\gamma(-\omega;\omega,0,0)$ and so forth.

The finite-field method is by far the easiest method to implement as long as the perturbations are real. Any program for the calculation of the property P can be used, as long as it is possible to include additional one-electron operators in the Hamiltonian.

The finite-field method can thus be applied at any level of approximation or correlation and even to approximations or methods for which a wavefunction or a ground-state energy is not defined. The latter approach was used for example for the calculation of the static second hyperpolarizability $\gamma(0;0,0,0)$ of Li$^-$ as second derivative of $\alpha(0;0)$ at the SOPPA(CCSD) level (Sauer, 1997).

Imaginary perturbation operators, like $\hat{O}_\alpha^{l\mathcal{B}}$, $\hat{O}_\alpha^{s\mathcal{B}}$, $\hat{O}_\alpha^{lm^K}$ and $\hat{O}_\alpha^{sm^K}$, require the use of complex arithmetic, which prevented a routine usage of the finite-field method for the calculation of magnetic properties. Nevertheless, finite-field approaches to the calculation of nuclear magnetic shielding constants (Fukui et al., 1992a) and nuclear spin-spin coupling constants (Fukui et al., 1992b) have been presented. In this method the paramagnetic contribution to $\sigma_{\alpha\beta}^K$, for example, was evaluated as numerical derivative of the expectation value of $\hat{O}_\alpha^{lm^K}$ with respect to $\mathcal{B}_\beta$. The expectation value itself was calculated to second order in electron correlation in the presence of the magnetic induction $\mathcal{B}_\beta$. The Fermi contact, spin-dipolar and orbital paramagnetic contributions to the coupling constants $J_{\alpha\beta}^{KL}$ were obtained as numerical derivatives with respect to m_α^K of an expression for the energy, which is second order in electron correlation and first order in $-\left(\hat{O}_\beta^{lm^L} + \hat{O}_\beta^{sm^L}\right)$ and is thus calculated in the presence of the perturbation $-\left(\hat{O}_\alpha^{lm^K} + \hat{O}_\alpha^{sm^K}\right)m_\alpha^K$. Mixed electric–magnetic properties, on the other hand, like nuclear magnetic shielding polarizabilities can easily be evaluated as numerical derivatives of electric-field-dependent nuclear magnetic shielding tensors without complex arithmetic.

A disadvantage of the finite-field method lies in the nature of numerical differentiation. Care must be taken in choosing the field strength, in our example $\mathcal{E}_\alpha$, which must not be too high, and in the number of different field strengths for which the property P is evaluated. For higher-order properties or multiple perturbations the method becomes cumbersome since the number of calculations to be performed increases rapidly. Secondly, adding the field to the Hamiltonian lowers the symmetry and therefore increases the computational cost of these calculations compared to the calculations without field. Finally, the method can obviously not be used for time-dependent perturbations and therefore for frequency-dependent properties.

A variation of this method is the finite point charge method, used by Maroulis and Thakkar (1988), in which the external electric field or field gradient is simulated by an appropriate arrangement of point charges. This method is even simpler to implement, since it only requires the option to include centres with a charge but no basis functions, rather than a modified one-electron Hamiltonian.

The finite-field method has been widely used but is becoming increasingly obsolete because of the advances in the analytical derivative methods: most electromagnetic properties can now be calculated analytically, obviating the need for a finite-field calculation.

However, there is one important exception, the calculation of geometrical derivatives of electromagnetic properties, i.e. derivatives with respect to a change in a nuclear coordinate or a normal coordinate. These derivatives, in particular first and second derivatives, are needed for the calculation of the vibrational averaging corrections to molecular properties as discussed in Chapter 8. Analytical derivatives are so far only implemented for a few, mainly first derivatives of first-order properties. Derivatives of other properties are therefore still done numerically nowadays, see e.g. (Bishop and Sauer, 1997; Sauer et al., 2001; Ruden et al., 2003).

12.2 The Analytic Derivative Method

In the analytic derivative method for the calculation of molecular properties, approximate expressions for P within a given method are differentiated analytically with respect to the perturbation. It is equally general as the finite-field method and does not suffer from the numerical problems of the latter method. However, it is much more difficult to apply to a new type of wavefunction, since expressions for the analytical derivatives have to be derived and implemented. Nevertheless, expressions for first- and second-order properties have been implemented for most *ab initio* methods following the derivation of analytical derivatives with respect to changes in the nuclear coordinates. Explicit expressions can be found in several reviews (Helgaker and Jørgensen, 1988; Amos and Rice, 1989; Gauss and Cremer, 1992).

12.2.1 First Derivatives

Let us consider again a system in the presence of a general perturbation $\mathcal{F}_{\alpha\cdots}$ that could be a component of the electric field $\mathcal{E}_\alpha$, field gradient $\mathcal{E}_{\alpha\beta}$, magnetic induction $\mathcal{B}_\alpha$ or nuclear moment m_α^K with the first- and second-order perturbation Hamiltonians $\hat{H}^{(1)} + \hat{H}^{(2)}$ as given in Eq. (2.101) and Eqs. (2.108)–(2.110).

The first derivative of the energy of such a perturbed system with respect to $\mathcal{F}_{\alpha\cdots}$ can then be written for most methods as

$$\left.\frac{dE(\vec{\mathcal{F}})}{d\mathcal{F}_{\alpha\cdots}}\right|_{|\vec{\mathcal{F}}|=0} = \sum_{pq} D_{pq} \langle \phi_p | \frac{\partial \hat{h}^{(1)}}{\partial \mathcal{F}_{\alpha\cdots}} | \phi_q \rangle = \sum_{\mu\nu} D_{\mu\nu} \langle \chi_\mu | \frac{\partial \hat{h}^{(1)}}{\partial \mathcal{F}_{\alpha\cdots}} | \chi_\nu \rangle \qquad (12.1)$$

or

$$\left.\frac{dE(\vec{\mathcal{F}})}{d\mathcal{F}_{\alpha\cdots}}\right|_{|\vec{\mathcal{F}}|=0} = \sum_{pq} D_{pq} \langle \phi_p | \hat{o}_{\alpha\cdots}^{\mathcal{F}} | \phi_q \rangle = \sum_{\mu\nu} D_{\mu\nu} \langle \chi_\mu | \hat{o}_{\alpha\cdots}^{\mathcal{F}} | \chi_\nu \rangle \qquad (12.2)$$

where D_{pq} and $D_{\mu\nu}$ are density matrices in the molecular and atomic orbital basis. The atomic orbitals χ_μ are here assumed to be independent of the perturbation. For variational wavefunctions, i.e. methods that fulfill the Hellmann–Feynman theorem, this result is equivalent to Eqs. (10.1) and (10.2) and the density matrices are identical.

However, for non-variational wavefunctions, as for example in the case of Møller–Plesset perturbation theory and the coupled cluster methods, the density matrices here are not consistent with the definition in Eq. (2.20) and thus Eqs. (9.107) and (9.108). The density matrices defined in Eq. (12.1) are for those methods therefore called **relaxed or response density matrix** (Trucks et al., 1988). In Section 9.7 it was discussed that the Hellmann–Feynman theorem can be fulfilled in non-variational methods such as coupled cluster theory (see e.g. Perera et al., 1996), when the first-order properties are defined as first derivatives of the Lagrangian Eq. (9.95) instead of the energy directly, meaning that the energy and first-order properties are evaluated as transition expectation values, Eqs. (9.95) and (9.103). The relaxed CC density can therefore also be calculated as the transition expectation value of the density operator, Eq. (2.21),

$$\rho^{CC,rel}(\vec{r}) = \langle \Phi_\Lambda | \hat{D}(\vec{r}) | \Phi_{CC} \rangle = \sum_{pq} \phi_p(\vec{r}) \phi_q^*(\vec{r}) D_{pq}^{CC,rel} \qquad (12.3)$$

The relaxed density matrix can in general be decomposed in an SCF and correlation part

$$\mathbf{D}^{rel} = \mathbf{D}^{SCF} + \mathbf{D}^{corr,rel} \qquad (12.4)$$

The SCF density is given in Eq. (9.112) and the correlation part consists of two parts

$$\mathbf{D}^{corr,rel} = \mathbf{D}^{amp,rel} + \mathbf{D}^{orb,rel} \qquad (12.5)$$

where $\mathbf{D}^{amp,rel}$ contains amplitudes or correlation coefficients and $\mathbf{D}^{orb,rel}$, obtained as a solution of the so-called Z-vector equations (Handy and Schaefer, 1984), arises because of the relaxation of the orbitals for non-variational wavefunctions. Only the occupied-virtual and virtual-occupied blocks are non-zero in $\mathbf{D}^{orb,rel}$, which is again a result of the Brillouin condition for the SCF ground state.

In the following, this will be illustrated for the MP2 energy correction, Eq. (9.68), whose straightforward differentiation gives

$$\left. \frac{dE^{MP2}(\vec{\mathcal{F}})}{d\mathcal{F}_{\alpha\cdots}} \right|_{\vec{\mathcal{F}}=0} = \sum_{ab} D_{ab}^{MP2} \langle \phi_a | \frac{\partial \hat{h}^{(1)}}{\partial \mathcal{F}_{\alpha\cdots}} | \phi_b \rangle + \sum_{ij} D_{ij}^{MP2} \langle \phi_i | \frac{\partial \hat{h}^{(1)}}{\partial \mathcal{F}_{\alpha\cdots}} | \phi_j \rangle$$

$$+ \sum_{ai} L_{ai}^{MP2} U_{\alpha\cdots,ai}^{(1)} \qquad (12.6)$$

or

$$\left. \frac{dE^{MP2}(\vec{\mathcal{F}})}{d\mathcal{F}_{\alpha\cdots}} \right|_{\vec{\mathcal{F}}=0} = \sum_{ab} D_{ab}^{MP2} \langle \phi_a | \hat{o}_{\alpha\cdots}^{\mathcal{F}} | \phi_b \rangle + \sum_{ij} D_{ij}^{MP2} \langle \phi_i | \hat{o}_{\alpha\cdots}^{\mathcal{F}} | \phi_j \rangle$$

$$+ \sum_{ai} L_{ai}^{MP2} U_{\alpha\cdots,ai}^{(1)} \qquad (12.7)$$

where D_{ij}^{MP2} and D_{ab}^{MP2} are the occupied-occupied and virtual-virtual blocks of the second-order correction to the density matrix as given in Eqs. (9.116) and (9.117). L_{ai}^{MP2} is called a Lagrangian although it differs from the Lagrangians discussed in Section 9.7 and is given as

$$L_{ai}^{\text{MP2}} = -4 \sum_{jbc} \left(2t_{ij}^{cb}[1] - t_{ji}^{cb}[1]\right) \left(\phi_a(\vec{r}_1)\,\phi_c(\vec{r}_1)\big|\phi_b(\vec{r}_2)\,\phi_j(\vec{r}_2)\right)$$

$$- 4 \sum_{jkb} \left(2t_{kj}^{ab}[1] - t_{jk}^{cb}[1]\right) \left(\phi_k(\vec{r}_1)\,\phi_i(\vec{r}_1)\big|\phi_b(\vec{r}_2)\,\phi_j(\vec{r}_2)\right)$$

$$+ \frac{1}{2} \sum_{bc} D_{bc}^{\text{MP2}} \left[4\left(\phi_a(\vec{r}_1)\,\phi_i(\vec{r}_1)\big|\phi_b(\vec{r}_2)\,\phi_c(\vec{r}_2)\right) - \left(\phi_a(\vec{r}_1)\,\phi_b(\vec{r}_1)\big|\phi_i(\vec{r}_2)\,\phi_c(\vec{r}_2)\right)\right.$$

$$\left. - \left(\phi_a(\vec{r}_1)\,\phi_c(\vec{r}_1)\big|\phi_b(\vec{r}_2)\,\phi_i(\vec{r}_2)\right)\right]$$

$$+ \frac{1}{2} \sum_{jk} D_{jk}^{\text{MP2}} \left[4\left(\phi_a(\vec{r}_1)\,\phi_i(\vec{r}_1)\big|\phi_k(\vec{r}_2)\,\phi_j(\vec{r}_2)\right) - \left(\phi_a(\vec{r}_1)\,\phi_k(\vec{r}_1)\big|\phi_i(\vec{r}_2)\,\phi_j(\vec{r}_2)\right)\right.$$

$$\left. - \left(\phi_a(\vec{r}_1)\,\phi_j(\vec{r}_1)\big|\phi_k(\vec{r}_2)\,\phi_i(\vec{r}_2)\right)\right] \tag{12.8}$$

Finally, $U_{\alpha\cdots,ai}^{(1)}$ are the solutions of the coupled Hartree–Fock equations, which enter in the expression for the derivative, because the molecular orbitals also depend on field. Inserting the expression for $U_{\alpha\cdots,ai}^{(1)}$ from Eq. (11.25) the third term can be written more explicitly as

$$\sum_{ai} L_{ai}^{\text{MP2}} U_{\alpha\cdots,ai}^{(1)} = -\sum_{ai}\sum_{bj} L_{ai}^{\text{MP2}} \left(\mathbf{A}^{(0,1)} - \mathbf{B}^{(1)}\right)^{-1}_{ai,bj} \langle\phi_b|\hat{o}_{\alpha\cdots}^{\mathcal{F}}|\phi_j\rangle \tag{12.9}$$

However, instead of solving the coupled Hartree–Fock equations for all components α of the field $\vec{\mathcal{F}}$, i.e. with the molecular orbital integrals $\langle\phi_b|\hat{o}_{\alpha\cdots}^{\mathcal{F}}|\phi_j\rangle$ as the right-hand side, one can solve one set of coupled Hartree–Fock equations for the so-called Z-vector (Handy and Schaefer, 1984), with this second type of Lagrangian, L_{ai}^{MP2}, as the right-hand side

$$\sum_{bj} \left(A_{ai,bj}^{(0,1)} - B_{ai,bj}^{(1)}\right) Z_{bj} = -L_{ai}^{\text{MP2}} \tag{12.10}$$

The last term in Eq. (12.6) can then be written as

$$\sum_{ai} L_{ai}^{\text{MP2}} U_{\alpha\cdots,ai}^{(1)} = \sum_{ai} Z_{ai} \langle\phi_a|\hat{o}_{\alpha\cdots}^{\mathcal{F}}|\phi_i\rangle \tag{12.11}$$

which is completely analogous to the other two contributions to the first derivative in Eq. (12.6), because it consists of a contraction of molecular orbital integrals of the perturbation operator $\hat{o}_{\alpha\cdots}^{\mathcal{F}}$ with another vector. The only difference is that now the summation is over pairs of occupied and virtual molecular orbitals. Defining therefore the occupied-virtual and virtual-orbital blocks $D_{ia}^{\text{MP2},rel}$ and $D_{ai}^{\text{MP2},rel}$ of the relaxed density matrix to be equal to the Z vector

$$D_{ai}^{\text{MP2},rel} = Z_{ai} \tag{12.12}$$

one can write the first derivative of the MP2 energy correction compactly as the contraction of all molecular orbital integrals of the perturbation operator with the corresponding elements of the relaxed density matrix

248 Derivative Methods

$$\left.\frac{dE^{\text{MP2,rel}}(\vec{\mathcal{F}})}{d\mathcal{F}_{\alpha\cdots}}\right|_{\vec{\mathcal{F}}=0} = \sum_{pq} D_{pq}^{\text{MP2}} \langle \phi_q | \hat{o}_{\alpha\cdots}^{\mathcal{F}} | \phi_p \rangle \quad (12.13)$$

as stated in Eq. (12.1) and in analogy to the expression in Eq. (10.1) for the unrelaxed density matrix consistent through second order in Eqs. (9.116)–(9.118). The difference between the unrelaxed density matrix consistent through second order (Jensen et al., 1988a; 1988b) and the relaxed density MP2 matrix (Gauss and Cremer, 1992; Cybulski and Bishop, 1994) is therefore in the occupied-virtual and virtual-orbital blocks.

12.2.2 Second Derivatives

The second derivative of the energy can be written in terms of atomic orbitals as

$$\left.\frac{d^2 E(\vec{\mathcal{F}})}{d\mathcal{F}_{\alpha\cdots} d\mathcal{F}_{\beta\cdots}}\right|_{|\vec{\mathcal{F}}|=0} = \sum_{\mu\nu} D_{\mu\nu} \langle \chi_\mu | \frac{\partial^2 \hat{h}^{(2)}}{\partial \mathcal{F}_{\beta\cdots} \partial \mathcal{F}_{\alpha\cdots}} | \chi_\nu \rangle + \sum_{\mu\nu} D^{(1)}_{\beta\cdots,\mu\nu} \langle \chi_\mu | \frac{\partial \hat{h}^{(1)}}{\partial \mathcal{F}_{\alpha\cdots}} | \chi_\nu \rangle$$

(12.14)

or

$$\left.\frac{d^2 E(\vec{\mathcal{F}})}{d\mathcal{F}_{\alpha\cdots} d\mathcal{F}_{\beta\cdots}}\right|_{|\vec{\mathcal{F}}|=0} = \sum_{\mu\nu} D_{\mu\nu} \langle \chi_\mu | \hat{o}_{\alpha\beta\cdots}^{\mathcal{F}\mathcal{F}} + \hat{o}_{\beta\alpha\cdots}^{\mathcal{F}\mathcal{F}} | \chi_\nu \rangle + \sum_{\mu\nu} D^{(1)}_{\beta\cdots,\mu\nu} \langle \chi_\mu | \hat{o}_{\alpha\cdots}^{\mathcal{F}} | \chi_\nu \rangle$$

(12.15)

where we have assumed again that the atomic orbitals are independent of the perturbation. The derivative of the relaxed density matrix, the so-called first-order relaxed density matrix, in the atomic orbital basis is given as

$$D^{(1)}_{\beta\cdots,\mu\nu} = \frac{\partial D_{\mu\nu}}{\partial \mathcal{F}_{\beta\cdots}} = \sum_{pq} c_{\mu p}^* \frac{\partial D_{pq}}{\partial \mathcal{F}_{\beta\cdots}} c_{\nu q} + \sum_{pq} D_{pq} \left(\frac{\partial c_{\mu p}^*}{\partial \mathcal{F}_{\beta\cdots}} c_{\nu q} + c_{\mu p}^* \frac{\partial c_{\nu q}}{\partial \mathcal{F}_{\beta\cdots}} \right) \quad (12.16)$$

The derivatives of the molecular orbital coefficients $\{c_{\nu q}\}$ are obtained by solving the coupled-perturbed Hartree–Fock equations (11.25). The first-order density matrix at the SCF level was given in Eq. (11.16). The occupied-occupied and virtual-virtual blocks of the correlated first-order density matrix contain derivatives of the amplitudes or correlation coefficients, which can be obtained by straightforward differentiation of the equations defining the amplitudes. The occupied-virtual and virtual-occupied part requires the solution of the first-order Z-vector equations, i.e. the derivative of the Z-vector equations. Explicit expressions for the relaxed density matrices and first-order relaxed density matrices for many methods can also be found in e.g. (Helgaker and Jørgensen, 1988; Amos and Rice, 1989; Gauss and Cremer, 1992; Gauss and Stanton, 1995).

12.3 Time-Dependent Analytical Derivatives

Several attempts have been made to extend the analytical energy derivative method also to the case of time-dependent perturbations. The pseudo-energy derivative (PED) method of Rice and Handy (1991), the quasi-energy derivative (QED) method of Sasagane, et al., (1993) and the time-dependent second-order Møller–Plesset method

(TDMP2) method of Hättig und Heß (1995) were steps in this direction. The most recent attempt that in a way summarises all previous developments is the time-averaged quasi-energy method of Christiansen et al. (Christiansen et al., 1998b).

All these methods define response functions as derivatives of a perturbed time-dependent quasi-energy (Löwdin and Mukherjee, 1972; Langhoff et al., 1972; Kutzelnigg, 1992)

$$Q(t,\vec{\mathcal{F}}) = \langle \Psi(t,\vec{\mathcal{F}}) | \hat{H}(t) - \imath\hbar \frac{\partial}{\partial t} | \Psi(t,\vec{\mathcal{F}}) \rangle \tag{12.17}$$

with the time-dependent perturbation Hamiltonian $\hat{H}(t)$ as defined in Eqs. (3.68) and (3.78).

While the linear response function $\langle\langle \hat{O}^\omega_{\alpha\dots} ; \hat{O}^\omega_{\beta\dots} \rangle\rangle_\omega$ of an operator $\hat{O}^\omega_{\alpha\dots}$ in the presence of a time-dependent field with Fourier components $\mathcal{F}_{\beta\dots}(\omega)$ was obtained in the PED method as

$$\langle\langle \hat{O}^\omega_{\alpha\dots} ; \hat{O}^\omega_{\beta\dots} \rangle\rangle_\omega = \frac{1}{2}\left(\frac{d^2 Q(t,\vec{\mathcal{F}}))}{d\mathcal{F}_{\beta\dots}(\omega) d\mathcal{F}_{\alpha\dots}(0)} \bigg|_{|\vec{\mathcal{F}}|=0} + \frac{d^2 Q(t,\vec{\mathcal{F}}))}{d\mathcal{F}_{\alpha\dots}(0) d\mathcal{F}_{\beta\dots}(\omega)} \bigg|_{|\vec{\mathcal{F}}|=0} \right) \tag{12.18}$$

and in the QED or TDMP2 method as

$$\langle\langle \hat{O}^\omega_{\alpha\dots} ; \hat{O}^\omega_{\beta\dots} \rangle\rangle_\omega = \frac{d^2 Q(t,\vec{\mathcal{F}})}{d\mathcal{F}_{\alpha\dots}(-\omega) d\mathcal{F}_{\beta\dots}(\omega)} \bigg|_{|\vec{\mathcal{F}}|=0} \tag{12.19}$$

it is finally in the time-averaged quasi-energy method given as

$$\langle\langle \hat{O}^\omega_{\alpha\dots} ; \hat{O}^\omega_{\beta\dots} \rangle\rangle_\omega = \frac{d^2 \{Q(t,\vec{\mathcal{F}})\}_T}{d\mathcal{F}_{\alpha\dots}(-\omega) d\mathcal{F}_{\beta\dots}(\omega)} \bigg|_{|\vec{\mathcal{F}}|=0} \tag{12.20}$$

where the time average is defined as (Langhoff et al., 1972)

$$\{Q(t,\vec{\mathcal{F}})\}_T = \frac{1}{T} \int_{-\frac{T}{2}}^{\frac{T}{2}} Q(t,\vec{\mathcal{F}}) \, dt \tag{12.21}$$

and T is a period of the perturbation Eq. (3.78)

$$\hat{H}^{(1)}(t+T) = \hat{H}^{(1)}(t) \tag{12.22}$$

The TDMP2 approach (Hättig and Heß, 1995) has only been applied to the linear response function. PED (Rice and Handy, 1991, 1992) and QED (Aiga et al., 1993) expressions for the frequency-dependent polarizability and first hyperpolarizability, i.e. linear and quadratic response functions, have been derived at the SCF and MP2 level, whereas QED expressions have also been presented for the coupled cluster level (Aiga et al., 1994). Furthermore, QED expressions for second and third hyperpolarizabilities, i.e. cubic and quartic response functions, have been presented for SCF, MCSCF, full and truncated CI wavefunctions (Sasagane et al., 1993). Finally, time-averaged QED expressions for linear and higher response functions, excitation energies and excited-state properties have been presented for general variational and non-variational theories and in particular for various coupled cluster methods (Christiansen et al., 1998b).

At the SCF level all methods lead to the same expressions for the response functions as obtained in the random phase approximation, in Section 10.3, with the time-dependent Hartree–Fock approximation, in Chapter 11.1, or with SCF linear response theory. The QED and time-averaged QED method for an MCSCF energy was also shown to yield the same expressions as obtained from propagator or response theory in Sections 10.4 and 11.2.

However, for non-variational wavefunctions and in particular at the MP2 level the various methods differ, despite the fact that they were constructed to give the correct static perturbation limit, meaning the same as obtained from taking derivatives of the time-independent MP2 energy in Section 12.2.2. The PED method started from the normal expression for the MP2 closed-shell energy, Eq. (9.68), but expressed with the time-dependent perturbed molecular orbitals from Eq. (11.32). In addition, it was required that the condition

$$\langle \Phi(t) | \frac{\partial \Phi(t)}{\partial \mathcal{F}_{\alpha\cdots}(0)} \rangle = 0 \tag{12.23}$$

is fulfilled for the first-order time-dependent MP wavefunction

$$|\Phi(t)\rangle = |\Phi_0^{\text{SCF}}(t)\rangle + |\Phi^{\text{MP1}}(t)\rangle \tag{12.24}$$

However, no time-dependent contribution was included in the first-order correlation coefficients apart from the time dependence of the molecular orbitals.

In the TDMP2 method, in the newest version of the QED method, called QED-MP2 Aiga and Itoh, 1996, and in the time-averaged QED method, the derivatives are taken of an MP2 time-dependent and relaxed quasi-energy Lagrangian

$$L_0^{\text{MP2,relax}}(t, \vec{\mathcal{F}}) = \langle \Phi_0^{\text{SCF}} | e^{-\hat{\kappa}(t)} \left\{ \hat{H}(t) - \imath\hbar \frac{\partial}{\partial t} \right\} e^{\hat{\kappa}(t)} \left\{ 1 + \hat{T}_2[1](t) \right\} | \Phi_0^{\text{SCF}} \rangle$$

$$+ \sum_{2\mu} \lambda_{2\mu}[1](t) \langle \Phi_0^{\text{SCF}} | {}^d\hat{h}_{2\mu} \, e^{-\hat{\kappa}(t)} \left\{ \hat{H}(t) - \imath\hbar \frac{\partial}{\partial t} \right\} e^{\hat{\kappa}(t)} | \Phi_0^{\text{SCF}} \rangle$$

$$+ \sum_{2\mu} \lambda_{2\mu}[1](t) \langle \Phi_0^{\text{SCF}} | {}^d\hat{h}_{2\mu} \left[e^{-\hat{\kappa}(t)} \left\{ \hat{F}(t) - \imath\hbar \frac{\partial}{\partial t} \right\} e^{\hat{\kappa}(t)}, \hat{T}_2[1](t) \right] | \Phi_0^{\text{SCF}} \rangle$$

$$+ \sum_{pq} \tau_{pq}(t) \langle \Phi_0^{\text{SCF}} | \left[q_{pq}^\dagger, e^{-\hat{\kappa}(t)} \left\{ \hat{H}(t) - \imath\hbar \frac{\partial}{\partial t} \right\} e^{\hat{\kappa}(t)} \right] | \Phi_0^{\text{SCF}} \rangle \tag{12.25}$$

The first term is the generalization of the normal MP2 energy, Eq. (9.68), to the case of time-dependent molecular orbitals and time-dependent first-order doubles correlation coefficients $t_{2\mu}[1](t)$. The second and third terms are the time-dependent version of the equations for the $t_{2\mu}[1](t)$ coefficients multiplied with their Lagrangian multipliers $\lambda_{2\mu}[1](t)$. It is these equations that were not included in the PED approach. Finally,

$\hat{\kappa}(t)$ is the time-dependent orbital rotation operator, defined in Eq. (11.37) with the time-dependent orbital rotation parameters $\kappa_{pq}(t)$ and $\kappa_{pq}^*(t)$, which are equal to the Fourier transforms of the TDHF expansion coefficients $U^{(1)}_{\alpha\cdots,pi}(\omega)$ and $U^{(1)}_{\alpha\cdots,pi}(-\omega)$ in Eq. (11.32). The last term is therefore the TDHF equations multiplied by corresponding time-dependent Lagrangian multipliers $\bar{\kappa}_{pq}(t)$. This Lagrangian is thus variational with respect to the TDHF coefficients, Eq. (11.35), the first-order doubles correlation coefficients as well as with respect to the Lagrangian multipliers for both types of coefficients. A constraint like Eq. (12.23) is not necessary in the QED method as a result of the fact that the second derivative is with respect to $\mathcal{F}_{\alpha\cdots}(-\omega)$. The TDHF coefficients have to be obtained by solving the TDHF equations, Eq. (11.35), as in the PED method. The first-order MP2 amplitudes as well as the Lagrangian multipliers for both the TDHF coefficients and the first-order MP2 amplitudes are obtained by solving appropriate response equations.

An important difference between the PED and QED method at the MP2 level as well as between these methods and the SOPPA or CC2 linear response functions concerns the poles of the response function, i.e. the values of the frequency of radiation for which the response function becomes singular. The poles of the exact response functions are equal to the vertical excitation energies of the system, as discussed in Section 7.4. The PED expression contains Hartree–Fock orbital energy differences as poles and thus excitation energies, whereas the QED method has the RPA or TDHF poles and double excitation poles coming from the time-dependent first-order doubles correlation coefficients $t_{2_\mu}[1](t)$. This is a major drawback of this approach because it is not in agreement with the pole structure of the exact response functions. It is a consequence of including orbital relaxation in the Lagrangian given in Eq. (12.25), which implies a two-step procedure, where first the TDHF equations are solved and afterwards equations for the time-dependent $t_{2_\mu}[1](t)$ amplitudes. The SOPPA and CC2 response functions, on the other hand, which are implicitly based on an unrelaxed formalism have the correct pole structure.

The same problem with the pole structure appears also for coupled cluster response functions, if one defines them as derivatives of a time-average quasi-energy Lagrangian including orbital relaxation. It is therefore preferable also in the analytical derivative approach like in Section 11.4 to derive coupled cluster response functions as derivatives of a time-dependent quasi-energy Lagrangian without orbital relaxation

$$L_0^{CC}(t,\vec{\mathcal{F}}) = \langle \Phi_0^{SCF} | \left\{ \hat{H}(t) - \imath\hbar\frac{\partial}{\partial t} \right\} e^{\hat{T}} | \Phi_0^{SCF} \rangle$$

$$+ \sum_{2\mu} \lambda_{2_\mu}(t) \langle \Phi_0^{SCF} | {}^d\hat{h}_{2_\mu} e^{-\hat{T}} \left\{ \hat{H}(t) - \imath\hbar\frac{\partial}{\partial t} \right\} e^{\hat{T}} | \Phi_0^{SCF} \rangle \quad (12.26)$$

Inserted in Eq. (12.20) one then obtains an expression for the coupled cluster response function that is essentially Eq. (11.91) but symmetrized with respect to the two property operators $\hat{P} = \hat{O}^\omega_{\alpha\cdots}$ and $\hat{O}^\omega_{\beta\cdots}$ and the sign of the frequency ω.

12.4 Further Reading

Time-Independent Derivatives

Y. Yamaguchi, Y. Osamura, J. D. Goddard and H. F. Schaefer, *A New Dimension to Quantum Chemistry: Analytic Derivative Methods in Ab Initio Molecular Electronic Structure Theory*, Oxford University Press, New York (1994).

Time-Dependent Derivatives

O. Christiansen, P. Jørgensen, and C. Hättig, *Response Functions from Fourier Component Variational Perturbation Theory Applied to a Time-Averaged Quasienergy*, Int. J. Quantum Chem. **68**, 1 (1998).

13
Examples of Calculations and Practical Issues

In the following sections some examples will be given of the calculation of the electromagnetic molecular properties introduced in Chapters 4 to 8 with some of the *ab initio* methods described in Chapters 10 to 12. The examples are neither meant to give an exhaustive overview of the performance of the different *ab initio* methods nor the molecular properties. But before doing so we have to discuss one important practical issue in all quantum chemical calculations, the one-electron basis set, and the more technical question of how the response functions or propagators are evaluated in actual calculations, i.e. the reduced linear equations algorithm.

13.1   Basis Sets for the Calculation of Molecular Properties

In all the quantum chemical methods described in Chapters 10 to 12 the approximate electronic wavefunctions are built from molecular orbitals, which are the solutions of the Hartree–Fock equations Eq. (9.11) or their multiconfigurational extension. In all modern methods these molecular orbitals are expanded in a set of one-electron functions $\{\chi_\mu\}$, called the basis set

$$\phi_p(\vec{r}) = \sum_\mu \chi_\mu(\vec{r})\, c_{\mu p} \tag{13.1}$$

In most cases these basis functions are what is called **Gaussian-type orbitals (GTO)**

$$\chi_{\zeta,k,m,n}(\vec{R},\vec{r}) = C(x - R_x)^k (y - R_y)^m (z - R_z)^n e^{-\zeta(\vec{r}-\vec{R})^2} \tag{13.2}$$

here given as cartesian Gaussian-type functions, where C is the normalization constant and the basis function is placed at point $\vec{R}$. The sum of the exponents k, m and n is the angular momentum quantum number l of this function. Basis functions always come in sets of functions having the same exponent ζ but all possible combinations of the exponents k, m and n whose sums give the same value of the angular momentum quantum number l.

Mathematically, the set of functions only has to be complete, apart from fulfilling the same boundary conditions as the molecular orbitals. However, this is not possible in practice, because it would require infinitely large basis sets. Often, one chooses therefore a more physical approach, where the basis functions are placed at the atoms and are meant to simulate atomic orbitals of the individual atoms in the molecule. The expansion in Eq. (13.1) then becomes a linear combination of atomic orbitals (LCAO). However, the GTOs are not an ideal approximation to atomic orbitals, which would be exponential functions $e^{-\zeta|\vec{r}-\vec{R}|}$ instead of Gaussian-type functions $e^{-\zeta(\vec{r}-\vec{R})^2}$ of the distance from the centre $\vec{R}$. This means that the GTOs fall off too quickly as compared to real atomic orbitals and that they have a maximum at the nucleus, i.e. at $\vec{r} = \vec{R}$, instead of the cusp of proper atomic orbitals. The most inner and outer parts of the atomic orbitals are therefore not well described by GTOs. The common solution to this problem is to combine several GTOs, which are called **primitive GTOs**, in a **contracted GTO**, which is then meant to simulate one atomic orbital. The coefficients of this combination are preset and kept fixed during the calculation of the molecular orbital coefficients $c_{\mu p}$, i.e. the solution of the Hartree–Fock equations Eq. (9.11).

Although the Gaussian-type orbitals (contracted or not) are not atomic orbitals but just basis functions, one still keeps the nomenclature and distinguishes between **valence orbitals**, which are meant to describe the electrons in the outermost shell, e.g. the 2s and 2p electrons in carbon or the 1s electron in hydrogen, and **core orbitals**, which are meant to describe the inner electrons, e.g. the 1s electrons in carbon. If each core and valence orbital of an atom is represented by a single primitive or contracted GTO one speaks of a **minimal basis set**.

However, such a basis set is not suitable for any quantitative calculations, even if it were to consist of the proper atomic orbitals. The reason is that proper atomic orbitals describe the electrons in an isolated atom with a spherically symmetric potential. This symmetry is broken in any molecule and the basis has to reflect this leading to two different kinds of extension of a minimal basis set. First, one represents each atomic orbital not by a single contracted or primitive GTO, but by two, three or more GTOs with different exponents ζ. Such basis sets are therefore called **double zeta (DZ)**, **triple zeta (TZ)**, and so forth, basis sets. Normally, this is in particular important for the valence functions, whereas the core orbitals are described by a single but then contracted GTO. A basis set with a single contracted GTO for each core orbital and two, three or more contracted or primitive GTOs for each valence orbital is then called a **valence double zeta (VDZ)**, **valence triple zeta (VTZ)**, and so forth, basis set. A typical example of a VDZ basis set is the Pople-style basis sets (Hehre et al., 1972) 6-31G, where each core orbital is represented by a contraction of 6 primitive GTOs while the valence orbitals are represented by two basis functions: one contraction of three primitive GTOs and a second but primitive GTO with a smaller exponent ζ.

However, this is often still not enough because it does not introduce the required asymmetry in the basis set. This can be achieved by also including basis functions with a higher angular momentum quantum number l than the valence orbitals, e.g. d-orbitals for carbon or p-orbitals for hydrogen. They are called **polarization functions** and are important for smaller molecules with high symmetry in particular, whereas in large molecules with low symmetry the atoms can partly "borrow" from each other. In the Pople-style basis sets this would lead to a 6-31G(d,p) or shorter 6-31G** basis set, which has an additional set of d-type GTOs for all second- and third-row atoms and an additional set of p-type GTOs for hydrogen (Francl et al., 1982).

A basis set can therefore be improved by both increasing the number of valence orbitals as well as adding more polarization functions. However, in a balanced basis set this should be done in such a way that both changes lead to changes of equal importance, which normally means changes in the electronic energy of the same order of magnitude. The series of correlation consistent polarized basis sets cc-pVXZ (with the cardinal number X = D, T, Q, 5, 6) by Dunning and coworkers (Dunning Jr., 1989; Woon and Dunning Jr., 1993; Wilson et al., 1999) and the series of polarization consistent pc-n (with n = 0, 1, 2, 3, 4) basis sets by Jensen (Jensen, 2001, 2002a, 2002b, 2002c, 2003, Jensen and Helgaker 2004, 2007) are two prominent examples of this. In the cc-pVXZ basis sets the criteria was the change in the energy of a correlated wavefunction calculation, whereas the pc-n basis sets are optimized for density functional theory calculations.

All these basis sets are essentially optimized for the calculation of electronic energies and are therefore able to represent the operators included in the field-free electronic Hamiltonian $\hat{H}^{(0)}$ reasonably well. However, in the calculation of molecular electromagnetic properties it is necessary also to represent other operators such as the electric dipole operator, the electronic angular momentum operator, the Fermi-contact operator and more. Most of these basis sets are *a priori* not optimized for this and have to be extended.

In the following, this will be discussed in more detail for electric properties like the dipole polarizabilities, for magnetic properties like the magnetizabilities and nuclear magnetic shielding tensors and for the indirect nuclear spin-spin coupling constants.

13.1.1 Electric Properties

The perturbation operator in the calculation of electric dipole moments, electric dipole polarizabilities and hyperpolarizabilities, i.e. the electric dipole moment operator Eq. (4.30), contains the position vector $\vec{r}$ of the electrons, which implies that the tail of the wavefunction becomes important. However, this is not well described in GTOs as discussed before and it is therefore essential to include additional valence functions with very small exponents ζ – so-called **diffuse basis functions**. In the Pople-style basis sets this is done in the 6-31G+ and 6-31G++ basis sets, where in the "+" basis set one diffuse function is added only for second- and third-row atoms, while in the "++" basis set one diffuse function is also added for hydrogen (Clark et al., 1983). In the series of correlation consistent and polarization consistent basis sets one set of diffuse functions of each type present in the basis set is added in the aug-cc-pVXZ (Kendall et al., 1992; Woon and Dunning Jr., 1993, 1994; Balabanov and Peterson, 2005) and aug-pc-n (Jensen, 2002c) version of these basis sets. In the series of correlation consistent basis sets it is also possible to add two or more sets of diffuse functions in the "d-aug", "t-aug" and so forth versions.

It is observed and can theoretically be supported that the requirements on the basis set increase with the order of the molecular property, i.e. from an electric dipole moment via the electric dipole polarizability to the first dipole hyperpolarizability and so forth. In particular, functions with a higher angular momentum quantum number than the valence orbitals, the polarization functions, become increasingly important and the required maximum angular momentum quantum number represented in the basis set increases along this series. A similar increase in the requirements on the basis set is observed, when one goes from the electric dipole moment to the electric quadrupole moment and so forth.

The correlation consistent basis sets contain a systematically increasing amount of polarization functions not only with respect to the number of functions but also to the highest angular momentum quantum number, which is always just one value smaller than the cardinal number X. This series of basis sets is therefore well suited for and frequently used in the calculation of polarizabilities and hyperpolarizabilities. Often, one can observe a monotonic convergence of the results, which offer the possibility to extrapolate to a complete basis set limit. However, these basis sets quickly become very large with increasing cardinal number X.

An often cheaper alternative are the medium-size polarized basis sets by Sadlej and co-workers (Sadlej, 1988, 1991a,b, 1992; Sadlej and Urban, 1991; Kellö and Sadlej, 1995; Neogrády et al., 1996; Cernusak et al., 2003). They are essentially triple zeta basis sets with polarization functions. The key to their success is the way the ζ exponents for the polarization functions are chosen: the polarization functions have the same exponents as some of the valence orbitals with the highest angular momentum quantum number, e.g. the d- and f-type functions on carbon have the same exponents as some of the p-type functions. The reasoning behind this choice is that the first-order correction to the perturbed orbitals of an atom in the presence of an electric field would approximately consist of atomic orbitals with the same exponent ζ but the angular momentum quantum number l one higher and one lower than the original function. This means that the basis set already includes also functions for the first-order correction to the atomic orbitals. However, for larger molecules this leads sometimes to linear dependencies and consequently convergence problems in the calculation of the molecular orbitals. The same problem can also be observed in calculations with the larger correlation consistent basis sets. Therefore, a modified version of the medium-size polarized basis sets, called reduced-size polarized (ZmPolX) basis sets, has recently been developed (Benkova et al., 2005a,b,c; Baranowska et al., 2007) that contains a smaller number of diffuse polarization functions. Very recently, Rappoport and Furche (2010) also developed a version of the Karlsruhe valence double-, triple- and quadruple-zeta basis sets (Schäfer et al., 1992, 1994; Weigend et al., 2003; Weigend and Ahlrichs, 2005) that was variationally optimised for the calculation of polarizabilities.

13.1.2 Magnetic Properties

The perturbation operators in the calculation of magnetisabilities and rotational g-factors are the magnetic dipole moment or angular momentum operators of the electrons, Eq. (5.21). Inclusion of enough diffuse functions is therefore equally important as for the electric analogue, the polarizability. However, while it was advisable to include polarization functions with high enough angular momentum in the calculation of polarizabilities, it is absolutely necessary for the case of the angular momentum operator as perturbation operator. Again, the correlation consistent basis sets in particular in their augmented form are frequently used, but it is often necessary for convergence to go to higher cardinal numbers than in the case of polarizabilities.

Furthermore, one is faced with the gauge-origin problem, as discussed in Section 5.10. A popular solution to this problem is to work with perturbation-dependent basis sets. In the case of an external magnetic field as perturbation such basis functions take the following form

$$\chi_{\zeta,k,m,n}(\vec{R},\vec{r},\vec{B}) = e^{\frac{ie}{2\hbar}\{(\vec{R}_{GO}-\vec{R})\times\vec{B}\}\cdot\vec{r}}\chi_{\zeta,k,m,n}(\vec{R},\vec{r}) \tag{13.3}$$

These are most often called **gauge including atomic orbitals** (GIAO) but sometimes also London orbitals (Helgaker and Jørgensen, 1991), because London (London, 1937) employed functions of this type for the first time. Their first application to the calculation of NMR nuclear magnetic shielding tensors, on the other hand, goes back to Hameka (Hameka, 1958) and Ditchfield (Ditchfield, 1974). Comparison with Eqs. (2.117) and (2.119) shows that the prefactor in a GIAO basis function corresponds actually to a unitary gauge transformation, which moves the global gauge origin $\vec{R}_{GO}$ to the centre $\vec{R}$ of this basis function. The GIAO basis functions are thus just one example for the general idea of distributed gauge origins. Application of the same prefactor to molecular orbitals is the idea behind the IGLO (Kutzelnigg, 1980; Schindler and Kutzelnigg, 1982; van Wüllen and Kutzelnigg, 1996) and LORG (Hansen and Bouman, 1985) approaches to the calculation of magnetizabilities and NMR nuclear magnetic shielding tensors.

Employing these GIAO basis functions removes the dependence on the global origin from property integrals necessary for the calculation of magnetizabilities and NMR nuclear magnetic shielding tensors but introduces a dependence on the external magnetic field also in the two-electron repulsion integrals Eq. (9.9). However, most important in the context of basis-set requirements is that the exponential prefactor in GIAOs introduces implicitly higher angular momentum functions in the basis set, as can be seen by expanding it

$$\chi_{\zeta,k,m,n}(\vec{R},\vec{r},\vec{B}) = \chi_{\zeta,k,m,n}(\vec{R},\vec{r}) + \frac{ie}{2\hbar}\left\{(\vec{R}_{GO}-\vec{R})\times\vec{B}\right\}\cdot\vec{r}\,\chi_{\zeta,k,m,n}(\vec{R},\vec{r})+\cdots \tag{13.4}$$

Calculation of magnetic properties with GIAO basis sets therefore show a significantly improved basis set convergence over regular basis sets.

The nuclear magnetic shielding tensor contains in addition to the magnetic dipole moment operator also the orbital paramagnetic operator, Eq. (5.58), with its $\frac{1}{|\vec{r}_i-\vec{R}_K|^3}$ dependence. Consequently, also the inner region of the electron density, i.e. closer to the nuclei, has to be described properly by the basis set. It is therefore necessary to add additional functions with large exponents, so-called **tight functions**. This can partly be achieved by using the correlation consistent core-valence cc-pCVXZ series of basis sets or their augmented version aug-cc-pCVXZ (Woon and Dunning Jr., 1995). Jensen has also developed a modified version of his polarization consistent basis sets, called pcS-n, which are optimised for DFT calculations of shielding constants (Jensen, 2008) by including tight p-functions.

In the CTOCD-DZ approach for solving the gauge-origin problem in shielding calculations, Section 5.10, the diamagnetic contribution to the shielding tensor is reformulated as a linear response function of the total electronic momentum operator, Eq. (3.65), and a new CTOCD-DZ operator, Eq. (5.117), which is essentially the electronic position vector times the orbital paramagnetic operator. Consequently, it becomes important to add not only p-type functions with large exponents, as in the pcS-n basis sets, but also d-type functions with large exponents (Sauer et al., 1994a,b). Sauer and coworkers (Ligabue et al., 2003; Bruun-Ghalbia et al., 2010) have therefore developed a modified version of the aug-cc-pCVTZ basis sets, called aug-cc-pCVTZ-CTOCD-uc, where such functions were added.

13.1.3 Electron-Spin-Dependent Properties

The NMR indirect nuclear spin-spin coupling constants as well as the ESR hyperfine coupling constants are probably the extreme cases of additional basis-set requirements in molecular property calculations. The Fermi-contact operator, Eq. (5.60), includes a Dirac δ function, which means that only the electron density at the coupled nuclei contributes. However, GTO basis functions, being Gaussian functions, have the fundamentally wrong behavior in that region: a maximum instead of a cusp. It is therefore ultimately important to have s-type GTOs with very large exponents in the basis set such that the cusp at the nucleus can at least approximately be simulated. This was already realized by Schulman and Kaufman (1970) or Kowalewski et al. (1979) and later by Oddershede and coworkers (Oddershede et al., 1988; Geertsen et al., 1991b) or Guilleme and San Fabián (Guilleme and San Fabián, 1998). Based on this several specialized spin-spin coupling constants basis sets have been developed that are modifications of the correlation consistent basis sets. In the cc-pVXZ-Cs series by Helgaker and coworkers (Helgaker et al., 1998) the cc-pVXZ basis sets were extended with the core-valence s-type functions of the cc-pCVXZ basis sets. In the cc-pVXZ-su$n$ basis sets by Helgaker and coworkers (Helgaker et al., 1998) the s-type functions of the cc-pVXZ basis sets were decontracted and augmented with a series of n s-type functions with increasingly larger exponents. In the aug-cc-pVTZ-J basis sets by Sauer and coworkers (Enevoldsen et al., 1998; Provasi et al., 2001; Barone et al., 2003; Rusakov et al., 2010; Provasi and Sauer, 2010) the aug-cc-pVTZ basis set was totally uncontracted, 4 s-type functions with increasingly larger exponents and for third-row atoms also 3 d-type functions with large exponents were added. Finally, the basis set was recontracted using the molecular orbital coefficients of suitable test molecules as contraction coefficients and the most diffuse second polarization function was removed. The resulting basis set was shown to reproduce results of DFT calculations for spin-spin coupling constants obtained with much larger basis sets (Peralta et al., 2003; Deng et al., 2006). Finally, Jensen and coworkers (Jensen, 2006; Benedikt et al., 2008; Jensen, 2010) generated modifications of the pc-n and cc-pVXZ basis sets, called, respectively, pcJ-n and ccJ-pVXZ, where not only tight s-type but also tight p- and d-type functions were added to the original basis sets. These basis sets are to be preferred over the basis sets with only tight s-type functions whenever the spin-dipolar Eq. (5.90) and orbital paramagnetic Eq. (5.88) contributions to the indirect spin-spin coupling constants are equally or more important than the Fermi-contact contribution Eq. (5.89).

13.2 Reduced Linear Equations

A feature common to all propagator or response function methods is that the response function is given as the product of a property gradient vector $\boldsymbol{T}^T(\hat{P}_\alpha)$ with the inverse of the principal propagator matrix $(\hbar\omega \boldsymbol{S} - \boldsymbol{E})$ and another property gradient vector $\boldsymbol{T}(\hat{O}^\omega_{\beta\ldots})$

$$\langle\langle \hat{P}_\alpha \,;\, \hat{O}^\omega_{\beta\ldots} \rangle\rangle_\omega = \boldsymbol{T}^T(\hat{P}_\alpha)(\hbar\omega \boldsymbol{S} - \boldsymbol{E})^{-1}\boldsymbol{T}(\hat{O}^\omega_{\beta\ldots}) \tag{13.5}$$

which can also be written as

$$\langle\langle \hat{P}_\alpha \,;\, \hat{O}^\omega_{\beta\ldots} \rangle\rangle_\omega = \boldsymbol{T}^T(\hat{P}_\alpha)\boldsymbol{X}(\hat{O}^\omega_{\beta\ldots}) \tag{13.6}$$

where $\boldsymbol{X}(\hat{O}^\omega_{\beta\ldots})$ is the so-called solution vector, defined as

$$\boldsymbol{X}(\hat{O}^\omega_{\beta\ldots}) = (\hbar\omega \boldsymbol{S} - \boldsymbol{E})^{-1}\boldsymbol{T}(\hat{O}^\omega_{\beta\ldots}) \tag{13.7}$$

In actual calculations, however, the inverse of the principal propagator is never evaluated. Instead, the solution vector is obtained as a solution of the corresponding set of coupled linear equations

$$(\hbar\omega \boldsymbol{S} - \boldsymbol{E})\,\boldsymbol{X}(\hat{O}^\omega_{\beta\ldots}) = \boldsymbol{T}(\hat{O}^\omega_{\beta\ldots}) \tag{13.8}$$

which are solved iteratively by expanding the solution vector in a basis of orthogonal trial vectors $\{\boldsymbol{b}_i\}$ (Pople et al., 1979)

$$\boldsymbol{X}(\hat{O}^\omega_{\beta\ldots}) = \sum_i \boldsymbol{b}_i c_i \tag{13.9}$$

The number of trial vectors required for a converged solution is normally orders of magnitude smaller than the dimension of the principal propagator.

In a given iteration n the expansion of the solution vector, Eq. (13.9), is inserted in the linear equations Eq. (13.8), which are then premultiplied with the trial vectors $\{\boldsymbol{b}_1 \cdots \boldsymbol{b}_n\}$. This transforms the original linear equations, Eq. (13.8), to the basis of the trial vectors $\{\boldsymbol{b}_1 \cdots \boldsymbol{b}_n\}$, which is called the reduced space,

$$\left(\hbar\omega \boldsymbol{S}^R - \boldsymbol{E}^R\right) \boldsymbol{X}^R(\hat{O}^\omega_{\beta\ldots}) = \boldsymbol{T}^R(\hat{O}^\omega_{\beta\ldots}) \tag{13.10}$$

where the matrices in the reduced space are defined as

$$S^R_{ij} = \boldsymbol{b}_i^T \, \boldsymbol{S} \, \boldsymbol{b}_j \tag{13.11}$$

$$E^R_{ij} = \boldsymbol{b}_i^T \, \boldsymbol{E} \, \boldsymbol{b}_j \tag{13.12}$$

$$T^R_i(\hat{O}^\omega_{\beta\ldots}) = \boldsymbol{b}_i^T \, \boldsymbol{T}(\hat{O}^\omega_{\beta\ldots}) \tag{13.13}$$

while the elements of the solution vector, $\boldsymbol{X}^R(\hat{O}^\omega_{\beta\ldots})$ in the reduced space are the optimal coefficients $\{c_i\}$ in the expansion of the trial vector $\boldsymbol{X}(\hat{O}^\omega_{\beta\ldots})$ in iteration n Eq. (13.9).

The linear equations in the reduced space are then solved with standard techniques, meaning by calculating the inverse of $(\hbar\omega \boldsymbol{S}^R - \boldsymbol{E}^R)$. In order to check whether the solution vector is already converged in this iteration one compares the norm of a residual vector, defined as,

$$\boldsymbol{R}_n = (\hbar\omega \boldsymbol{S} - \boldsymbol{E})\ \boldsymbol{X}_n(\hat{O}^{\omega}_{\beta...}) - \boldsymbol{T}(\hat{O}^{\omega}_{\beta...}) \tag{13.14}$$

with a preset threshold.

If the solution vector is not yet converged, one has to extend the set of trial vectors. A new trial vector $\boldsymbol{b}_{n+1}$ can be generated by a generalization of the conjugate gradient method

$$\boldsymbol{b}_{n+1} = (\hbar\omega \boldsymbol{S}^{diag} - \boldsymbol{E}^{diag})^{-1}\ \boldsymbol{R}_n\ , \tag{13.15}$$

where $\boldsymbol{S}^{diag}$ and $\boldsymbol{E}^{diag}$ are diagonal matrices consisting of the diagonal elements of $\boldsymbol{S}$ and $\boldsymbol{E}$ and the matrix $(\hbar\omega \boldsymbol{S}^{diag} - \boldsymbol{E}^{diag})^{-1}$ is called a preconditioner. Afterwards, one has to calculate the new elements of the matrices in the reduced space and therefore $(\hbar\omega \boldsymbol{S} - \boldsymbol{E})\ \boldsymbol{b}_{n+1}$, which is called a linear transformed trial vector. This can be done directly without ever calculating the $(\hbar\omega \boldsymbol{S} - \boldsymbol{E})$ matrix explicitly (Olsen and Jørgensen, 1985; Packer et al., 1996). For the RPA and SOPPA polarization propagators this can also be done directly from the two electron integrals in the basis of the atomic orbitals (Feyereisen et al., 1992; Bak et al., 2000; Christiansen et al., 1998a).

13.3 Examples of Electron Correlation Effects

In this section we will illustrate the calculation of electromagnetic properties taking the electric dipole moment $\vec{\mu}$ Eq. (4.40) in Table 13.1, the static electric dipole polarizability $\bar{\alpha}$ Eq. (4.52) in Table 13.2, the absolute nuclear magnetic shielding constant σ Eq. (5.67) in Table 13.3 and the indirect nuclear spin-spin coupling constants J

Table 13.1 Electric dipole moment (in a.u. $\approx 8.478358 \times 10^{-30}$ C m) of the hydrogen halides, HX, and methyl halides, CH_3X, calculated with an SCF as well as an unrelaxed and relaxed MP2 density matrix. Ab initio results taken from Packer et al. (1994), experimental equilibrium geometry data for HX from Ogilvie et al. (1980) and for CH_3X from Landolt-Börnstein (1976).

Molecule	SCF	unrelaxed MP2	relaxed MP2	Exp.
HF	0.7570	0.6994	0.7100	0.7094
HCl	0.4725	0.4455	0.4419	0.4305
HBr	0.3777	0.3375	0.3417	0.3219
CH_3F	0.8443	0.7521	0.7380	0.7312
CH_3Cl	0.8095	0.7271	0.7589	0.7461
CH_3Br	0.8456	0.7524	0.7462	0.7162

Table 13.2 Comparison of different polarization propagator (Dalskov and Sauer, 1998) and analytical derivative methods (McDowell et al., 1995) for the calculation of static dipole polarizabilities α (in units of $e^2 a_0^2 E_h^{-1}$) using the medium-size polarized basis sets (Sadlej, 1988, 1991a; Andersson and Sadlej, 1992).

	RPA SCF	SOPPA	SOPPA (CCSD)	CCSD	MP2	MP4	Exp.
HF	4.874	6.085	5.818	5.724	5.674	5.770	5.60
HCl	16.664	17.671	17.352	17.499	17.368	17.433	17.39
H_2O	8.492	10.319	9.939	9.824	9.792	9.866	9.64
H_2S	23.614	24.922	24.343	24.604	24.570	24.542	24.71
NH_3	12.926	14.736	14.366	14.411	14.432	14.411	14.56
PH_3	29.915	31.120	30.184	30.674	30.689	30.510	30.93
CH_4	16.120	16.853	16.520	16.709	16.754	16.704	17.27
SiH_4	29.960	31.414	30.742	31.467	31.035	31.216	31.90
F_2	8.593	8.903	8.525	8.550	8.219	8.662	8.38
Cl_2	29.886	31.346	30.556	30.905	30.556	30.707	30.42
C_2H_4	28.303	28.329	27.482	27.534	27.793	27.635	27.70
CO_2	15.841	19.444	18.726	18.013	17.884	17.846	17.51
SO_2	23.653	28.659	27.407	26.444	26.174	26.343	25.61

Eq. (5.75) in Tables 13.4 to 13.6 as examples. The emphasis is here on the comparison of some of the methods introduced in Chapters 10 to 12 and in particular on the effect of **electron correlation**, meaning the difference in the results obtained with methods based on the Hartree–Fock wavefunction, like SCF linear response (section 11.2) or RPA (section 10.3) and CHF (section 11.1) on one side and with methods based on multiconfigurational (sections 10.4 and 11.2), Møller–Plesset (sections 10.3 and 12.2) or coupled cluster wavefunctions (sections 10.3, 11.4 and 12.2) on the other side. Only results for small molecules are discussed here.

13.3.1 Electric Dipole Moment

In Table 13.1 some results for the electric dipole moment (Packer et al., 1994) of the hydrogen halides, HX, and methyl halides, CH_3X, are shown. They are calculated with the SCF density matrix, Eq. (9.112), with the unrelaxed second-order (MP2) density matrix in Eqs. (9.116)–(9.118) and with the relaxed second-order (MP2) density matrix in Eq. (12.5). The results for the dipole moments are clearly improved by the second-order correction to the MP density matrix. However, no clear trend is observable for the comparison of the relaxed and unrelaxed MP2 density matrix. Correlation at this level reduces the dipole moments on average by 9%. The root-mean-square percentage deviation of the unrelaxed MP2 results from the experimental equilibrium geometry values is 3.6% with a maximum and minimum deviation of 5.0% and −1.4%, respectively.

13.3.2 Isotropic Dipole Polarizability

In Table 13.2 results for the isotropic dipole polarizability are compared, calculated as polarization propagator at the RPA, SOPPA and SOPPA(CCSD) level, as CCSD linear response function and as second derivative of the SCF, MP2 and MP4 energy. The maximum difference between the results at the highest level, CCSD, and the experimental results is 3% or 0.83 a.u. for SO_2. However, one should remember that the calculated results are for a fixed equilibrium geometry, while the experimental results are for the vibrational ground state. As indicated in Table 13.2 the RPA polarization propagator results and SCF second derivative results are identical, illustrating the equivalence of these two approaches. Concerning the effect of electron correlation one can see that at the CCSD level the effect varies between 1% and 17% or −0.04 and 2.79 a.u. It is interesting to note that the difference between the unrelaxed CCSD linear response and the relaxed MP4 results is very small and the agreement with the CCSD linear response results is in most cases better in the SOPPA(CCSD) than in the SOPPA approach.

13.3.3 Nuclear Magnetic Shielding Constants

In Table 13.3 results for nuclear magnetic shielding constants σ Eq. (5.67) are compared, which were calculated (Gauss, 1992, 1993, 1994; Gauss and Stanton, 1995, 1996) as analytical second derivatives of the MP2, MP3, CCSD and CCSD(T) energies with GIAO basis functions. For hydrides of second-row atoms one observes that the differences between the MP3, CCSD and CCSD(T) results are rather small, whereas for the multiply bonded diatomic molecules there are still such large changes between the CCSD and CCSD(T) results that one cannot be sure that the results at the CCSD(T) level are converged. MP2, on the other hand, typically overestimates the correlation

Table 13.3 Comparison of different analytical derivative methods for the calculation of nuclear magnetic shielding constants σ (in ppm) (Gauss, 1992, 1993, 1994; Gauss and Stanton, 1995, 1996) The experimental data (Gauss, 1992; Wasylishen and Bryce, 2002) are extrapolated to the equilibrium geometry by subtracting calculated ro-vibrational corrections from the experimental values.

		SCF	MP2	MP3	CCSD	CCSD(T)	Experiment
HF	$\sigma^{19}F$	413.6	424.2	417.8	418.1	418.6	419.7 ±6
H_2O	$\sigma^{17}O$	328.1	346.1	336.7	336.9	337.9	337 ±2
NH_3	$\sigma^{15}N$	262.3	276.5	270.1	269.7	270.7	273.3 ±0.1
CH_4	$\sigma^{13}C$	194.8	201.0	198.8	198.7	198.9	198.4 ±0.9
F_2	$\sigma^{19}F$	−167.9	−170.0	−176.9	−171.1	−186.5	−192.8
N_2	$\sigma^{15}N$	−112.4	−41.6	−72.2	−63.9	−58.1	−59.6 ±1.5
CO	$\sigma^{13}C$	−25.5	20.6	−4.2	0.8	5.6	2.8 ±0.9
	$\sigma^{17}O$	−87.7	−46.5	−68.3	−56.0	−52.9	−56.79 ±0.59

correction and the MP2 results are sometimes in not much better agreement with higher-level methods or experimental results than the SCF results.

13.3.4 Indirect Nuclear Spin-Spin Coupling Constants

Finally, in Tables 13.4, 13.5 and 13.6 we compare results (Vahtras et al., 1992, 1993; Wigglesworth et al., 1998; Enevoldsen et al., 1998; Kaski et al., 1998; Åstrand et al., 1999; Jaszuński and Ruud, 2001; Sauer et al., 2001; Yachmenev et al., 2010) for the indirect nuclear spin-spin coupling constants J, Eq. (5.75), calculated with four polarization propagator methods: RPA, SOPPA, SOPPA(CCSD) and MCRPA. The results of the RPA, SOPPA and SOPPA(CCSD) calculations (Wigglesworth et al., 1998; Enevoldsen et al., 1998; Sauer et al., 2001; Yachmenev et al., 2010) for a given molecule were obtained with the same basis set and at the same nuclear geometry, whereas in some of the MCRPA calculations (Vahtras et al., 1992, 1993; Kaski et al., 1998;

Table 13.4 Comparison of different polarization propagator methods for the calculation of indirect nuclear spin-spin coupling constants J (in Hz) (Vahtras et al., 1992, 1993; Wigglesworth et al., 1998; Enevoldsen et al., 1998; Kaski et al., 1998; Åstrand et al., 1999; Jaszuński and Ruud, 2001; Sauer et al., 2001; Yachmenev et al., 2010). The experimental data are extrapolated to the equilibrium geometry by subtracting calculated ro-vibrational corrections from the experimental values.

		RPA	MCRPA	SOPPA	SOPPA (CCSD)	Experiment
N_2	$^1J^{15}N^{-14}N$	−14.9	0.8	2.7	2.1	1.4±0.6
CO	$^1J^{13}C^{-17}O$	−5.7	16.1	20.4	18.6	15.6±0.1
C_2H_2	$^1J^{13}C^{-13}C$	409.5	188.1	189.3	188.7	185.04
HF	$^1J^{1}H^{-19}F$	666.9	543.7	539.5	529.4	540
H_2O	$^1J^{1}H^{-17}O$	−103.4	−83.9	−82.4	−80.6	−83.04±0.02
NH_3	$^1J^{1}H^{-15}N$	−78.4	−62.2	−62.4	−62.1	
CH_4	$^1J^{1}H^{-13}C$	156.9	135.7	126.9	122.3	120.87±0.05
C_2H_2	$^1J^{1}H^{-13}C$	411.1	247.3	262.9	253.6	242.70
SiH_4	$^1J^{1}H^{-29}Si$	−241.2	—	−202.5	−192.1	−193.3±0.6
H_2O	$^2J^{1}H^{-1}H$	−22.4	−9.6	−9.1	−8.8	−7.8±0.7
NH_3	$^2J^{1}H^{-1}H$	−24.4	−11.2	−11.9	−11.3	
CH_4	$^2J^{1}H^{-1}H$	−27.0	−20.8	−15.3	−14.0	−11.878±0.004
C_2H_2	$^2J^{1}H^{-13}C$	−49.9	53.0	52.6	51.7	53.82
SiH_4	$^2J^{1}H^{-1}H$	−1.2	—	2.1	2.6	2.62±0.08
C_2H_2	$^3J^{1}H^{-1}H$	84.9	11.2	12.2	11.3	10.89

Åstrand et al., 1999; Jaszuński and Ruud, 2001) a slightly different basis set and nuclear geometry were employed than in the other calculations. However, this has no effect on the conclusions of the comparison.

The comparison of the three contributions in Tables 13.5 and 13.6 with the total coupling constants in Table 13.4 shows that one-bond C–C, C–H, N–H, Si–H and to a lesser degree also O–H coupling constants are dominated by the Fermi contact contribution. For two-bond H–H coupling constants the Fermi-contact contribution is still the largest contribution. However, the orbital paramagnetic and diamagnetic contributions are often of the same order of magnitude as the Fermi-contact contribution but partially cancel each other, because they often have opposite signs. Only in the case of coupling constants involving atoms with many lone pairs like F or couplings across double and triple bonds should one expect that the orbital paramagnetic

Table 13.5 Comparison of different polarization propagator methods for the calculation of the orbital paramagnetic contribution J^{OP} to the indirect nuclear spin-spin coupling constants (in Hz) (Vahtras et al., 1992, 1993; Wigglesworth et al., 1998; Enevoldsen et al., 1998; Kaski et al., 1998; Åstrand et al., 1999; Jaszuński and Ruud, 2001; Sauer et al., 2001; Yachmenev et al., 2010).

		J^{OP}		
		RPA	MCRPA	SOPPA (CCSD)
N$_2$	$^1J^{15}N-^{14}N$	0.43	2.69	3.00
CO	$^1J^{13}C-^{17}O$	11.81	12.89	14.11
C$_2$H$_2$	$^1J^{13}C-^{13}C$	15.05	6.67	6.34
HF	$^1J^{1}H-^{19}F$	195.05	182.0	189.82
H$_2$O	$^1J^{1}H-^{17}O$	−12.27	−11.45	−11.51
NH$_3$	$^1J^{1}H-^{15}N$	−3.07	−2.93	−2.97
CH$_4$	$^1J^{1}H-^{13}C$	1.47	1.48	1.50
C$_2$H$_2$	$^1J^{1}H-^{13}C$	−3.60	−0.84	−0.85
SiH$_4$	$^1J^{1}H-^{29}Si$	0.47	—	0.44
H$_2$O	$^2J^{1}H-^{1}H$	9.09	9.23	9.31
NH$_3$	$^2J^{1}H-^{1}H$	6.24	6.19	6.24
CH$_4$	$^2J^{1}H-^{1}H$	3.73	3.59	3.72
C$_2$H$_2$	$^2J^{1}H-^{13}C$	8.28	5.58	5.60
SiH$_4$	$^2J^{1}H-^{1}H$	2.35	—	2.34
C$_2$H$_2$	$^3J^{1}H-^{1}H$	5.54	4.80	4.81

Table 13.6 Comparison of different polarization propagator methods for the calculation of the spin-dipolar, J^{SD}, and Fermi contact J^{FC} contributions to the indirect nuclear spin-spin coupling constants (in Hz) (Vahtras et al., 1992, 1993; Wigglesworth et al., 1998; Enevoldsen et al., 1998; Kaski et al., 1998; Åstrand et al., 1999; Jaszuński and Ruud, 2001; Sauer et al., 2001; Yachmenev et al., 2010.)

		J^{SD}			J^{FC}		
		RPA	MCRPA	SOPPA (CCSD)	RPA	MCRPA	SOPPA (CCSD)
N_2	$^1J^{15N-14N}$	−7.84	−1.95	−1.76	−7.49	−0.53	0.79
CO	$^1J^{13C-17O}$	−9.07	−4.77	−4.37	−8.53	3.90	8.76
C_2H_2	$^1J^{13C-13C}$	29.06	9.04	8.46	365.35	172.34	173.92
HF	$^1J^{1H-19F}$	−11.73	−1.41	−0.94	483.62	363.2	340.50
H_2O	$^1J^{1H-17O}$	−0.01	−0.41	−0.47	−91.12	−72.08	−68.56
NH_3	$^1J^{1H-15N}$	−0.02	−0.18	−0.18	−75.22	−58.98	−58.87
CH_4	$^1J^{1H-13C}$	−0.21	0.02	0.03	155.42	123.53	120.58
C_2H_2	$^1J^{1H-13C}$	3.04	0.43	0.43	411.41	247.40	253.73
SiH_4	$^1J^{1H-29Si}$	0.05	—	−0.05	−241.78	—	−192.43
H_2O	$^2J^{1H-1H}$	1.25	1.03	0.89	−25.50	−12.70	−11.87
NH_3	$^2J^{1H-1H}$	0.91	0.68	0.67	−26.30	−12.82	−12.94
CH_4	$^2J^{1H-1H}$	0.46	0.35	0.36	−27.68	−15.73	−14.53
C_2H_2	$^2J^{1H-13C}$	−1.52	1.02	0.98	−55.25	47.77	46.47
C_2H_2	$^3J^{1H-1H}$	3.02	0.57	0.59	79.93	9.43	9.49
SiH_4	$^2J^{1H-1H}$	0.09	—	0.07	−1.22	—	2.59

and or the spin-dipolar contributions will make a significant contribution to the total coupling constant.

Turning to the comparison of the different methods we can see from Table 13.4 that the agreement of the SOPPA(CCSD) results with the experimental values is very good for some of the indirect nuclear spin-spin coupling constants. For most molecules the deviations are less than 4 Hz with a rms value of 2.3 Hz, apart from HF and $^1J^{1H-13C}$ in C_2H_2, where the deviations are about 10 Hz. Including also those couplings increases the rms to 5.1 Hz. The rms value of the percentage deviations is for all molecules 18% (10% without HF and $^1J^{1H-13C}$ in C_2H_2).

Based on these statistical data one can conclude that spin-spin coupling constants are more difficult to calculate than dipole polarizabilities and nuclear magnetic shielding constants. This can also be seen if one compares the correlation corrections to both properties as calculated at the SOPPA(CCSD) level. The correlation corrections for the molecules in Table 13.4 vary between 20% for HF and 426% for CO.

The calculation of spin-spin coupling constants is also complicated because it consists of four terms (see Section 5.7), of which the three linear response function contributions are shown in Tables 13.5 and 13.6. In Section 5.7, it was discussed that two of them, the spin-dipolar and the Fermi-contact contribution, depend on excited triplet states, which uncorrelated methods like RPA are often not able to describe properly. The large correlation effects in the spin-spin coupling constants are thus normally due to the Fermi-contact term.

13.4 Examples of Vibrational Averaging Effects

In Tables 13.7 and 13.8 some illustrative examples of the zero-point-vibrational corrections to nuclear magnetic shielding constants, σ, and indirect nuclear spin-spin coupling constants, J, are collected.

A few general conclusions can be drawn from these examples. Zero-point-vibrational corrections for properties, which describe an interaction with nuclear magnetic moments, like the nuclear magnetic shielding constant and the indirect nuclear spin-spin coupling constant, can amount to 10% and are typically larger for the coupling than for the shielding constants. Compared with the correlation effects discussed in Section 13.3 zero-point-vibrational corrections are smaller but not significantly smaller, which implies that they have to considered in a high-level calculation of NMR parameters. The large zero-point-vibrational corrections to the nuclear magnetic shielding constants of F_2 and CO are two extreme cases, well known in the literature. The corrections to geminal hydrogen-hydrogen couplings, $^2J^{1H-1H}$, are per cent wise larger than the corrections to other couplings in the same molecule.

Table 13.7 Calculated zero-point-vibrational corrections (ZPVC) to the nuclear magnetic shielding constant (in ppm) (Sundholm et al., 1996; Åstrand et al., 1999; 1999, 200a)

Molecule	Property	result at R_e	ZPVC	%
HF	σ^{19F}	419.68	−10.01	2.4
HF	σ^{1H}	29.01	−0.32	1.1
H_2O	σ^{17O}	343.94	−9.86	2.9
H_2O	σ^{1H}	30.97	−0.48	1.6
F_2	σ^{19F}	−187.84	30.90	16.5
C_2H_2	σ^{13C}	128.89	−3.78	2.9
C_2H_2	σ^{1H}	30.45	−0.80	2.6
CO	σ^{13C}	5.29	−1.82	34.5
CO	σ^{17O}	−53.5	−4.8	9.0
N_2	σ^{15N}	−58.7	−3.5	5.9

Table 13.8 Calculated zero-point-vibrational corrections (ZPVC) to the indirect nuclear spin-spin coupling constant (in Hz) (Wigglesworth et al., 1997, 1998; Åstrand et al., 1999; Wigglesworth et al., 2000b, 2001; Sauer et al., 2001; Yachmenev et al., 2010)

Molecule	Property	result at R_e	ZPVC	%
HF	$^1J^{1}H-^{19}F$	526.4	−26.9	5.1
H_2O	$^1J^{1}H-^{17}O$	−81.555	3.963	4.9
H_2O	$^2J^{1}H-^{1}H$	−8.581	0.653	7.6
NH_3	$^1J^{1}H-^{15}N$	−61.968	0.341	0.5
NH_3	$^2J^{1}H-^{1}H$	−10.699	0.537	5.0
CH_4	$^1J^{1}H-^{13}C$	123.846	5.030	4.1
CH_4	$^2J^{1}H-^{1}H$	−14.450	−0.686	4.7
SiH_4	$^1J^{1}H-^{29}Si$	−129.059	−7.585	3.8
C_2H_2	$^1J^{13}C-^{13}C$	189.995	4.861	1.9
C_2H_2	$^1J^{1}H-^{13}C$	254.906	−9.212	4.9
C_2H_2	$^2J^{1}H-^{13}C$	51.727	−3.237	6.3
C_2H_2	$^3J^{1}H-^{1}H$	11.311	−1.184	10.5

13.5 Further Reading

Basis Sets

F. Jensen, *Introduction to Computational Chemistry*, 2nd edn John Wiley & Sons, Chichester (2007): Chapter 5.

Ro-Vibrational Corrections to NMR Parameters

T. A. Ruden and K. Ruud, in M. Kaupp, M. Bühl and V. G. Malkin, ed. *Calculation of NMR and EPR Parameters Theory and Applications*, Wiley-VCH, Weinheim (2004): Chapter 10, pages 153–173.

Part IV

Appendices

Appendix A
Operators

In this appendix, explicit expressions for all the perturbation operators are collected. They were derived in Chapters 4 to 8 by expressing the scalar and vector potentials in the molecular electronic Hamiltonian, Eq. (2.101), in terms of electric fields and various magnetic inductions.

A.1 Perturbation Operators

The scalar potential of an external electric field with non-zero gradient is, Eq. (4.28),

$$\hat{\phi}^{\mathcal{E}}(\vec{r}_i) = -(\vec{r}_i - \vec{R}_O) \cdot \vec{\mathcal{E}}(\vec{R}_O)$$

$$-\frac{1}{2}\sum_{\alpha\beta}\left[(r_{i,\alpha} - R_{O,\alpha})(r_{i,\beta} - R_{O,\beta}) - \frac{1}{3}\delta_{\alpha\beta}(\vec{r}_i - \vec{R}_O)^2\right]\mathcal{E}_{\alpha\beta}(\vec{R}_O) \quad (A.1)$$

where the scalar potential at the origin of the coordinate system, $\phi^{\mathcal{E}}(\vec{R}_O)$ is set to zero. For the vector potential we want to consider three different cases:

- a uniform magnetic induction, Eq. (5.19),

$$\hat{\vec{A}}^B(\vec{r}_i) = \frac{1}{2}\vec{B} \times (\vec{r}_i - \vec{R}_{GO}) \quad (A.2)$$

where $\vec{R}_{GO}$ is the arbitrary gauge origin, defined in Section 5.10
- the magnetic induction due to the rotation of a molecule, Eq. (6.5),

$$\hat{\vec{A}}^J(\vec{r}_i) = -\frac{m_e}{e}\left(\mathbf{I}^{-1}\,\vec{J}\right) \times (\vec{r}_i - \vec{R}_{CM}) \quad (A.3)$$

- the magnetic dipole moment of nucleus K, Eq. (5.55),

$$\hat{\vec{A}}^K(\vec{r}_i) = \frac{\mu_0}{4\pi}\vec{m}^K \times \frac{\vec{r}_i - \vec{R}_K}{|\vec{r}_i - \vec{R}_K|^3} \quad (A.4)$$

whose magnetic induction is

$$\vec{B}^K(\vec{r}_i) = \frac{\mu_0}{4\pi}\left\{\frac{3\left[\vec{m}^K \cdot (\vec{r}_i - \vec{R}_K)\right](\vec{r}_i - \vec{R}_K)}{|\vec{r}_i - \vec{R}_K|^5} - \frac{\vec{m}^K}{|\vec{r}_i - \vec{R}_K|^3}\right\}$$

$$+ \frac{\mu_0}{4\pi}\frac{8\pi}{3}\delta(\vec{r}_i - \vec{R}_K)\vec{m}^K \quad (A.5)$$

Inserting these potentials in the molecular electronic Hamiltonian, Eq. (2.101), we obtain after some manipulation the following expressions for the first- and second-order perturbation Hamiltonian operators:

$$\hat{H}^{(1)} = -\sum_\alpha \left(\hat{O}_\alpha^{lB} + \hat{O}_\alpha^{sB}\right) \mathcal{B}_\alpha + \sum_\alpha \left(\hat{O}_\alpha^{lJ} + \hat{O}_\alpha^{sJ}\right) (\mathbf{I}^{-1} \vec{J})_\alpha \quad (A.6)$$

$$-\sum_K \sum_\alpha \left(\hat{O}_\alpha^{lm^K} + \hat{O}_\alpha^{sm^K}\right) m_\alpha^K$$

$$-\sum_\alpha \left(\hat{O}_\alpha^\mathcal{E} + \hat{\Omega}_\alpha^\mathcal{E}\right) \mathcal{E}_\alpha(\vec{R}_O) - \sum_{\alpha\beta}\left(\hat{O}_{\alpha\beta}^{\nabla\mathcal{E}} + \hat{\Omega}_{\alpha\beta}^{\nabla\mathcal{E}}\right) \mathcal{E}_{\alpha\beta}(\vec{R}_O)$$

$$\hat{H}^{(2)} = \sum_{\alpha\beta} \hat{O}_{\alpha\beta}^{BB} \mathcal{B}_\alpha \mathcal{B}_\beta + \sum_{\alpha\beta} \hat{O}_{\alpha\beta}^{JJ} (\mathbf{I}^{-1}\vec{J})_\alpha (\mathbf{I}^{-1}\vec{J})_\beta + \sum_{KL}\sum_{\alpha\beta} \hat{O}_{\alpha\beta}^{m^K m^L} m_\alpha^K m_\beta^L \quad (A.7)$$

$$+ \sum_{\alpha\beta} \hat{O}_{\alpha\beta}^{BJ} \mathcal{B}_\alpha (\mathbf{I}^{-1}\vec{J})_\beta + \sum_K \sum_{\alpha\beta} \hat{O}_{\alpha\beta}^{m^K B} m_\alpha^K \mathcal{B}_\beta + \sum_K \sum_{\alpha\beta} \hat{O}_{\alpha\beta}^{m^K J} m_\alpha^K (\mathbf{I}^{-1}\vec{J})_\beta$$

A.1.1 Field-Independent Perturbation Operators

The perturbation operators $\hat{O}$ are obtained as derivatives of the molecular electronic Hamiltonian, Eqs. (2.101) and (2.108), evaluated for zero fields or magnetic moments:

- **the electric dipole moment operators** coupling a molecule with a uniform electric field

$$\hat{\mu}_\alpha \equiv \hat{O}_\alpha^\mathcal{E}(\vec{R}_O) = \sum_i^N \hat{o}_{i,\alpha}^\mathcal{E}(\vec{R}_O) = \sum_i^N \hat{\mu}_{i,\alpha}(\vec{R}_O) = -e\sum_i^N (r_{i,\alpha} - R_{O,\alpha}) \quad (A.8)$$

$$\hat{\Omega}_\alpha^\mathcal{E}(\vec{R}_O) = \sum_K^M Z_K e (R_{K,\alpha} - R_{O,\alpha}) \quad (A.9)$$

- **the electric quadrupole moment operators** coupling a molecule with the gradient of an electric field

$$\hat{O}_{\alpha\beta}^{\nabla\mathcal{E}}(\vec{R}_O) = \sum_i^N \hat{o}_{i,\alpha\beta}^{\nabla\mathcal{E}}(\vec{R}_O) = \frac{1}{3}\sum_i^N \hat{\Theta}_{i,\alpha\beta}(\vec{R}_O)$$

$$= -\frac{e}{2}\sum_i^N \left[(r_{i,\alpha} - R_{O,\alpha})(r_{i,\beta} - R_{O,\beta}) - \frac{1}{3}\delta_{\alpha\beta}(\vec{r}_i - \vec{R}_O)^2\right] \quad (A.10)$$

$$\hat{\Omega}_{\alpha\beta}^{\nabla\mathcal{E}}(\vec{R}_O) = \frac{1}{2}\sum_K^M Z_K e \left[(R_{K,\alpha} - R_{O,\alpha})(R_{K,\beta} - R_{O,\beta}) - \frac{1}{3}\delta_{\alpha\beta}(\vec{R}_K - \vec{R}_O)^2\right] \quad (A.11)$$

- **the electric field operator** coupling a molecule with an electric dipole moment

$$\hat{O}^\mu_\alpha(\vec{R}) = \sum_i^N \hat{o}^\mu_{i,\alpha}(\vec{R}) = \frac{e}{4\pi\epsilon_0} \sum_i^N \frac{r_{i,\alpha} - R_\alpha}{|\vec{r}_i - \vec{R}|^3} \qquad (A.12)$$

$$\hat{\Omega}^\mu_\alpha(\vec{R}) = -\sum_K \frac{Z_K e}{4\pi\epsilon_0} \frac{R_{K,\alpha} - R_\alpha}{|\vec{R}_K - \vec{R}|^3} \qquad (A.13)$$

- **electric field gradient operators** coupling a molecule with an electric quadrupole moment located at $\vec{R}_K$

$$\hat{O}^\Theta_{\alpha\beta}(\vec{R}) = \sum_i^N \hat{o}^\Theta_{i,\alpha\beta}(\vec{R})$$

$$= \frac{e}{4\pi\epsilon_0} \sum_i^N \left[3\frac{(r_{i,\alpha} - R_{K,\alpha})(r_{i,\beta} - R_{K,\beta})}{|\vec{r}_i - \vec{R}_K|^5} - \frac{\delta_{\alpha\beta}}{|\vec{r}_i - \vec{R}_K|^3} \right] \qquad (A.14)$$

$$\hat{\Omega}^\Theta_{\alpha\beta}(\vec{R}) = -\frac{1}{4\pi\epsilon_0} \sum_{L \neq K} Z_L e \left[3\frac{(R_{L,\alpha} - R_{K,\alpha})(R_{L,\beta} - R_{K,\beta})}{|\vec{R}_L - \vec{R}_K|^5} - \frac{\delta_{\alpha\beta}}{|\vec{R}_L - \vec{R}_K|^3} \right] \qquad (A.15)$$

- **the orbital magnetic dipole moment operator** coupling the motion of the electrons in a molecule with a uniform magnetic induction

$$\hat{m}^l \equiv \hat{O}^{lB}_\alpha(\vec{R}_{GO}) = \sum_i^N \hat{o}^{lB}_{i,\alpha}(\vec{R}_{GO}) = \sum_i^N \hat{m}^l_{i,\alpha}(\vec{R}_{GO}) \qquad (A.16)$$

$$= -\frac{e}{2m_e} \sum_i^N \hat{l}_{i,\alpha}(\vec{R}_{GO}) = -\frac{e}{2m_e} \sum_i^N \left[(\vec{r}_i - \vec{R}_{GO}) \times \hat{\vec{p}}_i \right]_\alpha$$

where $\hat{\vec{l}}_i(\vec{R}_{GO})$ is the **orbital angular momentum operator** of electron i with respect to the gauge origin $\vec{R}_{GO}$
- **the spin magnetic dipole moment operator** coupling the spin of the electrons in a molecule with a uniform magnetic induction

$$\hat{m}^s \equiv \hat{O}^{sB}_\alpha = \sum_i^N \hat{o}^{sB}_{i,\alpha} = \sum_i^N \hat{m}^s_{i,\alpha} = -\frac{g_e e}{2m_e} \sum_i^N \hat{s}_{i,\alpha} \qquad (A.17)$$

where $\hat{\vec{s}}_i$ is the **spin operator** of electron i
- **the operator coupling the orbital angular momentum of the electrons in a molecule with the rotation of the molecule**

$$\hat{O}^{lJ}_\alpha(\vec{R}_{CM}) = \sum_i^N \hat{o}^{lJ}_{i,\alpha}(\vec{R}_{CM}) = -\sum_i^N \hat{l}_{i,\alpha}(\vec{R}_{CM}) \qquad (A.18)$$

- **the operator coupling the spin of the electrons** in a molecule with the rotation of the molecule

$$\hat{O}^{sJ}_\alpha = \sum_i^N \hat{o}^{sJ}_{i,\alpha} = \frac{2m_e}{e} \hat{O}^{sB}_\alpha \tag{A.19}$$

- **the orbital paramagnetic (OP) or paramagnetic nuclear spin-electron orbit (PSO) operator** coupling the orbital angular momentum of the electrons in a molecule with a magnetic dipole moment located at $\vec{R}_K$

$$\hat{O}^{OP}_{K,\alpha} \equiv \hat{O}^{lm^K}_\alpha = \sum_i^N \hat{o}^{lm^K}_{i,\alpha} = -\frac{e}{m_e} \frac{\mu_0}{4\pi} \sum_i^N \frac{\hat{l}_{i,\alpha}(\vec{R}_K)}{|\vec{r}_i - \vec{R}_K|^3} \tag{A.20}$$

$$= -\frac{e}{m_e} \frac{\mu_0}{4\pi} \sum_i^N \left(\frac{\vec{r}_i - \vec{R}_K}{|\vec{r}_i - \vec{R}_K|^3} \times \hat{\vec{p}}_i \right)_\alpha$$

where $\hat{\vec{l}}_i(\vec{R}_K)$ is the **orbital angular momentum operator** of electron i with respect to the position of the magnetic dipole moment $\vec{R}_K$

- **the Fermi-contact (FC) and spin-dipolar (SD) operators** coupling the spin of the electrons in a molecule with a magnetic dipole moment located at $\vec{R}_K$

$$\hat{O}^{FC}_{K,\alpha} + \hat{O}^{SD}_{K,\alpha} \equiv \hat{O}^{sm^K}_\alpha = \sum_i^N \hat{o}^{sm^K}_{i,\alpha} \tag{A.21}$$

$$= -\frac{g_e e}{2m_e} \frac{\mu_0}{4\pi} \frac{8\pi}{3} \sum_i^N \delta(\vec{r}_i - \vec{R}_K) \hat{s}_{i,\alpha}$$

$$- \frac{g_e e}{2m_e} \frac{\mu_0}{4\pi} \sum_i^N \left\{ \frac{3 \left[\hat{\vec{s}}_i \cdot (\vec{r}_i - \vec{R}_K) \right] (r_{i,\alpha} - R_{K,\alpha})}{|\vec{r}_i - \vec{R}_K|^5} - \frac{\hat{s}_{i,\alpha}}{|\vec{r}_i - \vec{R}_K|^3} \right\}$$

- **the diamagnetic magnetizability tensor operator** coupling the motion of the electrons in a molecule with the square of a uniform magnetic induction

$$\hat{O}^{BB}_{\alpha\beta}(\vec{R}_{GO}) = \sum_i^N \hat{o}^{BB}_{i,\alpha\beta}(\vec{R}_{GO}) \tag{A.22}$$

$$= \frac{e^2}{8m_e} \sum_i^N \left[(\vec{r}_i - \vec{R}_{GO})^2 \delta_{\alpha\beta} - (r_{i,\alpha} - R_{GO,\alpha})(r_{i,\beta} - R_{GO,\beta}) \right]$$

- **the operator coupling the orbital angular momentum of the electrons** in a molecule with the rotation of the molecule to second order

$$\hat{O}^{JJ}_{\alpha\beta}(\vec{R}_{CM}) = \sum_i^N \hat{o}^{JJ}_{i,\alpha\beta}(\vec{R}_{CM}) \tag{A.23}$$

$$= \frac{m_e}{2} \sum_i^N \left[(\vec{r}_i - \vec{R}_{CM})^2 \delta_{\alpha\beta} - (r_{i,\alpha} - R_{CM,\alpha})(r_{i,\beta} - R_{CM,\beta}) \right]$$

- **the orbital diamagnetic (OD) or diamagnetic nuclear spin-electron orbit (DSO) operator** coupling the orbital angular momentum of the electrons in a molecule with two magnetic dipole moments located at $\vec{R}_K$ and $\vec{R}_L$

$$\hat{O}^{m^K m^L}_{\alpha\beta} = \sum_i^N \hat{o}^{m^K m^L}_{i,\alpha\beta} \tag{A.24}$$

$$= \frac{e^2}{2m_e}\left(\frac{\mu_0}{4\pi}\right)^2 \sum_i^N \left[\frac{(\vec{r}_i - \vec{R}_L)}{|\vec{r}_i - \vec{R}_L|^3} \cdot \frac{(\vec{r}_i - \vec{R}_K)}{|\vec{r}_i - \vec{R}_K|^3}\delta_{\alpha\beta} - \frac{(r_{i,\alpha} - R_{L,\alpha})}{|\vec{r}_i - \vec{R}_L|^3}\frac{(r_{i,\beta} - R_{K,\beta})}{|\vec{r}_i - \vec{R}_K|^3}\right]$$

- **the diamagnetic rotational g tensor operator** coupling the orbital angular momentum of the electrons in a molecule with the rotation of the molecule and a uniform magnetic induction

$$\hat{O}^{BJ}_{\alpha\beta}(\vec{R}_{CM}, \vec{R}_{GO}) = \sum_i^N \hat{o}^{BJ}_{i,\alpha\beta}(\vec{R}_{CM}, \vec{R}_{GO}) \tag{A.25}$$

$$= -\frac{e}{2}\sum_i^N \left[(\vec{r}_i - \vec{R}_{CM}) \cdot (\vec{r}_i - \vec{R}_{GO})\delta_{\alpha\beta} - (r_{i,\alpha} - R_{CM,\alpha})(r_{i,\beta} - R_{GO,\beta})\right]$$

- **the diamagnetic nuclear magnetic shielding tensor operator** coupling the orbital angular momentum of the electrons in a molecule with a magnetic dipole moment located at $\vec{R}_K$ and a uniform magnetic induction

$$\hat{O}^{m^K B}_{\alpha\beta}(\vec{R}_{GO}) = \sum_i^N \hat{o}^{m^K B}_{i,\alpha\beta}(\vec{R}_{GO}) \tag{A.26}$$

$$= \frac{e^2}{2m_e}\frac{\mu_0}{4\pi}\sum_i^N \left[(\vec{r}_i - \vec{R}_{GO}) \cdot \frac{(\vec{r}_i - \vec{R}_K)}{|\vec{r}_i - \vec{R}_K|^3}\delta_{\alpha\beta} - (r_{i,\alpha} - R_{GO,\alpha})\frac{(r_{i,\beta} - R_{K,\beta})}{|\vec{r}_i - \vec{R}_K|^3}\right]$$

- **the diamagnetic spin-rotation tensor operator** coupling the orbital angular momentum of the electrons in a molecule with the rotation of the molecule and a magnetic dipole moment located at $\vec{R}_K$

$$\hat{O}^{m^K J}_{\alpha\beta}(\vec{R}_{CM}) = \sum_i^N \hat{o}^{m^K J}_{i,\alpha\beta}(\vec{R}_{CM}) \tag{A.27}$$

$$= -e\frac{\mu_0}{4\pi}\sum_i^N \left[(\vec{r}_i - \vec{R}_{CM}) \cdot \frac{(\vec{r}_i - \vec{R}_K)}{|\vec{r}_i - \vec{R}_K|^3}\delta_{\alpha\beta} - (r_{i,\alpha} - R_{CM,\alpha})\frac{(r_{i,\beta} - R_{K,\beta})}{|\vec{r}_i - \vec{R}_K|^3}\right]$$

A.1.2 Field-Dependent Perturbation Operators

For the derivation of magnetic second- or higher-order properties via response theory as discussed in Section 3.3 it is necessary to know the molecular magnetic moment and molecular magnetic induction operators in the presence of an external magnetic induction $\vec{\mathcal{B}}$, nuclear magnetic moments $\{\vec{m}^L\}$ or the perturbation due to the rotation of the molecule. These operators are also obtained as derivatives of the molecular electronic Hamiltonian, Eq. (2.101), but now for non-zero fields or magnetic moments:

- the orbital magnetic dipole moment operator in the presence of magnetic perturbations

$$\hat{m}_\alpha(\vec{R}_{GO}, \vec{\mathcal{B}}, \{\vec{m}^K\}, \vec{J}) = -\frac{\partial \hat{H}}{\partial \mathcal{B}_\alpha} \quad (A.28)$$

$$= \hat{O}_\alpha^{lB} + \hat{O}_\alpha^{sB} - 2\sum_\beta \hat{O}_{\alpha\beta}^{BB} \mathcal{B}_\beta - \sum_K \sum_\beta \hat{O}_{\beta\alpha}^{m^K B} m_\beta^K - \sum_\beta \hat{O}_{\alpha\beta}^{BJ} (\boldsymbol{I}^{-1}\vec{J})_\beta$$

- the operator for the molecular magnetic induction in the presence of magnetic perturbations

$$\hat{\mathcal{B}}_\alpha^j(\vec{R}_K, \vec{\mathcal{B}}, \{\vec{m}^L\}, \vec{J}) = -\frac{\partial \hat{H}}{\partial m_\alpha^K} \quad (A.29)$$

$$= \hat{O}_\alpha^{lm^K} + \hat{O}_\alpha^{sm^K} - \sum_\beta \hat{O}_{\alpha\beta}^{m^K \mathcal{B}} \mathcal{B}_\beta - 2\sum_L \sum_\beta \hat{O}_{\alpha\beta}^{m^K m^L} m_\beta^L - \sum_K \sum_\beta \hat{O}_{\alpha\beta}^{m^K J} (\boldsymbol{I}^{-1}\vec{J})_\beta$$

A.1.3 Sum-over-States Diamagnetic Contribution Operators

The reformulation of the diamagnetic contributions to the magnetizability, nuclear magnetic shielding and indirect nuclear spin-spin coupling tensor as linear response functions or sum-over-states (SOS) term discussed in Section 5.9 leads to new operators:

- the SOS diamagnetic magnetizability tensor operator

$$\hat{O}_\alpha^{\xi^\Delta}(\vec{R}_{GO}) = \left[\hat{\vec{\mu}}(\vec{R}_{GO}) \times \hat{\vec{m}}^l(\vec{R}_{GO})\right]_\alpha \quad (A.30)$$

- the SOS diamagnetic nuclear magnetic shielding tensor operator

$$\hat{O}_\alpha^{\sigma^{K,\Delta}}(\vec{R}_{GO}) = \left[\hat{\vec{O}}^\mu(\vec{R}_K) \times \hat{\vec{m}}^l(\vec{R}_{GO})\right]_\alpha \quad (A.31)$$

- the CTOCD-DZ diamagnetic nuclear magnetic shielding tensor operator

$$\hat{O}_{K,\delta\alpha}^{CTOCD-DZ}(\vec{R}_{GO}) = \frac{1}{4m_e}\left[\hat{\mu}_\delta(\vec{R}_{GO})\,\hat{O}_{K,\alpha}^{OP} + \hat{O}_{K,\alpha}^{OP}\,\hat{\mu}_\delta(\vec{R}_{GO})\right] \quad (A.32)$$

- the SOS orbital diamagnetic or diamagnetic nuclear spin-electron orbit operator

$$\hat{O}_\alpha^{K^{KL,\Delta}} = \left[\hat{\vec{O}}^\mu(\vec{R}_K) \times \hat{\vec{O}}_L^{OP}\right]_\alpha \quad (A.33)$$

A.2 Other Electronic Operators

- the density operator

$$\hat{D}(\vec{r}) = \sum_i^N \delta(\vec{r}_i - \vec{r}) \tag{A.34}$$

- the sum of the electronic position operators

$$\hat{O}_\alpha^r = \sum_i^N \hat{r}_{i,\alpha} \tag{A.35}$$

- the second moment operator

$$\hat{O}_{\alpha\beta}^{rr} = \sum_i^N \hat{r}_{i,\alpha} \hat{r}_{i,\beta} \tag{A.36}$$

- the total electronic canonical momentum operator

$$\hat{O}_\alpha^p = \sum_i^N \hat{p}_{i,\alpha} \tag{A.37}$$

- the total electronic angular momentum operator

$$\hat{L}_\alpha(R_{GO}) = \sum_i^N \left[(\vec{r}_i - \vec{R}_{GO}) \times \hat{\vec{p}}_i \right]_\alpha \tag{A.38}$$

Appendix B
Definitions of Properties

In the following tables all definitions of molecular response properties as derivatives of the energy or derivatives of other properties are collected.

Table B.1 Definitions of tensor components of the electric polarizabilities and hyperpolarizabilities as derivatives of components of the field-dependent electric dipole $\mu_\alpha(\vec{\mathcal{E}}, \mathcal{E})$ and quadrupole $\Theta_{\gamma\delta}(\vec{\mathcal{E}}, \mathcal{E})$ moments or of the field-dependent energy $E(\vec{\mathcal{E}}, \mathcal{E})$. All derivatives have to be evaluated at zero field and field gradient.

	$\mu_\alpha(\vec{\mathcal{E}}, \mathcal{E})$	$\Theta_{\gamma\delta}(\vec{\mathcal{E}}, \mathcal{E})$	$E(\vec{\mathcal{E}}, \mathcal{E})$
$\alpha_{\alpha\beta}$	$\dfrac{\partial}{\partial \mathcal{E}_\beta}$	—	$-\dfrac{\partial^2}{\partial \mathcal{E}_\beta \partial \mathcal{E}_\alpha}$
$\beta_{\alpha\beta\gamma}$	$\dfrac{\partial^2}{\partial \mathcal{E}_\gamma \partial \mathcal{E}_\beta}$	—	$-\dfrac{\partial^3}{\partial \mathcal{E}_\gamma \partial \mathcal{E}_\beta \partial \mathcal{E}_\alpha}$
$\gamma_{\alpha\beta\gamma\delta}$	$\dfrac{\partial^3}{\partial \mathcal{E}_\delta \partial \mathcal{E}_\gamma \partial \mathcal{E}_\beta}$	—	$-\dfrac{\partial^4}{\partial \mathcal{E}_\delta \partial \mathcal{E}_\gamma \partial \mathcal{E}_\beta \partial \mathcal{E}_\alpha}$
$A_{\alpha,\gamma\delta}$	$3\dfrac{\partial}{\partial \mathcal{E}_{\gamma\delta}}$	$\dfrac{\partial}{\partial \mathcal{E}_\alpha}$	$-3\dfrac{\partial^2}{\partial \mathcal{E}_{\gamma\delta} \partial \mathcal{E}_\alpha}$
$B_{\alpha\beta,\gamma\delta}$	$3\dfrac{\partial^2}{\partial \mathcal{E}_{\gamma\delta} \partial \mathcal{E}_\beta}$	$\dfrac{\partial^2}{\partial \mathcal{E}_\beta \partial \mathcal{E}_\alpha}$	$-3\dfrac{\partial^3}{\partial \mathcal{E}_{\gamma\delta} \partial \mathcal{E}_\beta \partial \mathcal{E}_\alpha}$
$C_{\gamma\delta,\alpha\beta}$	—	$\dfrac{\partial}{\partial \mathcal{E}_{\alpha\beta}}$	$-3\dfrac{\partial^2}{\partial \mathcal{E}_{\gamma\delta} \partial \mathcal{E}_{\alpha\beta}}$

Definitions of Properties

Table B.2 Definitions of various magnetic properties as derivatives of the perturbed energy $E(\vec{B}, \vec{m}^K, \vec{m}^L)$ or as derivatives of components of the perturbed magnetic dipole moment $m_\alpha(\vec{B}, \vec{m}^K)$ and molecular magnetic induction $B^j_\beta(\vec{R}; \vec{B}, \vec{m}^L)$. All derivatives are evaluated at zero magnetic field and zero nuclear magnetic moment.

	$m_\alpha(\vec{B}, \vec{m}^K)$	$B^j_\beta(\vec{R}; \vec{B}, \vec{m}^L)$	$B^j_\beta(\vec{R}_K; \vec{B}, \vec{m}^L)$	$E(\vec{B}, \vec{m}^K, \vec{m}^L)$
$\xi_{\alpha\beta}$	$\dfrac{\partial}{\partial B_\beta}$	—	—	$-\dfrac{\partial^2}{\partial B_\beta \partial B_\alpha}$
$\sigma_{\beta\alpha}(\vec{R})$	—	$-\dfrac{\partial}{\partial B_\alpha}$	—	—
$\sigma^K_{\beta\alpha}$	$-\dfrac{\partial}{\partial m^K_\beta}$	—	$\dfrac{\partial}{\partial B_\alpha}$	$\dfrac{\partial^2}{\partial m^K_\beta \partial B_\alpha}$
$K^L_{\beta\alpha}(\vec{R})$	—	$-\dfrac{\partial}{\partial m^L_\alpha}$	—	—
$K^{KL}_{\beta\alpha}$	—	—	$-\dfrac{\partial}{\partial m^L_\alpha}$	$\dfrac{\partial^2}{\partial m^K_\beta \partial m^L_\alpha}$

Table B.3 Definitions of the rotational g tensor and spin rotation as derivatives of the perturbed energy $E(\vec{B}, \vec{m}^K, \vec{J})$ or as derivatives of components of the perturbed magnetic dipole moment $m^J_\alpha(\vec{J})$ and molecular magnetic induction $B^{j,J}_\beta(\vec{R}; \vec{J})$. All derivatives have to be evaluated for zero perturbation.

	$m^J_\alpha(\vec{J})$	$B^{j,J}_\alpha(\vec{R}; \vec{J})$	$B^{j,J}_\alpha(\vec{R}_K; \vec{J})$	$E(\vec{B}, \vec{m}^K, \vec{J})$
$g_{J,\alpha\beta}$	$\dfrac{\hbar}{\mu_N}\dfrac{\partial}{\partial J_\beta}$	—	—	$-\dfrac{\hbar}{\mu_N}\dfrac{\partial^2}{\partial B_\alpha \partial J_\beta}$
$C_{\alpha\beta}(\vec{R})$	—	$\dfrac{\mu_N g_K}{2\pi}\dfrac{\partial}{\partial J_\beta}$	—	—
$C^K_{\alpha\beta}$	—	—	$\dfrac{\mu_N g_K}{2\pi}\dfrac{\partial}{\partial J_\beta}$	$-\dfrac{\mu_N g_K}{2\pi}\dfrac{\partial^2}{\partial m^K_\alpha \partial J_\beta}$

Appendix C
Perturbation Theory Expressions for Properties

Using time-independent perturbation theory from Section 3.2 or response theory as described in Sections 3.3 and 3.11 one can derive the following expressions for the first-order $P^{(1)}$

$$P^{(1)} = f_1 \langle \Psi_0^{(0)} | \hat{O}_1 | \Psi_0^{(0)} \rangle \tag{C.1}$$

and second-order $P^{(2)}$ molecular properties

$$P^{(2)} = f_1 \langle \Psi_0^{(0)} | \hat{O}_1 | \Psi_0^{(0)} \rangle + f_2 \sum_{n \neq 0} \frac{\langle \Psi_0^{(0)} | \hat{O}_2 | \Psi_n^{(0)} \rangle \langle \Psi_n^{(0)} | \hat{O}_3 | \Psi_0^{(0)} \rangle}{E_0^{(0)} - E_n^{(0)}}$$

$$+ f_2 \sum_{n \neq 0} \frac{\langle \Psi_0^{(0)} | \hat{O}_3 | \Psi_n^{(0)} \rangle \langle \Psi_n^{(0)} | \hat{O}_2 | \Psi_0^{(0)} \rangle}{E_0^{(0)} - E_n^{(0)}} \tag{C.2}$$

$$= f_1 \langle \Psi_0^{(0)} | \hat{O}_1 | \Psi_0^{(0)} \rangle + f_2 \langle\langle \hat{O}_2 ; \hat{O}_3 \rangle\rangle_{\omega=0} \tag{C.3}$$

for fixed nuclear positions. The operators $\hat{O}_1$, $\hat{O}_2$ and $\hat{O}_3$ and the prefactors f_1 and f_2 are collected in Tables C.1 and C.2.

Table C.1 Operators and prefactors for the exact first-order Rayleigh–Schrödinger perturbation theory expressions for molecular properties. See Eq. (C.1).

$P^{(1)}$	f_1	$\hat{O}_1$
$\mu_\alpha(\vec{R}_O)$	1	$\hat{O}_\alpha^E(\vec{R}_O) + \hat{\Omega}_\alpha^{\mathcal{E}}(\vec{R}_O)$
$\Theta_{\alpha\beta}(\vec{R}_O)$	3	$\hat{O}_{\alpha\beta}^{\nabla E}(\vec{R}_O) + \hat{\Omega}_{\alpha\beta}^{\nabla \mathcal{E}}(\vec{R}_O)$
$m_\alpha(\vec{R}_{GO})$	1	$\hat{O}_\alpha^{lB}(\vec{R}_{GO}) + \hat{O}_\alpha^{sB}$
$B_\alpha^j(\vec{R}_K)$	1	$\hat{O}_\alpha^{lm^K}(\vec{R}_{GO}) + \hat{O}_\alpha^{sm^K}$
$a_{\alpha\beta}^K$	$-\dfrac{g_K \mu_N}{2\pi} \dfrac{1}{\langle \Psi_0^{(0)} | \hat{S}_\alpha | \Psi_0^{(0)} \rangle}$	$\hat{O}_\beta^{sm^K}$

Table C.2 Operators and prefactors for the exact second-order Rayleigh–Schrödinger perturbation theory expressions for molecular properties. See Eqs. (C.2) and (C.3).

$P^{(2)}$	f_1	$\hat{O}_1$	f_2	$\hat{O}_2$	$\hat{O}_3$
$\alpha_{\alpha\beta}$			-1	$\hat{O}_\alpha^E$	$\hat{O}_\beta^E$
$A_{\alpha,\beta\gamma}(\vec{R}_O)$			-3	$\hat{O}_\alpha^E$	$\hat{O}_{\beta\gamma}^{\nabla E}$
$C_{\alpha\beta,\gamma\delta}(\vec{R}_O)$			-3	$\hat{O}_{\alpha\beta}^{\nabla E}$	$\hat{O}_{\beta\gamma}^{\nabla E}$
$\xi_{\alpha\beta}(\vec{R}_{GO})$	-2	$\hat{O}_{\alpha\beta}^{BB}(\vec{R}_{GO})$	-1	$\hat{O}_\alpha^{lB}(\vec{R}_{GO})$	$\hat{O}_\beta^{lB}(\vec{R}_{GO})$
$\xi_{\alpha\beta}^\Delta(\vec{R}_{GO})$			$-\dfrac{1}{2m_e}$	$\hat{O}_\alpha^{\xi^\Delta}(\vec{R}_{GO})$	$\hat{O}_\beta^p$
$\sigma_{\alpha\beta}^K(\vec{R}_{GO})$	1	$\hat{O}_{\alpha\beta}^{m^K B}(\vec{R}_{GO})$	1	$\hat{O}_\alpha^{lm^K}$	$\hat{O}_\beta^{lB}(\vec{R}_{GO})$
$\sigma_{\alpha\beta}^{K,\Delta}(\vec{R}_{GO})$			$\dfrac{1}{m_e c^2}$	$\hat{O}_\alpha^{\sigma^{K,\Delta}}(\vec{R}_{GO})$	$\hat{O}_\beta^p$
$\sigma_{\alpha\beta}^{K,\Delta}(\vec{R}_{GO})$			$\displaystyle\sum_{\gamma\delta}\epsilon_{\beta\gamma\delta}$	$\hat{O}_\gamma^p$	$\hat{O}_{K,\delta\alpha}^{CTOCD-DZ}(\vec{R}_{GO})$
$K_{\alpha\beta}^{KL}$	2	$\hat{O}_{\alpha\beta}^{m^K m^L}$	1	$\hat{O}_\alpha^{lm^K}+\hat{O}_\alpha^{sm^K}$	$\hat{O}_\beta^{lm^L}+\hat{O}_\beta^{sm^L}$
$K_{\alpha\beta}^{KL,\Delta}$			$\dfrac{1}{m_e c^2}$	$\hat{O}_\alpha^{K^{KL,\Delta}}$	$\hat{O}_\beta^p$
$g_{J,\alpha\beta}^{rig}(\vec{R}_{CM})$	$-\dfrac{2m_p}{eI_{\beta\beta}}$	$\hat{O}_{\alpha\beta}^{JJ}(\vec{R}_{CM})$			
$g_{J,\alpha\beta}^{ind}(\vec{R}_{CM},\vec{R}_{GO})$	$-\dfrac{2m_p}{eI_{\beta\beta}}$	$\hat{O}_{\alpha\beta}^{BJ}(\vec{R}_{CM},\vec{R}_{GO})$	$\dfrac{2m_p}{eI_{\beta\beta}}$	$\hat{O}_\alpha^{lB}(\vec{R}_{GO})$	$\hat{O}_\beta^{lJ}(\vec{R}_{CM})$
$g_{J,\alpha\beta}^{el}(\vec{R}_{CM})$			$\dfrac{4m_p m_e}{e^2 I_{\beta\beta}}$	$\hat{O}_\alpha^{lB}(\vec{R}_{CM})$	$\hat{O}_\beta^{lB}(\vec{R}_{CM})$
$C_{\alpha\beta}^{K,rig}(\vec{R}_K)$	$\dfrac{\mu_N g_K}{2\pi I_{\beta\beta}}$	$\hat{O}_{\alpha\beta}^{m^K J}(\vec{R}_K)$			
$C_{\alpha\beta}^{K,ind}(\vec{R}_{CM},\vec{R}_K)$	$-\dfrac{\mu_N g_K}{2\pi I_{\beta\beta}}$	$\hat{O}_{\alpha\beta}^{m^K J}(\vec{R}_{CM})$	$\dfrac{\mu_N g_K}{2\pi I_{\beta\beta}}$	$\hat{O}_\alpha^{lm^K}$	$\hat{O}_\beta^{lJ}(\vec{R}_{CM})$
$C_{\alpha\beta}^{K,el}(\vec{R}_K)$			$\dfrac{\mu_N g_K}{2\pi I_{\beta\beta}}\dfrac{2m_e}{e}$	$\hat{O}_\alpha^{lm^K}$	$\hat{O}_\beta^{lB}(\vec{R}_K)$

References

Aiga, F. and Itoh, R. (1996). Calculation of frequency-dependent polarizabilities and hyperpolarizabilities by the second-order Møller-Plesset perturbation theory. *Chem. Phys. Lett.*, **251**, 372–380.

Aiga, F., Sasagane, K., and Itoh, R. (1993). Frequency-dependent hyperpolarizabilities in the Møller-Plesset perturbation theory. *J. Chem. Phys.*, **99**, 3779–3789.

Aiga, F., Sasagane, K., and Itoh, R. (1994). Frequency-dependent hyperpolarizabilities in the Brueckner Coupled-Cluster Theory. *Int. J. Quantum Chem.*, **51**, 87–97.

Amos, R. D. and Rice, J. E. (1989). Implementation of analytic derivative methods in quantum chemistry. *Comput. Phys. Rep.*, **10**, 147–187.

Andersson, K. and Sadlej, A. J. (1992). Electric dipole polarizabilities of atomic valence states. *Phys. Rev. A*, **46**, 2356–2362.

Arponen, J. (1983). Variational principles and linked-cluster exp S expansions for static and dynamic many-body problems. *Ann. Phys.*, **151**, 311–382.

Åstrand, P.-O., Ruud, K., Mikkelsen, K. V., and Helgaker, T. (1999). Rovibrationally averaged magnetizability, rotational g factor and indirect nuclear spin-spin coupling of the hydrogen fluoride molecule. *J. Chem. Phys.*, **110**, 9463–9468.

Bak, K. L., Koch, H., Oddershede, J., Christiansen, O., and Sauer, S. P. A. (2000). Atomic integral driven second order polarization propagator calculations of the excitation spectra of naphthalene and anthracene. *J. Chem. Phys.*, **112**, 4173–4185.

Bak, K. L., Sauer, S. P. A., Oddershede, J., and Ogilvie, J. F. (2005). The vibrational g factor of dihydrogen from theoretical calculation and analysis of vibration-rotational spectra. *Phys. Chem. Chem. Phys.*, **7**, 1747–1758.

Balabanov, N. B. and Peterson, K. A. (2005). Systematically convergent basis sets for transition metals. I. All-electron correlation consistent basis sets for the 3d elements Sc-Zn. *J. Chem. Phys.*, **123**, 64107.

Baranowska, A., Siedlecka, M., and Sadlej, A. J. (2007). Reduced-size polarized basis sets for calculations of molecular electric properties. IV. First-row transition metals. *Theor. Chim. Acc.*, **118**, 959–972.

Barone, V., Provasi, P. F., Peralta, J. E., Snyder, J. P., Sauer, S. P. A., and Contreras, R. H. (2003). Substituent effects on scalar $^2J(^{19}F, ^{19}F)$ and $^3J(^{19}F, ^{19}F)$ NMR couplings: A comparison of SOPPA and DFT methods. *J. Phys. Chem. A*, **107**, 4748–4754.

Benedikt, U., Auer, A. A., and Jensen, F. (2008). Optimization of augmentation functions for correlated calculations of spin-spin coupling constants and related properties. *J. Chem. Phys.*, **129**, 64111.

Benkova, Z., Sadlej, A. J., Oakes, R. E., and Bell, S. E. J. (2005a). Reduced-size polarized basis sets for calculations of molecular electric properties. I. The basis set generation. *J. Comput. Chem.*, **26**, 145–153.

Benkova, Z., Sadlej, A. J., Oakes, R. E., and Bell, S. E. J. (2005b). Reduced-size polarized basis sets for calculations of molecular electric properties. II. Simulation of the Raman spectra. *J. Comput. Chem.*, **26**, 154–159.

Benkova, Z., Sadlej, A. J., Oakes, R. E., and Bell, S. E. J. (2005c). Reduced-size polarized basis sets for calculations of molecular electric properties. III. Second-row atoms. *Theor. Chim. Acc.*, **113**, 238–247.

Bethe, H. (1930). Zur Theorie des Durchgangs schneller Korpuskularstrahlen durch Materie. *Ann. d. Phys.*, **397**, 325–400.

Bishop, D. M. (1990). Molecular vibrational and rotational motion in static and dynamic electric fields. *Rev. Mod. Phys.*, **62**, 343–374.

Bishop, D. M. (1994). Aspects of non-linear-optical calculations. *Adv. Quantum Chem.*, **25**, 1–45.

Bishop, D. M. (1998). Molecular Vibration and Nonlinear Optics. *Adv. Chem. Phys.*, **104**, 1–40.

Bishop, D. M. and Cheung, L. M. (1980). Dynamic polarizability of H_2 and HeH^+. *J. Chem. Phys.*, **72**, 5125–5132.

Bishop, D. M., Cheung, L. M., and Buckingham, A. D. (1980). Dipole polarizability formulae. *Mol. Phys.*, **41**, 1225–1226.

Bishop, D. M. and Kirtman, B. (1991). A perturbation method for calculating vibrational dynamic dipole polarizabilities and hyperpolarizabilities. *J. Chem. Phys.*, **95**, 2646–2658.

Bishop, D. M. and Sauer, S. P. A. (1997). Calculation, with the inclusion of vibrational corrections, of the DC-electric-field induced second-harmonic-generation hyperpolarizability of methane. *J. Chem. Phys.*, **107**, 8502–8509.

Bruun-Ghalbia, S., Sauer, S. P. A., Oddershede, J., and Sabin, J. R. (2010). Mean excitation energies and energy deposition characteristics of bio-organic molecules. *J. Phys. Chem. B*, **114**, 633–637.

Buckingham, A. D. (1967). Permanent and induced molecular moments and long-range intermolecular forces. *Adv. Chem. Phys.*, **12**, 107–142.

Buckingham, A. D. and Stiles, P. J. (1972). Magnetic multipoles and the 'pseudocontact' chemical shift. *Mol. Phys.*, **24**, 99–108.

Bunker, P. R. and Moss, R. E. (1977). The breakdown of the Born-Oppenheimer approximation: the effective vibration-rotation Hamiltonian for a diatomic molecule. *Mol. Phys.*, **33**, 417–424.

Casimir, H. B. G. and Polder, D. (1948). The Influence of retardation on the London-van der Waals forces. *Phys. Rev.*, **73**, 360–372.

Cernusak, I., Kellö, V., and Sadlej, A. J. (2003). Standardized medium-size basis sets for calculations of molecular electric properties: Group IIIA. *Collect. Czech. Chem. Commun.*, **68**, 211–239.

Chang, Ch., Pelissier, M., and Durand, Ph. (1986). Regular two-component Pauli-like effective Hamiltonians in Dirac theory. *Phys. Scr.*, **34**, 394–404.

Chen, J. C. Y. (1964). Off-diagonal hypervirial theorem and its applications. *J. Chem. Phys.*, **40**, 615–621.

Christiansen, O., Bak, K. L., Koch, H., and Sauer, S. P. A. (1998a). A second-order doubles correction to excitation energies in the random phase approximation. *Chem. Phys. Lett.*, **284**, 47–62.

Christiansen, O, Jørgensen, P., and Hättig, C. (1998b). Response functions from Fourier component variational perturbation theory applied to a time-averaged quasienergy. *Int. J. Quantum Chem.*, **68**, 1–52.

Christiansen, O., Koch, H., and Jørgensen, P. (1995a). Response functions in the CC3 iterative triple excitation model. *J. Chem. Phys.*, **103**, 7429–7441.

Christiansen, O., Koch, H., and Jørgensen, P. (1995b). The second-order approximate coupled cluster singles and doubles model CC2. *Chem. Phys. Lett.*, **243**, 409–418.

Christiansen, O., Koch, H., and Jørgensen, P. (1996). Perturbative triple excitation corrections to coupled cluster singles and doubles excitation energies. *J. Chem. Phys.*, **105**, 1451–1459.

Clark, T., Chandrasekhar, J., Spitznagel, G. W., and Schleyer, P. (1983). Efficient diffuse function-augmented basis sets for anion calculations. III. The 3-21+G basis set for first-row elements, Li-F. *J. Comput. Chem.*, **4**, 294–301.

Cohen, H. D. and Roothaan, C. C. J. (1965). Electric dipole polarizability of atoms by the Hartree-Fock method. I. Theory for closed-shell systems. *J. Chem. Phys.*, **43**, S34–S39.

Cybulski, S. M. and Bishop, D. M. (1994). Theory of relaxed density matrices: Application to second-order response properties. *Int. J. Quantum Chem.*, **49**, 371–381.

Dalgaard, E. (1979). Expansion and completeness theorems for operator manifolds. *Int. J. Quantum Chem.*, **15**, 169–180.

Dalgarno, A. (1959). Perturbation theory for atomic systems. *Proc. R. Soc. Lond. A*, **251**, 282–290.

Dalskov, E. K. and Sauer, S. P. A. (1998). Correlated, static and dynamic polarizabilities of small molecules. A comparison of four black box methods. *J. Phys. Chem. A*, **102**, 5269–5274.

Deng, W., Cheeseman, J. R., and Frisch, M. J. (2006). Calculation of nuclear spin-spin coupling constants of molecules with first and second row atoms in study of basis set dependence. *J. Chem. Theory Comput.*, **2**, 1028–1037.

Diercksen, G. H. F. and McWeeny, R. (1966). Self-consistent perturbation theory. I. General formulation and some applications. *J. Chem. Phys.*, **44**, 3554–3560.

Dirac, P. A. M. (1958). *The principles of quantum mechanics* (4th edn) Chapter VII. Oxford University Press, Oxford.

Ditchfield, R. (1974). Self-consistent perturbation theory of diamagnetism I. A gauge-invariant LCAO method for N.M.R. chemical shifts. *Mol. Phys.*, **27**, 789–807.

Dunning Jr., T. H. (1989). Gaussian basis sets for use in correlated molecular calculations. I. The atoms boron through neon and hydrogen. *J. Chem. Phys.*, **90**, 1007–1023.

Enevoldsen, T., Oddershede, J., and Sauer, S. P. A. (1998). Correlated calculations of indirect nuclear spin-spin coupling constants using second order polarization propagator approximations: SOPPA and SOPPA(CCSD). *Theor. Chem. Acc.*, **100**, 275–284.

Eshbach, J. R. and Strandberg, M. W. P. (1952). Rotational magnetic moments of $^1\Sigma$ molecules. *Phys. Rev.*, **85**, 24–34.

Estermann, I. and Stern, O. (1933). Über die magnetische Ablenkung von Wasserstoffmolekülen und das magnetische Moment des Protons. II. *Z. Physik*, **85**, 17–24.

Falden, H. H., Falster-Hansen, K. R., Bak, K. L., Rettrup, S., and Sauer, S. P. A. (2009). Benchmarking second order methods for the calculation of vertical electronic excitation energies: Valence and Rydberg states in polycyclic aromatic hydrocarbons. *J. Phys. Chem. A*, **113**, 11995–12012.

Feyereisen, M., Nichols, J., Oddershede, J., and Simons, J. (1992). Direct atomic-orbital-based time-dependent Hartree-Fock calculations of frequency-dependent polarizabilities. *J. Chem. Phys.*, **96**, 2978–2987.

Feynman, R. P. (1939). Forces in molecules. *Phys. Rev.*, **56**, 340–343.

Flygare, W. H. (1964). Spin-rotation interaction and magnetic shielding in molecules. *J. Chem. Phys.*, **41**, 793–800.

Flygare, W. H. (1974). Magnetic interactions in molecules and an analysis of molecular electronic charge distribution from magnetic parameters. *Chem. Rev.*, **74**, 653–687.

Flygare, W. H. and Benson, R. C. (1971). The molecular Zeeman effect in diamagnetic molecules and the determination of molecular magnetic moments (g values), magnetic susceptibilities, and molecular quadrupole moments. *Mol. Phys.*, **20**, 225–250.

Francl, M. M., Pietro, W. J., Hehre, W. J., Binkley, J. S., Gordon, M. S., DeFrees, D. J., and Pople, J. A. (1982). Self-consistent molecular orbital methods. XXIII. A polarization-type basis set for second-row elements. *J. Chem. Phys.*, **77**, 3654–3665.

Frenkel, J. (1934). *Wave mechanics advanced general theory.* Clarendon Press, Oxford.

Frisch, R. and Stern, O. (1933). Über die magnetische Ablenkung von Wasserstoffmolekülen und das magnetische Moment des Protons. I. *Z. Physik*, **85**, 4–16.

Fuchs, C., Bonacic-Koutecky, V., and Koutecky, J. (1993). Compact formulation of multiconfigurational response theory. Applications to small alkali metal clusters. *J. Chem. Phys.*, **98**, 3121–3140.

Fukui, H., Miura, K., and Matsuda, H. (1992a). Calculation of nuclear magnetic shieldings. VIII. Gauge invariant many-body perturbation method. *J. Chem. Phys.*, **96**, 2039–2043.

Fukui, H., Miura, K., Matsuda, H., and Baba, T. (1992b). Calculation of nuclear spin-spin couplings. VII. Electron correlation effects on the five coupling mechanisms. *J. Chem. Phys.*, **97**, 2299–2303.

Gauss, J. (1992). Calculation of NMR chemical shifts at second-order many-body perturbation theory using gauge-including atomic orbitals. *Chem. Phys. Lett.*, **191**, 614–620.

Gauss, J. (1993). Effects of electron correlation in the calculation of NMR chemical shifts. *J. Chem. Phys.*, **99**, 3629–3643.

Gauss, J. (1994). GIAO-MBPT(3) and GIAO-SDQ-MBPT(4) Calculations of nuclear magnetic shielding constants. *Chem. Phys. Lett.*, **229**, 198–203.

Gauss, J. and Cremer, D. (1992). Analytical energy gradients in Møller-Plesset perturbation and quadratic configuration interaction methods: theory and application. *Adv. Quantum Chem.*, **23**, 205–299.

Gauss, J. and Stanton, J. F. (1995). Coupled-cluster calculations of nuclear magnetic resonance chemical shifts. *J. Chem. Phys.*, **103**, 3561–3577.

Gauss, J. and Stanton, J. F. (1996). Perturbative treatment of triple excitations in coupled-cluster calculations of nuclear magnetic shielding constants. *J. Chem. Phys.*, **104**, 2574–2582.

Geertsen, J. (1989). A solution of the gauge origin problem for the magnetic susceptibility. *J. Chem. Phys.*, **90**, 4892–4894.

Geertsen, J. (1991). A solution of the gauge-origin problem for the magnetic shielding constant. *Chem. Phys. Lett.*, **179**, 479–482.

Geertsen, J., Eriksen, S., and Oddershede, J. (1991a). Some aspects of the coupled cluster based polarization propagator method. *Adv. Quantum Chem.*, **22**, 167–209.

Geertsen, J., Oddershede, J., Raynes, W. T., and Scuseria, G. E. (1991b). Nuclear spin-spin coupling in the methane isotopomers. *J. Magn. Reson.*, **93**, 458–471.

Geertsen, J., Rittby, M., and Bartlett, R. J. (1989). The equation-of-motion coupled-cluster method: Excitation energies of Be and CO. *Chem. Phys. Lett.*, **164**, 57–62.

Gerratt, J. and Mills, I. M. (1968). Force constants and dipole-moment derivatives of molecules from perturbed Hartree-Fock calculations. *J. Chem. Phys.*, **49**, 1719–1729.

Goscinski, O. and Lukman, B. (1970). Moment-conserving decoupling of Green functions via Pade approximants. *Chem. Phys. Lett.*, **7**, 573–576.

Guilleme, J. and San Fabián, J. (1998). Basis sets and active space in multiconfigurational self-consistent field calculations of nuclear magnetic resonance spin-spin coupling constants. *J. Chem. Phys.*, **109**, 8168–8181.

Gunther-Mohr, G. R., Townes, C. H., and Van Vleck, J. H. (1954). Hyperfine structure in the spectrum of $N^{14}H_3$. II. Theoretical discussion. *Phys. Rev.*, **94**, 1191–1203.

Hald, K., Jørgensen, P., Olsen, J., and Jaszuński, M. (2001). An analysis and implementation of a general coupled cluster approach to excitation energies with application to the B_2 molecule. *J. Chem. Phys.*, **115**, 671–679.

Hameka, H. (1958). On the nuclear magnetic shielding in the hydrogen molecule. *Mol. Phys.*, **1**, 203–215.

Handy, N. C. and Schaefer, H. F. (1984). On the evaluation of analytic energy derivatives for correlated wave functions. *J. Chem. Phys.*, **81**, 5031–5033.

Hansen, Aa. E. (1967). Correlation effects in the calculation of ordinary and rotatory intensities. *Mol. Phys.*, **13**, 425–431.

Hansen, Aa. E. and Bouman, T. D. (1979). Hypervirial relations as constraints in calculations of electronic excitation properties: The random phase approximation in configuration interaction language. *Mol. Phys.*, **37**, 1713–1724.

Hansen, Aa. E. and Bouman, T. D. (1985). Localized orbital/local origin method for calculation and analysis of NMR shieldings. Applications to ^{13}C shielding tensors. *J. Chem. Phys.*, **82**, 5035–5047.

Hättig, C. and Heß, B. A. (1995). Correlated frequency-dependent polarizabilities and dispersion coefficients in the time-dependent second-order Møller-Plesset approximation. *Chem. Phys. Lett.*, **233**, 359–370.

Hehre, W. J., Ditchfield, R., and Pople, J. A. (1972). Self-consistent molecular orbital methods. XII. Further extensions of Gaussian-type basis sets for use in molecular orbital studies of organic molecules. *J. Chem. Phys.*, **56**, 2257–2261.

Helgaker, T., Jaszuński, M., Ruud, K., and Górska, A. (1998). Basis-set dependence of nuclear spin-spin coupling constants. *Theor. Chem. Acc.*, **99**, 175–182.

Helgaker, T. and Jørgensen, P. (1988). Analytical calculation of geometrical derivatives in molecular electronic structure theory. *Adv. Quantum Chem.*, **19**, 183–245.

Helgaker, T. and Jørgensen, P. (1991). An electronic Hamiltonian for origin independent calculations of magnetic properties. *J. Chem. Phys.*, **95**, 2595–2601.

Helgaker, T., Jørgensen, P., and Handy, N. C. (1989). A numerically stable procedure for calculating Møller-Plesset energy derivatives, derived using the theory of Lagrangians. *Theor. Chim. Acta*, **76**, 227–245.

Hellmann, H. (1937). *Einführung in die Quantemchemie*. Franz Deuticke, Leipzig, p. 285.

Herman, R. M. and Asgharian, A. (1966). Theory of energy shifts associated with deviations from Born-Oppenheimer behavior in $^1\Sigma$-state diatomic molecules. *J. Mol. Spectrosc.*, **19**, 305–324.

Herman, R. M. and Ogilvie, J. F. (1998). An effective Hamiltonian to treat adiabatic and nonadiabatic effects in the rotational and vibrational spectra of diatomic molecules. *Adv. Chem. Phys.*, **103**, 187–215.

Hettema, H., Wormer, P. E. S., Jørgensen, P., Jensen, H. J. Aa., and Helgaker, T. (1994). Frequency-dependent polarizabilities of O_2 and van der Waals coefficients of dimers containing O_2. *J. Chem. Phys.*, **100**, 1297–1302.

Hindermann, D. K. and Cornwell, C. D. (1968). Vibrational corrections to the nuclear-magnetic shielding and spin-rotation constants for hydrogen fluoride shielding scale for ^{19}F. *J. Chem. Phys.*, **48**, 4148–4154.

Jameson, C. J., Jameson, A. K., and Burrell, P. M. (1980). ^{19}F nuclear magnetic shielding scale from gas phase studies. *J. Chem. Phys.*, **73**, 6013–6020.

Jaszuński, M. (1987). A mixed numerical analytical approach to the calculation of non-linear electric properties. *Chem. Phys. Lett.*, **140**, 130–132.

Jaszuński, M. and Ruud, K. (2001). Spin-spin coupling constants in C_2H_2. *Chem. Phys. Lett.*, **336**, 473–478.

Jensen, F. (2001). Polarization consistent basis sets: Principles. *J. Chem. Phys.*, **115**, 9113–9125.

Jensen, F. (2002*a*). Erratum: Polarization consistent basis sets: Principles [J. Chem. Phys. 115, 9113 (2001)]. *J. Chem. Phys.*, **116**, 3502.

Jensen, F. (2002*b*). Polarization consistent basis sets. II. Estimating the Kohn-Sham basis set limit. *J. Chem. Phys.*, **116**, 7372–7379.

Jensen, F. (2002*c*). Polarization consistent basis sets. III. The importance of diffuse functions. *J. Chem. Phys.*, **117**, 9234–9240.

Jensen, F. (2003). Polarization consistent basis sets. IV. The basis set convergence of equilibrium geometries, harmonic vibrational frequencies, and intensities. *J. Chem. Phys.*, **118**, 2459–2463.

Jensen, F. (2006). The basis set convergence of spin-spin coupling constants calculated by density functional methods. *J. Chem. Theory Comput.*, **2**, 1360–1369.

Jensen, F. (2007). Polarization consistent basis sets. 4: The elements He, Li, Be, B, Ne, Na, Mg, Al, and Ar. *J. Phys. Chem. A*, **111**, 11198–11204.

Jensen, F. (2008). Basis set convergence of nuclear magnetic shielding constants calculated by density functional methods. *J. Chem. Theory Comput.*, **4**, 719–727.

Jensen, F. (2010). The optimum contraction of basis sets for calculating spin-spin coupling constants. *Theo. Chem. Acc.*, **126**, 371–382.

Jensen, F. and Helgaker T. (2004). Polarization consistent basis sets. V. The elements Si-Cl. *J. Chem. Phys.*, **121**, 3463–3470.

Jensen, H. J. Aa., Jørgensen, P., Ågren, H., and Olsen, J. (1988a). Second-order Møller-Plesset perturbation theory as a configuration and orbital generator in multiconfiguration self-consistent field calculations. *J. Chem. Phys.*, **88**, 3834–3839.

Jensen, H. J. Aa., Jørgensen, P., Ågren, H., and Olsen, J. (1988b). Erratum: Second-order Møller-Plesset perturbation theory as a configuration and orbital generator in multiconfiguration self-consistent field calculations. *J. Chem. Phys.*, **89**, 5354.

Jensen, M. Ø. and Hansen, Aa. E. (1999). The molecular magnetic shielding field: Response graph illustrations of the benzene field. *Adv. Quantum Chem.*, **35**, 193–215.

Jørgensen, P. and Helgaker, T. (1988). Møller-Plesset energy derivatives. *J. Chem. Phys.*, **89**, 1560–1570.

Jørgensen, P., Oddershede, J., and Beebe, N. H. F. (1978). Polarization propagator calculations of frequency-dependent polarizabilities, Verdet constant, and energy weighted sum rules. *J. Chem. Phys.*, **68**, 2527–2532.

Jørgensen, P. and Simons, J. (1981). *Second quantization-based methods in quantum chemistry*. Academic Press, New York.

Kaski, J., Lantto, P., Vaara, J., and Jokisaari, J. (1998). Experimental and theoretical ab initio study of the ^{13}C-^{13}C spin-spin coupling and ^{1}H and ^{13}C shielding tensors in ethane, ethene, and ethyne. *J. Am. Chem. Soc.*, **120**, 3993–4005.

Kellö, V. and Sadlej, A. J. (1995). Polarized basis sets for high-level-correlated calculations of molecular electric properties VIII. Elements of the group IIb: Zn, Cd, Hg. *Theor. Chim. Acta*, **91**, 353–371.

Kendall, R. A., Dunning, T. H., and Harrison, R. J. (1992). Electron affinities of the first-row atoms revisited. Systematic basis sets and wave functions. *J. Chem. Phys.*, **96**, 6796–6806.

Kern, C. W. and Matcha, R. L. (1968). Nuclear corrections to electronic expectation values: Zero-point vibrational effects in the water molecule. *J. Chem. Phys.*, **49**, 2081–2091.

Kjær, H. and Sauer, S. P. A. (2009). On the relation between the non-adiabatic vibrational reduced mass and the electric dipole moment gradient of a diatomic molecule. *Theo. Chem. Acc.*, **122**, 137–143.

Kjær, H., Sauer, S. P. A., and Kongsted, J. (2010). Benchmarking NMR indirect nuclear spin-spin coupling constants: SOPPA, SOPPA(CC2) and SOPPA(CCSD) versus CCSD. *J. Chem. Phys.*, **133**, 144106.

Koch, H., Christiansen, O., Kobayashi, R., Jørgensen, P., and Helgaker, T. (1994). A direct atomic orbital driven implementation of the coupled cluster singles and doubles (CCSD) model. *Chem. Phys. Lett.*, **228**, 233–238.

Koch, H. and Jørgensen, P. (1990). Coupled cluster response function. *J. Chem. Phys.*, **93**, 3333–3344.

Kowalewski, J., Laaksonen, A., Roos, B., and Siegbahn, P. (1979). Finite perturbation-configuration interaction calculations of nuclear spin-spin coupling constants. I. The first row hydrides and the hydrogen molecule. *J. Chem. Phys.*, **71**, 2896–2902.

Kutzelnigg, W. (1980). Theory of magnetic susceptibility and NMR chemical shift in terms of localized quantities. *Israel J. Chem.*, **19**, 193–200.

Kutzelnigg, W. (1992). Stationary perturbation theory I. Survey of basic concepts. *Theo. Chim. Acta*, **83**, 263–312.

Landolt-Börnstein (1976). *New series, structure data of free polyatomic molecules*. Volume 2.7., 2.2.1. Springer, Berlin.

Langhoff, P. W., Epstein, S. T., and Karplus, M. (1972). Aspects of time-dependent perturbation theory. *Rev. Mod. Phys.*, **44**, 602–644.

Langhoff, P. W. and Karplus, M. (1970). Application of Padé approximants to dispersion force and optical polarizability computations. In *The Padé Approximant in Theoretical Physics* (ed. G. A. Baker Jr. and J. L. Gammel), Chapter 2, pp. 41–97. Academic Press, New York.

Lazzeretti, P. (1989). General connections among nuclear electromagnetic shieldings and polarizabilities. *Adv. Chem. Phys.*, **75**, 507–549.

Lazzeretti, P., Malagoli, M., and Zanasi, R. (1994). Computational approach to molecular magnetic properties by continuous transformation of the origin of the current density. *Chem. Phys. Lett.*, **220**, 299–304.

Ligabue, A., Sauer, S. P. A., and Lazzeretti, P. (2003). Correlated and gauge invariant calculations of nuclear magnetic shielding constants using the continuous transformation of the origin of the current density approach. *J. Chem. Phys.*, **118**, 6830–6845.

Linderberg, J. and Öhrn, Y. (1973). *Propagators in quantum chemistry*. Academic Press, London.

London, F. (1937). Theorie quantique des courants interatomiques dans les combinaisons aromatiques. *J. Phys. Radium*, **8**, 397–409.

Löwdin, P.-O. (1965). Studies in perturbation theory. IX. Connection between various approaches in the recent development. Evaluation of upper bounds to energy eigenvalues in Schrödinger's perturbation theory. *J. Math. Phys.*, **6**, 1341–1353.

Löwdin, P.-O. and Mukherjee, P. K. (1972). Some comments on the time-dependent variation principle. *Chem. Phys. Lett.*, **14**, 1–7.

Lutnæs, O. B., Teale, A. M., Helgaker, T., Tozer, D. J., Ruud, K., and Gauss, J. (2009). Benchmarking density-functional-theory calculations of rotational g tensors and magnetizabilities using accurate coupled-cluster calculations. *J. Chem. Phys.*, **131**, 144104.

Maroulis, G. and Thakkar, A. J. (1988). Multipole moments, polarizabilities, and hyperpolarizabilities for N_2 from fourth-order many-body perturbation theory calculations. *J. Chem. Phys.*, **88**, 7623–7632.

McDowell, S. A. C., Amos, R. D., and Handy, N. C. (1995). Molecular polarizabilities - a comparision of density functional theory with standard ab initio methods. *Chem. Phys. Lett.*, **235**, 1–4.

McLachlan, A. D. and Ball, M. A. (1964). Time-dependent Hartree-Fock theory for molecules. *Rev. Mod. Phys.*, **36**, 844–855.

Mills, I., Cvitaš, T., Homann, K., Kallay, N., and Kuchitsu, K. (1993). *Quantities, units and symbols in physical chemistry*. Blackwell Scientific Publications, Oxford.

Møller, C. and Plesset, M. S. (1934). Note on an approximation treatment for many-electron systems. *Phys. Rev.*, **46**, 618–622.

Neogrády, P., Kellö, V., Urban, M., and Sadlej, A. J. (1996). Polarized basis sets for high-level-correlated calculations of molecular electric properties VII. Elements of the group Ib: Cu, Ag, Au. *Theor. Chim. Acc.*, **93**, 101–129.

Nielsen, E. S., Jørgensen, P., and Oddershede, J. (1980). Transition moments and dynamic polarizabilities in a second order polarization propagator approach. *J. Chem. Phys.*, **73**, 6238–6246.

Noga, J. and Bartlett, R. J. (1987). The full CCSDT model for molecular electronic structure. *J. Chem. Phys.*, **86**, 7041–7050.

Noga, J. and Bartlett, R. J. (1988). Erratum: The full CCSDT model for molecular electronic structure [J. Chem. Phys. 86, 7041 (1987)]. *J. Chem. Phys.*, **89**, 3401.

Norman, P., Jiemchooroj, A., and Sernelius, B. E. (2003). Polarization propagator calculations of the polarizability tensor at imaginary frequencies and long-range interactions for the noble gases and n-alkanes. *J. Chem. Phys.*, **118**, 9167–9174.

Oddershede, J., Geertsen, J., and Scuseria, G. E. (1988). Nuclear spin-spin coupling constant of HD. *J. Phys. Chem.*, **92**, 3056–3059.

Ogilvie, J. F. and Liao, S. C. (1994). Electric and magnetic molecular properties from analysis of vibration-rotational spectral data of samples measured without applied fields - application to GaH $X^1\Sigma^+$. *Chem. Phys. Lett.*, **226**, 281–288.

Ogilvie, J. F., Rodwell, W. R., and Tipping, R. H. (1980). Dipole moment functions of the hydrogen halides. *J. Chem. Phys.*, **73**, 5221–5229.

Olsen, J. and Jørgensen, P. (1985). Linear and nonlinear response functions for an exact state and for an MCSCF state. *J. Chem. Phys.*, **82**, 3235–3264.

Olsen, J., Jørgensen, P., Helgaker, T., and Oddershede, J. (2005). Quadratic response functions in a second-order polarization propagator framework. *J. Phys. Chem. A*, **109**, 11618–11628.

Packer, M. J., Dalskov, E. K., Enevoldsen, T., Jensen, H. J. Aa., and Oddershede, J. (1996). A new implementation of the second order polarization propagator approximation (SOPPA): The excitation spectra of benzene and naphthalene. *J. Chem. Phys.*, **105**, 5886–5900.

Packer, M. J., Dalskov, E. K., Sauer, S. P. A., and Oddershede, J. (1994). Correlated dipole polarizabilities and dipole moments of the halides HX and CH_3X (X = F, Cl and Br). *Theor. Chim. Acta*, **89**, 323–333.

Packer, M. J. and Pickup, B. T. (1995). An analysis of the magnetizability tensor at CHF level. *Mol. Phys.*, **84**, 1179–1192.

Peng, H. W. (1941). Perturbation theory for the self-consistent field. *Proc. Roy. Soc. London A*, **178**, 499–505.

Peralta, J. E., Scuseria, G. E., Cheeseman, J. R., and Frisch, M. J. (2003). Basis set dependence of NMR spin-spin couplings in density functional theory calculations: First row and hydrogen atoms. *Chem. Phys. Lett.*, **375**, 452–458.

Perera, S. A., Nooijen, M., and Bartlett, R. J. (1996). Electron correlation effects on the theoretical calculation of nuclear magnetic resonance spin-spin coupling constants. *J. Chem. Phys.*, **104**, 3290–3305.

Pickup, B. T. (1992). The propagator theory of non-linear response properties. In *Methods in computational chemistry: Theory and computation of molecular properties* (ed. S. Wilson), pp. 107–265. Plenum Press, New York.

Pickup, B. T. and Goscinski, O. (1973). Direct calculations of ionization energies I. Closed shells. *Mol. Phys.*, **26**, 1013–1035.

Pople, J. A., Binkley, J. S., and Seeger, R. (1976). Theoretical models incorporating electron correlation. *Int. J. Quantum Chem. Symp.*, **10**, 1–19.

Pople, J. A., Krishnan, R., Schlegel, H. B., and Binkley, J. S. (1979). Derivate studies in Hartree-Fock and Møller-Plesset theories. *Int. J. Quantum Chem. Symp.*, **13**, 225–241.

Pople, J. A., McIver, J. W., and Ostlund, N. S. (1968). Self-consistent perturbation theory. I. Finite perturbation methods. *J. Chem. Phys.*, **49**, 2960–2964.

Provasi, P. F., Aucar, G. A., and Sauer, S. P. A. (2001). The effect of lone pairs and electronegativity on the indirect nuclear spin-spin coupling constants in CH_2X (X = CH_2, NH, O, S): Ab initio calculations using optimized contracted basis sets. *J. Chem. Phys.*, **115**, 1324–1334.

Provasi, P. F. and Sauer, S. P. A. (2009). Analysis of isotope effects in NMR one-bond indirect nuclear spin-spin coupling constants in terms of localized molecular orbitals. *Phys. Chem. Chem. Phys.*, **11**, 3987–3995.

Provasi, P. F. and Sauer, S. P. A. (2010). Optimized basis sets for the calculation of indirect nuclear spin-spin coupling constants involving the atoms B, Al, Si, P, and Cl. *J. Chem. Phys.*, **133**, 54308.

Puzzarini, C., Cazzoli, G., Harding, M. E., Vazquez, J., and Gauss, J. (2009). A new experimental absolute nuclear magnetic shielding scale for oxygen based on the rotational hyperfine structure of $H_2^{17}O$. *J. Chem. Phys.*, **131**, 234304.

Rabi, I. I., Ramsey, N. F., and Schwinger, J. (1954). Use of rotating coordinates in magnetic resonance problems. *Rev. Mod. Phys.*, **26**, 167–171.

Raghavachari, K., Trucks, G. W., Pople, J. A., and Head-Gordon, M. (1989). A fifth-order perturbation comparison of electron correlation theories. *Chem. Phys. Lett.*, **157**, 479–483.

Rappoport, D. and Furche, F. (2010). Property-optimized Gaussian basis sets for molecular response calculations. *J. Chem. Phys.*, **133**, 134105.

Rice, J. E. and Handy, N. C. (1991). The calculation of frequency-dependent polarizabilities as pseudo-energy derivatives. *J. Chem. Phys.*, **94**, 4959–4971.

Rice, J. E. and Handy, N. C. (1992). The calculation of frequency-dependent hyperpolarizabilities including electron correlation effects. *Int. J. Quantum Chem.*, **43**, 91–118.

Rijks, W. and Wormer, P. E. S. (1988). Correlated van der Waals coefficients for dimers consisting of He, Ne, H_2, and N_2. *J. Chem. Phys.*, **88**, 5704–4714.

Romero, H. R. and Aucar, G. A. (2002). QED approach to the nuclear spin-spin coupling tensor. *Phys. Rev. A*, **65**, 53411.

Roothaan, C. C. J. (1951). New developments in molecular orbital theory. *Rev. Mod. Phys.*, **23**, 69–89.

Rosenblum, B., Nethercot, A. H., and Townes, C. H. (1958). Isotopic mass ratios, magnetic moments and the sign of the electric dipole moment in carbon monoxide. *Phys. Rev.*, **109**, 400–412.

Rowe, D. J. (1968). Equations-of-motion method and the extended shell model. *Rev. Mod. Phys.*, **40**, 153–166.

Ruden, T. A., Lutnæs, O. B., and Helgaker, T. (2003). Vibrational corrections to indirect nuclear spin-spin coupling constants calculated by density-functional theory. *J. Chem. Phys.*, **118**, 9572–9581.

Rusakov, Y. Yu., Krivdin, L. B., Sauer, S. P. A., Levanova, E. P., and Levkovskaya, G. G. (2010). Structural trends of ^{77}Se-^{1}H spin-spin coupling constants and conformational behavior of 2-substituted selenophenes. *Magn. Reson. Chem.*, **48**, 633–637.

Sadlej, A. J. (1988). Medium size polarized basis sets for high level correlated calculations of molecular electric properties. *Collect. Czech. Chem. Commun.*, **53**, 1995–2016.

Sadlej, A. J. (1991a). Medium-size polarized basis sets for high-level-correlated calculations of molecular electric properties II. Second-row atoms: Si through Cl. *Theor. Chim. Acta*, **79**, 123–140.

Sadlej, A. J. (1991b). Medium-size polarized basis sets for high-level-correlated calculations of molecular electric properties IV. Third-row atoms: Ge through Br. *Theor. Chim. Acta*, **81**, 45–63.

Sadlej, A. J. (1992). Medium-size polarized basis sets for high-level-correlated calculations of molecular electric properties V. Fourth-row atoms: Sn through I. *Theor. Chim. Acta*, **81**, 339–354.

Sadlej, A. J. and Urban, M. (1991). Medium-size polarized basis sets for high-level-correlated calculations of molecular electric properties III. Alkali (Li, Na, K, Rb) and alkaline-earth (Be, Mg, Ca, Sr) atoms. *J. Mol. Struct. (Theochem)*, **234**, 147–171.

Sasagane, K., Aiga, F., and Itoh, R. (1993). Higher-order response theory based on the quasienergy derivatives: The derivation of the frequency-dependent polarizabilities and hyperpolarizabilities. *J. Chem. Phys.*, **99**, 3738–3778.

Saue, T. (2001). Post Dirac-Hartree-Fock methods – Properties. In *Relativistic electronic structure theory, Part 1. Fundamentals* (ed. P. Schwerdtfeger), Chapter 7, pp. 332–467. Elsevier, Amsterdam.

Sauer, S. P. A. (1993). A sum-over-states formulation of the diamagnetic contribution to the indirect nuclear spin-spin coupling constant. *J. Chem. Phys.*, **98**, 9220–9221.

Sauer, S. P. A. (1997). Second order polarization propagator approximation with coupled cluster singles and doubles amplitudes - SOPPA(CCSD): The polarizability and hyperpolarizability of Li^-. *J. Phys. B: At. Mol. Opt. Phys.*, **30**, 3773–3780.

Sauer, S. P. A. (1998). A relation between the rotational g-factor and the electric dipolar moment of a diatomic molecule. *Chem. Phys. Lett.*, **297**, 475–483.

Sauer, S. P. A., Enevoldsen, T., and Oddershede, J. (1993). Paramagnetism of closed shell diatomic hydrides with six valence electrons. *J. Chem. Phys.*, **98**, 9748–9757.

Sauer, S. P. A., Paidarová, I., and Oddershede, J. (1994a). Correlated and gauge origin independent calculations of magnetic properties. I. Triply bonded molecules. *Mol. Phys.*, **81**, 87–118.

Sauer, S. P. A., Paidarová, I., and Oddershede, J. (1994b). Correlated and gauge origin independent calculations of magnetic properties II. Shielding constants of simple singly bonded molecules. *Theor. Chim. Acta*, **88**, 351–361.

Sauer, S. P. A. and Provasi, P. F. (2008). The anomalous deuterium isotope effect in the NMR spectrum of methane: An analysis in localized molecular orbitals. *ChemPhysChem*, **9**, 1259–1261.

Sauer, S. P. A., Raynes, W. T., and Nicholls, R. A. (2001). Nuclear spin-spin coupling in silane and its isotopomers: ab initio calculation and experimental investigation. *J. Chem. Phys.*, **115**, 5994–6006.

Sauer, S. P. A., Schreiber, M., Silva-Junior, M. R., and Thiel, W. (2009). Benchmarks for electronically excited states: A comparison of noniterative and iterative triples corrections in linear response coupled cluster methods - CCSDR(3) versus CC3. *J. Chem. Theory Comput.*, **5**, 555–564.

Schäfer, A., Horn, H., and Ahlrichs, R. (1992). Fully optimized contracted Gaussian basis sets for atoms Li to Kr. *J. Chem. Phys.*, **97**, 2571–2577.

Schäfer, A., Huber, C., and Ahlrichs, R. (1994). Fully optimized contracted Gaussian basis sets of triple zeta valence quality for atoms Li to Kr. *J. Chem. Phys.*, **100**, 5829–5835.

Schindler, M. and Kutzelnigg, W. (1982). Theory of magnetic susceptibilities and NMR chemical shifts in terms of localized quantities. II. Application to some simple molecules. *J. Chem. Phys.*, **76**, 1919–1933.

Schleyer, P., Maerker, C., Dransfeld, A., Jiao, H., and van Eikema Hommes, J. R. (1996). Nucleus-independent chemical shifts: A simple and efficient aromaticity probe. *J. Am. Chem. Soc.*, **118**, 6317–6318.

Schrödinger, E. (1926). 1. Quantisierung als Eigenwertproblem Dritte Mitteilung: Störungstheorie, mit Anwendung auf den Starkeffekt der Balmerlinien. *Ann. d. Phys.*, **80**, 437–490.

Schulman, J. M. and Kaufman, D. N. (1970). Application of many-body perturbation theory to the hydrogen molecule. *J. Chem. Phys.*, **53**, 477–484.

Scuseria, G. E. and Schaefer, H. F. (1988). A new implementation of the full CCSDT model for molecular electronic structure. *Chem. Phys. Lett.*, **152**, 382–386.

Sekino, H. and Bartlett, R. J. (1984). A linear response, coupled-cluster theory for excitation energy. *Int. J. Quantum Chem. Symp.*, **26**, 255–265.

Silva-Junior, M. R., Sauer, S. P. A., Schreiber, M., and Thiel, W. (2010). Basis set effects on coupled cluster benchmarks of electronically excited states: CC3, CCSDR(3) and CC2. *Mol. Phys.*, **108**, 453–465.

Smith, C. M., Amos, R. D., and Handy, N. C. (1992). Theoretical calculations of the nuclear magnetic shielding tensors for the ethylenic carbon atoms in cyclopropenes. *Mol. Phys.*, **77**, 381–396.

Stanton, J. F. and Bartlett, R. J. (1993). The equation of motion coupled-cluster method. A systematic biorthogonal approach to molecular excitation energies, transition probabilities, and excited state properties. *J. Chem. Phys.*, **98**, 7029–7039.

Stevens, R. M., Pitzer, R. M., and Lipscomb, W. N. (1963). Perturbed Hartree-Fock calculations. I. Magnetic susceptibility and shielding in the LiH molecule. *J. Chem. Phys.*, **38**, 550–560.

Sundholm, D., Gauss, J., and Schäfer, A. (1996). Rovibrationally averaged nuclear magnetic shielding tensors calculated at the coupled cluster level. *J. Chem. Phys.*, **105**, 11051–11059.

Sutter, D. H. and Flygare, W. H. (1976). The molecular Zeeman effect. *Top. Curr. Chem.*, **63**, 89–196.

Toyama, M., Oka, T., and Morino, Y. (1964). Effect of vibration and rotation on the internuclear distance. *J. Mol. Spectrosc.*, **13**, 193–213.

Trucks, G. W., Salter, E. A., Sosa, C., and Bartlett, R. J. (1988). Theory and implementation of the MBPT density-matrix - an application to one-electron properties. *Chem. Phys. Lett.*, **147**, 359–366.

Vaara, J., Lounila, J., Ruud, K., and Helgaker, T. (1998). Rovibrational effects, temperature dependence, and isotope effects on the nuclear shielding tensors of water: a new ^{17}O absolute shielding scale. *J. Chem. Phys.*, **109**, 8388–8397.

Vahtras, O., Ågren, H., Jørgensen, P., Helgaker, T., and Jensen, H. J. Aa. (1993). The nuclear spin-spin coupling in N_2 and CO. *Chem. Phys. Lett.*, **209**, 201–206.

Vahtras, O., Ågren, H., Jørgensen, P., Jensen, H. J. Aa., Padkjær, S. B., and Helgaker, T. (1992). Indirect nuclear spin-spin coupling constant from multiconfigurational linear response theory. *J. Chem. Phys.*, **96**, 6120–6125.

van Lenthe, E., Baerends, E. J., and Snijders, J. G. (1993). Relativistic regular two-component Hamiltonians. *J. Chem. Phys.*, **99**, 4597–4610.

van Wüllen, C. and Kutzelnigg, W. (1996). Calculation of nuclear magnetic resonance shieldings and magnetic susceptibilities using multiconfigurational Hartree-Fock wave functions and local gauge origins. *J. Chem. Phys.*, **104**, 2330–2340.

Visser, F. and Wormer, P. E. S. (1984). The non-empirical calculation of second order molecular properties by means of effective states. I Application to time-dependent coupled Hartree-Fock method. *Mol. Phys.*, **52**, 923–937.

Visser, F., Wormer, P. E. S., and Stam, P. (1983). Time-dependent coupled Hartree-Fock calculations of multipole polarizabilities and dispersion interactions in van der Waals dimers consisting of He, H_2, Ne, and N_2. *J. Chem. Phys.*, **79**, 4973–4984.

Visser, F., Wormer, P. E. S., and Stam, P. (1984). Erratum: Time-dependent coupled Hartree-Fock calculations of multipole polarizabilities and dispersion interactions in van der Waals dimers consisting of He, H_2, Ne, and N_2 [J. Chem. Phys. 79, 4973 (1983)]. *J. Chem. Phys.*, **81**, 3755.

Wasylishen, R. E. and Bryce, D. L. (2002). A revised experimental absolute magnetic shielding scale for oxygen. *J. Chem. Phys.*, **117**, 10061.

Watson, J. K. G. (1973). The isotope dependence of the equilibrium rotational constants in $^1\Sigma$ states of diatomic molecules. *J. Mol. Spectrosc.*, **45**, 99–113.

Watson, J. K. G. (1980). The isotope dependence of Diatomic Dunham coefficients. *J. Mol. Spectrosc.*, **80**, 411–421.

Weigend, F. and Ahlrichs, R. (2005). Balanced basis sets of split valence, triple zeta valence and quadruple zeta valence quality for H to Rn: Design and assessment of accuracy. *Phys. Chem. Chem. Phys.*, **7**, 3297–3305.

Weigend, F., Furche, F., and Ahlrichs, R. (2003). Gaussian basis sets of quadruple zeta valence quality for atoms H to Kr. *J. Chem. Phys.*, **119**, 12753–12762.

Wick, G. C. (1933a). Sul momento magnetico di una molecola d'idrogeno. *Nuovo Cimento*, **10**, 118–127.

Wick, G. C. (1933b). Über das magnetische Moment eines rotierenden Wasserstoff-moleküls. *Z. Physik*, **85**, 25–28.

Wick, G. C. (1948). On the magnetic field of a rotating molecule. *Phys. Rev.*, **73**, 51–57.

Wigglesworth, R. D., Raynes, W. T., Kirpekar, S., Oddershede, J., and Sauer, S. P. A. (2000a). Nuclear magnetic shielding in the acetylene isotopomers calculated from MCSCF shielding surfaces. *J. Chem. Phys.*, **112**, 736–746.

Wigglesworth, R. D., Raynes, W. T., Kirpekar, S., Oddershede, J., and Sauer, S. P. A. (2000b). Nuclear spin-spin coupling in the acetylene isotopomers calculated from ab initio correlated surfaces for $^1J(C,H)$, $^1J(C,C)$, $^2J(C,H)$ and $^3J(H,H)$. *J. Chem. Phys.*, **112**, 3735–3746.

Wigglesworth, R. D., Raynes, W. T., Kirpekar, S., Oddershede, J., and Sauer, S. P. A. (2001). Erratum: Nuclear spin-spin coupling in the acetylene isotopomers calculated from ab initio correlated surfaces for $^1J(C,H)$, $^1J(C,C)$, $^2J(C,H)$ and $^3J(H,H)$. *J. Chem. Phys.*, **114**, 9192.

Wigglesworth, R. D., Raynes, W. T., Oddershede, J., and Sauer, S. P. A. (1999). Calculated nuclear shielding surfaces in the water molecule; prediction and analysis of $\sigma(O)$, $\sigma(H)$ and $\sigma(D)$ in water isotopomeres. *Mol. Phys.*, **96**, 1595–1607.

Wigglesworth, R. D., Raynes, W. T., Sauer, S. P. A., and Oddershede, J. (1997). The calculation and analysis of isotope effects on the nuclear spin-spin coupling constants of methane at various temperatures. *Mol. Phys.*, **92**, 77–88.

Wigglesworth, R. D., Raynes, W. T., Sauer, S. P. A., and Oddershede, J. (1998). Calculated spin-spin coupling surfaces in the water molecule; prediction and analysis of J(O,H), J(O,D) and J(H,D) in water isotopomeres. *Mol. Phys.*, **94**, 851–862.

Wilson, A. K., Woon, D. E., Peterson, K. A., and Dunning Jr., T. H. (1999). Gaussian basis sets for use in correlated molecular calculations. IX. The atoms gallium through krypton. *J. Chem. Phys.*, **110**, 7667–7676.

Woon, D. E. and Dunning Jr., T. H. (1993). Gaussian basis sets for use in correlated molecular calculations. III. The atoms aluminum through argon. *J. Chem. Phys.*, **98**, 1358–1371.

Woon, D. E. and Dunning Jr., T. H. (1994). Gaussian basis sets for use in correlated molecular calculations. IV. Calculation of static electrical response properties. *J. Chem. Phys.*, **100**, 2975–2988.

Woon, D. E. and Dunning Jr., T. H. (1995). Gaussian basis sets for use in correlated molecular calculations. V. Core-valence basis sets for boron through neon. *J. Chem. Phys.*, **103**, 4572–4585.

Yachmenev, A., Yurchenko, S. N., Paidarová, I., Jensen, P., Thiel, W., and Sauer, S. P. A. (2010). Thermal averaging of the indirect nuclear spin-spin coupling constants of ammonia: The importance of the large amplitude inversion mode. *J. Chem. Phys.*, **132**, 114305.

Yeager, D. L. and Jørgensen, P. (1979). A multiconfigurational time-dependent Hartree-Fock approach. *Chem. Phys. Lett.*, **65**, 77–80.

Zubarev, D. N. (1974). *Nonequilibrium statistical thermodynamics*. Consultants Bureau, New York.

Index

6-31G, 254
6-31G(d,p), 254
6-31G**, 254
6-31G++, 255

ab initio, 186, 189, 191, 210, 245
adiabatic, 143
angle of rotation, 159
angular momentum
  orbital, 98, 132, 140, 142, 147
    quenching, 99
  rotation, 126–128, 137, 138, 142
approximation
  Born–Oppenheimer, 5, 7, 26, 174–177
    breakdown, 126, 127, 130, 137, 141, 144
  dipole, 45, 155, 158, 162
  electric quadrupole, 94
  harmonic, 181
  minimal coupling, 5, 13, 17
  polarization propagator
    first-order, 186, 214
    multiconfigurational, 225, 233, 234
    second-order, 216, 235, 241, 251
    third-order, 222
    zeroth-order, 216
  random phase, 186, 212, 214, 215, 222, 224, 230–232, 250, 261–263, 266
    doubles corrected, 223, 224
    multiconfigurational, 225, 233, 234, 263
  Tamm–Dancoff, 216
  zeroth-order regular, 23
atomic orbital, 190, 193, 232, 254
  gauge including, 204, 257
aug-cc-pCVTZ-CTOCD-uc, 258
aug-cc-pCVXZ, 257
aug-cc-pVTZ-J, 258
aug-cc-pVXZ, 255
aug-pc-n, 255

basis set, 253
  6-31G, 254
  6-31G(d,p), 254
  6-31G**, 254
  6-31G++, 255
  aug-cc-pCVTZ-CTOCD-uc, 258
  aug-cc-pCVXZ, 257
  aug-cc-pVTZ-J, 258
  aug-cc-pVXZ, 255
  aug-pc-n, 255
  cc-pCVXZ, 257
  cc-pVXZ, 255
  cc-pVXZ-Cs, 258
  cc-pVXZ-su$n$, 258
  ccJ-pVXZ, 258
  correlation consistent core-valence, 257
  correlation consistent polarized, 255, 256
  diffuse functions, 255
  double zeta, 254
  Gaussian type, 253
    contracted, 254
    primitive, 254
  GIAO, 257
  medium size polarized, 256
  minimal, 254
  pc-n, 255
  pcJ-n, 258
  pcS-n, 257
  polarization consistent, 255, 256
  polarization functions, 254, 255
  reduced size polarized, 256
  tight functions, 257
  triple zeta, 254
  valence double zeta, 254
  valence triple zeta, 254
Bethe formula, 169
birefringence
  circular, 158
Born interpretation, 9
Born–Oppenheimer approximation, 5, 7, 26, 174–177
  breakdown, 126, 127, 130, 137, 141, 144
Brillouin theorem, 198, 200, 202, 206, 246

Casimir-Polder formula, 172
Cauchy moment, 167
cc-pCVXZ, 257
cc-pVXZ, 255
cc-pVXZ-Cs, 258
cc-pVXZ-su$n$, 258
CC2 model, 202, 241
  amplitude, 203, 222
  Jacobian, 241
CC3 model, 203, 241
ccJ-pVXZ, 258
CCLR, 236, 240
CCSD, 201, 202, 241
CCSD model
  amplitude, 202, 222
  Jacobian, 241
CCSD(T) model, 203, 262

CCSDPPA, 222
CCSDR(3) model, 242
CCSDT model, 203, 241
centre of mass, 130, 139, 140, 142, 145, 150
charge
  density, 10, 72
  distribution, 71, 75
    polarizable, 83, 100, 101
  rotating, 126
  total, 73, 135
chemical shift, 100, 110, 111
  nucleus independent, 110
CHF, 105, 186, 227, 231, 232, 247, 248, 261
chirality, 158
CI, 190, 197, 203, 204
  matrix, 198, 216
  mono-excited, 216
clamped-nucleus, 177
classical electromagnetism, 12
cluster operator, 201, 236
commutator, 42, 50, 57, 60, 61, 120, 134, 149, 150, 163
configuration
  coefficient, 196, 197, 203
  interaction, 190, 197
    matrix, 198, 216
    mono-excited, 216
  state function, 196
correlation coefficient, 195, 200, 206, 218, 222, 250
correlation consistent core-valence basis set, 257
correlation consistent polarized basis set, 255, 256
Coulomb
  gauge, 16, 24
  potential, 22
coupled cluster, 190, 203, 204, 245
  amplitude, 195, 201, 222
    doubles, 201
    equation, 202, 203
    first-order, 237, 238
    λ, 196, 205, 237
    singles, 201
    time-dependent, 237
  bra wavefunction, 205
    time-dependent, 236
  energy, 201, 202
  Jacobian, 238, 240, 241
    CC2, 241
    CCSD, 241
  Lagrangian, 205, 251
  linear response function, 236, 240
    CC2, 241, 251
    CC3, 241
    CCSD, 241, 262
  response equation, 239
  singles and doubles model, 201
    Jacobian, 241

singles, doubles and triples model, 203
vector function, 202, 203, 238
wavefunction, 201, 222
  Λ, 205, 236
  linearised, 222
  time-dependent, 236
coupled Hartree–Fock, 105, 186, 227, 231, 232, 247, 248, 261
coupling
  minimal, 5, 13, 17
  nuclear motion–electron motion, 126, 137, 141, 144, 273–276
coupling tensor
  hyperfine, 106, 280, 281
  nuclear quadrupole, 90
  nuclear spin-spin, 263
    basis set, 258
    direct, 112
    Fermi contact term, 116, 244, 258, 264, 266
    indirect, 110–113, 117, 118, 244, 266
    spin-dipolar term, 116, 244, 258, 265, 266
CTOCD-DZ, 124, 257
current
  density, 10, 11, 93, 105, 126, 136
    continuous transformation of origin, 124
    induced, 100
    polarizable, 101
  steady, 94

density
  charge, 10, 72
  current, 10, 11, 93, 105, 126, 136
    continuous transformation of origin, 124
    induced, 100
    polarizable, 101
  electron, 9
    perturbed, 40
  flux, 12
  matrix, 39, 207
    atomic orbital, 207, 210, 245
    first-order, 41, 54, 209, 211, 229, 248
    molecular orbital, 207, 210, 245
    perturbed, 40, 229
    reduced one-electron, 10, 40, 210
    relaxed, 246, 261
    response, 246
    SCF, 208, 210, 229, 246, 261
    second-order correction, 208, 218
    unrelaxed MP2, 208, 210, 218, 261
  operator, 10
density functional theory, 186, 255, 257, 258
  time-dependent, 212, 214, 215
DFT, 186, 255, 257, 258
diamagnetic, 104, 114, 119, 121, 131–133, 136, 139, 140

expectation value, 105, 116, 118, 124, 131, 132, 140, 258
linear response function, 105, 116, 118, 124, 131, 132, 140, 258
sum-over-states, 105, 116, 118, 124, 131, 132, 140, 258
diffuse functions, 255
dipole approximation, 45, 155, 158, 162
dipole moment
  electric, 73, 77, 78, 133, 135, 147, 150, 151, 170, 178, 243, 255, 261, 278–281
    induced, 81, 158, 169
    time-dependent, 156, 157
  gradient, 151
  magnetic, 95, 97, 98, 280, 281
    induced, 100, 127
    nuclear, 96, 100, 106, 137
    rotating, 126, 127
  time-dependent, 160
dipole oscillator strength, 164–169
  sum rules, 166, 167
  sums, 166, 172
Dirac
  δ function, 9, 11, 57, 58, 78, 258
  equation, 21
    time-dependent, 19, 20
  Hamiltonian operator, 17
  representation, 43
dispersion, 153
  coefficient, 172
  energy, 169, 171, 172
  force, 169
divergence theorem, 94
DSO, 114
DZ, 254

Ehrenfest theorem, 41, 233, 235
eigenvalue, 224
  problem, 65, 165, 223, 241
  second-order correction, 66
eigenvector, 65, 165
  RPA, 224
electric moment
  first, 73, 77
  second, 73, 77
electric quadrupole approximation, 94
electromagnetism
  classical, 12
electron
  density, 9
electron spin resonance, 5, 96, 106
energy
  m-th order, 36
  adiabatic, 143
  charge distribution, 75, 83, 101
  derivative, 32, 243
    first, 31, 79, 97, 204, 206, 245, 246, 278, 279

  second, 38, 53, 85, 102, 112, 175, 207, 231, 243, 248, 262, 278, 279
  dispersion, 169, 171, 172
  electronic, 177
  first-order, 36
  induction, 170
  kinetic, 16
  loss, 168
  mean excitation, 169
  non-adiabatic, 143
  perturbed, 34, 75, 96, 106, 112, 127, 137, 243
  potential, 75, 96
    anharmonic, 181, 183
    nuclear, 181
    surface, 8
  rotational, 128, 138, 141
  second-order, 36, 37, 143
  vertical excitation, 53, 64, 161, 165, 171, 211, 212, 224, 232, 242, 250, 251
  vibrational, 141, 177, 181
EOM-CCSD, 242
equation-of-motion
  coupled cluster method, 242
  expectation value, 41
  interaction picture, 45
  Lagrange's, 14
  polarization propagator, 58, 59, 134
ESR, 5, 96, 106
Euler–Lagrange equations, 14
excitation energy
  vertical, 53, 64, 161, 165
expectation value, 6, 7, 33, 39, 98, 131, 134, 139, 206, 210
  equation-of-motion, 41
  field-dependent, 37, 54, 87, 104, 117, 211, 244
  first-order, 41
  gauge invariant, 26
  relation to polarization propagator, 59
  second-order, 54
  time-dependent, 31, 41, 49–51, 53, 56, 156, 159, 240
  time-derivative, 41
  transition, 190, 201, 203, 205, 236, 240, 246

FC, 108
Fermi contact, 108, 116
field
  electric, 12, 71, 75, 76, 78, 80, 81, 83–85, 158, 175, 177
    electromagnetic wave, 153–155
    molecular, 89
    time-dependent, 158
  magnetic, 96, 100
field gradient
  electric, 75, 76, 78, 81, 83, 85, 88
  molecular, 89
finite field, 243

finite point charge, 244
first-order polarization propagator
    approximation, 186, 214
fluctuation potential, 199, 213, 235
flux density, 12
Fock matrix, 198
    perturbed, 228
FOPPA, 186, 214
force
    dispersion, 169
    intermolecular, 169, 170
    London, 169
    Lorentz, 12, 15, 17
force constant
    cubic, 181
    harmonic, 181
form invariance, 25, 26
frequency
    angular, 153
    Larmor, 110
    resonance, 111
    vibrational
        harmonic, 181

g factor
    electron, 5, 22
    nuclear, 106, 111, 137
    rotational, 145–147, 149, 151, 256
        irreducible non-adiabatic, 151
        isotopically independent, 148, 151
    vibrational, 145–147, 149, 151
        irreducible non-adiabatic, 151
        isotopically independent, 148, 151
g tensor
    rotational, 127–136, 141, 145,
        278–281
gauge
    Coulomb, 16, 24
    function, 13, 27, 28
    length, 27, 44, 155, 156
    Lorenz, 28
    origin, 27, 94, 95, 105, 121, 122, 130,
        132, 133, 139, 257
    transformation, 13, 25, 26, 257
    velocity, 28, 155
gauge including atomic orbital, 204,
    257
GIAO, 204, 257
gradient
    electric
        field, 75, 76, 78, 81, 83, 85, 88
        property, 62, 165, 213, 215, 240, 259
Green's function
    double-time, 51
GTO, 253

Hamiltonian, 5
    classical, 13, 15, 18, 19
    derivative, 32, 39, 80, 103
    Dirac, 17
    electronic, 7, 24
    gauge transformed, 26
    Hartree-Fock, 191–193, 199
    operator, 16
    perturbation, 24, 33, 45, 78, 97, 102, 107,
        113, 130, 138, 142, 156, 159,
        272, 273
    perturbed, 24, 31, 33, 43, 44, 203
    rotational, 128, 138
    spin, 106, 112
    superoperator, 60
    time-dependent, 43, 44, 156
    unperturbed, 7, 24, 33, 142
    vibration-rotational, 143, 145, 147
Hartree-Fock, 191
    coupled, 227
    energy, 191–193, 199
        perturbed, 231
    equation, 191, 192, 254
        first-order, 229, 230
        perturbed, 227, 228
        time-dependent, 232
    Hamiltonian, 191, 192, 199
    potential, 191, 199
    Roothaan, 193
    time-dependent, 214, 232
    uncoupled, 211
    wavefunction, 191–193, 197, 198, 207
Heaviside step function, 50, 57
helicity, 155
Hellmann-Feynman theorem, 31, 33, 38, 77,
    80, 98, 186, 203, 204, 206, 245
Hessian matrix
    electronic, 63, 165, 214, 223, 234, 240
hyperfine coupling, 106, 280, 281
hyperfine splitting, 90
hyperpolarizability, 212, 255
    first, 174
        dipole, 81, 249, 255, 278, 279
        dipole-quadrupole, 81, 278, 279
    second, 174
        dipole, 81, 244, 249, 278, 279
hypervirial theorem, 42, 121
    momentum operator, 43
    off-diagonal, 42, 121, 124, 162, 164
    transition moment, 43, 216

IGLO, 257
induction
    energy, 170
    magnetic, 12, 96, 127, 128
        electromagnetic wave, 153–155
        induced, 109
        local, 110
        molecular, 105, 280, 281
        rotational, 136, 137
        time-dependent, 158
        time-derivative, 160

inner projection, 62
interaction
    electric, 170
    intermolecular, 71, 169
interaction picture, 43
    equation-of-motion, 45
intermediate normalization, 34
intermolecular interactions, 71, 169
irrotational, 17
isotope shift, 174
isotopologue, 133, 135, 148

Jacobian
    CC2 model, 241
    CCSD model, 241
    coupled cluster, 238, 240, 241
Jahn–Teller theorem, 99

Lagrangian, 205
    classical, 14, 15, 17
    coupled cluster, 205, 246, 251
    Møller–Plesset perturbation theory, 206, 246, 250
    quasi-energy
        time-dependent, 250, 251
Lamb shift, 5
Larmor
    frequency, 110
    theorem, 127
LCAO, 254
length
    gauge, 27, 44, 155, 156
    representation, 164, 166–168
London
    force, 169
    orbital, 257
longitudinal, 17
Lorentz force, 12, 15, 17
LORG, 257

Mössbauer spectroscopy, 90
magnetizability, 100–102, 118, 122, 130, 133, 136, 145, 164, 278–281
    basis set, 256
magneton
    nuclear, 106
mass
    centre of, 130, 139, 140, 142, 145, 150
    reduced, 142, 144–146, 148
        non-adiabatic, 143, 145
    relativistic, 17
    rest, 17
Maxwell's equations, 12, 28, 153, 158
MCRPA, 225, 233, 234, 263
MCSCF, 190, 196, 197, 203, 204
    linear response, 233, 234
    wavefunction, 186, 196, 225
        time-dependent, 233
mean excitation energy, 169

medium-size polarized basis set, 256
minimal coupling, 5, 13, 17
mixed representation, 164, 166, 168
molecular orbital, 190, 191, 193, 210, 221, 253
    coefficient, 190, 193, 196, 197, 203, 204, 254, 258
    derivative, 248
    first-order, 215, 229, 230, 232
    localised, 212
    time-dependent, 232, 250
moment
    electric, 77
        dipole, 73, 77, 78, 81, 133, 135, 147, 150, 151, 156, 160, 170, 178, 243, 255, 261, 278–281
        first, 73, 77
        gradient, 151
        induced, 80, 81, 158, 169
        quadrupole, 73, 77, 78, 81, 135, 278–281
        second, 73, 77
    magnetic
        dipole, 95, 97, 280, 281
        first, 95
        induced, 100, 127
        monopole, 95
        nuclear, 96, 100, 106, 137
        rotating, 127
        rotational, 126
    of inertia, 126, 133
    transition, 49, 53, 161, 162, 165, 171, 211, 212, 232
        dipole, 162
        magnetic dipole, 164
        quadrupole, 164
moment expansion
    Cauchy, 167
    polarization propagator, 60, 61
    polarizability, 167
momentum
    angular
        orbital, 98, 99, 132, 140, 142, 147
        rotation, 126–128, 137, 138, 142
    canonical, 14, 15, 17, 26, 134
    gauge invariant, 27
    generalised, 15
    kinematical, 16, 26
    mechanical, 16, 23, 26
mono-excited CI, 216
Mulliken notation, 191
multiconfigurational polarization propagator approximation, 225, 233, 234
multiconfigurational self-consistent field, 190, 196, 197
    wavefunction, 186, 196
multipole expansion
    electric, 71, 74, 169
    magnetic, 93

Møller–Plesset perturbation theory
  correlation coefficient, 200, 201, 222, 250
  energy, 199
    second-order, 200, 246, 250, 262
  Lagrangian, 206, 246
  polarization propagator, 239
  relaxed density matrix, 247, 261
  time-dependent, 249
  wavefunction, 199, 204, 207, 222, 235, 245
    first-order, 200, 217, 218
    second-order, 200, 217
    time-dependent, 235

Newton's second law, 14, 17
NICS, 110
NMR, 5, 90, 96, 100, 109, 110, 141
non-adiabatic, 143–145
normal
  coordinate, 180, 183
  mode, 180, 181
normalisation
  intermediate, 34
nuclear magnetic resonance, 5, 90, 96, 100, 109, 110, 141
nuclear magneton, 106

OP, 107
operator
  canonical momentum, 42, 134, 148, 162
  cluster, 201
  complete set, 61, 63, 196, 212
  CTOCD-DZ, 124, 276, 277
  Darwin, 22
  de-excitation, 62, 195, 212
    double, 195
    single, 195, 225
    time-dependent, 236
  density, 10, 277, 278
  diamagnetic magnetizability, 103, 274, 275, 276, 277
  diamagnetic nuclear magnetic shielding, 114, 275–277
  diamagnetic spin-orbit, 114, 275–277
  electric dipole, 79, 136, 155, 156, 160, 162, 255, 272, 273
  electric field, 91, 273, 274
  electric-field gradient, 91, 273, 274
  electric quadrupole, 79, 155, 272, 273
  electron position, 42
  electron spin, 22
  excitation, 62, 194, 201, 212
    double, 194, 201, 216, 219, 222
    single, 194, 201, 214, 225
    time-dependent, 236
  Fermi contact, 108, 109, 113, 255, 258, 274, 275
  field-dependent, 32, 116
  Fock, 191, 192
  Hamiltonian, 5, 16
    derivative, 32, 39, 80, 103
    Dirac, 17
    electronic, 7, 24
    gauge transformed, 26
    Hartree–Fock, 191–193, 199
    perturbation, 24, 33, 45, 78, 97, 102, 107, 113, 114, 130, 138, 142, 156, 159, 272, 273
    perturbed, 24, 31, 33, 43, 44, 203
    rotational, 128, 138
    spin, 106, 112
    time-dependent, 43, 44, 156
    unperturbed, 7, 24, 33, 142
    vibration-rotational, 143, 145, 147
  hermitian, 53
  interaction, 25, 79
  Λ, 195, 205
    time-dependent, 236
  magnetic dipole, 98, 116, 155, 163, 256, 273, 274, 276, 277
  magnetic induction, 116, 276, 277
  mass-velocity, 22
  orbital angular momentum, 98, 108, 132, 140, 142, 147, 163, 255, 256, 273–275
  orbital diamagnetic, 114, 276, 277
  orbital paramagnetic, 107, 113, 257, 274–276
  orbital rotation, 194, 197, 206, 225, 233, 236, 251
  paramagnetic spin-orbit, 107, 274, 275
  perturbation, 25, 79, 97, 103, 107, 108, 129, 130, 138
  position
    electron, 277, 278
  second moment, 277, 278
  spin, 98
    electron, 116, 273, 274
  spin-dipolar, 108, 109, 113, 274, 275
  spin-orbit, 22
  state transfer, 63, 196, 225, 233, 235
  T, 194, 201
    time-dependent, 236
  T_1 transformed, 202
  tensor, 25
  total electronic angular momentum, 277, 278
  total electronic canonical momentum, 277, 278
  vector, 25
  Zeeman, 22, 24
optical rotation, 158
orbital
  atomic, 190, 193, 232, 254
    gauge including, 204, 257
  core, 254
  energy, 192, 193, 199, 211, 251

perturbed, 228
Gaussian type, 253
   contracted, 254
   primitive, 254
London, 257
molecular, 190, 191, 193, 210, 221, 253
   coefficient, 190, 193, 196, 197, 203, 204, 254, 258
   derivative, 248
   first-order, 215, 229, 230, 232
   localised, 212
   occupied, 192–194, 197
   time-dependent, 232, 250
   unoccupied, 192–194, 197
   virtual, 192–194, 197
relaxation, 206, 246, 251
spatial, 192
   molecular, 189–192
spin, 189–192, 198, 215, 228
   perturbed, 227, 229
valence, 254–256
oscillator strength, 164–169
   sum rules, 166, 167
   sums, 166, 172
overlap matrix, 63

Padé approximant, 172
paramagnetic, 105, 114, 121, 131, 133, 139, 140, 244
paramagnetism, 96
   temperature independent, 105
Pauli spin matrix, 19, 22
pc-n, 255
pcJ-n, 258
pcS-n, 257
perturbation theory
   $2m+1$ rule, 36
   mth order equation, 35
   Møller–Plesset, 190, 199, 202, 203, 235, 243, 261
      correlation coefficient, 200, 201, 222, 250
      energy, 199
      Lagrangian, 206
      relaxed density matrix, 247, 261
      time-dependent, 249
      wavefunction, 199, 204, 207, 212, 222, 235, 245
   pseudo, 64, 223
   Rayleigh-Schrödinger, 33, 199
   time-dependent, 31, 44
   time-independent, 30, 33
phase factor
   time-dependent, 6, 20
polarizability
   atomic, 175
   dipole, 81, 85, 174, 175, 243, 255, 278–281
      frequency-dependent, 156, 157, 164, 167–169, 172, 249
   dipole-quadrupole, 81, 85, 278–281
   electric dipole-magnetic dipole
      frequency-dependent, 158, 160, 165
   electronic-vibrational, 176
   quadrupole, 81, 85, 278–281
      frequency-dependent, 164
   vibrational, 175
   vibrational averaged, 176
polarization consistent basis set, 255
polarization functions, 254, 255
polarization propagator, 88, 157, 160, 215
   approximation
      first-order, 186, 214
      multiconfigurational, 225, 233, 234
      second-order, 216, 235, 241, 251
      third-order, 222
      zeroth-order, 216
   derivative, 168
   eigenvalue problem, 165, 223
   eigenvector, 165
   equation-of-motion, 134
   matrix representation, 213
      partitioned, 213, 216, 223
      moment expansion 59, 60
   Møller–Plesset, 239
   pole, 161, 165, 171, 212
   residuum, 161, 164, 165, 171, 212
   spectral representation, 161, 211
   static, 52
   symmetry, 52
   time domain, 50
   time-derivative, 56
potential
   Coulomb, 22
   electrostatic, 71, 78, 89
   fluctuation, 199, 213, 235
   generalised, 14, 15
   harmonic, 181
   Hartree–Fock, 191, 199
   scalar, 12, 25, 71, 75
      electric field, 78, 271, 272
      time-dependent, 13, 27, 28
   vector, 12, 16, 23, 25, 93, 96, 105
      electromagnetic wave, 153, 154
      magnetic moment, 95, 107, 271, 272
      rotating nuclei, 127, 271, 272
      time-dependent, 12, 27, 28, 153, 154
      uniform magnetic induction, 27, 97, 271, 272
probability, 9
   amplitude, 47
   density, 9
   transition, 47, 49
propagator
   principal, 63, 213, 259
property
   first-order, 206, 246, 280, 281

gradient, 62, 165, 213, 215, 219, 240, 259
  second-order, 87, 102, 231, 280, 281
PSO, 107

quadrupole moment
  electric, 73, 77, 78, 135, 278–281
  induced, 81
  nuclear, 90
quadrupole splitting, 90
quantum electrodynamics, 5, 22
quasi-energy
  time-averaged, 249
  time-dependent, 249

radiation
  circular polarized, 155, 157
  linear polarized, 45, 153, 155, 157
  plane polarized, 45, 153, 155, 157
  polychromatic, 45, 154
random-phase approximation, 186, 212, 214, 215, 222, 224, 231, 232, 250, 261–263, 266
  doubles corrected, 223, 224
  matrix, 215, 230
  multiconfigurational, 225, 233, 234, 263
reduced mass, 142, 144–146, 148
  non-adiabatic, 143, 145
reduced size polarized basis set, 256
reduced space, 259
refractive index, 153, 155–159
renormalisation, 218
representation
  Dirac, 43
  interaction, 43
  length, 164, 166–168
  mixed, 164, 166, 168
  Schrödinger, 43
  velocity, 164, 166, 168
resolution of the identity, 52, 59, 61, 120, 124, 196
response function
  linear, 50, 88, 100, 104, 118, 131, 132, 134, 139, 141, 145, 157, 160, 249
    CC2, 241, 251
    CC3, 241
    CCSD, 262
    complex, 172
    coupled cluster, 236, 240, 251
    eigenvalue problem, 65, 165, 223, 241
    frequency domain, 52
    matrix representation, 62
    pole, 53, 64, 161, 165, 171, 212, 241, 251
    self-consistent field, 186, 214, 233, 234, 250
    spectral representation, 52
    static, 53
    symmetry, 52
    time domain, 51

    time-derivative, 57
  quadratic, 235, 249
    frequency domain, 55
    time domain, 55
response theory
  time-dependent, 31, 49
  time-independent, 30, 37
Roothaan Hartree-Fock, 193
rotating frame, 44, 127
rotational g factor, 145–147, 149, 151, 256
  irreducible non-adiabatic, 151
  isotopically independent, 148, 151
rotational g tensor, 127–136, 141, 145, 278–281
rotational spectroscopy, 90
rotational strength, 165
RPA, 186, 212, 214, 215, 222, 224, 231, 232, 250, 261–263, 266
  matrix, 215, 230
RPA(D), 223, 224

scalar potential, 12, 25
  electric field, 78
  time-dependent, 13, 27, 28
SCF, 186, 190, 203, 204, 249, 250, 260, 261
  energy, 186, 190
  linear response, 214, 250
  wavefunction, 186
Schrödinger equation
  electronic, 7, 142
  nuclear, 8, 178, 180, 181
  time-dependent, 6, 8, 14, 25, 44
    field-dependent, 30, 237
  time-independent, 6
    field-dependent, 30
  unperturbed, 33
  vibrational, 181
Schrödinger picture, 43
Schrödinger–Pauli equation, 22
SD, 108
second-order polarization propagator approximation, 216, 235, 241, 251, 262, 263
second-order property, 87, 102, 231
self-consistent field, 186, 190–192
  energy, 186
  linear response, 214
  multiconfigurational, 190, 196, 197
    wavefunction, 186, 196
  wavefunction, 186, 191–193, 197, 198, 207
shielding
  nuclear magnetic, 100–102, 109, 111–113, 117, 118, 122, 139, 141, 244, 257, 262, 266, 278–281
    basis set, 257
    CTOCD-DZ, 124, 257
    IGLO, 257
    LORG, 257
    polarizability, 244

shielding field
  magnetic, 109, 278–281
Slater determinant, 189–192, 196–198, 210
  excited
    doubly, 194, 200, 201
    quadruply, 201
    singly, 194, 198, 200, 201, 211, 222
    triply, 201
  linear combination, 194
Slater–Condon rules, 198, 200, 208, 212, 217
small component
  elimination, 20
solenoidal, 17
solution vector, 214, 216, 259
SOPPA, 216, 235, 241, 251, 262, 263
SOPPA(CC2), 222
SOPPA(CCSD), 222, 244, 262, 263, 265
spatial orbital, 192
  molecular, 189–192
special relativity, 17
spectroscopy
  electron spin resonance, 5, 96, 106
  Mössbauer, 90
  nuclear magnetic resonance, 5, 90, 96, 100, 109, 110, 141
  rotational, 90
  vibration-rotational, 143
spin
  electron, 22, 98, 103, 116
  function, 189
  nuclear, 90, 106, 111
  Pauli matrix, 19, 22
  photon, 155
spin rotation tensor, 137–141, 278–281
spin-dipolar, 108, 116
spin-orbital, 189–192, 198, 215, 228
  perturbed, 227, 229
spin-spin coupling tensor
  nuclear, 263
    basis set, 258
    direct, 112
    Fermi contact term, 116, 244, 258, 264, 266
    indirect, 110–113, 117, 118, 244, 266
    spin-dipolar term, 116, 244, 258, 265, 266
spinor
  four-component, 19
  two-component, 19
splitting
  hyperfine, 90
  quadrupole, 90
stationary state, 7, 11
  electronic, 8
stopping power, 168
substitution rule, 14, 19
sum rule
  dipole oscillator strength, 166, 167
  Thomas–Reiche–Kuhn, 166

sum-over-states, 86, 103, 114, 118, 131, 139, 141, 144, 145, 211
superoperator, 60
  binary product, 60
  Hamiltonian, 60
  resolution of the identity, 61
  resolvent, 61

Tamm–Dancoff approximation, 216
TD-DFT, 212, 214, 215
TDHF, 214, 232
temperature
  averaging, 183
  dependence, 174
  independent paramagnetism, 105
theorem
  Brillouin, 198, 200, 202, 206, 246
  divergence, 94
  Ehrenfest, 41, 233, 235
  Hellmann–Feynman, 77, 80, 98, 186, 203, 204, 206, 245
  hypervirial, 42, 121
    momentum operator, 43
    off-diagonal, 42, 121, 124, 162, 164, 216
    transition moment, 43, 216
  Jahn–Teller, 99
  Larmor, 127
third–order polarization propagator approximation, 222
Thomas–Reiche–Kuhn sum rule, 166
time–dependent Hartree–Fock, 214, 232
TIP, 105
TOPPA, 222
transition
  moment, 49, 53, 161, 162, 165, 171, 211, 212, 232
    dipole, 162, 177
    magnetic dipole, 164
    quadrupole, 164
    vibrational, 177
  probability, 47, 49
  rate, 49, 162
transverse, 17
two-electron repulsion integral
  Mulliken notation, 191
TZ, 254

uncoupled Hartree–Fock, 216
unrelaxed, 206

van der Waals coefficient, 172
VDZ, 254
vector potential, 12, 16, 23, 25, 93, 96, 105
  electromagnetic wave, 153, 154
  magnetic moment, 95, 107
  time-dependent, 12, 27, 28, 153, 154
  uniform magnetic induction, 27, 97

velocity
　angular, 126
　gauge, 28, 155
　representation, 164, 166, 168
vibrational
　averaging, 174
　correction, 133, 135, 136, 174, 180, 266
　　clamped-nucleus, 177, 180
　　sum-over-states, 175
　　zero-point, 180, 183
vibrational g factor, 145–147, 149, 151
　irreducible non-adiabatic, 151
　isotopically independent, 148, 151
VTZ, 254

wave
　electromagnetic
　　circular polarized, 155, 157
　　linear polarized, 45, 153, 155, 157
　　plane polarized, 45, 153, 155, 157
　　polychromatic, 45, 154
　vector, 153
wavefunction
　mth–order, 36
　Born interpretation, 9
　configuration interaction, 197
　electronic, 7, 142, 176, 177
　first-order, 36
　harmonic oscillator, 182
　Hartree–Fock, 191–193, 197, 198, 207

large component, 20
many-electron, 189, 191, 194, 199, 201
　bra, 195
nuclear, 8, 174
orthogonal complement, 196
perturbed, 34, 203
self-consistent field, 191–193, 197, 198, 207
　multiconfigurational, 196, 225, 233
small component, 20
time-dependent, 6, 26, 43, 45, 233
　first-order, 47
　second-order, 47
time-derivative, 41
variational, 190, 197, 204
vibrational, 176–178, 180, 182
vibronic, 174–177
wavelength, 153

Z vector, 247, 248
Zeeman effect
　rotational, 128
Zeeman term, 22, 24
zero point vibrational correction, 180, 183
zeroth-order polarization propagator approximation, 216
ZOPPA, 216
ZORA, 23
ZPVC, 180, 183